Quo Vadis: Evolution of Modern Navigation

F.G. Major

Quo Vadis: Evolution of Modern Navigation

The Rise of Quantum Techniques

Springer

F.G. Major
Severna Park, MD, USA

ISBN 978-1-4614-8671-8 ISBN 978-1-4614-8672-5 (eBook)
DOI 10.1007/978-1-4614-8672-5
Springer New York Heidelberg Dordrecht London

Library of Congress Control Number: 2013945533

Figure 1.2 © The Smithsonian Institution
Figure 2.2 © The British Library Board
Figure 2.10 © Trustees of the British Museum
Figure 5.12 Courtesy of Celestaire Corp
Figure 14.9 Courtesy of Boeing Corp under contract to NASA
Figures 17.1, 17.6 and 17.7 Courtesy of NASA-Hqrs Still Photography Library, Washington, DC
Figures in Chaps. 7–10 taken from the author's Quantum Beat, Springer sec. ed. 2007
Original art work inside by Christopher Major. chris@chmajor.com

Printed on acid-free paper

Springer is part of Springer Science+Business Media (www.springer.com)

Preface

The fields of global navigation and geodesy have undergone revolutionary advances in recent years, in regard to both precision and immediacy. The rapid "conquest of space" starting in the middle of the last century has led, in conjunction with the development of ultra-stable atomic clocks, to the realization of satellite-based navigational systems of intrinsically high precision and continuous global coverage. In the USA the original motivation and support for the system commonly referred to as GPS (Global Positioning System) came from the military, but later made available to the general public, while no doubt retaining to the military future refinements in resolution. Other satellite systems include the European GALILEO, the Russian GLONASS, and the Chinese BEIDOU.

The crucial elements in these systems are the spacecraft atomic clocks that can maintain, with ground support, a continuous sub-microsecond synchronization over the satellite network. The GPS has matured into a reliable, universally available facility accessible with a relatively inexpensive handheld receiver. It has been absorbed into our culture much like handheld wireless communication. This accelerating pace of technical assimilation has inevitably widened the gap between the level of technical understanding by the public and the level of sophistication of the new technology. This is exacerbated by the development of microelectronics and computer sophistication, reducing the needed knowledge to operate devices to simply knowing what button to push.

It is the purpose of this book to help bridge that gap by presenting an intelligent and intelligible account of the essential principles underlying the design of satellite navigational systems, with introductory chapters placing them in context with the early development of navigational methods. While the book makes no pretense at being a history of navigation, it nevertheless attempts to establish a historical perspective in its treatment.

The book is intended for nonspecialists with an inquiring turn of mind, who have some knowledge of physics or engineering, typically at a junior college level. It covers in an intuitive manner a broad range of topics relevant to the evolution of surface and space navigation, with minimum mathematical formalism. The first two introductory chapters are devoted to navigation in the natural world and among the ancient mariners, including of course the Phoenicians, Vikings, and Pacific Islanders. The third and fourth chapters deal with the elements of star navigation, the constellations, star coordinates, etc. The next three chapters are devoted to precise mechanical chronometers, the determination of longitude at sea, and the eventual development of precise quartz-controlled clocks. A summary introduction to quantum ideas is contained in the next chapter as a prerequisite for a discussion of microwave and optical interactions with atoms. The subject of atomic clocks is dealt with at some length, including the hydrogen maser, the cesium beam, rubidium cell, and the latest development, the cesium fountain and single ion frequency standards. A chapter on gyrocompasses is introduced next leading to a discussion of the Sagnac laser gyro. The thirteenth chapter deals with time-based navigation: the former ground-based networks LORAN C and OMEGA, followed by the next three chapters dealing with satellite-based systems, including orbit theory, the early satellite navigation systems SECOR, TRANSIT, etc., culminating in an extended treatment of the global navigation satellite system, the Global Positioning System (GPS), a major technical achievement of our time. The subject of GPS naturally divides itself into descriptions of the three principal areas: the space, control, and user segments. An attempt is made to convey the extraordinary degree of coordination and precision required to make such a system work. The list of applications of GPS, particularly Differential GPS, is long: in addition to the obvious air and sea navigation and missile guidance, GPS finds applications in geodesy, major construction and mining projects, agriculture, ecology, emergency location, and of course navigating on the highway. Finally a chapter is devoted to the future of navigation in which atom interferometers may be used as gyrocompasses, cesium fountains as atomic time standards, and quantum computers to update satellite ephemerides.

The author is deeply grateful to the many who have contributed in different ways to the completion of this book. Special thanks are due to Prof. Fançois Barlier, Astronome émirité of the Observatoire de la Côte d'Azur, for his early support. Thanks are also due to those whose criticism of portions of the manuscript he had sought, and who gave so generously of their time. These include in no particular order Prof. Neil Ashby, emeritus professor University of Colorado; Prof. Hans Schüssler of Texas A&M; Prof. emeritus Günter Werth of the University of Mainz; Flt. Capt. Darrin Colegrave; Flt. Captain Sean Lee; Dr. Glenn Major; and electronics engineer Barry Colegrave. Special thanks are also due to Christopher Major, who so ably and patiently prepared all the inside original artwork, to William Carter for providing useful material, and especially to Christopher Coughlin, my editor at Springer, for his early and continued encouragement.

Contents

Navigation in Nature

1

1.1 Animal Navigation

Life on earth displays myriads of forms and behaviors, but none so mystifies us as the ability of many creatures to find their way over immense distances spanning the globe, often crossing wide expanses of open sea to perform their migrating ritual. It is still largely a mystery that reaches to the very genetic basis of animal behavior. The seemingly infinite paths of evolution are here again evident in the many ways that animals have developed to find their way according to their different environments. Some animals live and move on the earth's surface, over widely varying terrain, others high in the atmosphere through wind and cloud, while yet others through the great ocean depths. In one notable instance, the monarch butterfly is able to migrate thousands of miles across the North American continent, a journey so long that it incredibly requires at least two generations to complete it! No less baffling is the migratory behavior of some other animals: for example, the Manx shearwater, a fast-flying seabird, that can glide and plane its way to breeding grounds across the Atlantic Ocean, and the Pacific salmon that returns to the very stream where it was hatched after 2 years of wandering thousands of miles at sea.

For many migratory species, our present lack of knowledge is such that it is not even certain just how to analyze their behavior in terms of basic navigational functions. There is a complexity which arises from the many strategies animals have evolved to deal with the vagaries of their great migrations. This makes it hard to analyze the way we see animals behave, because the navigational strategy they adopt depends on the particular environmental conditions they happen to face. Nevertheless to begin to understand the migratory and homing behavior of animals, we must distinguish some basic navigational skills in terms of which we can describe the behavior we observe. The most rudimentary type of navigation has been called by animal behaviorists [1] *piloting*, a term borrowed from marine navigation. In this, the movement is guided by consecutive familiar *landmarks*, with possibly some random or systematic searching between them. Here, the term *landmark* is used broadly to include not only cues that are recognizable physical

F.G. Major, *Quo Vadis: Evolution of Modern Navigation: The Rise of Quantum Techniques*, DOI 10.1007/978-1-4614-8672-5_1,

features that can be seen but also cues based on the senses of hearing and smell. Wayfinding to destinations beyond the current sensory range requires a conceptual map and a bodily sense of space–time and direction to seek the next landmark. Such a higher level of navigational ability, exhibited by many species, is acquired through *compass orientation*. In this, an animal has the ability to maintain a particular compass bearing, that is, direction with respect to the earth's north–south axis independently of any preknowledge of landmarks. And finally, the highest level, which has been attributed to only a few species, is *true navigation* where an animal is able to direct its movement toward a destination without any prior sensory knowledge of any part of the path it takes. It is not implied of course that these animals must have some supernatural perception, a sixth sense, to navigate under these conditions, but merely that it is possible that some animals are genetically programmed with an internal "map" which they are able to follow using their ability to orient their movement and sense the passage of time, as well as other cues. How else to explain the migration of the monarch butterfly?

What serves as useful landmarks for the purposes of navigation, of course, depends on the space within which the animal moves and may include odors or sounds in addition to the more common recognition of visible objects or topographical features. An animal's behavior is based on all the cues that it recognizes in its surroundings. While earthbound insects may use visual landmarks or scents on the ground as backup to other cues, birds migrating during the day may recognize geographical features, such as coastlines, far below. Such landmarks are important to migrating birds, allowing them to correct for any deviation from their intended course, such as may be caused by a cross wind, for example. Pilotage plays some role in the homing success of some animals but appears however to be of secondary importance in the case of homing pigeons. This was made evident by experiments in which manipulating their ability to correctly orient themselves caused them to deviate from reaching their home loft, even while approaching full view of it.

1.2 Orientation by the Sun

By far the most remarkable, and surprisingly common feature of animal behavior, is the ability of some animals to orient themselves with respect to the earth's geographical axis, that is, to achieve compass orientation. For many who have focused vision, or at least a sufficient degree of light sensing ability, the sun can provide an orientation cue, provided the animal can take into account the changing position of the sun throughout the day. Indeed, there are many animals that *can* do precisely that: using their *biological clock*, they are able to compensate exactly for the movement of the sun across the sky and deduce a constant compass orientation. This biological clock regulates the inner workings of animals and plants, giving rise to what is called a *circadian rhythm* (Latin: circa = about, dies = day), a term applied to changes that repeat regularly on a daily basis. Such periodic behavior that follows the day–night cycle has been noticed since ancient times. In the eighteenth century, the great naturalist Carolus Linnaeus, best known for his classification of animals and plants, noted that the petals of different species of flowers opened and closed at different specific times during the day. This occurred with such regularity that he is said to have cultivated a garden having flowers that opened at various times of the day to form a *flower clock.* Recent research has revealed the biochemical pathway connecting light stimulus with the chemical processes driving the circadian rhythm. Light-sensitive substances called *cryptochromes,* first discovered in plants, serve as transducers of light in the blue to ultraviolet region of the spectrum, setting in train the 24-h cyclical biochemical processes. More recently, it has been shown that cells in the retinas of mice and humans express genes similar to those that code *cryptochromes* in plants, suggesting that these genes play a role in the circadian light-sensing mechanism. Detailed studies on fruit flies have revealed the specific sequence of biochemical reactions that control the circadian rhythm.

The ability to compensate for the movement of the sun has been demonstrated quantitatively in honey bees and certain birds. In a classical series of experiments on caged starlings carried out in the 1950s by Gustav Kramer [2], it was shown that by using mirrors to alter the apparent position of the sun as seen by the birds, their sense of orientation could be altered accordingly. This effect was made evident by the change in direction of the caged birds' urge or "restlessness" (*Zugunruhe* in German) to begin their migration. A more quantitative demonstration of the same ability was achieved by systematically shifting the phase of the starlings' internal clock [3] by gradually conditioning them to an altered day–night cycle. It was found that indeed the birds compensated for the movement of the sun at the proper rate of 15° per hour. Similar experiments have been conducted on many other animal species both vertebrate and invertebrate including homing pigeons, amphibians, etc., with similar results. For many species, it is not necessary for the sun to be directly visible. Many have acquired the ability to sense the direction of *polarization* of sunlight as seen indirectly through scattering by clouds and the atmosphere. The same atmospheric scattering of sunlight that produces the blueness of the sky results in the light becoming polarized in a pattern which depends on the relative position of the sun and the direction of observation. One can demonstrate this by using a polarizing filter, such as a pair of "polaroid" sunglasses, and noting that the brightness of the sky, as seen through the filter, varies as it is rotated about an axis along the line of sight. This confers on members of a species equipped to detect this polarization, the special ability to orient themselves even under overcast skies, or when the sun is below the horizon, during the hours of dawn and dusk. More will be said in the next chapter on this subject in the context of the Vikings' ability to determine the sun's position in the presence of fog. What is truly remarkable is that in the case of invertebrates such as bees, ants, and spiders, there is evidence that it is the light in the *ultraviolet* region of the spectrum (380–410 nm) whose polarization direction is detected and used for orientation. In experiments on a certain desert ant [4], for example, a creature that is able to return to its nest after foraging over wide areas without a scent trail, individuals were allowed to roam across an especially prepared sandy test area extending over several hundred meters from their nest site. Light polarization filters mounted on a movable frame exposed the ants to sunlight whose polarization direction could be manipulated, while depriving them as far as possible the ability to see any landmarks as they tried to find their way to their nest. It was found that the ants did indeed appear to respond to the polarization of the ultraviolet component of sunlight in guiding their successful return to the nest. The ultraviolet peculiarity has been demonstrated also in a species of spider, which is endowed with secondary optic sensors sensitive to the direction of polarization of sunlight falling on them.

The possibility of using the polarization of sunlight even extends to an underwater environment, where scattering by the water itself leads to a polarization of the light, a polarization whose orientation depends on the position of the sun. This polarization can be sensed at depths beyond where the sun itself would be visible.

1.3 Orientation by the Stars

The ability of certain species to migrate at night obviously presents a case where compass orientation, where it exits, must clearly be achieved by some means other than the sun. It would be natural to assume that the star patterns must be used, even though such a cue is obviously a good deal more subtle than the bright sun. For this to be possible, the animal must compensate for the apparent rotation of the star patterns due to the earth's rotation, just as orientation by the sun requires it. Again a circadian internal clock must come into play. Indeed, some experiments on night migrating birds do suggest that such is the case. These experiments were done on night migrating birds again exploiting the technique of observing their preferred direction of migratory *Zugunruhe*. In a series of classic experiments on the indigo bunting [5], the birds were put in special circular cages and taken outside after dark under open skies. On observing that their migratory urge was correctly oriented when the skies were clear and the stars clearly visible, but that under overcast skies there was an unmistakable loss of direction and vigor in the birds' drive, it was evident that the simplest explanation was that the birds take their directional cue from the stars. Of course, it is difficult with a living organism, such as a bird, to reach definitive conclusions concerning their behavior since it is difficult to identify and control all conceivable factors that may affect the observation; perhaps on seeing an overcast sky, the birds become depressed, or sense the approach of a storm. In a more convincing series of experiments on indigo buntings [6], in which a planetarium was used to simulate the night sky, it was shown that these birds oriented their migratory urge with respect to the star positions as projected onto the planetarium dome, even when the direction of the planetarium Pole Star was set far from its true compass direction. If the planetarium stars are turned off and the dome dimly lit with diffuse light, the birds' nocturnal migratory behavior was noticeably affected becoming more random or ceasing altogether. To test whether the birds need early exposure to the night sky in order to develop the correct migratory orientation, indigo bunting hatchlings were raised in isolation until the time of first migration, at which time they were tested in the planetarium. After an initial randomness of orientation, they were found to show normal migratory behavior.

Further experiments using two groups of young buntings conditioned to different orientation of the celestial pole reinforced the idea that they indeed use star patterns to guide their migration.

Only birds were once thought to have this remarkable ability with any degree of certainty; however, very recently researchers Marie Dacke and Eric Warrant affiliated with Lund University reported on experiments [7] carried out in South Africa on dung beetles. These beetles mold dung into perfect spheres larger than themselves and escape the crowd around the dung pile by a hurried departure along a *straight* line, even at night. This requires them to keep a constant bearing, and must therefore have some form of navigational aid at night. It had previously been assumed that only the Moon could provide such aid, but then it was discovered that many beetles were still able to navigate under a clear sky in the absence of the Moon! This led to speculation as to whether the stars play a role. To test this, observations were made in a planetarium where the beetles could be subjected to controlled displays of the heavens. The results showed unequivocally that the stars provided the necessary cue and that in particular only that linear concentration of bright stars we call the Milky Way determines their path.

It might be thought that the moon, whose presence in the night sky is generally much more obvious than the stars, would be a natural guide for nocturnal migrants, but that is not expected to be the case for long range migration. The primary reason has to do with the fact that the moon's motion is such that its apparent position and appearance vary considerably from day to day throughout the lunar cycle, even disappearing at times. To be able to use the moon to fix a compass bearing, an organism would have to keep track of two incoherent cycles, the daily and the lunar cycles. There are cases however where evidence exists suggesting lunar orientation, based on mirror experiments, in certain insects such as sand hoppers, and in more recent experiments, dung beetles.

1.4 Magnetic Orientation

From the earliest times since the discovery of magnetism, the action of a magnetic field on animals was conceived as an all pervading force that can mystically exert an influence for good or evil. It persists even today in the revival of the ancient Chinese practice of *Feng Shui* (pronounced "fung shway") which uses a magnetic "compass" for the propitious alignment of objects in space, such as monuments and buildings, and even the arrangement of living spaces and furniture. From this flowed the early observation that a lodestone (a form of the iron ore *magnetite*), if free to rotate, naturally sought a fixed direction with respect to the *earth's axis*. It was very much later in the sixteenth century that this remarkable property of lodestone was the subject of serious study by William Gilbert, who in his book *De Magnete* put forth his revolutionary idea that the explanation for the behavior of the lodestone was to be found in the earth itself. His theory of terrestrial magnetism describes the magnetic field that surrounds the earth, as if arising from a great magnet imbedded at the center of the earth, aligned more or less along its axis of spin. This provided the theoretical basis for constructing a *marine compass* consisting of a magnetic "needle" so mounted as to be free to rotate. More will be said about the history of the magnetic compass in the next chapter.

In the biological sphere, the earth's magnetic field does indeed play an important role in guiding the path of migratory and homing animals. The ability of some animals to travel great distances over trackless oceans or deserts, or under overcast skies, where visual cues are nonexistent, clearly implies that a nonvisual backup system of navigation is involved. In fact, indisputable evidence now exists that such a nonvisual capability derives from an ability to sense the direction of the earth's magnetic field. The initial suggestion that this may be so dates back almost a century. In a comprehensive series of experiments on European robins [8], a uniform magnetic field was applied over the robins' cage using pairs of current-carrying circular coils, erected coaxially on opposite sides of the cage: one pair along a horizontal axis and the other along a vertical axis. In this way, the vertical and horizontal components of the magnetic field sensed by the birds could be independently varied. Again observing the birds' average directionality in its *Zugunruhe* activity led to the startling discovery that the robins not only sensed the magnetic field but also responded to it in a rather intricate way. They did not simply use the direction of the horizontal component, as one does with a magnetic compass, but rather the horizontal component only defines an axis, without defining a particular *direction* along it; that is determined by the slope of the *total* magnetic field with respect to the vertical direction. This means that a reversal of the vertical component of the field will reverse the direction of the birds' migration for a given horizontal component as shown in Fig. 1.1.

Magnetic orientation has also been observed to be an important capability of homing pigeons [9], one to which they have recourse under cloudy skies. More space will be devoted to homing pigeons later in this chapter. The manner in which magnetic information is processed and determines the pigeons' behavior remains to be studied. It is difficult to design and interpret unequivocally experiments intended to discover the role played by the environment; nevertheless, observations do indicate an extraordinary sensitivity of these birds to variations in the magnetic field about them. There have been recent reports of finding magnetic substances in the heads of homing pigeons and the abdomens of honey bees, but further research is needed to establish whether these are relevant to orientation.

Another important example of animals sensitive to ambient magnetic fields is among the invertebrates, in particular honey bees. These creatures of course have a special place among humans as producers of an important commodity and have therefore been the subject of a great deal of interest, particularly their social behavior and organization. Especially fascinating is the way in which a "field-worker" bee returning to a hive after having found a source of nectar communicates to fellow bees in the hive exactly where this source is to be found. It was an Austrian biologist Karl von Frisch [10] who, using specially designed beehives equipped with viewing windows, observed returning bees execute an elaborate dance, in which the bees "*waggle*" their bodies from side to side as they followed a figure-of-

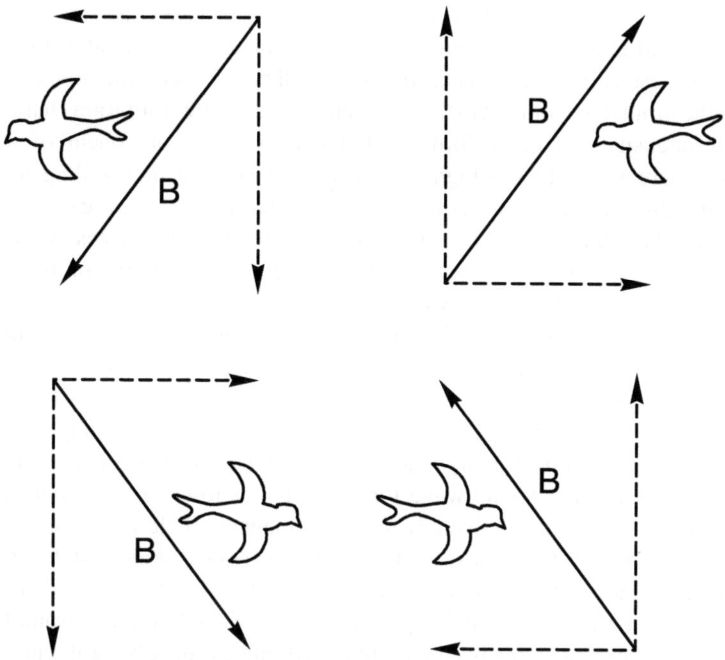

Fig. 1.1 Magnetic field effect on European robins

eight path. The dance is performed on the vertical comb surface in the hive. What is perhaps the most remarkable example of animal communication is the fact that this motion conveys *quantitatively* the direction and distance of the food source. The angle of the path along which the waggle occurs with respect to the vertical is equal to the angle of the direction of the food source with respect to the direction of the sun, and the distance is commensurate with the frequency or number of waggles executed. It has been reported that applying magnetic fields of varying intensity over the region occupied by honey bees can affect the orientation of the path of the waggle dance.

The use of the earth's magnetic field by migratory birds, among other cues, to navigate over intercontinental distances, raises the question of the effects on bird migrations of the transitory disturbances in the earth's magnetic field called *magnetic storms* [11]. These magnetic storms have been the subject of study in the wider context of terrestrial magnetism, and their association with *solar flares* that also produce spectacular aurora displays and interfere with radio communications. Magnetic observatories as early as the eighteenth century made systematic observations of the variations in the earth's magnetic field over time. During some days the recorded intensity of the magnetic field showed a comparatively smooth and regular quasi-periodic graph; other days the record showed large irregular transient fluctuations. The regular magnetic variations repeated every half hour and had a maximum variation of around 100 gamma (1 gamma $= 10^{-9}$ T). The origin of this smooth variation is thought to be atmospheric "tides" at ionospheric altitudes, causing a flow of electrically conducting gas across the steady magnetic field of the earth, thereby inducing currents as in an electric generator: the *dynamo effect*. It is these currents which contribute a small periodic addition to the magnetic field on the ground. In contrast to these variations, magnetic storms are characterized by their intensity and pulse form, with the horizontal component of the field suddenly increasing, staying high for an hour or so, and then falling to a deep minimum before returning to its normal value after several days. Experiments on radio-tracked homing pigeons have shown that magnetic disturbances over the order of 100 gamma are sufficient to noticeably degrade the ability of the birds to accurately determine their homeward direction.

While a great deal has been learned about the *behavioral* effects of magnetic fields on animal navigation, very little progress has been made at understanding the neural and biochemical basis of this sensory ability. Other sensory systems such as visual, olfactory, auditory all have identifiable receptors, but no magnetoreceptor has been identified with certainty in any animal. What sets a magnetic field apart is the fact that, whereas a sound wave for example has to come into contact with a specific part of the organism, namely the ear, a magnetic field freely permeates the whole organism; therefore, an interaction could take place at any point or points in the organism. Furthermore, the interaction may not be localized in an identifiable organ at all, but rather in certain dispersed regions of the brain, for example.

In recent years, there has been an intensified interest in solving the problem of magnetoreception in animals. This has been encouraged by recent promising approaches to solving the mystery. For creatures that live in the sea, a plausible way in which they may be sensitive to the earth's magnetic field is through *electromagnetic induction*. Thanks to Michael Faraday we know that an electrical conductor moving across a magnetic field will have induced along it an electromotive force (emf). It has been postulated that animals such as sharks and skates have organs whose motion in the earth's magnetic field would produce such a perceptible voltage and function as magnetoreceptors. The main difficulty with this postulate is that the animals are surrounded by a moving conductor, the sea, producing its own electric field. Several reports have been published of experiments designed to clarify the basis of the sensitivity of rays to magnetic fields, but no consensus has emerged. However, that still leaves magnetoreception in birds and insects to explain, although it is not inconceivable that some conductive part of an animal's body could develop an emf across it due to motion through a magnetic field.

A more promising approach posits that the earth's magnetic field, or its variation, affects *biochemical* processes in the animal. It is clear that if indeed this is true, it must be through some *very subtle* process. It certainly cannot be a simple exchange of energy, since it requires orders of magnitude more energy to affect a biological reaction than the earth's field can provide. However, there are examples in atomic physics where weak magnetic fields can affect the absorption/emission of much more energetic photons through quantum *selection rules* ultimately based on quantum symmetry requirements. One proposed mechanism for magnetoreception that has generated a good deal of interest depends in a similar way on the symmetry of quantum states. It involves *coherent* electron spin states of *paired radical ions*. The concept of *coherent states* is an intrinsically quantum one: unlike classical mechanics in which particles have individual, independent coordinates and momenta, a quantum description of a collection of electrons for example must obey a certain symmetry principle, which in effect restricts the behavior of the system. We will resist here a diversion into quantum theory; some quantum concepts will be described in a later chapter in the context of atomic clocks. Suffice it to say that for a collection of electrons, as in a molecule, the mathematical representation of their quantum state must be *antisymmetric,* that is an exchange of the coordinates and spin direction of any two electrons must yield the same function with the opposite sign. The electron spin direction, which has only two values, "up" or "down", is regarded as just another coordinate. In the case of just two electrons α and β, the spin part of the function must have the symmetrized forms:

$$\alpha(\uparrow)\beta(\uparrow), [\alpha(\uparrow)\beta(\downarrow) + \alpha(\downarrow)\beta(\uparrow)], \alpha(\downarrow)\beta(\downarrow) \quad \text{symmetric}$$
$$[\alpha(\uparrow)\beta(\downarrow) - \alpha(\downarrow)\beta(\uparrow)] \quad \text{antisymmetric.}$$

A symmetric spin function must be associated with an antisymmetric function of the space coordinates, while the reverse is true for the antisymmetric spin function. The symmetric spin functions form a triplet, while the antisymmetric spin functions form a singlet. Getting back to the subject of radical ion pair reactions, these may result in an outcome strongly affected by a weak magnetic field as a consequence of quantum symmetry conditions. A radical ion pair is produced for example by the optical excitation of one molecule into a state from which it is energetically favorable to transfer one electron to another molecule acting as an acceptor, resulting in a positive–negative ion pair. The positive ion is left with an unpaired electron spin which couples to the acquired electron spin on the acceptor molecule. These electrons interact magnetically with any external magnetic field as well as any nuclear magnetic moment, the so-called *hyperfine interaction.* The important feature is that the probability of the electron being transferred back depends on the time evolution of the quantum state of the coupled electron spins of the two ions. This time evolution is sensitive to the magnetic field.

1.5 Winds and Current

For migratory birds and insects whose living space is mainly in the atmosphere, successful migration critically depends on their ability to navigate under unpredictably variable conditions of air currents and cloud cover. It is not surprising that such species should have evolved enhanced awareness of such conditions as are likely to be favorable to ease their exhausting journey or could prove fatal if contrary on a long journey over an expanse of water. Their flight path and timing are affected by conditions that produce *thermals*, rising currents of air sought after by large birds of prey, and the direction of seasonal prevailing winds. Since wind direction and speed are variable in space and time, it is inevitable that there may be on occasion crosswinds that will cause the flight path to drift from the intended course. To correct for this, the migrant must rely on visual landmarks such as coastlines or mountains to sense how far it has been driven off course. Clearly, the relatively slow flying insects are more at the mercy of winds than large species such as birds, and their progress can be totally determined by wind patterns. A number of field studies have been made on the factors that determine the precise timing that birds choose to initiate their migration. It is well established, not surprisingly, that they favor a tail wind; however, it is impossible to be certain whether birds have somehow a prior sense of the direction in which they should start their migration and simply wait until the wind direction is favorable, or their migration path evolved originally from the prevailing wind direction at the time of year when they now feel the *Zugunruhe*. In any event a persistent head wind would obviously be avoided if at all possible, as it could spell disaster if a large body of water must be crossed. As already mentioned wind is a particularly crucial factor for insects, which are at its mercy because they are slow flying and unable to resist it. The desert locusts for example, the biblical scourge that periodically plagues areas of North Africa and the Middle East, swarm in huge clouds that block out the sun and denude all vegetation along a path determined by where the wind carries them. Because of the catastrophic destruction of crops they cause, locust swarms have long been a subject of study on an international scale. Researchers at Cambridge University in the UK have discovered that higher levels of the hormone serotonin are correlated with the transformation of the locust from a solitary creature to an aggressively gregarious one forming the destructive swarms [12]. The swarms are tracked by satellite imagery and their progress closely monitored. The phenomenon of swarming, which also occurs in birds, raises interesting questions as to how the individuals in a swarm seem to coordinate their flight so as to produce intricate maneuvers of the swarm as a whole. The wind pattern alone cannot explain how the swarm avoids being dispersed and remains a coherent mass. It appears that individual insects on the outer reaches of the swarm consistently direct their flight toward the center of the cloud.

1.6 Electric Orientation

For animals whose environment is in the depths of the sea, sensitivity to the electric potential (voltage) gradients in the water may serve to guide their migration. The motion of the electrically conducting sea water through the earth's magnetic field produces voltage gradients, on the same principle as an electrical generator. The direction of these electric potential gradients is perpendicular to both the magnetic field and the direction of the ocean current. Given that these currents often follow consistent patterns over wide areas of ocean, it follows that those creatures sensitive to potential gradients may be able to orient themselves with respect to the direction of such currents, providing them with navigational cues. It has been shown that skates and sharks are sensitive to electrical potential gradients in the water around them, and can therefore presumably orient themselves with respect to ocean currents.

A most remarkable example of the use of electrical fields to achieve *electrolocation* to navigate in the murky waters of some South American and African rivers are certain "weak" electric fish, among them *Gymnotus carapo* [13]. These creatures have evolved a highly specialized electrical means of "navigating" over short distances and avoiding obstacles and other fish under conditions of low visibility. A specialized organ in their tails can discharge a relatively weak electrical pulse (unlike the lethal electric eel), called an *electric organ discharge* (EOD) which, because of the electrical conductivity of the water, produces around the fish an electrical field whose distribution is distorted by variations in the conductivity of the water and any solid objects in it. By sensing the distribution of the electric field along its own anterior skin surface, this fish is able to "see" objects as well as detect the presence of other fish of the same species, similarly probing their surroundings. Since the electric field strength falls off very rapidly with distance away from the source (as much as the inverse cube of the distance for an ideal dipole in a uniform medium), it follows that this sensing method is effective over only the immediate surroundings of the animal, and not for long-distance navigation. The principle of this mode of sensing, though reminiscent of the echo-location used by bats, has little in common with it, either in the type of energy radiated or its maximum range.

1.7 The Homing Pigeon

The unique talent of the homing pigeon (*Columba livia*) of being able to fly to its home loft when released from distant points with which it may be totally unfamiliar has been known since ancient times. When Noah, according to the Old Testament, released those doves to verify the end of the Great Flood, he fully expected them to return, if they could! The use of pigeons as rapid message carriers continued through the middle ages and in times of war into the twentieth century. It is said that the financier Rothschild, the banker to kings and emperors, made substantial profit by receiving news by carrier pigeon in London of Napoleon's defeat at Waterloo, long before anyone else did. In a similar vein, during the Franco-Prussian war of 1870–1871, Julius Reuter saw an opportunity to fill an urgent need for communication between Paris, which was under siege, and the outside world. He established a carrier pigeon service that ultimately blossomed into the present day Reuters. At the time, the science of photography was rapidly advancing and in particular microphotography became feasible, making it possible to send thousands of messages by courier pigeon. During the Great War (1914–1918), both sides used carrier pigeons on a large scale, since radio was still in its infancy and not sufficiently sophisticated, at least at the beginning of the war, to be useful under battlefront conditions. The U.S. Navy department issued in 1918 a manual for *The Reception, Care and Training of Homing Pigeons*. One of the most illustrious

Fig. 1.2 Cher Ami, the pigeon hero (Armed Forces History/The Smithsonian Institution, with permission)

heroes of the Great War was a pigeon named *Cher Ami*. It saved the lives of an entire battalion, the so-called *Lost Battalion,* of the New York 77th division. Briefly, this battalion had advanced too far into enemy territory and came under devastating friendly artillery fire and would no doubt have perished if word did not reach the command post immediately. The only means available was by carrier pigeon, and that is how *Cher Ami* had fame thrust upon it. Unfortunately, the Germans spotted it taking off and it came under intense fire; it took a bullet in the breast, lost one eye and part of the leg holding the message dangling by a tendon! Undeterred, it continued to the command post and saved the battalion. It was awarded *La Croix de Guerre* by the French government for extraordinary valor. A photograph of a statuette depicting Cher Ami is reproduced in Fig. 1.2.

It is natural to regard this event in an anthropomorphic light, that the pigeon did it in the spirit of devotion to duty. The fact is of course, it really did not have much of a choice but to continue flying to its "loft", which in this case was at the command post. In World War II, another carrier pigeon named *G.I. Joe* was honored for exceptional service in rescuing the lives of 1,000 British soldiers in an Italian town that was under imminent threat of being bombed by the American army air force. Immediate communication was essential, and *G.I. Joe* provided it. For that he received the *Dickin Medal*, a British award given to any animal that has done extraordinary service, typically to dogs taking part in search and rescue missions.

The many aspects of homing behavior in pigeons have been parts of a popular field of study, since, unlike migratory behavior, it can be studied in a local environment, making it easier to observe and control. There is still considerable controversy about the way the various environmental factors come into play in the act of homing from an unfamiliar location. It is generally assumed that the pigeon uses a map and compass method, that is, it can construct a conceptual map with respect to which it can extrapolate its position and then direct its flight in a direction according to some orientation mechanism. There is good evidence that the bird senses its location on the map through environmental stimuli, principally olfactory (smell), magnetic and visual, while the directional sense relies on the sun and the earth's magnetic field. Although homing success in general is the result of a number of redundant capabilities and the interpretation of the results of experiments is often difficult because of their variability; nevertheless, there is little question that the earth's magnetic field plays some role in the navigation ability of homing pigeons as well as migrating birds. Their ability to navigate under totally overcast skies first prompted a series of experiments to test directly whether the birds can use magnetic orientation [14]. To that end, small magnets were actually attached to one group of pigeons' backs and a control group were similarly burdened with an identical load of nonmagnetic brass and released. It was found that provided the sun was visible, there was no loss of ability to head off in the right homeward direction (the so-called *vanishing direction*), but when released under dense cloud cover, those whose magnetic field was disturbed by the small magnets were more likely to deviate randomly than the control group. A more refined approach [15] consists of replacing the small magnets with miniature field coils, which enabled the direction of the magnetic field passing through the birds' heads to be reversed simply by reversing the direction of the energizing electric current in the coils. Two groups of homing pigeons carrying such coils, with one group having the field direction set in an upward vertical direction, while the other in a downward direction were released under overcast sky conditions. It was observed that the two groups tended to fly off in *opposite* initial directions when released. With the advent of the general availability of the *Global Positioning System* (GPS), the subject of later chapters, it is now possible to track the movement of migrating and homing animals with real-time precision measured in meters. Recently, GPS was applied to study in detail the progress of homing pigeons under controlled conditions; later in this chapter, we describe the long-distance GPS tracking of migratory birds such as the *Manx shearwater*. The recent application of GPS allowed the detailed study of the movement of the homing pigeon [15] over its entire journey. Previous studies relied on only the *vanishing bearing,* the compass direction in which the bird was traveling as it disappeared. It was found that if the pigeon was allowed a 5 min preview of its surroundings prior to release, it would reach its home loft faster. Otherwise it may circle an unknown release point and reconnoiter up to a distance of about 1,000 m, before it continues its homeward flight as if it had completed a preview of its point of departure. Recent research [16] on the nature and location of the actual sensor involved in the homing ability of pigeons suggests that *magnetoreception* occurs in the upper beak area of the bird. It was shown that local anesthesia of that area, or the placement of a magnet there, impairs the birds' ability to detect a magnetic field or sense its true direction. The manner in which magnetic information is processed and determines the pigeons' behavior remains a mystery. It is difficult to design and interpret unequivocally experiments intended to discover the role played by the environment. Nevertheless, observations do indicate an extraordinary sensitivity of these birds to variations in the magnetic field about them. There have been recent reports of finding magnetic substances in the heads of homing pigeons and the abdomens of honey bees, but further research is needed to establish whether these are relevant to orientation.

A radically new area of research was brought to light very recently [17] as a result of attempting to resolve what appeared to be an anomaly in the homing patterns of pigeons released from certain particular sites. In one case, the birds invariably took a wrong turn when released, and from another location the birds always went off in more or less random directions; except on one particular day, when they all successfully returned home. Since no magnetic anomaly had been reported which might explain this strange behavior, it remained a deep mystery until a geologist by the name of Hagstrum learned of the extraordinary ability of pigeons to detect infrasound. This is vibration of matter at such a low frequency, typically below 20 Hz, that it is inaudible to humans. It occurred to Hagstrum that perhaps the pigeons are guided by the character of the infrasound emanating from their loft area like an acoustic beacon. Remarkably, he was able to gather sufficient weather data for the date of the releases and using a special computer program make a prediction of the wind directions for that date and therefore the infrasound signal that is carried by the wind. The results were totally consistent with the observations.

Recent interest has been directed to discovering what areas of the bird's *brain* are involved in the act of successful homing. In particular much interest has been shown in the relevance of the part of the bird's brain called *hippocampus* to the function of homing. It has been experimentally shown that damage to the hippocampus degrades spatial learning and causes memory loss, but it is the further loss of the ability to *associate* compass direction with other navigational cues, such as visual and olfactory cues, that result in the disruption of a pigeon's homing ability [18].

1.8 The Monarch Butterfly

Of all the migratory species of animals, the monarch butterfly (*Danaus Plexippus*) is in many respects the most enigmatic: there is the existence in the same species of both migratory and nonmigratory generations, the narrowly defined destination of their migration and the extraordinary length of the journey to reach it. Furthermore, their individual ability to navigate to their destination with no outward help from previous generations who had completed the journey clearly points to some preprogrammed activity in the miniscule brain of a butterfly!

During the fall, monarch butterflies begin an unremitting journey of over 3,500 km from the northeast of the Rocky Mountains of North America to wintering

Fig. 1.3 Monarch butterfly migration routes

grounds at certain *specific* fir groves in the mountains of central Mexico. Those originating west of the Rocky Mountains tend to migrate to scattered points south along the Pacific coast, not nearly as impressive as the Mexican migration. The approximate migration routes are illustrated in Fig. 1.3.

The fall migrant butterflies, as distinct from the spring and summer butterflies, have a deficiency in a certain *juvenile hormone* which results in *diapause*, that is, reproductive inactivity, and a marked increase in longevity extending to months, compared with the weeks of summer butterflies. There is also buildup of abdominal fat that prepares them for the long migration south. This reproductive passivity lasts throughout the overwintering period in Mexico until the arrival of spring when they break the diapause and become reproductively active. They mate then fly north to lay their fertilized eggs on milkweed plants in the south western regions of the USA. There they develop through the normal life cycle from egg to larva to pupa, emerging as relatively reproductively active adults. The life span of these fecund nonmigratory butterflies is relatively short, a matter of only weeks, during which time they spread northward, possibly following the abundance of the poisonous milkweed plant that the larvae feed on, repopulating the northern reaches of their range as far as southern Canada. As many as four generations of nonmigratory butterflies may be completed before the time of the migration to Mexico in the fall. This means that migratory butterflies are several generations removed from the last members of their kind to make the journey.

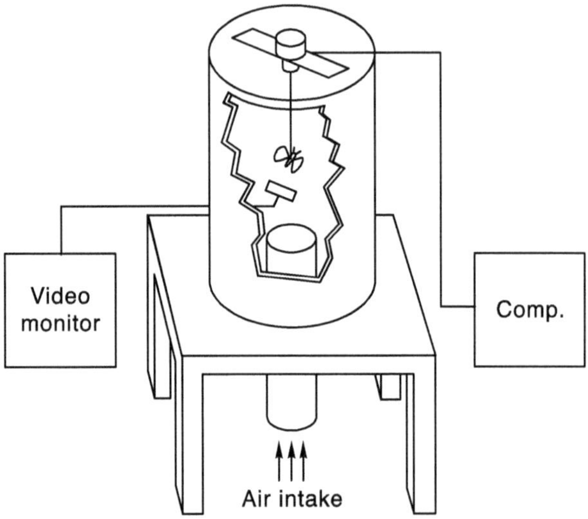

Fig. 1.4 Monarch butterfly flight simulator

Adding to the mystery that there are two kinds of Monarch butterflies is the question of how they are able to congregate from widely separated points in the USA to such a precise location in central Mexico, with several generations separating one migration from the last! Of course, many animal species habitually congregate at certain geographical locations to breed, but generally adults return to where they themselves were given life, not some previous generation! Tracking the migration route of Monarch butterflies using such powerful techniques as GPS is difficult because of the weight constraint; GPS receivers have been miniaturized to a remarkable degree but still would be ill adapted to be carried by a delicate butterfly. Nevertheless, their migratory route has been established through field observation; in fact, there is so much public interest in their progress that groups have been organized that are dedicated to keep track of these beautiful insects.

The central question of how they navigate has been addressed by laboratory experiments that test the sensitivity of the butterflies to the earth's magnetic field, and the direction of the sun as possible means of navigation. These methods of navigation are known to play a role in the navigation of so many other animals, but up until recently there were conflicting observations about their use by the Monarchs. Definitive answers were obtained by Mouritsen and Frost [19] using a specially designed flight simulator in which monarchs are tethered, but free to "fly". By simulating the free flight of the butterflies under controlled environmental conditions, these authors were able to settle some of the questions about the navigational cues on which the monarchs rely. A schematic drawing of the simulator is reproduced in Fig. 1.4.

The flight simulator chamber consisted of a translucent plastic cylinder about the size of a large bird cage with a 6 in. diameter vertical pipe admitting air into it through a hole at the center of the base directed at a suspended butterfly. The vertical air flow was controlled by a variable fan and made as free of turbulence as possible by filling the inside of the pipe with parallel narrow tubes (drinking straws) to collimate the flow. Miniature cameras were mounted outside four small holes drilled 90° apart near the bottom of the cylinder. The cameras could be moved to a different angular position to avoid any directional bias. The butterfly was "tethered" from above by a very thin, straight tungsten wire glued with beeswax to its dorsal thorax. This suspension wire was coupled rigidly to the vertical shaft of a commercial optical encoder in such a way that the butterfly body was constrained to stay in a horizontal plane but could turn with very little friction. The optical encoder was connected to a computer that recorded headings at timed intervals, producing a derived flight path based on an assumed constant speed.

Experiments were run on over 50 wild monarch butterflies captured at the northern shore of Lake Ontario during the fall migration period. Most butterflies were observed for at least 1 h, and some as long as 4 h. Assuming their normal air speed of around 18 km/h this corresponds to a virtual flight of 72 km. Control experiments were undertaken to make sure there were no directional biases, including purposely turning butterflies into other directions, which would be equivalent to turning the apparatus, and observing that the butterflies always returned to the proper geographic direction. To test their use of the sun for orientation, the butterflies were first kept indoors for at least 5 days, then subjected them to three different light/dark sequences to reset their *circadian* clock: (a) 6 h time-advanced sequence; (b) 6 h time-retarded sequence, and (c) normal day/night sequence. At the time and place where the experiments were run, the 6 h azimuth movement of the sun varied from 91° to 115°. Therefore, if the butterflies use the sun for direction then the 6 h shifted butterflies should fly in directions displaced in equal and opposite amount relative to the normal group. For this part of the experiment, the time-displaced butterflies were "flown" in a clear *Lucite* simulator outdoors on a clear sunny day. The time-shifted results clearly verify that the directions of flight are shifted by the predicted amount, both in magnitude and direction, relative to the direction of normal migration, proving that they take their directional cue from the sun. For the magnetic test series, the butterflies were "flown" in a dense translucent simulator to prevent any residual cues from sunlight. No evidence was found that rotating an applied magnetic field over the simulator produced any change in the directional behavior of the butterflies, showing conclusively that monarch butterflies do not respond to the earth's magnetic field for navigation. In summary, this series of experiments using a flight simulator under tightly controlled conditions yielded excellent data over long flight paths, and convincingly establishes the role of the sun; it lays to rest the contention that magnetic orientation is involved in the monarch butterfly migration.

Recent studies have been conducted to throw light on the molecular basis for the behavioral differences between summer (nonmigratory) and migratory monarch butterflies. Haizun Zhu et al. [20] identified 23 genes in the brain of monarch butterflies, genes that correlate with a juvenile hormone that separate reproductive from nonreproductive monarchs. Genes involved in longevity, fatty acid metabolism and innate immunity are enhanced in nonreproductive migratory butterflies. Other recent studies have dealt with the circadian clock of the monarch butterfly and the effects at the genetic level of exposure to the light/dark cycle and ultraviolet radiation. Froy et al. [21] report that continued exposure of the butterflies to *constant* light intensity disrupts the basic circadian clock function at the molecular level, with a consequent change in behavior. The time marker used to define the beginning of the circadian cycle is the time when the adult butterfly emerges from the *pupa* (*eclosion*). A 6 h shift in the light/dark cycle in the laboratory produced a corresponding shift in the adult emergence. This rhythm is unaffected if it is followed by constant darkness. On the other hand, constant *light* disrupts the timing of the circadian clock, and therefore the butterfly's ability to properly correct for the movement of the sun in directing its flight. Curiously, exposure to constant light was found to cause a tendency to direct the monarch's flight toward the sun itself. Froy et al. also investigated whether the ultraviolet part of the solar spectrum played any observable part in monarch butterfly navigation, as it is known to do in some other insects. To investigate this question, the butterflies were allowed to firstly reach a steady flight pattern outdoors under a clear sunny sky, then an ultraviolet filter was put over the simulator to block all radiation of wavelength shorter than 394 nm from reaching the butterflies. It was found that the butterflies simply *stopped* their directed flight and resumed flying only after the UV filter was removed!

1.9 Long-Distance Migratory Birds

Of the many examples of animal navigation, none are more wide ranging or more challenging to track than the transoceanic migrations of sea birds such as shearwaters and arctic terns. The shearwaters are relatively small seabirds (less than 1 kg) that fly fast with much gliding and planning close to the water's surface or diving beneath it. The Sooty shearwater has the distinction of migrating the farthest of any bird. It flies the full length of the Pacific Ocean starting from New Zealand all the way to the north Pacific, in constant pursuit of good feeding grounds. The Manx shearwater (the prefix Manx refers to Isle of Man, where presumably a colony once existed) also has a respectable distance of migration. It nests in deep burrows on Skomer, a small island on the southwest coast of Wales, and every fall it embarks on an annual long-distance migration that takes it south along the west coast of Africa, across the Atlantic to South America where it winters on the coast of Argentina. The return journey begins in a northerly direction toward the Caribbean and then northeasterly across the North Atlantic Ocean to its burrows at Skomer where it nests.

The navigational ability of the Manx shearwater first captured the interest of the Cambridge biologist Geoffrey Matthews in the early 1950s [22]. His pioneering work was mainly a systematic study to establish the extent of the bird's ability to navigate to its home burrows over unfamiliar *land* routes from different points in Britain. There had previously been homing experiments in which incubating birds were taken to distant points, in one case as far as Italy, with inconclusive results.

A unique transatlantic crossing by one Manx shearwater reported in the *The Times* in 1952 [23] briefly captured public interest.

The difficulties of tracking transoceanic bird migrations have, in recent years, been much alleviated with the development of miniature GPS receivers. These have already been used to study the flight of the Manx shearwater on its feeding excursions to the colony on Skomer Island during its incubating period. In a recently published article, Guilford et al. [24] describe the application of a different device: a miniature archival (recording) light logger weighing only 2.5 g to track the activities of Manx shearwaters on their annual long-distance migration from Wales to South America. These miniature light loggers, also called *geolocators* to emphasize their ability to estimate geographical position, carry miniature photometric and other types of detectors with a time stamp that records the exact time of sunrise, sunset, and noon, as well as other events, such as diving under water. They have a computer chip with integrated software to determine geographical position based on the input from the light detectors. The latitude calculation is based on the length of daylight, and the precise time of noon with respect to Greenwich is used to estimate longitude.

The tracking project carried out by Guilford et al. used six established breeding pairs of Manx shearwaters on Skomer Island, whose coordinates are approximately 52°N, 5°W. The miniature geolocators were attached to the legs of the 12 birds and released in late August. All 12 birds returned to the Skomer breeding grounds the following season and the geolocators retrieved successfully. The recorded data showed the path of migration to be southward along the west coast of Africa, across the Atlantic Ocean approximately in a southwesterly direction to the Brazilian coast, and finally south to the overwintering grounds south of the Rio de la Plata in Argentina, around 40°S. This is an area where the mixing of ocean currents is known to produce an abundance of fish. The return journey to their breeding grounds in Wales takes them on a northerly track veering toward the west in the direction of the eastern Caribbean, continuing north to the east coast of the USA, and finally east across the north Atlantic to Wales. The shortest recorded travel time during the southward migration of 7,750 km was achieved by a male in 6.5 days, including time spent in the water, presumably feeding. Allowing for the recorded times of immersion into the water, it is estimated that this bird averaged a flight speed of 55 km/h. The recorded data showed that the birds often made stopovers, spending days not making any migratory progress. Upon return to their breeding grounds, the females are observed to go through an egg-forming stage involving a "prelaying exodus" to forage in the Irish Sea.

The other long-range migratory seabird is the *sooty shearwater*, an abundant species that through sightings at sea is known to migrate over great distances in the Pacific Ocean. It is only recently that with the availability of the miniature geolocators of the type described earlier, that *individual* birds can be tracked and their migratory routes and behavior accurately recorded. Using this technology, they have been tracked by S.A. Shaffer et al. [25] from their breeding grounds in New Zealand for about 260 days covering a total distance of about 64,000 km, making it the longest animal migration ever recorded electronically. Their flight paths would naturally be expected to follow the most favorable prevailing wind patterns. Indeed, they begin their migration from New Zealand in the easterly direction of the local prevailing winds, until they leave the southern ocean and head north, at which point they appear to take advantage of the trade winds blowing in a westerly direction. Once in the cool waters of the North Pacific, they head for one of three feeding grounds, making lengthy stopovers there: Japan, Alaska, or California Each shearwater was recorded as having flown the entire Pacific Ocean in a figure eight, before returning south through a narrow region in the middle of the Pacific Ocean. The shearwaters exhibit well-defined seasonal behavior: during the breeding season in the southern spring/summer they migrate to colder Antarctic waters where they engage in frequent diving to depths of tens of meters, foraging for food. In the southern fall, they begin their long migration to the northern Pacific, where again they gravitate around the three areas previously mentioned to feed.

1.10 Pacific Salmon

Just as remarkable as the sooty shearwaters' migratory feats are those of the Pacific salmon. These fish are famous for their migrations from the rivers or lakes where they are born into the wide expanses of the Pacific Ocean to feed, mature, and then return to the very place in inland waters where they were born, to spawn and die. These migrations that may range thousands of miles over open ocean must pose extraordinary challenges to the fish's ability to find its way to the mouth of its natal river mouth, and then to further travel up the river to its spawning grounds. These migrations have been so consistently successful that spawning populations have evolved adaptations to their specific habitats.

Of particular interest is the way in which the timing and *imprinting* is accomplished in the juvenile salmon prior to their migration. Imprinting is a well-known phenomenon in animal behavior that has been much studied, for example, in ducklings. A.H. Dittman and T.P. Quinn of the University of Washington School of Fisheries [26] have studied imprinting and homing by wild salmon, in particular the freshwater habits experienced by juvenile salmon. Because of the difficulties of tracking salmon in the ocean, research has been limited principally to tagging and recovery studies, and orientation tests on juveniles in inland waterways. A highly successful method of tagging fish was developed in the early 1960s by Keith Jefferts [27]. It involves injecting into the head of the fish a fine magnetic steel wire 1 mm in length on which had been etched a digital code giving pertinent information, such as date and place of release, etc. Homing salmon tagged and recovered have shown that they converge on their natal estuary with remarkable consistency. The fundamental question of the mechanism by which salmon orient themselves at sea is not entirely resolved. Some claim that in effect a map and compass are used; others contend that only a compass is necessary. Juvenile salmon studied in large lakes have shown an ability to orient themselves with respect to the sun, polarization of sunlight, and the earth's magnetic field. Indeed, magnetite crystals have been isolated from the head of several salmon species. Their relevance to magnetoreception is yet to be proved.

The transition from open saltwater to inland rivers faced by returning salmon poses special adaptation problems requiring perhaps different orientation techniques. The experimental evidence is that navigation in freshwater to home in on spawning grounds is predominantly by olfactory cues. The olfactory basis of imprinting in salmon was originally proposed by Hasler and his colleagues [28]. This was tested by them experimentally by exposing coho salmon juveniles to one of two synthetic (presumably odoriferous) chemicals, and finding that upon their return to spawn, they were diverted to different streams scented with one of these chemicals, 1.5 years later!

1.11 Human Navigation

According to the eminent behaviorist Piaget [29] in his book on a child's conception of geometry, human spatial knowledge has a hierarchical structure that evolves in stages. A fundamental stage in this development is the construction of representations of the environment beginning with elementary sensorimotor experiences, and progressing to the inclusion of spatial and directional information to form a coherent *cognitive map*. There is, however, a danger of taking the cartographic metaphor too far. Revealing experiments have been conducted by Scholl [30] that distinguish between spatial representations formed by actual physical experience in the environment and those derived from a study of a cartographic map. Results on the ability of test subjects to point to different unseen objects led him to the conclusion that streams of egocentric views observed during movement provide the basis of human cognitive representations of the spatial environment. The truly amazing fact is that the representation cannot be a static one: it must be automatically transformed to any perspective depending on the way the subject is facing! This requires a significant part of the human brain to be dedicated to a sophisticated navigational system in which vision plays a crucial role. The vision system continually monitors optical flow and sends data to the brain where spatial relationships in three dimensions are processed and recognized from all points of view. This ability is crucial to the formation of a coherent reality in three-dimensional space.

Navigational impairment in humans can result from many different brain deficiencies, from poor neural encoding of visual information to impaired memory of spatial images. In the absence of orienting cues, a natural left–right asymmetry in the human body leads to movement in a circle. This explains the veering tendency of blind individuals, who lack visual cues; in this case, any cognitive representation must be constructed from tactile or other contact with the environment. To fully appreciate the complexity of the human cognitive apparatus, we need only consider the innumerable ways it can go wrong, but usually does not. For example, damage to some areas of the brain can cause a condition called *unilateral neglect* in which patients can respond to a left or right stimulus individually, but when presented with simultaneous stimuli, one side is consistently neglected. There are visual integration deficiencies that lead an individual to perceive an eye, a nose, a mouth, etc. but cannot construct a face, or see a chair, a desk, a window and not be able to conclude that it is a body.

Recent advances in virtual reality technology have made it possible to study human navigation under controllable conditions that closely duplicate actual conditions of travel. That is, the perspective of the test subject is of one immersed in his surroundings with an *allocentric* frame of reference, that is, one outside one's self, as opposed to *egocentric*. Tests by a number of researchers have confirmed that cognitive maps constructed with virtual reality are similar to those acquired in the real environment.

References

1. R.G. Golledge (ed.), *Wayfinding Behavior* (Johns Hopkins University Press, New York, 1999), p. 127
2. G. Kramer, Ibis **101**, 399 (1959)
3. W.T. Keeton, Adv. Study Behav. **5**, 47–132 (1974)
4. R. Wehner, M. Mueller, Naturwissenschaften **80**, 331–333 (1993)
5. S.T. Emlen, Auk **84**, 309–342 (1967)
6. S.T. Emlen, in *Avian Biology*, ed. by D.S. Farner and J.R. King, vol 5 129–219 (1975)
7. M. Dacke, E.J. Warrant, *Current Biology* (Elsevier, New York, 2013)
8. W. Wiltschko, R. Wiltschko, J. Comp, Physiology **109**, 91–100 (1976)
9. C. Walcott, J. Exp, Biology **70**, 105–123 (1977)
10. K. von Frisch, *The dance language and orientation of bees* (Harvard University Press, Cambridge, MA, 1967)
11. J.J. Love, Phys. Today **61**, 31–37 (2008)
12. M.L. Anstey et al., Science **323**(5914), 627–630 (2009)
13. H. W. Lissmann, J. Exp. Biol. **35** 451–486 (1958)
14. W.T. Keeton, Proc. Natl. Acad. Sci. **68**(1), 102–106 (1971)
15. F. Papi et al., J. Exp. Biol. **166**, 169–179 (1992)
16. D. Biro et al., J. Exp. Biol. **205**, 3833–3844 (2002)
17. J. Hagstrum, J. Exp. Biol. **216**, 687 (2013)
18. V.F.P. Bingman, T.J. Jones, J. Neuroscience **14**, 6687–6694 (1994)
19. H. Mouritsen, B.J. Frost, Proc. Natl. Acad. Sci. USA **99**, 10162 (2002)
20. Z. Haisun et al., BMC Biol. **7**, 14 (2009)
21. F. Oren et al., Science **300**, 1303 (2003)
22. G.V.T. Matthews, J. Exp. Biol. **30**, 370 (1953)
23. The Times (of London) June 28, (1952)
24. T. Guilford et al., Proc. R. Soc. B **276**, 1215–1223 (2009)
25. S.A. Shaffer et al., Proc. Natl. Acad. Sci. USA **103**, 12799–12802 (2006)
26. A.H. Dittman, T.P. Quinn, J. Exp. Biol. **199**, 83–91 (1996)
27. P.K. Bergmann, K.B. Jefferts, et al., Washington Department of Fisheries, Res. Paper **3**, 63 (1968)
28. A.D. Hasler, W.J. Wisby, Am. Nat. **85**, 223–238 (1951)
29. J. Piaget et al., *Children's Conception of Geometry* (Basic Books, New York, 1960)
30. M.J. Schell, J. Exp. Psych. **13**, 615 (1987)

2 Early Navigators

2.1 Desert Nomads

From the earliest times, long before the advent of the tools of navigation, humans had acquired path finding skills relying solely on their senses: whether it be nomads traversing expanses of forbidding deserts, or aborigines making tracks through the bush in search of game, or Pacific islanders sailing their canoes over vast stretches of ocean, humans developed the ability to find their way over increasingly wider areas on the earth's surface. Unlike animals that have evolved with certain programmed sensitivities to the environment enabling them for example to orient themselves with respect to the sun, or the earth's magnetic field, the natural human pathfinders acquired a heightened awareness of the physical surroundings that helped define their path. This goes far beyond the general level of awareness that would be typical of a casual observer with an untrained eye; it entails sharp observation, in the sense of seeing with great clarity the significance of subtle nuances in the changing environment as they move, and to remember these in a time-ordered way. The early human navigators developed their navigational lore through oral tradition over generations of men who ventured out on trackless expanses of sea and sand.

The navigational challenges attending a camel caravan crossing the dust blown shifting dunes of a desert are as daunting as those faced by Pacific islanders crossing the ocean to a distant island beyond the sight of land; getting lost in either case would truly seal their fate. The realization that their very survival depends on reaching their destination must surely have served as a powerful inducement to concentrate their attention with deadly earnestness on every possible cue to ascertain their path at every point along their journey. Such detailed observation, deeply impressed on the memory and accumulated over many generations, built up a body of empirical knowledge that extends from the motions of the heavenly bodies to the properties of shifting sand dunes in the case of the desert nomads, to the study of wind and wave in the case of the island seafarers. Before graphical maps came into existence, or even the art of writing, finding one's way depended on having a mental

F.G. Major, *Quo Vadis: Evolution of Modern Navigation: The Rise of Quantum Techniques*, DOI 10.1007/978-1-4614-8672-5_2,

map of significant cue patterns and waypoints stored in the memory, their relative directions, and the expected time of travel between them. Since major landmarks may be separated by long stretches of uncertain travel, the ancient traveler, on leaving one waypoint, would be guided by all the cues he could discern along the way to the next waypoint, interpreting them with reference to his mental map, which he had developed by having traveled the same route previously, or from detailed descriptions from those who had. In the case of desert nomads, it is generally assumed that because they live in the desert in the open under starry skies at night, they are therefore guided by the stars. While the desert inhabitants cannot fail to be in awe of the stars, their sole nightly companions, and to recognize patterns among them, nevertheless the evidence is, however, that among some nomads way finding is very much a down to earth activity. J.J. Hobbs [1], who lived among the Ma'aza nomads in the eastern Egyptian desert, found them to be remarkably ignorant of the stars. Similarly, the English traveler Tristan Gooley writes of a similar experience with a Tuareg guide in the North African desert. They relied mainly on the Sun and prevailing wind as well as topographical features such as cliffs, rocks, wadis, etc. Most striking about these nomads is their intimacy with their environment, giving names as mnemonics to subtle features that are remarkable only in some tribe-related way, perhaps where some incident occurred in the past. For example, the name *Jebel el Dibbaah* (slaughter mountain) is given to a place where the tribe's goats inexplicably succumb to some mysterious disease [1]. The Ma'aza sense of geographic orientation, that is bearing, is based primarily on the observed positions of the sun by day, and possibly the star patterns at night. Other secondary cues include the prevailing wind direction, the direction and shape of sand dunes, the growth pattern of any vegetation, etc. For the seafarer, the cues include the direction of swells and the wind, the color of the water, the presence of cloud formations, and the possible observation of sea birds and their flight paths.

2.2 Seafarers of the Pacific

Navigating on the open sea out of sight of land by ancient seafarers was generally limited to achieving a straight passage between the point of origin and landfall, if not driven off course by a storm.

The Pacific islanders who colonized the many islands spread out over much of the Pacific Ocean, could, by crossing the relatively short distance from one island to the next, arrive at the far reaches of the ocean. The longest stretch was to New Zealand, which was achieved by a hero of Maori folk lore named Kupe in the thirteenth century A.D. [2]. Polynesians and Melanesians, whose lives were intimately controlled by the ocean, sailed their outriggers and catamarans over open water without the benefit of instruments, or indeed a system of writing. They developed over many generations a star-based system of navigation passed down and elaborated as an oral tradition. Through a combination of accidental, storm-driven paths to discovery and purposeful explorations, they managed, hundreds of years ago, to colonize almost all the islands in the various archipelagos of the Pacific Ocean, stretching from the Hawaiian chain to New Zealand.

In order that a voyager be able just to traverse a trackless sea to reach a desired destination, without regard to its being the shortest way, it is sufficient to know what direction must be followed throughout the journey, from his starting point to the desired end point, and to have the means to maintain that direction. For this, he must use a frame of reference that has constant orientation, one ideally defined independently of his own particular position and immediate environment. There *is* such a global frame of reference: it is built around the natural invariance in space of the axis of the earth's rotation. Apart from an extremely slow, and for all practical purposes negligible, motion of the axis describing a cone, called *precession*, the axis is very nearly fixed relative to the "fixed" stars as well as with respect to the figure axis of the earth, which is defined by its rotational symmetry. To be useful, this frame of reference must be accessible from all points on the earth, with observable indicators of known fixed directions with respect to this frame. Just such a set of directional reference points are the fixed stars, fixed in the sense that unlike the planets, they appear not to move in *relative* position. Because of the enormous distances that separate them from the earth compared with the size of the earth, stars may be assumed to lie along fixed directions in space, no matter from what point on the earth they are observed. We shall take up the phenomenon of deviation from that property called stellar parallax in Chap. 4. To an earth-bound observer, the stars appear as points of light mounted on the inside surface of a huge black spherical dome, the *celestial sphere*, with the earth at its center. Since the earth spins on its axis at a constant rate, completing one turn in 24 h, to an earthling this celestial sphere appears to turn about the same axis in the opposite direction to the earth's rotation. This means that if we assume for a moment that the sun is somehow extinguished, we on earth would see the stars move in concentric circles, some completing the circles in full view, while others would disappear below the horizon and then reappear on the other side to complete the circle. In the northern hemisphere, the point about which the stars appear to turn is near *Polaris,* the North Star. Of course, in reality, when the sun is above the horizon, its brilliant rays are scattered by the earth's atmosphere to produce the bright blue sky of day that completely blots out the stars. If the earth were completely devoid of any atmosphere that can scatter the sun's rays, we would be able to see the sun among the stars as it appeared to move over the course of a year because of the earth's motion in its orbit around it. A line from the center of the earth to the sun rotates at a nearly constant rate sweeping out a plane called the *ecliptic*, completing the orbit in 1 year. It follows that in the actual presence of the earth's atmosphere, at any particular time of the year when the sun's direction points to a particular part of the celestial sphere, stars in that region will be above the horizon during the day and therefore invisible against the bright blue sky, while those in the opposite direction will be observable in the night sky. Thus at different times of the year, different star patterns, or constellations, become visible in the night sky. If one scans the horizon on a clear night, one sees that there are among the myriad stars some that are just above the horizon, of which some are setting on the western horizon and some are rising above the eastern horizon. Just which stars are on the horizon will of course depend on the precise hour and the time of year. But what is fundamental to the whole way-finding system known from ancient times as *navigating by the stars* is the fact that any given star will rise and set at points on the horizon that define directions making precisely the same angles to the direction of the earth's axis, no matter *when* (solar time) it actually rises and sets throughout the year. Because of the orbital motion of the earth around the sun, a given star will rise and set on the horizon about 4 min solar time earlier each night. This nightly shift will accumulate to 12 h in 6 months so that a star that rose just before sunrise at (say) 6 A.M. will, during the course of 6 months, rise earlier and earlier until it rises at 6 P.M. the day before. Since the stars appear to trace out concentric circular arcs, those whose directions are nearer the N–S polar axis will have smaller radii and will set (if at all) at a smaller angle to the horizon. The obliqueness of the crossing angle also depends on the latitude of the observer, as does in fact the exact point on the horizon where the crossing occurs. It follows that the closer a desired course is to the N–S axis, the faster the steering star's bearing changes as it rises, further limiting how long it can be used for steering without introducing serious error, thus making it necessary to have a sequence of many more stars to serve over a long passage.

For the ancient voyagers, the practice of *following a star* to reach a particular destination from a given starting point did not involve any astronomical instruments or knowledge beyond a total familiarity with the star patterns (constellations) visible in the night sky at different times of the year, and awareness of the apparent constant rotation of these star patterns during the course of the night. Over generations of seafaring by the Pacific islanders, a system of navigation evolved in which, for each given island of departure, the desired destination island is reached by steering by a particular star, or more commonly a sequence of stars, whose common point on the horizon where they rise or set indicates the direction which must be followed to arrive at that destination. Selection of the guiding stars depended on both the points of departure and destination. Consider such a passage between two islands. Since the surface of the sea conforms to the curvature of the earth, which is very nearly a sphere, the shortest distance between them is along what is called the *geodesic* line passing through them. This is the circular arc where the spherical surface of the earth intersects a plane imagined to be drawn through those points and passing through the center of the earth. Unfortunately, a course set at a constant bearing, having a constant direction relative to (say) north, does not, in general, follow a *geodesic*, but rather another curve, called a *rhumb*, and is therefore not the shortest distance. However, if the distance traveled is small compared with the radius of the earth, then the earth's surface over that distance is very nearly a plane. It follows that in this approximation the geodesic is a straight line. But a defining characteristic of a straight line is having a constant direction in space at all points along its length. Hence, in the immediate neighborhood of any point on the earth's surface (which may in fact extend over a hundred miles), the distance measured along a rhumb is approximately the same as along the geodesic and is therefore nearly the shortest distance between any two given points. If the distance between the points is so large that the curvature of the earth's surface is no longer negligible, then clearly a *rhumb* course would no longer be the shortest one to follow between the given points. This subject will be taken up again in Chap. 5.

Any lateral deviations from the intended course, whether intentional as in tacking, or drifts due to uncontrolled disturbances such as winds and ocean currents, required sufficient skill, if not to anticipate them at least to be able to detect them, so that timely actions can be taken to compensate for them. In the absence of the means to fix position at sea, the island navigator must rely on cues in his immediate environment, such as the strength and direction of prevailing winds, the pattern of ocean swells and currents, all of which may be subject to seasonal variation. Furthermore, he must have a thorough knowledge of the performance of his craft and its motion in response to these factors. This can come only after many years spent on the water, and assiduous teaching by master navigators of the tribe. Not only must the sequence of guiding stars for each destination be memorized (a sequence which differs for different starting points), but also the navigator must be able to tack against adverse winds, which requires repeated changes of course. This means he cannot simply keep following rigidly a particular sequence of stars, but must estimate how far has been traversed on each tack and correct his course accordingly. Considering that no instruments were used either to measure the passage of time or the speed of the canoe, it is indeed remarkable that these "primitive" people were such successful navigators.

2.3 The Star Compass

Considering the vagaries of the weather, a serious problem is the possibility of cloud cover obscuring one or more of the steering sequence of stars. In such cases, or when there may be a lapse in the presence of suitable bright stars on the horizon, it seems natural that other stars, not in direct line with the destination, would be used as reference by setting a course at some appropriate fixed angle to them. Thus, it happens at times that the steering is not directed toward a guiding star but aims to keep a particular star or stars possibly in the same constellation, aligned with different particular features on the catamaran, such as spars, sheets, etc. This is particularly likely when tacking, where numerous changes of course are required. This amounts to viewing the stars in the abstract as indicators of direction, rather than being unique markers identified with particular islands. The next logical step to flow from this crude concept of relative angular relationships is to construct what may be called by analogy with the mariner's compass, a *star compass*, in which the location (but not the time) of the rising and setting points of useful stars are marked around a closed figure representing the total sweep of the horizon. The inhabitants of the Caroline archipelago developed just such a star compass with the star positions, however, unevenly distributed around the figure of a square or circle, and like the mariner's compass, divided into 32 points [3]. This number owes nothing to Western ideas but arises naturally from subdividing the square into quarters and then bisecting these repeatedly into eighths, sixteenths, and finally 30 s. Since the Caroline archipelago is slightly north of the equator, the star *Polaris* is visible to its inhabitants who recognized its special property, calling it the *star that does not move*. It does not, however, play the same role with regard to the cardinal points as in the Western compass; rather the cardinal point is the rising position of the star *Altair* (in the constellation *Aquila, Eagle*) which has a declination around 8° making its point of rising slightly north of due east on the Western compass. This special role for Altair can be argued arises from the fact that the Caroline Islands stretch along the sixth parallel of latitude and therefore Altair rises and sets nearly vertically and passes overhead at the zenith. Aside from *Polaris* and its neighbors that the Carolinians observed to rotate above the horizon, every other star will rise and set at points on the horizon that are symmetric with respect to the N–S line. Of these, the stars that have come to have the status of identifying the points of the compass are primarily those whose positions, rather than their brightness, suitably cover the entire sweep of the horizon, albeit not at uniform intervals. Their N–S line, like the Western compass, is defined by *Polaris* and *Crux*, the Southern Cross. The latter having a declination around only −60° will therefore appear to follow a much larger arc around the South Pole than *Polaris* does around the North Pole. The principal arm of the cross remains pointing nearly toward the South Pole as it describes its arc from a point somewhat east of the pole, through a vertical position, to set at a point as far to the west. Among the northerly points are named groupings of stars, corresponding to *Ursa Minor*, and *Ursa Major*, for example, while toward the south the constellation *Crux* serves to define five separate points along its arc corresponding to the positions where it rises, sets, major arm vertical, and the two positions when *Alpha Centauri* rises and sets. It would be natural to assume that the island peoples of the Pacific, like other peoples, instinctively imagined patterns among the stars organized in certain groupings, to which names were given, and which to the trained navigators became instantly recognizable, even when overcast conditions permitted only one or two members of the group to be clearly visible. It has been remarked that the compass points are closer together for directions toward the east and west than for north and south; this probably reflects the fact that stars near the celestial equator are more suited for horizon point navigation, since they rise and set more nearly in a vertical direction than ones of higher declination. Lacking the concept of a navigational chart, or indeed a system of writing, the Carolinians nevertheless were able to develop their star compass as a useful navigational aid which brought together and organized what was known of stars that may be followed to sail from one island to another. For each island destination a line is drawn to the appropriate star point, which had to be memorized. A given line may correspond to other pairs of islands sharing a common relative bearing. As previously remarked, though each line is identified by the name of a principal star, in fact a sequence of stars appearing at the same point on the horizon had to be memorized. Knowledge of the named star points, the axes connecting pairs of islands and their star sequences was passed on from the master navigator to the young apprentices using seashells arranged around a closed figure, usually a circle or preferably a rectangle, since it is believed the corners serve to provide natural reference points to aid the memory in learning the star positions. Figure 2.1 gives a complete listing of stars that make up a Carolinian star compass, with Western names.

Fig. 2.1 The Carolinian star compass ↑= rising, ↓= setting

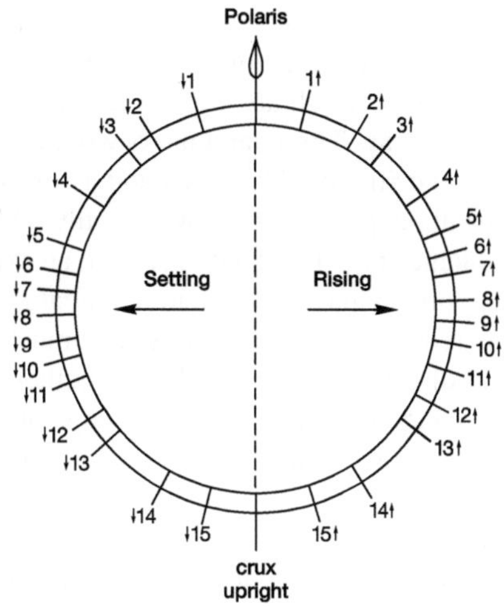

1. Ursa min.	6. Aldabran	11. γ-Corvi
2. Ursa maj.	7. γ-Aquil	12. Antares
3. α-Cassio	8. Altair	13. λ-Scorpi
4. Vega	9. β-Aquil	14. Crux
5. Pleiades	10. Orion	15. α-Centauri

As already remarked, the exact positions of the points of the star compass, apart from those at the poles and on the celestial equator, vary with the latitude of the observer, which would seem to severely limit the utility of the star compass as a navigational aid. But since the star sequence leading to a particular destination was arrived at through experience, it must already have taken into account the variation with latitude of the star point's absolute geographical bearing. Another effect of sailing over a wide range of latitude is that some stars may no longer appear above the horizon, and the visible star pattern continuously changes. Of particular importance is the Pole star which is not visible south of the equator; however, the island navigators have learned to estimate its position from the "pointer" stars of *Ursa major*. Sailing in an E–W direction, that is keeping a constant latitude, has no effect on the bearing of the star compass points.

Undoubtedly, the most challenging problem that faces a Pacific island navigator venturing out upon the ocean is to correct for the displacement of his vessel from the intended course by ocean currents and wind and then to be able to estimate what correction to his heading he must make to reach his destination. The exact bearing of the star points on the horizon vary somewhat with latitude; hence, it would be impossible to discern the insignificant change that would result from the drift, and therefore of no help in determining even the need for a correction. The ocean currents are particularly insidious, because the ocean water moves bodily carrying the canoe with it, and out of sight of land, there is no reference to indicate its existence. The ancient navigators, of necessity, evolved over many generations ways of allowing for ocean currents. One important technique is to set ashore landmarks which are sighted astern of the canoe while visible in the distance as it sailed away. These landmarks would serve as a reference with respect to which any lateral drift would be noted and assessed. Of course, this yielded essential information about the current flow, but only within a few miles from shore, where it may be influenced by the island itself. For an extended journey, there was no recourse but to rely on navigational lore to anticipate certain patterns of current flow on particular interisland crossings. A judgment must be made of the drift speed based on the rate and direction of current flow, together with the observed strength and direction of the wind. It still requires knowledge of the duration of the drift to obtain an estimate of the actual *course made good* during the drift. The problem is further aggravated by the relative slowness of the canoe, making the relative significance of the current greater.

Finally, we might mention what might be considered as a reenactment of a primitive Pacific voyage: the remarkable voyage in 1947 of the Kon-Tiki, a raft of balsa wood logs equipped with a sail that the Norwegian Thor Heyerdahl and his crew sailed 8,000 km from Peru across the Pacific Ocean. Since the object of the expedition was to show that early native South Americans could sail westward across the Pacific ocean using rafts of materials at hand, carried by the Humboldt current, Heyerdahl did not depend on any modern instruments to navigate; he relied strictly on the stars and ocean current.

2.4 The Phoenicians

Among the ancients, none so impressed their contemporaries by their seafaring skills as the Phoenicians, whose home was in the eastern Mediterranean, roughly coinciding with modern day Lebanon. From their cities of Byblos, Sidon, and Tyre, as early as the second millennium B.C., they sailed to the extremities of the known world, establishing colonies and trading outposts throughout the Mediterranean region and beyond the Pillars of Hercules, named by the Moors "Jabal Tareq" (Arabic for Mount Tareq) present day Gibraltar. Their principal colony in North Africa, Carthage, equaled Rome in power and culture, but ironically, though they were universally credited by their contemporaries with the invention of alphabetic writing, or at least for having introduced it throughout the Mediterranean, no significant literature of their own survives, aside from numerous fragments of inscriptions of a funerary or official nature. As Fredrico Mazza, an Italian historian [4] described it: "it is their singular destiny" that their cultural history, and magnificent achievements in navigation, trade and the useful crafts should be available to later generations only through the eyes of their enemies, the Greeks and Romans. The ancient Egyptians and Israelites were on more friendly terms, with exchange of trade and skilled artisans and seamen. It was the Romans who sacked the city of Carthage in the Punic wars and are guilty of the imperial destruction of its libraries, the entire intellectual patrimony of the Phoenician people. Therefore, in order to

Fig. 2.2 Bas relief of Phoenician merchant vessel (© The British Library Board, with permission)

discover the maritime legacy of these people, one is forced to rely on secondary sources: the writings of such Greeks as Homer, Herodotus, Plutarch, Strabo, and later Roman writers such as Cicero and Pliny. The description of the Phoenicians that emerges, even from the earliest writers, is an ambivalent one of a people whose reputation as skilled artisans and seafarers was unequaled, but who, according mostly to Roman writers, lacked morals, were devious, cruel, and cowardly. Any examples of a more positive Roman appraisal of these successful "barbarians," the Carthaginians, can be assumed to be either to mitigate a Roman defeat or inflate a victory by exaggerating the strength of the enemy. Their reputation however as skilled and innovative craftsmen was undisputed; it has been recorded that King Solomon selected Phoenician artisans to build his temple in Jerusalem. Their art in textile weaving and dyeing the Tyrian purple was much sought after in the ancient world; so much so that it became a mark of royalty. According to the Greek geographer Strabo, it was on the beaches of Phoenicia near Sidon that fine sand was found with which to produce their excellent glass. The Roman Pliny also confirms that glass was manufactured from sands found there, as were the first glass mirrors. The Phoenicians cannot claim to have discovered the vitrification process; it was known in other parts of the world, including Egypt and most likely China, but it is very probable that the Phoenicians produced a superior, more uniformly transparent glass, far beyond simple beads and bowls. Archeological discoveries, both underwater and on land, have yielded an expanding body of knowledge concerning Phoenician culture and commerce, including the design and manner of construction of their ships. On the evidence of the classical writers, there would seem to be no limit to the Phoenicians' skill in ship construction and the extent of their voyages of exploration and trade in distant lands. Figure 2.2 shows a bas-relief detail from a Phoenician sarcophagus found in Sidon; it shows a Phoenician merchant vessel dated around the second century B.C.

If the ancient historians are to be believed, the Phoenicians circumnavigated the continent of Africa almost 2,000 years before the Portuguese Vasco da Gama managed to round the Cape of Good Hope. They routinely sailed the length of the Mediterranean Sea, establishing trading posts and colonies in what is now Spain, Corsica, Sardinia, Malta, Southern France, and North Africa. They established sea routes beyond the Strait of Gibraltar into the Atlantic Ocean, one of which followed the coast of Africa as far as the Gulf of Guinea, and the other north across the Bay of Biscay, possibly as far as modern day Brittany, or even the tin mines of Wales. At the eastern end of the Mediterranean, they are known to have sailed in the employ of the Egyptian Pharaohs up and down the Red Sea to the Indian Ocean.

Since the Phoenicians are not known to have developed navigational instruments, it had been assumed until recently by scholars that the Phoenicians could sail only within sight of land and did so principally in daylight. Since the speed under sail can be presumed to be no more than two or three knots, it would mean that their staging posts for taking on provisions and taking refuge during the night could not be more than about 25 nautical miles apart. But a look at the map of the Mediterranean immediately shows that there are many passages between Phoenician trading posts that are in excess of 100 nautical miles in length, which could not possibly be covered in one day. Naturally, there were also coastal ships that engaged in fishing or local transfer of goods, which could layover during the night. The technique of coastal navigation is essentially based on visual cues on shore to determine bearing and to stay far enough out to minimize the risk of such hazards as going aground on shoals or breaking up on submerged rocks.

It is a testament to the skill of the Phoenicians as seamen that they were able to traverse long distances on what is at times a dangerously unpredictable sea, the Mediterranean. Violent squalls can erupt unpredictably with shipwrecking force. They were undoubtedly familiar through bitter experience with a number of more predictable seasonal winds that have, through the ages, become part of the vocabulary with names such as *Mistral*, *Scirocco*, and *Khamsin*. The *Mistral* is a very cold wind blowing in winter from the north, sometimes with gale force through the Rhone valley in France, bringing heavy seas. In contrast, the *Scirocco* and *Khamsin* are strong hot dust-laden southerly winds that blow in the spring from North Africa and the Red Sea in advance of a low-pressure system. As experienced sailors, the Phoenicians undoubtedly were alert to the vagaries of sailing the Mediterranean. Figure 2.3 shows the named prevailing winds of the Mediterranean Sea.

The fact that they could traverse the open sea from, for example, present day Tunis in North Africa to Marsala in Sicily, a distance that exceeded what can reasonably be covered in daylight, means that they had the means of preserving their bearing at night. The simplest explanation of how they accomplished this is to assume they used at least a rudimentary form of *star navigation*. Indeed, *Polaris*, the North Star in the constellation Ursa Minor, is said to have been called *the Phoenician Star*. Open sea navigation, however, was extremely hazardous for the ancients, fraught with the dangers of sudden changes in the strength and direction of the wind and heavy swells. Although the Phoenicians and later their descendants

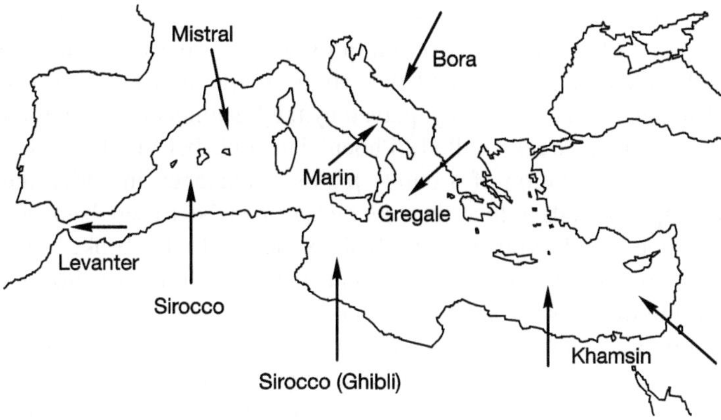

Fig. 2.3 The prevailing winds of the Mediterranean Sea

the Carthaginians had attained a remarkable degree of proficiency in the design and construction of seaworthy ships, to the envy of the Romans, nevertheless, these were light craft at the mercy of wind and sea. It would not be surprising therefore if coastal navigation was chosen wherever possible. The speed and maneuverability of Carthaginian vessels were so superior to that of the Roman vessels, that according to Herodotus, the preeminent Greek historian of antiquity, the Romans reverse engineered a wrecked Carthaginian man-of-war to design and build their own fleet.

Herodotus in his *History* also gives an interesting account of the Egyptian Pharaoh Necho (610–594 B.C.) and his desire to find a sea route from the Red sea to the Mediterranean by *circumnavigating Africa*! Apparently, it did not occur to him to dig a canal through the Suez isthmus, although according to Herodotus a channel to join the Nile River and the Red Sea was begun, but soon abandoned. He entrusted the expedition to sail around Africa to a Phoenician captain who was to fit out the ships with provisions and Phoenician crews. Obviously, the Pharaoh's advisers had no idea of the size of the African continent; they possibly assumed that the known land of Ethiopia was at its southern extremity. The expedition was to sail south on the Red Sea round the southern end of Africa, sail back north on the west side of Africa, and enter the Mediterranean through the *Pillars of Hercules* (Gibraltar). Many have accepted this account on the strength of Herodotus' standing as an ancient historian, often dubbed *the father of history*. The contention that this expedition succeeded, while theoretically possible since it relies only on coastal navigation, and the prevailing winds would favor the chosen direction over the one taken by Vasco da Gama; nevertheless, there is considerable skepticism as to the authenticity of this account.

A similar controversy surrounds the notion that the Phoenicians introduced the alphabet to the Greeks. Again Herodotus explicitly states in his *History* that "These Phoenicians who came with Cadmus .... introduced to Greece on their arrival a great variety of arts, among the rest, that of the alphabet, which up to that time, I believe, was unknown to the Greeks." Again out of respect to Herodotus, such eminent writers of antiquity as Pliny the Elder unequivocally state that the Phoenicians invented the letters of the alphabet. However, now scholars disagree on the role, if any, played by the Phoenicians in the introduction of what the ancient Greeks called "Phoenician alphabet" to the Greeks. The beginnings of the evolution of a purely *phonetic* alphabet, in which the letters represent simple sounds rather than *ideogram*, can be traced back to around 1700 B.C. when the Egyptian hieroglyphics began to be broken down and simplified; concurrently, in Mesopotamia the cuneiform writing was also evolving in the direction of simplified characters assuming phonetic values. On Crete a phonetic system of writing also contributed to the development of a phonetic alphabet. The confusion over the origins of the alphabet arises partly from differences in nomenclature, that is, differences in precisely to whom the name "Phoenician" and "Canaanite" apply, and from the existence of many interacting influences and cultures. The consensus seems to be that the alphabet was not "invented" but rather evolved from the melding of these sources to produce "Syrian" (Canaanite) or Phoenician writing. The form in which the alphabet arrived in Greece represents the last Phoenician phase.

2.5 The Vikings

The Vikings were sea-borne marauders that in the period between about 793 A.D. and 1066 A.D., plied the North Atlantic from points in the south and west of present day Norway to the east coast of England and beyond. The date taken to mark the end of these unprovoked attacks on England is 1066, the fateful year that William I, Duke of Normandy (commonly given the sobriquet "William the Conqueror") defeated the Anglo-Saxon King Harold at the battle of Hastings and brought French cuisine to Britain. The Normans had the Vikings to thank for their victory at Hastings because Harold had been busy fending them off from northeast England when William invaded the south, and Harold's battle-weary army had to regroup and march a considerable distance across England to face the Norman army. Contrary to heroic accounts of their exploits, the Vikings were bloodthirsty and destructive, who laid waste defenseless monasteries, and stole anything of value. Particularly atrocious was the sacking in 793 of the monastery associated with the English Saint Cuthbert.

But our only interest in the Vikings is in how they managed to make their way across the North Atlantic in what are often overcast and foggy conditions that may last for days. According to the Icelandic sagas, the Vikings not only invaded England wreaking destruction there, but under Eric the Red further sailed to Iceland, establishing a colony there. Eric's son, Leif Ericson, who was born in Iceland, continued his father's sea exploration, sailing farther west to Greenland, and some would say, all the way to North America. Very little substantiated knowledge has been unraveled about the navigational knowledge of the Norsemen, but there can be no doubt about their courage and determination to press ahead in the face of the unknown. One can speculate that they were driven by some mystical desire to follow the sun. It is known that if they kept the sun at noon at the same elevation in the initial exploratory journey from Iceland to Greenland, thereby maintaining a course along a roughly constant latitude, their ship was bound to hit Greenland, since it is so big! This contrasts with the exploits of the Pacific islanders who must find small islands scattered in the middle of a vast ocean. The principal shipping route the Vikings took from Norway to Greenland was along a constant latitude 61° N; however, once a particular point on Greenland had become the desired destination for a return journey, then of course navigational accuracy becomes necessary. The magnetic compass was still far in the future. Under clear skies, the Vikings relied on the sun; they undoubtedly would have observed that the arc traced by the sun differs between summer and winter and is south of the overhead (*zenith*) position at midday, passing *due south* of it at that hour. Archeologists have in fact found what is clearly a primitive sundial with the gnomon (the part that casts the shadow) perpendicular to the plane on which the shadow is cast. The refinement introduced by Arab astronomers of setting the gnomon at an angle to point toward the pole star was to come much later. The wooden face of the Viking sundial had carved on it curves presumably indicating the paths taken by the tip of the gnomon's shadow as the day progresses at different times of the year. If the length of the shadow at noon is plotted for different times of the year, the result is known to be an elongated figure of eight with the esoteric name *analemma*. The essential fact is that the direction of the shadow when the sun is at its highest would indicate the direction of north.

Of course, an important prerequisite to using a sundial is a visible sun! The question of how the Vikings avoided getting lost in a fog which may last for days has still to be answered. It was the Danish archeologist Thorkild Ramskou [5] who in 1967 first suggested that the Vikings may have discovered a certain crystalline mineral with the remarkable directional property of changing transparency when turned while viewing the sky at different angles to the sun. In modern terms, that amounts to being able to analyze the *polarization* of sunlight reaching their craft after it had been scattered by fog or a thin layer of cloud [6]. The suggestion brings to mind the studies on the use of daylight polarization by ants and bees, particularly the work of the Austrian biologist Karl von Frisch, whom we met in Chap. 1. Intrigued by the descriptions of the *Solstenen* (sunstone) in the Viking sagas, Ramskou pondered what possible use, if any, the Vikings had for what was evidently a very special kind of gem stone with unusual optical properties. If the sunstone was actually a crystalline mineral such as calcite or tourmaline, it may very well have been birefringent or even dichroic, that is, doubly refracting with two sets of optical properties: one involving what are called *normal* rays and the other *abnormal*. These rays have mutually perpendicular polarization directions and in the case of tourmaline differ greatly in their absorption by the crystal, resulting in only one able to pass through with its unique polarization direction: as such it would act as a polarization filter. Very recently [7], Guy Ropars and his colleagues at the University of Rennes in France reported on experiments using crystals of calcite to demonstrate the ability to use the polarization of scattered

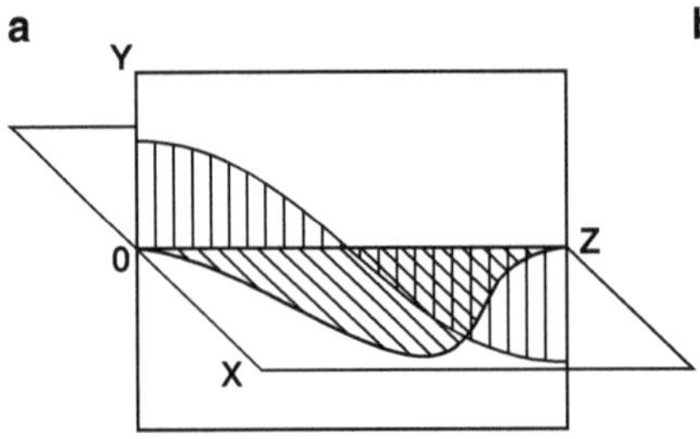

Fig. 2.4 Illustrations of different polarizations of light: (**a**) transverse field components in circular polarization and (**b**) linear polarization

sunlight to deduce the direction of the sun. Their work showed that by rotating the crystal it is possible to completely depolarize the light at a unique angle from which the direction of the sun can be accurately deduced. To add credence to the theory that the Vikings used the solstenen as a navigation device, a block of mineral found in a ship wreck 200 years before the advent of crystal optics was found to be calcite.

It might be well to pause here to recall what is meant by "polarization" in this context. Since Clark Maxwell's work in the nineteenth century, we have known that light under certain conditions manifests itself as a *transverse* electromagnetic (em) wave, by which we mean that the direction of its oscillating electromagnetic field is *perpendicular* to the direction in which the wave is traveling. The polarization state is simply the direction of the field in a plane perpendicular to the beam. An unpolarized beam has the field direction randomly directed to points on a circle around the beam as axis, whereas in a polarized beam the field direction may vary coherently along the axis, as in Fig. 2.4a for circular polarization, or Fig. 2.4b, linear polarization. A polarization filter that selects one type of polarization from an unpolarized beam can be constructed using any one of a number of birefringent or dichroic crystals. The result is a (colored) polarization filter that finds applications in many optical systems and is most familiar as "Polaroid" sunglasses.

Perhaps Ramskou was cognizant of the fact that in modern times an airline flying in the arctic region had made use of what is called a *Twilight compass*, which is one based on the analysis of the *polarization* of sunlight scattered by particles in the lower atmosphere. The fact that sunlight reaching the earth is scattered by air molecules and other particles in the atmosphere has been studied at least since the nineteenth century, but particularly in recent times because of concern about air pollution; so there is a very extensive literature on the subject. The most obvious effect of this scattering is the very blueness of the sky, and the redness at sunset. It was Lord Rayleigh who showed that the degree of scattering is proportional to the frequency of the light wave raised to the fourth power. This means that light at the higher frequency blue end of the solar spectrum is scattered far more than the red. Rayleigh's theory also explains the fact that sunlight is partially polarized, that is, the transverse electromagnetic field at different points along the beam points preferentially in one direction. It remains to apply Rayleigh's theory to determine what is required in order to uniquely find the direction of the sun using a polarizing

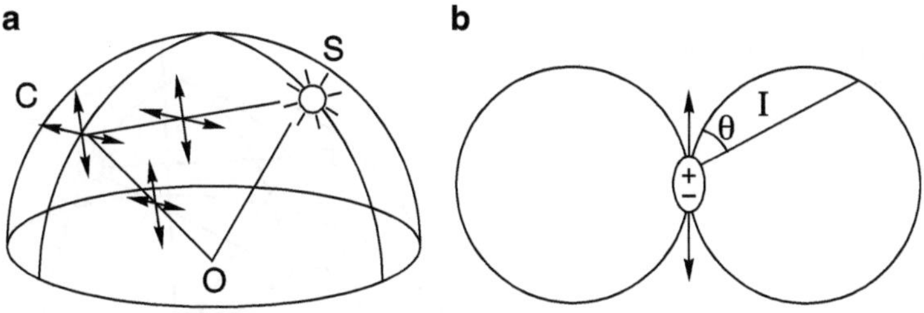

Fig. 2.5 (**a**) The polarization of light by scattering from atmospheric cloud cover; (**b**) radiation pattern of induced electric dipole oscillation

filter. In his scattering theory, the assumption is made that light is scattered from particles so small compared with the wavelength of light that their shape is irrelevant and they may be considered to be structureless. They are stretched by the electric component of the incident light wave to produce an oscillating electric dipole which reradiates an electromagnetic wave that constitutes the scattered light (see Fig. 2.5b). We will not bother with the mathematics but simply state in Eq. (2.1) the result for the intensities of the light polarized in a direction *perpendicular* to the scattering plane I_T, and *parallel* to that plane I_p. The scattering plane is simply the plane that passes through the sun, the scattering air particle, and the observer.

$$I_p = I_0 k^4 \alpha^2 \cos^2\theta / r^2 \quad I_T = I_0 k^4 \alpha^2 / r^2, \tag{2.1}$$

where α is proportional to the polarizability of the air molecule, k is the wave number of the light in wavelengths per meter, and θ is the scattering angle. The maximum polarization is observed when the direction of observation is at right angles to the direction from the sun to the point being observed in the sky, that is, when $\theta = 90°$, as shown in Fig. 2.5a.

One way in which the polarization filter may be used to locate the obscured sun is to first establish the direction of polarization of the filter by looking through it at some point in the sky away from a clearly visible sun and rotating it about the line of sight as axis. The brightness of the sky seen through the filter should vary between a maximum and minimum depending on the angle of rotation. At the angle when the brightness is at a minimum, a line is marked on the body of the filter to the direction of the sun. This line according to Rayleigh's theory should indicate the direction of polarization of the filter. With this "calibration" we are ready to find the direction of the sun when it is obscured by fog. Simply repeat the same kind of observation aiming the polarizer at two distinct widely separated points in the sky, each time noting the direction of the mark on the filter. If all goes well, the two directions to the sun should intersect at the position of the invisible sun. Mystery solved. Whether the Vikings actually did all that, while plausible, can never be certain. The procedure is illustrated in Fig. 2.6.

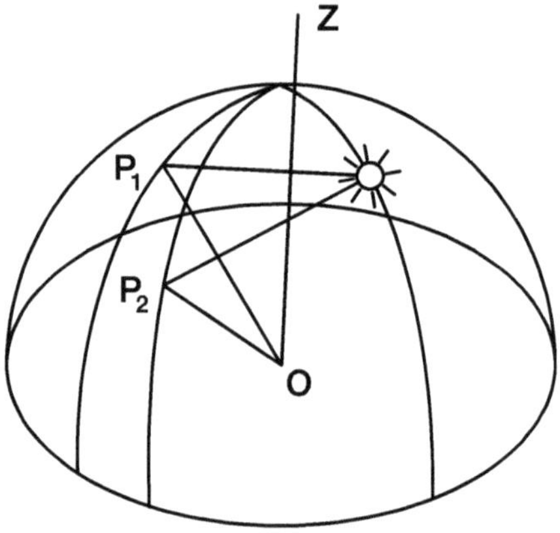

Fig. 2.6 Postulated Viking method of fixing the position of the sun under cloudy conditions

Recent studies [8] have been conducted on the polarization of daylight under different conditions of cloud cover and sun elevations to evaluate its suitability for what has been dubbed sky-polarimetric navigation. This has included statistical studies of the ability of persons to discern the position of the sun under somewhat overcast conditions. Curiously, some of the polarization studies have also been stimulated by the study of navigation by desert ants!

2.6 Ancient Instruments

2.6.1 Sundial and Nocturnal

The earliest devices for measuring the passage of time were naturally derived from the basic notion that time is defined by the natural cycle of day/night and the changing seasons. In order to keep track of the sun's journey across the sky, the simple *shadow clock* came into general use, with a vertical stick called a *gnomon* by the Greeks, casting the shadow. In its most primitive form, the shadow is cast on a horizontal plane marked with lines corresponding to subdivisions of the day. The principle was applied in many different forms: one of the earliest ancient Egyptian shadow clocks used the shadow of a *horizontal* bar placed in a N–S direction above a horizontal scale running E–W. These early shadow clocks did not indicate the time in hour intervals but larger significant subdivisions of the day.

The division of the day into 24 h is traceable back to the ancient Sumerians, who inhabited the region known in classical times as Babylonia [9]. They used a sexagesimal (base 60) system of numbers, reflected in the way the hour is divided into 60 min and the minute into 60 s. The same applies to angular measure: 360° in a complete circle, 60′ in a degree, and 60″ in a minute. More will be discussed about Sumerian contributions to astronomy in the next chapter. The Sumerian version of the shadow clock suffered from one critical flaw: it could not serve as a standard of time because the shadow of a vertical shaft clearly moves at a variable rate over the span of a day, and what is worse the variability itself changes with the season and latitude. To correct for this would require a sophisticated knowledge of celestial mechanics; it was Arab astronomers in the Middle Ages that hit upon an elegant solution. This consisted simply of setting the direction of the gnomon not in the vertical direction but inclined in the direction of the north celestial pole, that is, in the direction of the axis of rotation of the earth. This direction is currently within about 1° of the direction of *Polaris*, the North Star. This seemingly simple change transformed the shadow stick into a serious instrument for the measurement of time, the *sundial*.

An instrument somewhat analogous to the sundial was the marine instrument called *the nocturnal* which was used to measure time at *night* by observing the angle of apparent rotation of the night sky (*the celestial sphere*) about the north celestial pole. It consisted of two concentric circular dials, one larger than the other, rotatable about their common center. On their outer rims were engraved equal divisions, twelve main divisions on the outer rim corresponding to the months of the year and 24 on the inner rim corresponding to the hours of the day. An arm extending beyond the discs with its straight edge lined up with the center was pivoted to slide over the circular scales. On a clear night in the northern hemisphere where *Polaris* is visible, the instrument was held at arm's length and the North Star sighted through a hole in the center of the instrument provided for that purpose. The index arm was then turned to line up with two stars in the constellation *Ursa Major* called *the pointers* because a straight line joining them when extended comes within 1° of *Polaris*. As time passes, the rotation of the earth makes that line appear to rotate, and the angle of rotation is a measure of the *sidereal* time elapsed. The difference between sidereal and solar time will be discussed in Chap. 4.

2.6.2 Magnetic Compass

The recorded history of the use of the magnetic behavior of a certain mineral begins in ancient China perhaps as early as the fourth century B.C. but certainly by the time of the Han Dynasty in the second century B.C., when instruments of divination resembling compasses were in use. These consisted of pieces of the magnetic mineral *lodestone* finely carved into the shape of a spoon resting on a disc representing heaven, which was itself centered on a square bronze plate representing the earth. On the bronze plate, the so-called *divining board*, were inscribed various symbols of I Ching, the ancient Chinese work of divination, as well as markings indicating the azimuth angles of constellations, probably of astrological significance. Initially, these Chinese "compasses" had nothing to do with navigation in the sense we understand the term. It had more to do with present day *Feng Shui* (pronounced "feng shway" meaning wind and water) which has seen a revival even in modern times to determine the most propitious arrangement of objects in space, or time to embark on some activity. By the time of the T'ang dynasty in the seventh century A.D., the Chinese were able to magnetize iron needles by simply rubbing them with lodestone. In the eleventh century, the records show that it had been learned that an iron needle could be magnetized by heat treating it while positioned in a N–S direction. About that time also appeared a new design of "compass" in which a magnetized needle, which may in fact have the decorative form of a fish, is floated on a wooden platform in a bowl of water. The 24 markings of the "heaven" disc of the ancient "compass" were retained on the rim of the bowl.

Lodestone is a specific form of *magnetite,* a mineral with various complex chemical and crystalline structures, involving among other things, of course, oxides of iron. Magnetite in general is not uncommon; however, lodestone is rare. The attraction of iron to lodestone was known to Thales of Miletus, founder of the modern scientific method around 650 B.C. The question of why lodestone is found naturally magnetized raises the broader question concerning the magnetism of the earth as a whole. However, according to Wasilewski [10], the earth's magnetic field alone could not account for the magnetism of lodestone. He put forth the hypothesis that the magnetization occurs as a result of a lightning strike in which an intense transient high temperature and magnetic field would be produced simultaneously.

The origin of the earth's magnetism is conventionally assumed to be swirling currents of molten metal in the earth's core. Molten metal is pictured as surrounding an inner solid core which provides the energy for convective motion in the surrounding liquid metal [11]. If conditions exist in which a small fluctuation in the magnetic field, however small, can be reinforced by the magnetic field produced by induced electric currents in the moving molten metal, a *self-sustaining dynamo* can be produced. The theoretical challenge is to show that such a self-sustaining condition is possible, consistent with what is known or at least plausible about the interior of the earth. A new development in the theory of geomagnetism recently appeared [12], which shows how ocean currents may cause some secular (that is, long term) variations in the magnetic field at the earth's surface.

The gross distribution of the earth's magnetic field on its surface and beyond is as though a very large bar magnet was at the earth's center aligned very nearly with its axis of rotation, as shown in Fig. 2.7.

Of course, the detailed distribution of the strength and direction of the magnetic field varies over the surface due to variations in the magnetic properties of the local earth formations. As is obvious from Fig. 2.7, the direction of the local magnetic field at different points on the surface is generally not horizontal, except at the magnetic equator. The angle that the field direction makes with the horizontal is called the *angle of dip*. This angle can be measured by having a magnetized needle free to rotate about a horizontal axis. This dip in the magnetic field direction has the obvious effect of requiring a compass needle to be mounted in such a way as to prevent it dipping down as it tries to align itself with the field in a horizontal plane. The geographical location of the magnetic poles, where the angle of dip is 90°, is not exactly coincident with the geographical poles defined by the axis of the earth's spin; moreover, over the years, it can drift in position relative to it. Since the magnetic "meridians" do not coincide with the geographical meridians, that is, the N–S lines, the compass reading at any given point on the earth's surface differs from the true geographic north by what is called the *magnetic variation*. Although there is much less reliance on magnetic compasses now, navigational charts are

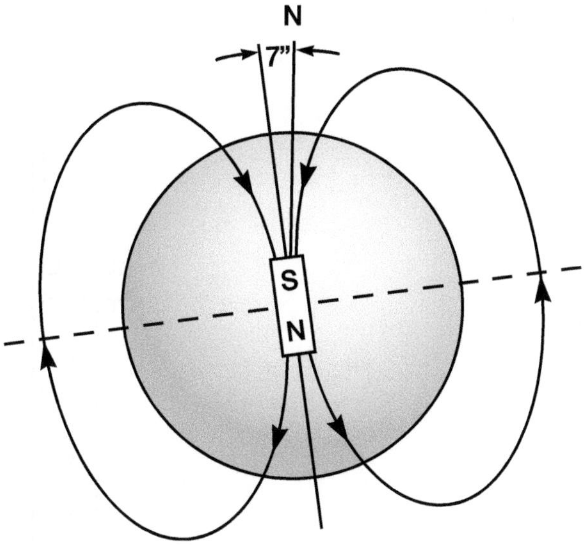

Fig. 2.7 Gilbert's view of the origin of the Earth's magnetic field

published in which the magnetic variation and amount of *secular* (long term) change that occurs in it per year may be shown. There are regions of the world where the magnetic variation changes over an anomalously short distance, detracting from the usefulness of a magnetic compass. Thus, in the southern Atlantic Ocean, there is a region of anomalously weak magnetic field at the earth's surface, a region that is subject to continuous change.

2.6.3 The Sandglass

The sandglass is a form of interval timer, familiar in the form of the hourglass, which is based on the time required for a given quantity of fine powder to flow through a constriction. Its European history evidently goes back only to the fourteenth century, although its liquid analog the *clepsydra*, the water clock of classical times, works on the same principle. It is hard to believe that a solid form of the clepsydra that avoids the messiness of a liquid would not have occurred to an inventive Greek or Phoenician. Of course, the clepsydra does have the advantage of continuous operation as long as there is a water supply. In fact, the finiteness of the time span an hourglass encompasses has become a symbol of the fleeting nature of man's life. Old father time is pictured carrying an hourglass and a scythe. Sand was not always the solid used: the choice is judged by the constancy of the flow; this means that the substance must not have a tendency to absorb moisture and change the mobility of its grains. The rate of flow of the grains through the hole clearly depends on grain size and shape as well as friction between grains. Generally, it was uncommon to have graduations on the outside of the glass to mark fractions of the

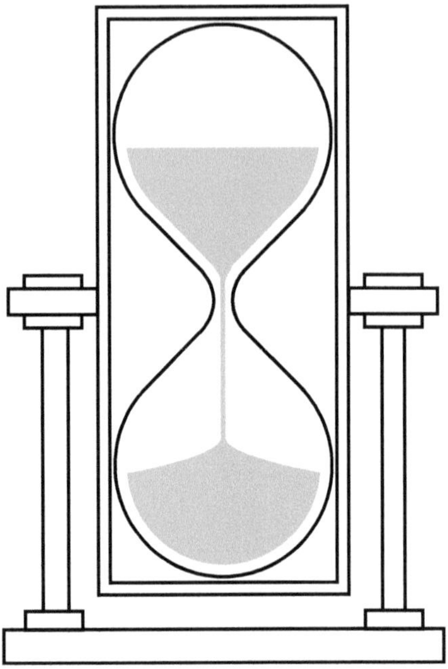

Fig. 2.8 Schematic drawing of a sandglass

hour; however, different size hourglasses were made that ran out after different chosen times. A typical sandglass is illustrated in Fig. 2.8.

The dynamics of flow of solid grains in a sandglass differ from liquid flow through an orifice, where the flow depends on the "head," to use a plumbers' term, that is, the hydrostatic pressure at the orifice. A *clepsydra* requires a supply of water in order to keep that head constant and therefore constant rate of flow. In the case of the sandglass, the rate of flow of grains through the constriction is expected to be more complicated. The theory has been worked out by A.A. Mills et al. [13]. They point out an essential difference from a water clock, namely that in the case of grains the rate of flow does not depend simply on the level remaining, except toward the end. Uniformity of size and shape of grains and low friction between them are important properties; the most reproducible results were obtained with glassy spherical grains of a material called *ballotini*. The theory shows that the time interval to complete one cycle is proportional to the bulk volume of *ballotini* and to $1/(D-d)^{2.5}$, where D is the diameter of a circular orifice and d is the maximum diameter of a grain.

Sandglasses have been used aboard ships as recently as the nineteenth century to mark the timing of the *watch*. A watch is a 4 h period during which one group of seamen is on duty; its end is marked by the sounding of a bell to signal another group to relieve them. A bell is rung every time the sand glass runs out and is flipped. Sandglasses measuring 1 h, one-half hour and one-half minute were carried on ships. One-half hour is the duration of one *bell*, 1 h that of two bells, etc.; a 4 h watch would be referred to as "*eight bells*."

2.6.4 The Chip and Patent Logs

The chip log was a contraption that in earlier times was used to measure a ship's speed in the water. It consisted essentially of a weighted wooden panel attached to a long rope that was knotted at regular intervals along its length and wound on a free turning reel. It was thrown overboard astern and as the ship moved forward, the rope is paid out with its knots moving relative to the ship. One crewman mans the half-minute sandglass, while another crewman yells out "mark!" as a knot passes him. At that instant, the second crewman flips the sandglass over, and then yells "mark" when the sand glass has run out. The first crewman having counted the knots that went by him between the two "marks," and knowing the distance between successive knots can calculate the speed of the ship relative to the water. The spacing of the knots in the rope is chosen so that the number of knots that pass a fixed point on the ship in one-half minute is numerically equal to the speed in *nautical* miles hour or *knots*.

The chip log was superseded in the nineteenth century by the *patent log* which consists of a screw rotator with four radial blades, towed from the stern of a ship at the end of a line, causing it to rotate. From the rate of rotation, mechanically displayed on a dial, the speed of the rotator relative to the water is deduced, and the speed integrated over time yields the distance traveled.

2.6.5 The Davis Quadrant

The precursor of the Davis Quadrant was the cross-staff. This ancient instrument is designed to measure the angle between the directions of two distant objects: in this context for example the altitude of the sun or star above the horizon. As the name suggests, it consisted of a long graduated staff with a sliding rod across it constrained to remain perpendicular to the long staff as it is moved along the length of it. For an altitude measurement, the crosspiece must also be held in a vertical direction. By placing the eye at the end of the staff and adjusting the position of the crosspiece along the staff, its two ends are sighted along the two directions under study. For example, one direction might be the horizon, and the other the sun. Knowing the length of the cross piece and reading its adjusted position off the linear scale on the staff, the tangent of the desired angle can be computed, and the angle found from a trigonometric table of tangents.

The cross-staff suffers from at least three serious drawbacks: first, the observer must look directly at the sun, second, the observer must alternate his gaze between two directions, and third his eyeball must be right up to the beginning of the linear scale on the staff. Nevertheless, it is very simple and was used for centuries; it was not until the seventeenth century that a more refined instrument was developed known as the Davis Quadrant, which is illustrated in Fig. 2.9.

Instead of viewing the sun directly, the direction of the sun is derived from the *shadow* that is cast by a shadow vane mounted on a small 60° circular scale above the main staff. Below the staff is a much larger 30° circular scale on which a sliding

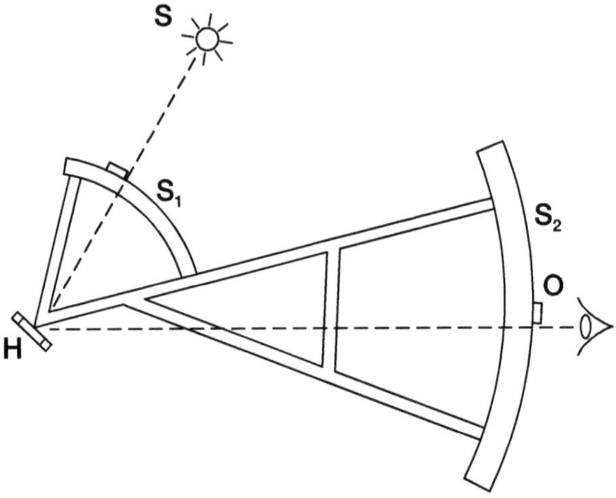

Fig. 2.9 Schematic design of the Davis quadrant

sighting vane is mounted. At the common center of the two circular scales is a horizon vane. By lining up the edge of the shadow with the horizon, the desired angle is read from the two circular scales. The use of reflective elements as in modern forms of a quadrant/sextant, an innovation attributed to Isaac Newton, will be discussed in a later chapter.

2.6.6 The Astrolabe

The *astrolabe*, an astronomical instrument of ancient origin was perfected by Arab astronomers in the Middle Ages. It is in essence an analog computational aid useful for tracking the constellations and determining latitude and time. In Fig. 2.10 is shown a photograph of an astrolabe similar to the one described in the *Treatise on the Astrolabe*, written by Geoffrey Chaucer for the benefit of his son's education [14]; the same Geoffrey Chaucer who wrote *Canterbury Tales*.

The astrolabe consisted of a disc on whose rim was engraved a uniform scale with 24 divisions, surmounted by a plate engraved with a projection of the celestial sphere as seen by the observer. The projection is specific to the particular latitude of the observer; another observer at a different location would have to change this plate. Pivoted at the center was a spidery (the Arabic name is "*Ankaboot*" which means spider) cutout star pattern of the night sky called the *rete*, on which was also inscribed the path of the sun among the stars as it appears to an earthly observer: the zodiac. Also, pivoted at the center is a straight pointer called the *rule*. On the back of the astrolabe were concentric circles consisting of a scale graduated in degrees, the houses of the zodiac, and calendar months. A pointer bar, called an *alidade*, spanning the full diameter was pivoted at the center. We will spare the reader the intricacies of using the Arab astrolabe. Sufficient to say that it is assumed the calendar month and day are known at the time it is to be used. The altitude of the sun is first observed using the angular scale on the back of the instrument to sight the sun. From that side also the user reads the position of the sun along the

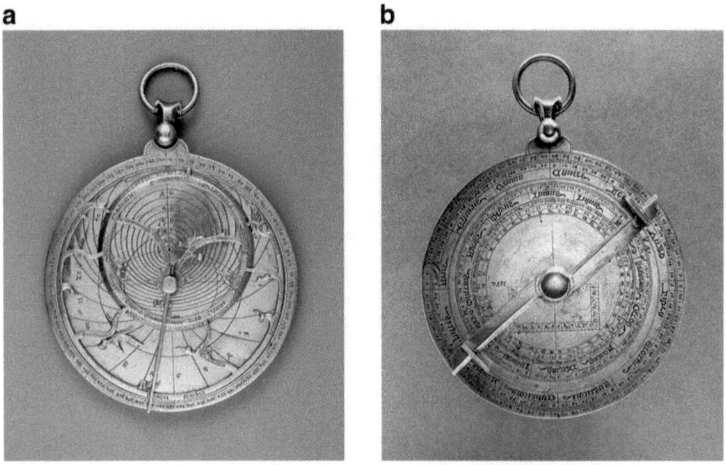

Fig. 2.10 An astrolabe from the time of Chaucer (1391): (**a**) front and (**b**) back (© The Trustees of the British Museum, with permission)

zodiac. This information is used on the front side to turn the star pattern with respect to the projection of the celestial sphere until the sun's position in the zodiac agrees with the observed altitude. The time is read by appropriately setting one end of the rule and reading the other end on the 24-h scale.

It should be pointed out that the Arab astrolabe was an *astronomical* instrument, far too sophisticated to appeal to the average mariner. Hence, the mariners' astrolabe was a radically stripped down version consisting mainly of the back of the Arab astrolabe that could be used to determine latitude.

References

1. J.J. Hobbs, *Bedouin Life in the Egyptian Desert* (University of Texas Press, Austin, TX, 1989)
2. M. King, *History of New Zealand* Penguin Books (2003); also A. Sharp, *Ancient Voyagers in the Pacific*, Penguin Books (1957)
3. D. Lewis, *We the Navigators*, secondth edn. (University of Hawaii Press, Hawaii, 1994)
4. F. Mazza, in *The Phoenicians* p. 628 Moscatti Ed Rizzoli Intl. Publ. N. Y. (1999): Auber, M.E., The Phoenicians and the West Cambridge University Press, 1996
5. B. Almgren et al., *The Viking* (Barnes and Noble, New York, 1995)
6. R. Hegedus et al., Proc. R. Soc. **A463**, 1081 (2007)
7. R. Guy et al., Proc. R. Soc. **468**, 671 (2012)
8. G. Horvath et al., Philos. Trans. R. Soc. B **366**, 772 (2011)
9. S.N. Kramer, *The Sumerians, Their History Culture and Character* (University of Chicago Press, Chicago, 1963)
10. P.J. Wasilewski, Phys. Earth Planet. Int. **15**, 349 (1977)
11. D.P. Stern, Rev. Geophys. **40**(3), 1–30 (2002)
12. G. Ryskin, New J. Phys. **11**, 063015 (2009)
13. A.A. Mills et al., Eur. J. Phys. **17**, 97–109 (1996)
14. St. John's College, University of Cambridge, Library website

3 Historical Background to Astronomy

3.1 Ancient Astronomy

The genealogy of modern astronomy is traceable back to the records of celestial events deciphered by archeologists working on the site of ancient Mesopotamia [1], so named by the Greeks because it lies between the two great rivers: the Tigris and the Euphrates, in what is now Iraq. These records consist of wedge-shaped cuneiform characters imprinted on clay tablets, the earliest of which date back to the third millennium BC. The Mesopotamians could not have chosen a more durable medium. Their number system is sexagesimal, that is, it is a place-value system with base 60; however, the representation of the numbers 0,1,2,.....59 is reminiscent of the Roman numerals in that these numbers are constructed by repetitions in just two distinct cuneiform characters: one vertical and the other horizontal, as shown in Fig. 3.1. They lacked a symbol for zero (whose origin is thought to be Indian) so the presence of a zero is shown simply by an empty space. Our division of hours into 60 min and minutes into 60 s, with similar base 60 divisions of degrees of arc into minutes and seconds of arc, is undoubtedly of Babylonian origin [2].

The ancient Mesopotamians displayed an extraordinary interest in matters astronomical. There is no surviving record of comparable fascination with the heavens among, for example, the great navigators of the ancient world, the Phoenecians. That is probably due to the thoroughness of the Romans in destroying Carthage. The most plausible explanation for the ancients' interest in the stars is the belief, particularly among some Mesopotamians, that the movement and patterns of stars in the night sky were very much relevant to everyday life on earth: the belief that what will happen on earth can be foreseen in the stars. Eclipses were considered particularly ominous; in other words, the detailed observations of the planets in relation to the stars as recorded in those ancient tablets were really about astrology.

Like other cultures, the Mesopotamians placed great importance on the construction of a calendar to regulate the timing of activities of the society in accordance with the cyclical changes of the seasons of the year. Clearly a calendar is essential for the timely performance of such mundane functions as the planting of

F.G. Major, *Quo Vadis: Evolution of Modern Navigation: The Rise of Quantum Techniques*, DOI 10.1007/978-1-4614-8672-5_3,

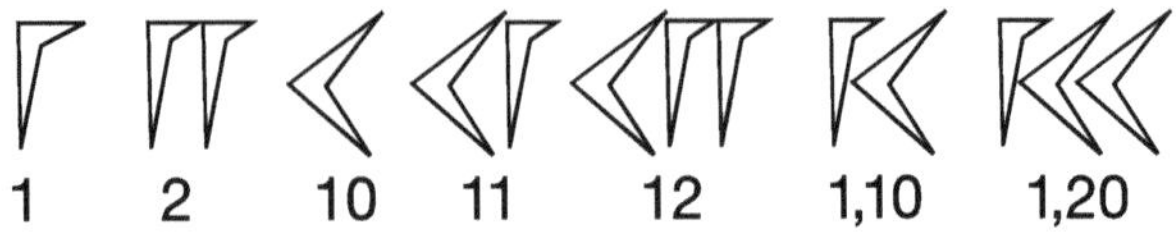

Fig. 3.1 Sexagesimal cuneiform number

seeds and harvesting of crops, as well as religious observances. The difficulty in constructing an accurate calendar is ultimately due to the inconvenient physical fact that the periods (times to complete a cycle) of the earth's spin on its axis, the earth's revolution around the sun, and the orbit of the moon around the earth are not in simple ratios. Thus the period of the moon is about 29.5 days and the period of the earth around the sun is about 365.25 days. To further complicate matters we have the conventional 7-day week, so there are not a whole number of weeks in a month, nor lunar months in a year. In ancient Mesopotamia the calendar was based on the motions of the sun and moon, but necessarily in a complicated way: The orbital periods of the moon (month) and the earth (year) are used, each month beginning on the evening when the crescent new moon is just visible for the first time on the western horizon. Since a month is not exactly a whole number of days, such a sighting may be judged to occur with roughly equal probability on the 29th or 30th day. This ambiguity was deemed unsatisfactory for a period in the third millennium BC and an administrative calendar was defined as having all months 30 days long. This practice survived in later astronomical applications. With this definition of the lunar month, a period of 12 months would be short of a year by about 5.25 days. Therefore in order to correct for this, intercalary days were inserted to bring the calendar into alignment with the seasons, on the same principle as the present day introduction of an extra day in February in leap years.

The Mesopotamians recognized the difference between the "fixed" stars and the planets, the wanderers, or according to modern scholars what the Mesopotamians called biblu, meaning wandering sheep! The Babylonians during the first millennium BC made extensive and detailed observations of the movements of the sun, moon, and the planets relative to certain prominent stars as reference. These observations were organized in a collection of clay tablets called the Babylonian Astronomical Diaries. Around the middle of the first millennium BC Babylonian astronomers reorganized the records in the Diaries to reflect a change in the reference system used to specify the positions of celestial bodies. In place of using prominent stars as a reference, it was noticed that the apparent motions of the sun and planets were confined to a narrow circular zone around the earth along the ecliptic, the apparent arc traced by the sun with respect to the fixed stars. We now know of course that since the earth and the planets revolve around the sun in planes that are nearly coincident, it follows that as the earth travels in its orbit, the sun and the planets appear to move within a narrow band relative to a background of distant stars. The angular position with respect to a chosen origin in this circular band, called the Zodiac, was used as the reference system. In modern astronomy the origin of the Zodiac is chosen as the position of the sun at the vernal equinox. This is the point in the earth's orbit where the earth's axis of spin is at right angles to the sun–earth direction. The Zodiac was divided into 12 equal segments, "houses" or

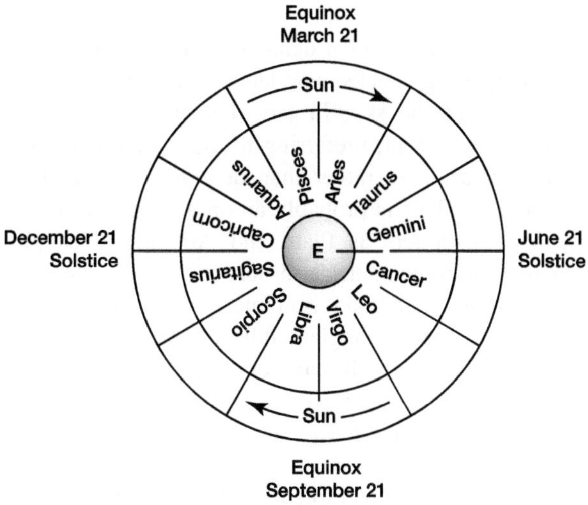

Fig. 3.2 The constellations ("houses") of the Zodiac

Table 3.1 Angular intervals of the houses of the Zodiac

Sign	Range (degs.)	Name	Sign	Range (degs.)	Name
Aries	0–30	Ram	Libra	180–210	Scale4s
Taurus	30–60	Bull	Scorpio	210–240	Scorpion
Gemini	60–90	Twins	Sagittarius	240–270	Archer
Cancer	90–120	Crab	Capricorn	270–300	Goat
Leo	120–150	Lion	Aquarius	300–330	Water carrier
Virgo	150–180	Virgin	Pisces	330–360	Fish

"signs", each named after the most prominent star or constellation in that direction, as shown in Fig. 3.2.

Special significance was given to the alignment of the sun and planets with particular constellations along the Zodiac: that is the basis of astrology. Meticulously detailed observations were made to determine the disposition of the sun, moon, and planets at the time of the birth of a child in order to cast a horoscope. In Table 3.1 are shown the constellations in the 12 signs of the Babylonian Zodiac.

Through the invention and refinement of an accurate calendar, and a systematic way of recording the motions of the sun and planets, the Babylonians contributed profoundly to the establishment of a science called astronomy. The fact that the original motivation may have been more to do with astrology than science does not dull the luster of their accomplishment. By making systematic time-marked observations and recording them over extended periods of time, they made it possible to discern order and periodicity in what would otherwise be incoherent and sporadic observations. This not only made possible the prediction of events, but also provided the basis for forming hypotheses as to the interrelationship between the phenomena observed, that is, creating models.

The expansion of Alexander's conquests to include Mesopotamia in 331 BC brought the Babylonian achievements in astronomy reluctantly into the arms of the conquering Greeks, who lacked a comparable tradition of systematic recording of events in the sky. In fact the ancient Greeks are reputed to have a more philosophical bent, preferring to make theoretical constructs of the universe and man's place in it, rather than deigning to actually make observations to verify hypotheses. This is probably an unfair characterization born of the dominance of such philosophers as Plato and Aristotle. This philosophical attitude may be appropriate to geometrical proofs such as the theorem of Pythagoras, where it can safely be assumed that it is not necessary to actually use a ruler to measure the sides of a right angle triangle—although, with modern distance measurement made possible with lasers, it may very well turn out that for a sufficiently large triangle it is not true—but we digress. The melding of Greek thought with the empiricism of the Mesopotamians led to important advances in the mathematical modeling of planetary motion based on Babylonian data, advances associated principally with the name of Claudius Ptolemy, a Greek living in Roman Alexandria [2].

Prior to the interaction with the Babylonians, Greek astronomers were preoccupied with questions which we would describe as cosmology: speculations about the universe and man's place in it. As early as the sixth century BC a Greek by the name of Pythagoras, immortalized for every pupil as the author of the *theorem of Pythagoras,* believed on philosophic grounds that the earth is a sphere because that figure is "perfect". A more rational reason for the earth's shape was given by another towering figure of Greek science: Aristotle (384–322 BC) whose writings describe the moon as not being luminous in itself, but that its changing shape is a result of the changing direction of the sun's rays illuminating its spherical form. He also postulated that the earth is spherical on the basis of *physical observations*, a departure from Greek tradition. Another Greek astronomer who deserves to be better known from the island of Samos (the same island where Pythagoras was born) was Aristarchus (310–230 BC). He believed that the earth revolves around the sun, a notion that was to cost Bruno his life more than one and a half thousand years later. The reason this was rejected by Aristotle and others at the time was the absence of observable parallax, the shift in apparent direction of an object as a result of a shift in the observer. This same argument was made by the Dane Tycho Brahe in the late sixteenth century. The argument in principle is perfectly correct, but it happens that even the nearest stars are so far away that the parallax is so small that it is observable only with modern telescopes.

During the third century BC Alexandria, Egypt, was a center of learning famous for its library, a repository of the ancient world's great works in the arts and sciences of Greek, Egyptian, Indian, and Mesopotamian origin. It was a major center of scholarship with vast holdings of scrolls which were destroyed by fire c. 48 BC under circumstances about which later historians do not agree. Alexandria was the home of a number of prominent astronomers referred to as belonging to the Alexandrian School. Among them was Aristarchus of Samos c. 310–230 BC, whom we have already mentioned, Eratosthenes (c. 276–196 BC), and of course the most famous of all, Claudius Ptolemy.

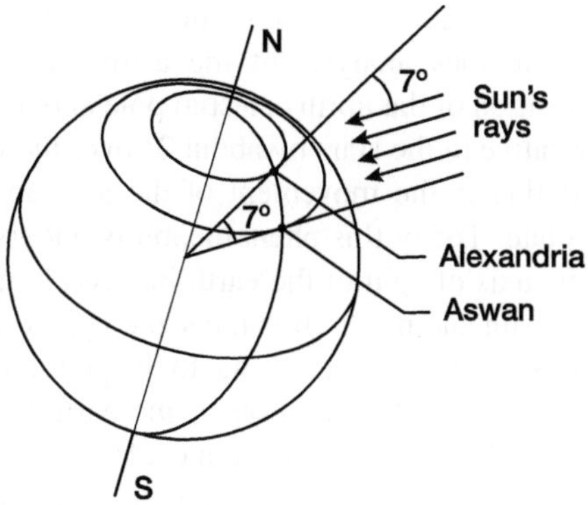

Fig. 3.3 The Eratosthenes experiment to measure the radius of the earth

It was Eratosthenes who devised a way to measure the curvature of the earth's surface [3]. It is based on the fact that the sun is so far from the earth that its rays reaching the earth are very nearly parallel; in fact the angle of divergence is only about one-third of one minute of arc! The circumstance that the rays of light approaching the earth from the sun, stars, and planets are, for all intents and purposes, parallel is essential for the art of celestial navigation, as we will see in a later chapter. The principle of Eratosthenes' experiment is simply to measure the difference in the direction, with respect to the local vertical, of the sun's rays reaching two widely separated points on the earth's surface at the same time, as illustrated in Fig. 3.3. The two points he used were Syene, Egypt (near modern Aswan), and Alexandria, 5,000 stadia north of it. (A stadium is an ancient unit of length about 1/6 km.)

It happens that Syene's latitude is about 23 degrees north (the latitude of the tropic of Cancer) and as he noted, at noon on the 22nd June (summer solstice), the sun's rays penetrated a deep *vertical* well, indicating that the sun was by definition at the zenith. But at Alexandria, at the same time the sun's rays were measured to be slanted a little over 7° from the vertical, or about 1/50 of a circle. Therefore, he argued that the surface of the earth is curved, confirming his preconceived notion that the earth is a sphere. The Alexandria–Syene distance (5,000 stadia) must therefore be 1/50 of the circumference of the earth, and the whole circumference must be 250,000 stadia. There is some uncertainty regarding the unit he used, so it is difficult to assess the accuracy of the result; nevertheless it represents a remarkable achievement of Greek empirical science.

The last and most prominent Greek astronomer of pre-Christian antiquity was unquestionably Hipparchus who was born in Nicea, Asia Minor, early in the second century BC, and lived most of his life on Rhodes, just off the coast of modern Turkey. Unfortunately all but one of his books are lost, and that one only a work of criticism. However his contributions to astronomy are reflected in the work of later astronomers, notably Ptolemy. Hipparchus is known to have compiled a star catalog in which a multitude of stars are listed with their celestial coordinates and magnitudes (brightness). A fuller description of these quantities is given later in the chapter. Through careful comparisons of his star coordinates with more ancient ones available to him, he made a truly remarkable discovery, which attests to the meticulous analysis of the astronomical data before him. He noticed that the position of the north celestial pole about which the stars appear to rotate had shifted relative to the stars by about 2° over the last 150 years! This he correctly explained as due to the movement of the axis about which the celestial sphere appears to rotate. Today this phenomenon is understood as due to the slow conical sweep of the axis of spin of the earth induced by gravitational forces acting on the earth by the sun and moon. It is the same gyroscopic effect as seen in a spinning top whose axis sweeps out a cone due to the gravitational pull of the earth trying to make it fall on its side. This motion of the earth's axis is called *precession* and takes about 26,000 years to complete a cycle.

Perhaps the one astronomer of antiquity that had the greatest impact on the development of astronomy was Claudius Ptolemy (or Ptolemaeus) who lived in Alexandria in Roman Egypt during the second century AD [2]. His actual birth date is uncertain, but assumed to be c. 100 AD. His writings ranged over many subjects including mathematical geography, astrology, and musical theory. But he is best known for his work in mathematical astronomy, embodied in a 13-volume treatise that goes under the Latinized Arabic title of Almagest. This work does not deal solely with Ptolemy's own work, but contains also a compilation of the achievements of earlier astronomers, mainly Hipparchus. It is so comprehensive in scope that it contains virtually all of Greek astronomical knowledge of the time. Building on the theories and observations of Hipparchus, Ptolemy's most important contribution was a geometrical model of the solar system that conformed to the Greek tradition of circular motion and yet predicted the observed motions of the moon and planets with considerable accuracy. By elaborating Ptolemy's geometrical construct, subsequent Islamic astronomers were able to refine the model into one which essentially stood until the time of Copernicus over 1,300 years later.

The source of the difficulty in describing mathematically the motions of the celestial bodies is ultimately due to the frame of reference being fixed on the earth, itself in motion with respect to the stars. As we now know, it is much easier to describe the motions of the planets if our coordinate system is fixed in the sun, and not in the earth. Of course at the time of Bruno, the argument was not about the choice of coordinate systems, but about the location of the center of the universe; his heresy went considerably beyond the motion of the planets. Bruno died to defend his right to speak out against church doctrine. Although the motions of the earth and planets with respect to the sun are relatively tractable, the motion of a planet relative to an earth that is itself moving is considerably more complicated. In order to describe it, Hipparchus and Ptolemy introduced eccentric circular motion and epicycles. By doing so they were able to explain the vexing phenomenon of retrograde planetary motion, in which a planet appears at times to go backwards along its orbit. As might be imagined, in ancient times this phenomenon inspired a great deal of awe and dread that it may be an omen related to some occurrence on earth. Thus to find a rational explanation was considered a great feat.

In attempting to construct a theoretical model of the solar system, Ptolemy was influenced by the Greek philosophical tradition established by Aristotle in which

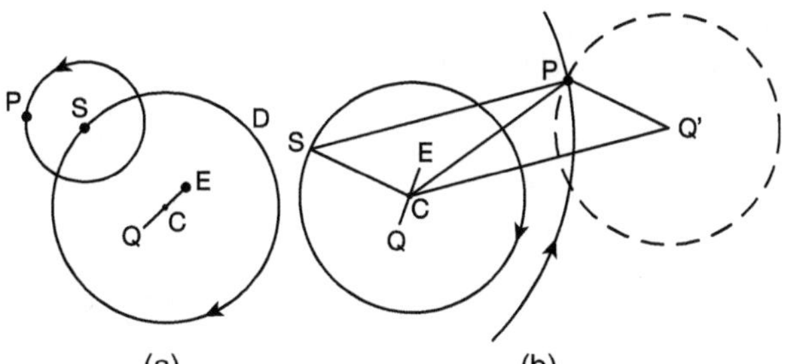

Fig. 3.4 Epicycles: (**a**) inner planet and (**b**) transformed outer planet motion

the universe beyond the moon is made up of a quite different *essence* from the earthly four: earth, air, fire, and water. It is a fifth essence or *quintessence* which obeyed different laws of motion. While the "natural" state of motion of the earthly four is rectilinear, the fifth element's natural motion is in circles. Therefore the problem of planetary motion he set out to solve was: what *circular* motions do the sun and planets execute that results in their observed positions from earth? The basic planetary model he described, before refinements were introduced to improve the accuracy of predictions, is based on the use of epicycles as illustrated in Fig. 3.4a, b.

An inner planet, such as Mercury or Venus, is postulated to move on an epicycle, that is, a circle around the sun whose center moves on a larger circle called the deferent with earth as center, as shown in Fig. 3.4a. The radius of the epicycle is the distance of the planet from the sun, and the radius of the deferent is the distance of the earth from the sun. It is not hard to prove that this corresponds merely to a mathematical transformation from a heliocentric system with the earth and planet moving in circular orbits around the sun. For the outer planets: Mars, Jupiter, and Saturn, the geometry is a little different, as shown in Fig. 3.4b where the planet P is shown following an orbit around the sun S that is larger than the earth's orbit. By applying a small geometrical trick we can transform the problem into one of epicyclic motion similar to the case of inner planets. We do this by defining a point Q' such that PQ' is equal and parallel to SC, thereby constructing the parallelogram SPQ'C. We note that since CS = Q'P throughout the motion it follows that P lies on a circle with center Q' and radius equal to that of the earth's orbit. Moreover since SP = CQ' at all times, the point Q' must move on a circle centered on the earth, point C, with radius equal to the planetary orbital radius. Thus the diagram for the outer planets Fig. 3.4b is made to involve an epicycle similar to the inner planet diagram Fig. (3.4a).

The predictions of the above planetary theory based on simple epicycles explained in a direct way the occurrence of retrograde planetary motion. This occurs, we may recall, when a planet appears to reverse its direction of motion along its orbit for some time and then resume its original direction. The Ptolemaic theory readily explains this phenomenon for an inner planet traveling faster in its orbit than the earth in its orbit; it is evident that during part of its orbit the planet moves in the same direction as the sun, and part opposed, with a resultant

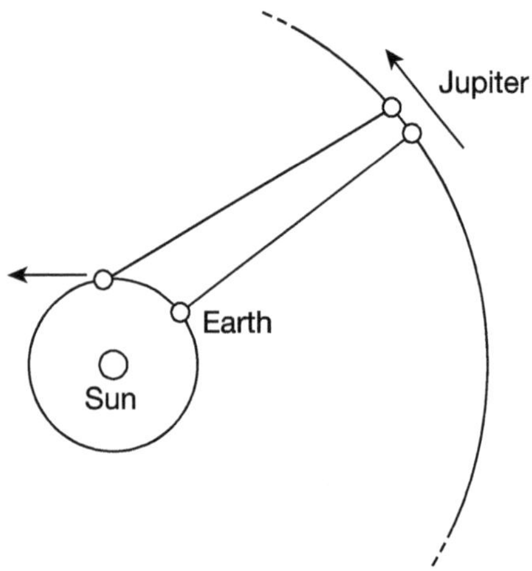

Fig. 3.5 Retrograde planetary motion in modern theory

movement in the reverse direction. In terms of the modern heliocentric orbital theory, retrograde motion is readily explained. In Fig. 3.5 are depicted the positions of the earth and a hypothetical outer planet with respect to a coordinate system centered on the sun. Retrograde motion appears in the outer planet if it is assumed that its orbital motion about the sun is slower than that of the earth.

We now know, thanks to Johannes Kepler, that the planetary orbits are not circular but rather elliptical, and that it was Ptolemy's preconceptions that the motion must be compounded of circular elements that forced him to attempt ever more complex systems to better predict the observed data. Following Hipparcus he assumed that the sun is in an eccentric position by an amount that could be adjusted to fit the observed variation in the apparent velocity of a planet in its orbit. This would help explain the phenomenon, well known at the time that the sun travels in its apparent orbit faster in winter, than in summer. This was given quantitative expression in Kepler's second law of planetary motion, about which more will be said later. In attempts to further improve the agreement of his theory with observation he also assumed that the center of the epicycle moves at a uniform angular rate not about the earth, nor about the center of the deferent, but about a third point called the equant on the opposite side of the deferent center from the earth. Although the ad hoc introduction of the equant into the model improved the agreement between theoretical predictions and observations, it nevertheless violated the philosophical world view that all heavenly motions are compounded of uniform circular motions. A more prosaic but serious problem remained however in the case of the moon's motion, namely according to the model the moon's distance from the earth varies by a factor of about two. This implies that the apparent size of the moon to an observer on earth varies by the same factor, contrary to what is observed. Clearly this is a fatal flaw, one that Ptolemy left unresolved, nor was it successfully addressed until 11 centuries later by a Persian astronomer named Nasir al-Din al-Tusi (c. 1201–1274 AD).

3.2 Islamic Astronomy

The work of Ptolemy embodied in the Almagest was so exhaustive and mathematically so sophisticated that it brought astronomy based on the classical Aristotelian universe to such a settled state that succeeding astronomers turned to such applications as astrology. It is true that there were a number of difficulties that later astronomers tried to resolve, but, working with the same philosophical constraints, that effort was no more than elaborating a flawed theory. Many authors wrote commentaries on the Almagest, but over the subsequent centuries, as the Roman Empire crumbled and Europe ultimately descended into the "Dark Ages", much of Greek scientific literature was lost. The emergence of the Arab Islamic empire in the seventh century AD stretching across North Africa and southern Europe and the Middle East ultimately saw a renewed interest in science and astronomy. Libraries were built in cities such as Baghdad, Damascus, and Granada, and observatories observed the movements of the moon and planets, and cataloged the stars. Scientific and astronomical treatises in Greek, Babylonian, and Hindu were translated into Arabic or copied and preserved. In fact some important writings of Ptolemy were available in later times only as Arabic translations.

Some of the impetus that drove the Islamic rulers to promote the learning of astronomy derives from the three important religious observances that good Moslems must keep. The first is to pray five times a day at prescribed times determined by the position of the sun or reading of a sundial. This of course requires knowledge of the precise time of sunrise and sunset. The second requirement is that the prayers should be performed facing Mecca, a direction called *kibla.* Strict adherence to this observance is of course difficult and requires the religious authorities to provide the compass direction to Mecca for any given location. Presumably the faithful lost in the desert can expect some dispensation. Lastly the faithful must observe the month of Ramadan during which there is total fasting during daylight hours. Since it is the lunar month that is involved, the beginning depends on observing the new moon's first crescent. The duty to inform the muezzin when to announce from the minaret the time for prayer, and the beginning of the fast, falls on a designated official of the mosque called the *muwaqqit*, which means in Arabic naturally enough, the timer.

Since Islam, as a religion, is such an integral part of the adherents' daily life, it is not surprising that so much emphasis was placed on the accurate observance of prayers and fasting at the proper time. It prescribes the use of the lunar calendar; hence the problem of lunar crescent visibility is of central importance. Notable among the Moslem mathematicians/astronomers who devoted themselves to this problem was Muhammed ibn Musa al-Khwarizmi (c. 780 AD). He was born in Baghdad and worked at *Dar el Hikma* (the House of Wisdom) in Baghdad, an institution set up for the advancement of learning. It was here that Greek, Egyptian, Hindu, and Andalusian writings were copied and translated into Arabic. Khwarizmi is best known for his major mathematical work given the Anglicized title *Algebra* which introduced to Europe the concepts of algebra [4]. His name gave us the word algorithm. Though his reputation rests on the invention of algebra, he made significant contributions in determining the precise setting times of the sun and moon throughout the year. This is not trivial, as we saw with the modeling of the moon's motion, which was left by Ptolemy in less than a satisfactory state.

The Ptolemaic system came under detailed questioning by another Islamic scholar named al-Hassan ibn al-Haytham, known in the west as Al Hazen, who was born in Basra (c. 965 AD). He is best known for his work on the optics of the eye, refuting the Greek theory of vision and reputedly inventing a form of the *camera obscura*. His criticism of the Ptolemaic theory was contained in a book whose translated title is *Doubts Concerning Ptolemaeus*. He believed that the theory of Ptolemy must physically conform to the Aristotelian cosmology, and therefore the contrivance of an eccentric equant point introduced *ad hoc* to improve the agreement with observation was just not acceptable. Of course Ptolemy realized this also, but could not find a more consistent alternative. Since the equant gave good agreement with observations, the implication is that to remove it requires a reworking of the entire model. As we shall see the continued dissatisfaction with the model eventually led to the heliocentric model of Copernicus 400 years later.

There was another eminent Moslem astronomer/mathematician who refined Ptolemy's theory of the moon [5]. His name was Nasir ed-Din al-Tusi [5], from which we deduce that he came from the city of Tus in northeastern Persia, where he was born in c. 1201 AD. This was a time when Mongolian hordes swept across Asia causing widespread destruction and massacre of indigenous populations. In 1214 Ghengis Khan led the Mongols in an invasion of the Islamic world, wreaking much devastation in their path, reaching the region around Tus in 1220. Al-Tusi was able to take refuge in nearby Nishapur, at the time an important center of learning. For a period he lived with the Assassins, a murderous, heretical offshoot of Shi'a Islam who had seized mountain fortresses in Persia. The world was not rid of the Assassin menace until they were eradicated by the Mongols under Hulagu in c. 1256 AD. To al-Tusi's good fortune, Hulagu was interested in astrology, and being informed of al-Tusi's talents in astronomy, he appointed him as a personal one-man brains trust, presumably on matters astrological. After the Mongols captured Baghdad in 1258 AD Hulagu returned to Persia to set up his capital at Maragheh in northwestern Persia. It was there that Hulagu authorized the building of a grand observatory using al-Tusi's design. Working at this observatory al-Tusi oversaw the design and construction of many sophisticated astronomical instruments, including astrolabes, and the computation of extensive mathematical and astronomical tables. He "invented" a mathematical device, now known as the *Tusi couple* which combines two counter-rotating circular motions to produce a linear motion, making it possible to more accurately model the vexing problem of lunar motion using only combinations of circular motions. In fact, al-Tusi made the most significant improvements in Ptolemy's theory of planetary motion until the heliocentric model of Copernicus, about 270 years later.

The last Muslim astronomer we will mention in this brief summary is Ibn al-Shatir (Arabic for "son of the Smart One"!) [6], the most prominent Muslim astronomer of the fourteenth century. He was born in Damascus, Syria c. 1305, and rose to be the *muwaqqit* at the famous Umayyed mosque in Damascus, responsible for astronomically determining the exact times of daily prayer and fasting during Ramadan. His reputation rests not on his time-keeping but on his work in planetary theory. It was not until the 1950s that Ibn Al-Shatir's writings on planetary theory were studied by modern scholars, studies that showed that Nicolaus Copernicus used some of the same mathematical techniques to improve the Ptolemaic model. Al-Shatir put forth a new planetary theory that did away with the eccentric deferent and the equant of the Ptolemaic model, replacing them with secondary epicycles instead. While no improvement was achieved with respect to the sun, the additional epicycles did improve the lunar theory by reducing the variation in the lunar distance from the earth, a major objection to Ptolemy's model. Ibn Al-Shatir was also responsible for the construction of a beautiful marble sundial of extraordinary proportions measuring 2 m by 1 m in size and mounted on a minaret of the Umayyed Mosque in Damascus. It represented the earliest example of a sundial having the gnomon set parallel to the earth's polar (spin) axis, in order for the shadow to turn uniformly with equal hour divisions.

While advances in the sciences of astronomy and mathematics were being made by Islamic scholars during what has been called the "golden age" of Islamic science, European science was encumbered by a preoccupation with religious disputation among the intellectuals. Of course there was a good deal of that in the Islamic societies too, but astronomers there were needed to serve the ritual needs of Islam. The Islamic role in the history of science is often portrayed as serving only to preserve the intellectual accomplishments of the ancient Greeks, whereas in fact notable contributions were made by Arab mathematicians in algebra and trigonometry as well as the science of optics. The communication of scientific knowledge between the Islamic centers and Christian Europe occurred mainly through Spain in the west and Byzantium in the east. Materials such as astronomical and mathematical tables as well as sophisticated instrument design such as the astrolabe found their way to Christian Europe.

3.3 European Astronomers

The best known European astronomer to build on the planetary theories of Ptolemy and his Islamic successors was Nicolas Copernicus (1473–1543) who was born in Toran, Poland. He studied medicine and cannon law and was called on by the Lateran Council to give his opinion on calendar reform. He was a true "renaissance man" having studied at universities in Italy and become proficient in many languages, mathematics, medicine, religion, and cosmology. He served for many years as civil administrator in Varmia, Poland. His heliocentric cosmology on which his place among the principal figures in the history of astronomy rests was not even his main preoccupation. In 1539 Copernicus happened to discuss his heliocentric idea for the solar system with a mathematician by the name of Georg Rheticus who published it in a book within an astrological context. With the encouragement of Rheticus, Copernicus wrote a comprehensive treatise on the subject, known as *De Revolutionibus Orbium Coelestium* (On the Revolutions of

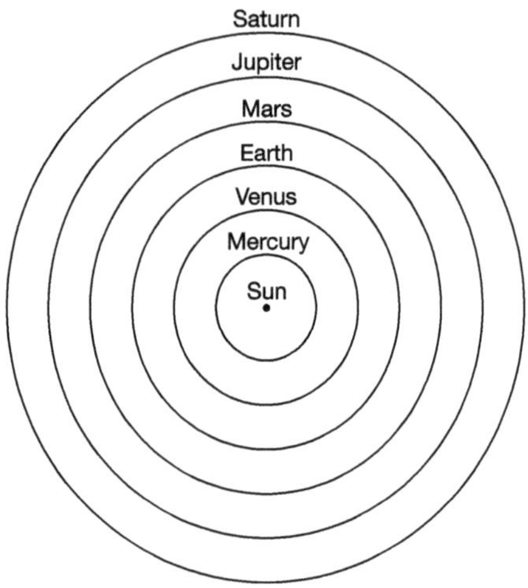

Fig. 3.6 Copernican solar system

the Celestial Sphere). It was published in 1543, inscribed with a dedication to the Pope. He died the same year. While his heliocentric model of the solar system was certainly a break from the traditional geocentric model, it would be wrong to suggest that it affected a great simplification. He made liberal use of the mathematical techniques involving multiple epicycles and Tusi couples derived from Arab astronomers. The complexity of the models that are constrained to be constructed from pure circular motion is not much reduced by a different choice of reference system. However in one important respect the heliocentric theory was able to account for the phenomenon of retrograde motion of the outer planets as all resulting from the motion of the earth around the sun. There is little to suggest that he was revolutionary in trying to oppose the religious orthodoxy of the time, but rather that he held deep-seated beliefs about the cosmos, and was looking for a simpler, more elegant planetary model that conforms to it. Figure 3.6 illustrates the concentric orbits of the planets as described by Copernicus.

Three years after the death of Copernicus, another great astronomer was born of a noble family in 1546 in Skane, a town at the time in Denmark, now in Sweden. His Latinized name is Tycho Brahe. He studied in Denmark and Germany and as a university student he developed a strong interest in astronomy and astronomical instruments. It appears that he was also involved in dueling, a fine old tradition (now forbidden at most German universities), which cost him the tip of his nose! His reputation as an accomplished astronomer must have been well established when, after a tour in Germany, King Frederick II of Denmark and Norway offered to build him an observatory on an island near Copenhagen, called *Hveen*. Thanks to astronomical instruments of old and original designs that he had built to exacting specifications, it became the preeminent observatory in Europe. It should be noted that the idea of glass lens systems for astronomical observations had not been "reduced to practice" until a decade later, around 1608. The observatory became a center for astronomical observation, seminars, and compilations of astronomical tables. In 1597 Tycho Brahe left Denmark when his sponsor Frederick II died, and eventually settled in Prague in 1599 where he died 2 years later. He hired Johannes Kepler as an assistant in Prague to calculate the planetary orbit of Mars based on his continuous careful record of observations. These observations were never formally published during his lifetime.

Johannes Kepler (1571–1630) was born near Weil der Stadt, Württemberg, in southwestern Germany, part of the Holy Roman Empire. It was his good fortune to have gained the position of assistant to the famous Tycho Brahe, in which capacity, as previously noted, he was assigned the job of using the detailed continuous observations on the planet Mars to calculate its precise orbit. In the course of this work he made the seminal discovery that the best fit to the data was not a circle but another member of the conic sections, a figure called an ellipse. This work was completed around 1605, and the result was published in his *Astronomia Nova*. In it he put forth what are now known as Kepler's first two laws of planetary motion. Kepler took over as Imperial Mathematician upon Tycho Brahe's death in 1601. Although his fame rests on his accomplishments in the science of astronomy, he, like Isaac Newton of a later generation, also studied the science of optics, making contributions to the theory of image formation in lens systems and the human eye; in fact one design of a telescope bears his name. In 1621 Kepler published the treatise *Astronomiae* in which a systematic treatment is given of heliocentric astronomy, followed later by the publication of the *Rudolphine Tables*, a compilation of Tycho Brahe's comprehensive and detailed star and planetary tables.

Kepler enunciated three planetary laws:

First Law: "*Each planet moves about the sun in an orbit that is an ellipse, with the sun at one focus*"

Second Law: "*The straight line joining a planet and the sun sweeps out equal areas in space in equal time intervals*"

Third Law: "*The squares of the (sidereal) periods of the planets are in direct proportion to the cubes of the semi-major axes of their orbits*"

The first law states that the shape of the orbit is that of an ellipse, which is a closed curve belonging to the general class of conic sections, among which is the simplest, the circle. To generate the members of this class, imagine two right circular conical shells mounted coaxially opposite to each other as shown in Fig. 3.7.

Now consider the curved line(s) of intersection between the cone(s) and a plane sheet. The intersection curve is determined by the relative size of the angle A that the plane makes with respect to the axis of the cone compared with the angle B that the cone's face makes with that axis: For a circle $A = 90$, for an ellipse $A > B$, for a parabola $A = B$, and for a hyperbola $A < B$. The study of the conic sections goes back to the geometers of ancient Greece, principally Pollonius c. 200 BC.

The geometry of the ellipse is what concerns us here. The large diameter is called the major axis while the smaller diameter is the minor axis. The ellipse has two characteristic points along the major axis equidistant from the center called the foci. The ellipse has the interesting property that the sum of the distances from any

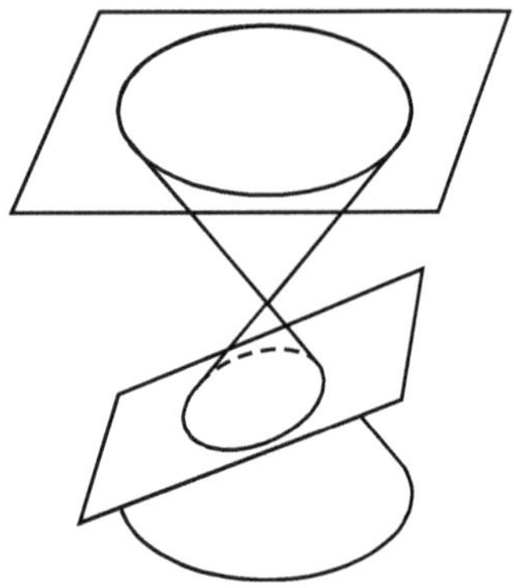

Fig. 3.7 The conic sections

point on it to the two foci is constant, no matter where the point is on the ellipse. This property implies a striking optical property, namely, if the reflecting surface a mirror conforms to the shape of an ellipsoid, which is a surface resulting from spinning an ellipse about its major axis, then a point source of light placed at one focus will converge precisely at the other focus. The ratio of the distance between the two foci to the length of the major axis is called the eccentricity and is a measure of how far it departs from the shape of a circle. If ε represents the eccentricity then for a circle $\varepsilon = 0$ and for a straight line segment $\varepsilon = 1$.

As Isaac Newton was to show later, when two bodies exert a force on each other as defined by Newton, and that force depends inversely on the square of the distance between them, that is,

$$F = \frac{k}{d^2}, \tag{3.1}$$

where F is the force, d is the distance between the bodies, and k is a constant, then the path the bodies will follow will be a conic section. The second of Kepler's laws has to do with the velocity of the planet in its orbit. It had been known since classical times that the progression of the sun along the zodiac does not occur at a constant rate. In modern terms this means the earth does not travel at a constant rate along its orbit. By analyzing in detail the precise observations available to him from Tycho Brahe, Kepler was able to quantify that variation in the earth's movement and relate it to the elliptical geometry of the orbit. Figure 3.8 illustrates the meaning of the second law, namely that the areas of the sectors swept out by the earth are equal for equal intervals of time. Since the radius varies, it follows that the angles of the sectors must also vary, and hence the apparent speed of the earth varies, since the time interval is the same.

A contemporary of Kepler of equal importance in the history of science is the Italian Galileo Galilei (1564–1642) who was born in Pisa, Italy, home of the famous leaning tower where he is reputed to have done experiments on falling bodies. In recent times there were fears that the tower itself may be falling! The invention of the telescope plays an important part in Galileo's contribution to astronomy.

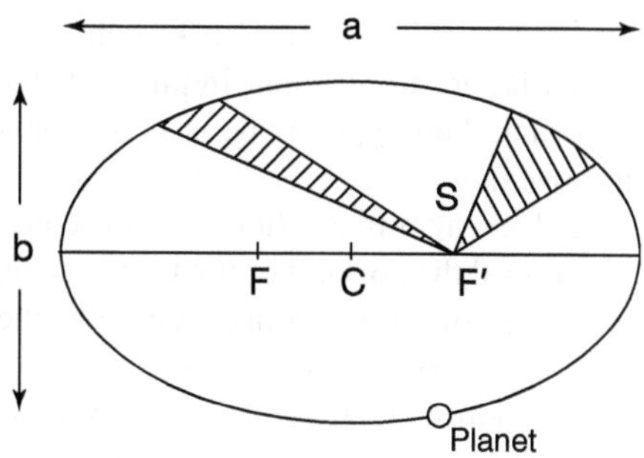

Fig. 3.8 The constant areal velocity law of Kepler

The basic optical elements that make up a (refractor) telescope, the glass lens, had reached Europe probably from Islamic Spain or Byzantium as early as the 1300s when single eye glasses had become familiar in Italy. However it was not until 1608 that the patent for a device using *two* lenses to enhance the ability to see *distant* objects was awarded to a Dutch spectacle maker named Hans Lippershey. It is almost certain that what Lippershey had built was more akin to a spy glass of the kind that the explorer Abel Tasman might have used. In any event Galileo soon learned about the invention and built his own telescope in 1609. He was the first to point a telescope up at the night sky and open up the whole field of observational astronomy. He designed and perfected many telescopes that reached a magnification of over 20 diameters. That was enough to enable him to see for the first time details of the mountains and craters on the moon's surface, the phases of Venus, four satellites of Jupiter, and resolve into individual stars what had appeared to the naked eye as clouds in the region of the Milky Way. The publication of these observations in a book with the (translated) title *Sidereal Messenger* in 1610 was received like a thunderbolt! The accepted notions of the celestial sphere, the fifth essence, had been opened up to human inspection and found to be much like the other of Aristotle's four essences. Galileo's monumental work entitled *Dialogue Concerning the Two Chief World Systems* was published in 1632. In an English edition by S. Drake of this classic work, there is a foreword written by Albert Einstein, which reads in part:

> *"A man is here revealed who possesses the passionate will, the intelligence, and the courage to stand up as the representative of rational thinking against the host of those who, relying on the ignorance of the people and the indolence of teachers in priest's and scholar's garb, maintain and defend their positions of authority…".*

The book has the format of a dialogue between three men named Salviati, Sagredo, and Simplicio, who converse on all manner of scientific questions, all designed to bring out fallacies and contradictions in the accepted view of the way physical objects move and behave. Salviati is the protagonist for Galileo's arguments and Simplicio represents an Aristotelian philosopher. The Catholic Church in the person of Cardinal Bellarmine had warned Galileo not to publicly teach the Copernican model since it contradicted church dogma. An appeal to Pope Urban VIII to permit the publication of the *Dialogue* was rebuffed. Nevertheless it was published in Italian in Florence, an infraction which precipitated his being called before the Roman Inquisition. Galileo was publicly humiliated and forced to recant. Apocryphally he is said to have muttered *Eppure si muove* (nevertheless it moves).

Galileo also made fundamental contributions to the science of mechanics. He introduced the concept of the inertia of matter. He contended that the "natural" state of motion of a body is not a circle as the Aristotelians would have it, but either at rest or constant motion in a straight line. This became the basis for Newton's first law of motion. Galileo also did the famous experiments on the accelerated motion of bodies falling freely or constrained to move on smooth inclined planes. He measured the value of that acceleration for different objects and as a function of the angle of inclination of the plane. Another, perhaps less fundamental discovery for which Galileo is given credit is noticing the isochronism of the pendulum, that is, a suspended mass, swinging freely through a small arc, takes the same time to complete an oscillation no matter how widely it swings. The story is that around 1583 when Galileo was a medical student he timed with his own pulse the swinging of a chandelier at the cathedral in Pisa (when presumably he should have been paying attention to the Mass!). Actually the period (time to complete an oscillation) of a simple pendulum is only approximately constant provided the angle of the swing is small. To have true isochronism the arc would have to be part of a cycloid rather than a circle. It was not until a few months before his death in 1642 that Galileo suggested the use of a pendulum to regulate a clock. He had become blind in 1638 and no longer able to put his idea into practice. He dictated a design to his son Vincenzo who made drawings but did not actually complete a working model. That honor went to Christiaan Huygens, of Huygens Principle fame in physical optics.

3.4 Isaac Newton

Another prominent figure in the history of mathematics and astronomy is Isaac Newton, who was born in 1643, less than one year after the Dutch explorer, Abel Tasman, sailed around what was to become New Zealand. Newton was born in the rural English county of Lincolnshire at Woolthrope Manor of a hardly literate rustic family. Isaac Newton was no Amadeus Mozart—he showed little early precociousness in mathematics or science—not until his twenties did his genius become manifest. The place where he was born and where he contributed so much to science and mathematics remains today as a historic site complete with the old apple tree made famous by its falling fruit, an occasion witnessed by the young professor from Cambridge and said to have inspired his concept of gravity. Cuttings from that parent tree have been acquired by many institutions around the world! The outbreak of the bubonic plague in 1665 caused general panic and the closure of Cambridge University, where Newton was on the faculty. When the university decided to release the students and faculty, Newton went home to Woolthrope

Manor, and spent two intensely creative years there until it was safe to return to Cambridge.

In that short period at Woolthrope Manor Newton broke new ground in the following areas:

1. The decomposition of white light into the colors of the rainbow.
2. The laws of motion.
3. A new branch of mathematics: what he called *fluxions* (calculus).
4. The law of universal gravitation.

The decomposition of white light into the colors of the rainbow was analyzed exhaustively by Newton in his book *Optiks*, the first edition of which was published in English in 1704. The first edition included an allusion to his discovery of the calculus of fluxions independently of the German mathematician Leibnitz. The latter challenged that claim and remained an antagonist throughout much of Newton's career. His contribution to optics is not simply observing that when white light is refracted at an air–glass interface the rays refracted in different directions have different colors, which had been known at least as far back as 1300 AD. In fact it had been noted by the Persian Kamal ad-Din al-Farisi, and later the German monk Theodoric of Freiberg had given a valid explanation of the formation of the rainbow. What Newton proved was that if one of the colors of the rainbow is refracted through a *second* prism, the light remains the same color. That is, the act of refraction does not produce a *change* of color; all the colors were present in the white light, and it is a physiological phenomenon in the eye that gives the perception of "white" when all the colors of the rainbow are perceived simultaneously in the eye in the proportion present in sunlight. He also investigated what are now called Newton's rings, the interference (often colored) rings that are formed when a convex optical lens is placed on an optically flat glass plate, and the related phenomenon seen in thin films. He also studied the diffraction of light at sharp boundaries. In order to explain these wavelike behaviors he found it necessary to reconcile his "corpuscular" theory of light with what is obviously wavelike behavior. His theory of light as consisting of corpuscles obeying his mechanics would require, in order to agree with the observed change in direction of light rays at an air–glass interface, that particles of light travel faster in glass than in air, a result that was demonstrably wrong. Now that modern physics has shown that light does indeed have a "particle aspect" people inclined to hero-worship might be tempted to say that Newton was half right! Actually the light particle, the photon, is nothing like a particle obeying Newton's laws of motion. . ..for one thing it always travels with the (constant) speed of light!

Newton's laws of motion, the foundations of Newtonian mechanics, were set forth in his monumental *Philosophia Naturalis Principia Mathematica*. They may be simply translated to read as follows:

1. Every body continues in its state of rest or uniform motion in a straight line unless it is compelled to change that state by force impressed upon it.
2. The rate of change of motion is proportional to the force impressed and takes place in the direction of the straight line along which the force acts.

3. To every action there is an equal but opposite reaction; or the mutual action of two bodies on each other is always equal and acts in opposite directions.

As already indicated, the first law is a statement of Galileo's postulate that the "natural" state of motion is uniform motion in a straight line. In dynamical terms it states that a material body resists change from its natural state of motion: it is said to have *inertia*. In the second law, Newton understands the quantity of motion to be the momentum, defined as the velocity V multiplied by a constant factor m which he called the inertial mass, a measure of the ability of a body to resist a change in its motion. An important consequence of this law is that for a body to be continually changing its direction of motion, as in a circular orbit for example, it must be continually subjected to a force toward the center. The third law of action and reaction is the basis of the law of conservation of momentum in a system of interacting bodies. For example if we have two colliding bodies of masses m_1 and m_2 then since $F_{12} = -F_{21}$ at all times during the interaction, it follows that the time rates of change of the momenta are also equal and opposite at all times: thus,

$$F_{12} = \frac{m_1 \Delta V_1}{\Delta t} = -F_{21} = -\frac{m_2 \Delta V_2}{\Delta t}, \tag{3.2}$$

and therefore the total change in momentum $(m_1\Delta V_1 + m_2\Delta V_2) = 0$. That is, there is no change in *total* momentum. This is the principle on which a jet engine depends to give thrust to an aircraft. The momentum carried by the continuous ejection of the hot gases from the engine results in the reactionary forward thrust.

An equally important conservation law that is derived from Newton's laws is the conservation of angular momentum or spin. This can be illustrated with a gyroscope by mounting the rotor on gimbals so that its support can rotate in a plane at right angles to the rotor axis. As the rotor is set in motion, its support will rotate in the opposite direction. The principle of the gyrocompass, once the prime instrument of navigation, is that it has so much angular momentum that it preserves its orientation in spite of small perturbations. Ideally, if there are no torques acting on the spinning rotor, its angular momentum will remain constant indefinitely, that is, it will remain pointing in the same direction. The same is true of the rotation of the planets such as the earth.

Newton's major contribution to mathematics is no less than the independent creation of a whole new branch dealing with what he called fluxions and fluents. This has to do with the concepts of mathematical functions and variables and the treatment of variations in the value of a mathematical function involving quantities that may be variable. He introduced the concepts of what we now know as calculus. Although this work is mentioned in his *Principia*, since that work was not published until 1687, after the German mathematician and philosopher Gottfried Leibnitz published similar ideas in 1684, Leibnitz claimed the prior invention of differential calculus. Newton bitterly disputed that claim having copious notes made when he originally conceived the idea in 1666 at Woolthrope Manor at the time of the plague.

Finally in our discussion of Isaac Newton we come to his greatest contribution: the law of universal gravitation. This was a brilliant example of inspired inductive reasoning. Starting with the knowledge, thanks to Galileo, that objects of whatever composition fall to the earth with a constant acceleration he deduced according to his own law that a force was acting on it, attracting it to the ground. And he had no doubt that it would continue falling down a hole in the ground if one were provided for it. He also knew that according to his first law a planet such as the moon must be subject continuously to a force towards the earth since it is constantly accelerating towards the earth, similarly for planets orbiting the sun. From this he hypothesized that there must be a universal attractive force between all bodies anywhere in space. To deduce how this force should depend on the distance separating them, he required the so-called *centripetal* acceleration of a mass towards the center, when moving uniformly in a circle. To a modern student imbued with the methods of calculus, deriving the formula for the acceleration would not be much of a challenge, but probably it was not trivial even for Newton. In any event, using the following expression for α, the acceleration towards the center:

$$\alpha = \omega^2 r, \tag{3.3}$$

where ω is the angular velocity in radians per second, and r is the radius of the orbit, a mass m would require a force according to Newton given by

$$F = m\alpha = m\omega^2 r. \tag{3.4}$$

But according to Kepler's third law of planetary motion, we have

$$\left(\frac{2\pi}{\omega}\right)^2 = T^2 = kr^3, \tag{3.5}$$

where T is the period, and k a constant. It follows on substituting for ω^2 in the expression for F, and simplifying we get

$$F = m\frac{(2\pi)^2}{k}\frac{1}{r^2}, \tag{3.6}$$

which shows that if indeed the gravitational force follows a simple power law, it must be an *inverse square* law in order to comply with Kepler's third law. This is a highly reasonable result if the gravitational force is pictured as radiating out like Faraday's "lines of force". Newton's law for the gravitational force between two bodies of mass m_1 and m_2 is usually written as follows:

$$F = \frac{Gm_1m_2}{r^2}, \tag{3.7}$$

Table 3.2 Conic section as a function of particle energy

ε	Conic section	Total energy E
0	Circle	$-\frac{mk^2}{2L^2}$
1	Parabola	0
<1	Ellipse	$0\rangle E\rangle -\frac{mk^2}{2L^2}$

where G is the gravitational constant which in the International System of units has the numerical value $G = 6.673\times10^{-11}\ \mathrm{Nm}^2\ (\mathrm{Kg})^{-2}$. This result applies to the force between theoretical point masses. To apply it to planetary motion it remains to compute the total force between two large uniform spheres. To accomplish this Newton had to solve a problem in integral calculus. He was able to show that the force between two homogeneous spheres is the same as if their entire mass was concentrated at their centers; this of course greatly simplified the problem of solving for the possible orbits. Newton was able to integrate the (differential) equation of motion of a particle subject to a central force obeying the inverse square law. In terms of plane polar coordinates $(r,\ \theta)$ the solution for a planet orbiting the sun is of the following form:

$$\frac{1}{r} = C\left[1 + \varepsilon\cos\left(\theta - \theta'\right)\right], \tag{3.8}$$

which represents a conic section with *eccentricity* ε, which is the following function of the *total* energy E and *angular momentum* L, both of which are constants of the motion:

$$\varepsilon = \sqrt{1 + \frac{2EL^2}{mk^2}}, \tag{3.9}$$

where $k = GMm$, and M the mass of the sun, assumed infinitely large compared with m, the mass of the planet (a good approximation for the earth, but not Jupiter). The form of the conic section, whether a circle, ellipse, parabola, or hyperbola, depends on the value of the eccentricity, ε, and therefore the total energy of the system, as shown in Table 3.2.

The total energy E is the sum of the kinetic energy and the gravitational potential energy. If a body starts with some kinetic energy from a great distance where the potential energy is zero, then the total energy is positive and the resulting motion will be hyperbolic. If however it approaches the origin and somehow loses enough kinetic energy to make E *negative*, then it would be captured in an elliptical orbit, as shown in Fig. 3.9 for the earth.

The second of Kepler's laws of planetary motion, namely the law of constant areal velocity, actually applies not only to an inverse square law force such as gravitation, but to *any central force*, that is, one directed always toward a central

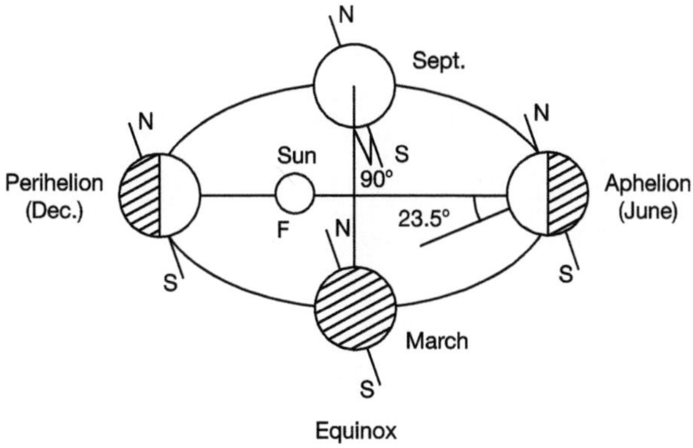

Fig. 3.9 Earth's elliptical orbit and the obliquity of its axis to the plane of the orbit

point. In fact it is a statement of the conservation of angular moment, which must hold for a central force since it does not exert any torque on the moving body. The conservation of angular momentum law can be stated as follows:

$$mr^2\omega = L = \text{const.} \tag{3.10}$$

It can be seen from Fig. 3.8 that $r^2\omega/2$ is in fact just rate at which area is swept out by the radius vector r to a point on the orbit, and since the mass m is constant the second law follows.

The third of Kepler's laws is a little more complicated to prove for an elliptical orbit, so the interested reader is referred to any textbook on classical mechanics. It is much simpler of course to prove for a circular orbit and we have indicated already the part it played in arriving at an inverse square law for the force of gravitation.

References

1. M.S. John, *A Brief Introduction to Astronomy in the Middle East* (Publ. Saqi, Beirut, 2008)
2. O. Neugebauer, *The Exact Sciences in Antiquity* (Dover Publications, New York, NY, 1969)
3. G.O. Abell, *Realm of the Universe*, 5th edn. (Saunders Publications, Philadelphia, MD, 1994)
4. O. Neugebauer, *The Exact Sciences in Antiquity* (Dover Publications, New York, NY, 1969), p. 146
5. M.S. John, *A Brief Introduction to Astronomy in the Middle East* (Publ. Saqi, Beirut, 2008), p. 131
6. O. Neugebauer, *The Exact Sciences in Antiquity* (Dover Publications, New York, NY, 1969), p. 197

Elements of Modern Astronomy 4

4.1 The Earth as a Planet

The observed eccentricity of the earth's orbit around the sun is small, amounting to about 0.017. This means that the apsides, that is, the maximum and minimum distances from the sun, differ by only about 3.4 %; this would contribute a negligible amount to the seasonal climatic changes on the earth's surface. Rather the seasons are due to the *obliquity* of the earth's rotation axis relative to the orbital plane, *the ecliptic*. The angle of obliquity is about 23.5°, as shown in Chap. 3 (Fig. 3.9). The direction of the rotation axis remains fixed in space aside from small perturbations that cause a very slow *precession*, in which the axis sweeps out a cone, like a spinning top. The consequence of the obliquity of the axis is that during different segments of the earth's orbit, the northern hemisphere, for example, will be inclined *towards* the sun increasing the duration of daylight, while in the southern hemisphere the opposite is true. It so happens that the orientation of the spin axis is such that the northern hemisphere is inclined *away* from the sun at the *perihelion* point. That point on the major axis of the orbit is nearest to the sun, where by Kepler's second law, the earth has the highest orbital velocity; at the opposite extreme point, the *aphelion*, the orbital velocity is minimum. The points in the orbit where the direction of the earth's spin axis is perpendicular to the *radius* vector, that is, the line joining the sun to the earth, are called the *equinoctial points*, because there day and night are of equal length.

The model of the earth as a uniform hard sphere spinning freely in the vacuum of space is unrealistic on many counts: first of all it is not a perfect sphere, it has a topographically complicated surface structure with large bodies of water, and it has a molten interior. It also has an equatorial bulge with a slight north to south asymmetry, making it a somewhat imperfect *oblate ellipsoid* [1]. Because the earth is not a uniform sphere, the moon can therefore exert a torque on it that tends to turn its axis in the direction of the perpendicular to the moon's orbit. This torque causes gyroscopic motion in which, instead of simply turning the axis in the direction of the force, it causes the earth's axis instead to precess, sweeping out a

F.G. Major, *Quo Vadis: Evolution of Modern Navigation: The Rise of Quantum Techniques*, DOI 10.1007/978-1-4614-8672-5_4,

cone, much like a spinning top tipping under gravity. The result in the case of the Earth is a precession of the equinoxes, that is, a movement of the points where the equator and ecliptic cross (these terms are defined in the next section); this causes the seasons to occur at different times of the year. The rate of precession is extremely slow amounting to about one cycle in 26,000 years.

The moon is also mainly responsible for the ocean tides. The tidal action in the world's oceans involving the massive movement of water can create a drag on the earth's rotation. The viscosity of the water, that is, its internal resistance to flow, will cause the kinetic energy of the earth's rotational motion to be dissipated slowly and irreversibly in the form of heat. The main cause of the tidal action is the gradient in the gravitational pull of the moon with a minor contribution from the sun. The connection of the tides with the moon was recognized even in early times because of the evident correlation of the neap and spring tides with the phases of the moon. The flow of the tide is *not* caused simply by the moon drawing to itself the waters of the oceans, as might naively be thought; the explanation is much more subtle. It is due to the strong *variation* across the earth in the relatively weak gravitational pull of the moon, with a weaker variation in the much *greater* pull of the sun. If we consider the dynamic equilibrium of a volume element of the ocean water, we note that the resultant gravitational pull of the sun and moon at the *surface* of the earth must provide the centripetal force to keep it moving with the earth as it pursues its orbit. When the moon is new, for example, so that it is in line with the Earth and sun, the gradient of the gravitational field across the earth is accentuated. But the orbital radius of the earth is determined by the gravitational pull of the field *near the center* of the earth, not at the surface. Since the surface ocean water is constrained to move with the earth, it follows that on the moon side of the earth the gravitational pull is greater than the centrifugal force, while on the other side of the earth the centrifugal force is greater than the gravitational pull. The result is as if there were a stretching of the ocean water, with high (spring) tide on opposite sides of the earth.

The rotation of the earth has significant effect on the surface wind and ocean currents principally through what is called the *coriolis force*. It is a fictitious force introduced when referring motion to a rotating frame of reference; in the case of a particle stationary with respect to the rotating frame it reduces to the familiar centrifugal force. Thus a plumb line at rest will not line up exactly with the earth's radius since it is deflected by the centrifugal force. However, since that force is so small relative to the force of gravity, the deflection angle is very small. In fact, no correction is made for it since *by definition* the plumb line *is* vertical. However the value of g the acceleration of gravity will, as a consequence, vary over the surface of the earth. In the case of moving systems such as in the ocean currents and winds, this coriolis force acts at all points perpendicular to their velocity. In the northern hemisphere where the angular velocity vector points generally in a northerly direction, the coriolis force always deflects the direction of the currents to the right. The opposite is true of the southern hemisphere. A popular demonstration of the coriolis effect is the *Foucault pendulum*, once a proud exhibit at a museum in Washington, DC. This consists of a massive c.900 kg metal sphere suspended by a long (perhaps 60 m) wire over a clock face, around which the arc of the swing would rotate.

The size and shape of planet earth have been a subject of speculation and study since ancient times. As we saw in an earlier chapter a fair measurement of the circumference of the earth as a sphere was accomplished in classical times by Eratosthenes. His result for the circumference, 39,000 km, came remarkably (or more likely, fortuitously) within 2 % of the modern value, if it is assumed that he used the length unit the *stade*, as commonly defined in Egypt at that time. It was generally accepted that the earth was a sphere by the first century A.D. notwithstanding the occasional metaphysical absurdities in the Middle Ages.

The theory of the shape of the earth has engaged the minds of all the great mathematicians for at least the last 350 years. These include Newton, Euler, Laplace, Legendre, Gauss, Jacobi, Riemann, and Poincaré, to mention the more familiar names. It was Newton who first postulated that the earth could not be spherical, and advanced an argument that suggested it must be an oblate ellipsoid of revolution, the figure produced by spinning an ellipse about its minor axis. He imagined two hypothetical shafts being drilled all the way to the center of the earth, one along the minor axis, and one along the major axis, and the two being filled with water. He reasoned that because the water in the equatorial shaft has the gravitational pull reduced by the presence of the centrifugal force, it must be longer than the axial shaft, if the water is to be in equilibrium at the center of the earth. Therefore he argued, the earth must have the form of an oblate spheroid, and he calculated a *flattening*, that is, $(a - b)/a = 1/230$ where a and b are the major and minor axes of the ellipse. However, the question of the shape was not settled experimentally until the beginning of the eighteenth century. The observational method for determining the surface geometry of the earth is fundamentally similar to that used by Eratosthenes. The object is to measure the distance one must travel along a meridian (*the meridian arc*) to observe a change in the latitude of a given amount, say 1°. The meridian arc yields the curvature of the earth's surface because it is a measure of the change in direction of the zenith with respect to a fixed direction in space as determined astronomically, say by the North Star. The challenge is to determine the exact distance between widely separated points on the earth's surface where the curvature measurements are to be made. The geodetic method of triangulation was used to make a survey of France by Picard and Cassini over the period 1684–1718 which established the distance between Dunkerque (51.02°N, 2 20°E) on the English channel and Collioure (42.5°N, 3.08°E) on the Mediterranean coast, chosen to be on the same meridian as Paris. Unfortunately the results for the curvature at the two locations were inconclusive, seeming to favor a *prolate* ellipsoidal shape! This accorded with the expectation of no less a philosopher than Rene Descartes, so de Maupertuis, a member of the French Academy and supporter of Newton's theory, persuaded Louis XV to approve expeditions to points as far apart as Peru on the equator and Lapland near the arctic circle. De Maupertuis reported that the 1° meridian arc at Lapland was 57,437.9 toise (an old unit equal to 1.949 m), while near Paris it was known to be somewhat less at 57,060 toise, confirming Newton's oblate ellipsoid prediction. Taken together with the results from the Peru expedition, the distance between the equator and the North Pole along the Paris meridian was calculated to be 5,130,762 toise. This distance was taken in 1791 soon after the French Revolution to *define* a new standard meter as being 10,000,000 m. Parenthetically we note that the current definition of the meter is based on the standard unit of time, the atomic second, through the constancy of the velocity of light: The meter is the length of path traveled by light in vacuo during the time interval of 1/299,792,458 of a second. Getting back to the shape of the earth, a definitive mathematical description of the figure of the earth was published in 1743 by Clairaut, in which he gave a general formula for the acceleration of gravity g at any point on the surface of any ellipsoid of revolution, as follows:

$$g = g_0\left[1 + \left(\frac{5}{2}m - f\right)\sin^2\lambda\right], \tag{4.1}$$

where g_0 is the value at the equator, λ the latitude, m the ratio of the centrifugal force to gravity at the equator, and $f = (a - b)/a$ where a and b are the semimajor and semiminor axes of the ellipsoid [1]. This result allows the shape of the earth to be determined, in principle, by an accurate measurement of g at different latitudes, for example, by a suitably designed pendulum. We know of course that the surface of the earth is not a smooth ellipsoid, but has irregular features such as mountains and valleys and oceans. The study of the earth's surface topography and gradients in the acceleration of gravity has long been of interest in fields such as mineral prospecting, and today it is of particular interest in the satellite-based global positioning systems. Historically the most prominent name in the study of gravity gradients is that of the Hungarian Roland Eötvös who in 1896 published his description of a gravity gradiometer that goes by his name. He designed and performed the most sensitive test of whether the gravitational mass of a body equals its inertial mass. The unit of acceleration gradient is called the Eotvos in his honor and has a value of $10^{-9}\ \mathrm{s}^{-2}$. Recent research in the area of gravity gradiometers has progressed far beyond torsion balances and suspended weights! Most notable among recent developments are the *superconducting* gravity gradiometer and the *quantum optical* gravity gradiometer. The former uses a superconducting proof mass whose minute movement is detectable using a sensing coil and an ultrasensitive detector, the SQUID (superconducting quantum interference device). The quantum gradiometer is a more sophisticated quantum optical system that uses *cold individual atoms* as proof mass in an optical quantum state interferometer using what are called Raman transitions. These transitions involve simultaneous changes in the quantum energy state of an atom as well as its state of motion. Since atomic interference is involved not only in the accurate mapping of the earth's gravitational field, but also in the construction of an atomic gyroscope, we will take up the subject in some detail in a later chapter.

4.2 The Moon and Planets

Thanks to the Apollo program initiated by President Kennedy in 1961 our satellite, the moon, has been walked upon and so thoroughly photographed and studied that its surface features and even subsurface composition are probably as well known and familiar as any part of the earth. But the moon is certainly more than just a satellite, given its long cultural, not to mention romantic significance.

Because of its relative proximity to the earth at a distance of 384,000 km, it appears to move more quickly in the sky as compared with any other celestial body, apart from the sun, of course. Next to the sun it is the brightest object in the sky; however, its reflected brightness is as little as 2.5×10^{-6} times the brightness of the sun because its *albedo* is only about 0.07. Traditionally we distinguish eight phases of the moon, which correspond to eight positions it occupies in its orbit around the earth. They are new moon, waxing crescent, first quarter, waxing gibbous, full, waning gibbous, third quarter waning crescent, and back to new moon. These phases are illustrated in Fig. 4.1.

The orbit of the moon can be tracked with extreme accuracy, again thanks to the Apollo program. Besides being the first to put man on the Moon, among the accomplishments of the famous Apollo 11 mission launched in 1969 was the deployment of a panel on which are mounted an array of 100 *corner cube reflectors*, each measuring 3.8 cm. These are essentially three mutually perpendicular plane reflecting surfaces forming a concave corner reflector. They have the property that a parallel beam of light incident at *any angle* will be exactly reversed in direction; hence they are called *retroreflectors*. Subsequently such panels were deployed also on Apollo 14 and 15. The Soviet Union also deployed in 1973 a laser corner reflector array in addition to other observational equipment on the robotic *Lunokhod 2* lunar rover, as part of their Luna program. The technique of using corner reflectors on artificial satellites for the purpose of ranging and tracking them using powerful pulsed lasers was initiated by H. Plotkin at The Goddard Space Flight Center in1965. It was not until corner reflectors were placed on the moon by Apollo 11 that lunar ranging experiments could be conducted. The principle of the technique is identical to *radar*; in fact, it has matured into a widely used technology called *LIDAR* [2]. It is essentially a pulse-echo technique in which a high-energy ultrashort pulse of laser radiation is directed, with appropriate optics, as a parallel beam at an object of interest, and the reflected or scattered radiation is received by the same optics as the transmitter. There is an intrinsic practical problem in this technique: a powerful pulse of radiation is in close proximity to a radiation detector designed to be as sensitive as possible. The time elapsed between the sending of the pulse and its reception gives directly twice the distance to the object in meters. In this case a very short (about 1 ns) powerful laser pulse is directed by a telescope with a large diameter mirror toward the moon. From 1969 to 1985 lunar ranging tests were conducted using available time on the 107-in. Smith telescope at the McDonald Observatory in Texas. A dedicated 30-in. telescope was commissioned in 1985. Using a large telescope ensures a small diffraction angle and therefore relatively small divergence of the beam; nevertheless the distance to the moon is so

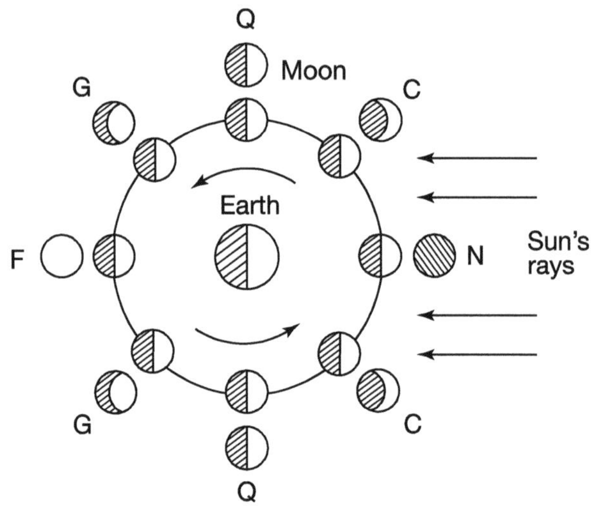

Fig. 4.1 Phases of the moon: *N* new, *C* crescent, *Q* quarter, *G* gibbous, *F* full

large that the beam is about 7 km in diameter when it reaches there, and the reflected beam arrives at the earth with a diameter of about 20 km. The return signal is so weak that detection requires long integration times. The time resolution achieved corresponds to a range error of less than one part in 10^{10}! This is a level of resolution that permits accurate mapping of the moon's surface topography. Spectacular photographs were obtained in the years 1964–1967 of the lunar surface by the Ranger and Surveyor probes and the Lunar Orbiter. In 1966 the Soviet Luna 9 lunar probe made the first soft landing on the far side of the moon and transmitted close-up photographs to earth. The lunar surface is characterized by the presence of what are called *maria* (Latin for *seas*).

The planets that constitute the solar system are the *inferior* planets: Mercury and Venus with orbital radii less than that of the Earth, and the *superior* planets: Mars, Jupiter, Saturn, Uranus, and Neptune that are in orbits outside that of the Earth. The former outermost planet Pluto was reclassified in 2006 and no longer counted as a planet. The order of the planetary orbits and their relative size were correctly deduced by Copernicus, whose arguments introduced the concept of *elongations.* In the context of the planets and the Sun, the elongation of a planet is the angle between the observed direction of the planet and that of the sun as seen from Earth. For an inferior planet there are two possible times when the planet is in line with the sun and the elongation is zero. These are called *inferior* and *superior conjunctions* of the planet and the Sun. It can easily be verified by a drawing that the elongation of an inferior planet cannot exceed 90° (See Fig. 4.2). A superior planet on the other hand can have any elongation between 0° (conjunction) and 180° (opposition). The time intervals between successive oppositions or superior conjunctions are called *synodic periods* [3].

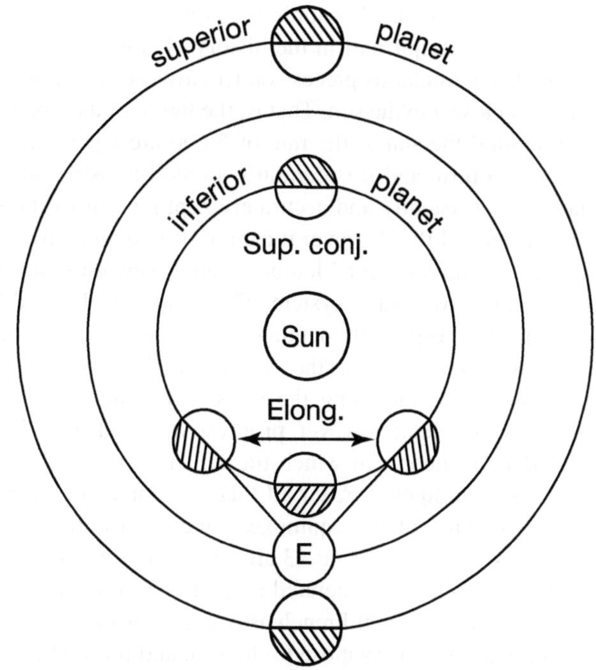

Fig. 4.2 Elongation (**a**) inferior planet (**b**) superior planet

4.3 The Inferior Planets

Of the inferior planets, Mercury is the smallest and closest to the sun; it is rarely seen by the unaided eye because of the sun's glare, but when it is visible at twilight it has the appearance of a bright star. It has a highly eccentric orbit inclined by 7° to the ecliptic, an orbit which it completes in 88 days. It has a synodic period of 116 Earth days, which means that the periods during which it is near conjunction with the sun occur about three times a year. It is then that it may be visible as a "morning star" or "evening star". The planet takes about 59 Earth days to complete a rotation about its axis with respect to the fixed stars; this is in the simple ratio 2:3 with respect to its period of revolution around the sun, hinting that the rotation may be driven by tidal action due to the sun's gravity. Thanks to NASA missions to Mercury beginning with the Mariner 10 flyby as early as 1974, followed by MESSENGER (a clever acronym) launched in 2004 and successfully put into orbit around the planet in March 2011, its surface has been thoroughly photographed and analyzed. Temperature measurements of the surface were recorded showing an astounding change in temperature between daylight hours and after sunset. At noon the temperature can reach 427°C and then after sunset can plunge below −123°C and continue falling until midnight! The high daytime temperatures would mean that any gas molecules on the surface would have thermal velocities that far exceed the gravitational escape velocity and therefore no substantial atmosphere is expected on Mercury.

To add to Mercury's impressive properties, particularly the temperature range, it has played a major part in the history of science. The reason has to do with what seemed an anomalous precession (rotation) of its perihelion, the point in its elliptical orbit closest to the sun. That is, the figure of its orbit has been observed to slowly turn around the sun at the rate of 5,600 arc s per century! Admittedly that is not much of a turn, and it is certainly a credit to astronomers throughout history who have so consistently and accurately kept records of their observations as to notice such a small shift. However it cannot be explained by Newton's law of gravitation applied to the sun and Mercury alone: Newton's law unambiguously predicts a fixed ellipse for such a system. The inclusion of possible perturbations caused by other planets, especially neighboring Venus, requires a new level of mathematical sophistication to treat the so-called three-body problem. Many prominent mathematicians including the likes of Lagrange and Laplace have attempted to solve this problem; however, progress was made only in developing some forms of perturbation theory, in which the effect of *small* changes in the parameters of a problem is computed. After all this was done and allowance made for the Earth's own precession of the equinoxes, since that defines the frame of reference, still there was left a significant 43 arc s per century to explain.

This discrepancy generated a great deal of debate about possible explanations. Leverrier, a prominent French astronomer proposed the idea that there was a small planet in an orbit very close to the sun, and for a while it was thought that one had been observed. But later observations failed to confirm this. The failure to find an explanation based on Newton's inverse square law led some to begin theorizing that perhaps Newton's theory is only an approximation and that instead of a $1/r^2$ dependence it is an inverse power law, but the exponent is not exactly 2 but $2 + \varepsilon$ where ε is a small number. In fact Hall did show that the theory will predict the exact precession rate if $\varepsilon = 0.00000016$. However this was not accepted because it would have broader implications about other basic theory. Many such ad hoc explanations were proffered, but they would not only have to correctly predict the precession of Mercury, but many other cases of precession, such as those of Venus, Earth, and the asteroids. The one answer that came naturally out of a consistent general theory of gravitation was put forward by Albert Einstein in 1915. His correct prediction of the size of the precession of Mercury provided a strong physical confirmation of the validity of his *General Theory of Relativity* [4].

Venus is a spectacularly bright object in the night sky, a planet which at its brightest can be seen even in the daytime. As an inferior planet it appears, like Mercury, to travel back and forth from one side of the sun to the other as it executes its synodic period. It is plainly seen as a "morning star" and as an "evening star". Its mass, which is a little less than that of the Earth, is fifteen times the mass of Mercury and two and a half times as big in diameter, from which we can conclude that they have approximately the same *average* density Venus has a very nearly circular orbit about twice that of Mercury, and an eccentricity of only 0.007 with a sidereal orbital period of 224.7 days. The inclination of the plane of its orbit relative to that of the Earth (the ecliptic) is about 3°24, compared with 7° for Mercury. The most singular fact about Venus is that the direction of it rotation about its axis is opposite to most of the other planets, namely from *east to west* with a period of 243 days! A consequence of this very slow rotation is that its shape is very nearly spherical rather than having a pronounced oblateness like the Earth.

The *transit of Venus* [5], that is, the passage of Venus in front of the sun's disc which would constitute an eclipse if it were the moon and not Venus, appears as a small black spot moving across the bright disc of the sun. It occurs *twice* every 113 years approximately, with the two transits separated by about 8 years. One transit occurred in 2004 followed by another in 2012. Observations of this occurrence have played an important role in the history of astronomy. The English astronomer Edmund Halley, for whom a comet is named, proposed a method of using the transit of Venus to measure the distance from the Earth to the sun, a distance that constitutes a fundamental unit in astronomy, the *astronomical unit* (AU). Unfortunately he died before the next transit of Venus was expected to occur. The prospect of being able to measure that fundamental distance generated great interest; international expeditions were organized and sent out to remote parts of the world to observe and make measurements on the transit. Halley's proposed method is based on the phenomenon of *parallax* [6]. This is simply the apparent difference in the *relative* directions of two distant objects when viewed from different angles because of their different lateral positions. This effect is dramatically seen when sitting in a speeding train watching the landscape as near objects seem to sweep around distant objects. The method requires at least two observers on the Earth to be as widely separated in latitude as possible, preferably on the same meridian, and be able to observe as much of the transit as possible. The two observers must time exactly the duration of the transit across the sun's disc. The time to complete the transit will be different for the two observers because the lines of transit across the sun's disc as seen by them are displaced because of parallax. The difference in the timed duration of the transit is then used to compute the parallax displacement of the line of transit. From this parallax displacement Venus's distance can be deduced. The value of AU, the distance from the Earth to the sun, can then be deduced using Kepler's third law. It is presumed of course that the distance between the two observers can be determined exactly in meters. In summary the transit of Venus provides primarily a precise way to determine the Earth–Venus distance. Presumably a parallax measurement relative to the fixed stars would not be as accurate. Of particular historical interest is the fact that it was that great navigator and explorer Cpt. James Cook who, as an important part of his first journey of discovery in 1769, was assigned the observation of the transit of Venus on the newly discovered island of Tahiti. On the same journey, of course, he also happened to circumnavigate New Zealand and produce a remarkably accurate map!

Venus has a very inhospitable dense atmosphere. At altitudes up to about 50 km it is principally carbon dioxide, accounting for 85 % with the remainder nitrogen and other gases. However the upper atmosphere is composed of nasty stuff like sulfuric acid and other corrosive gases. Thanks to the space mission Magellan, launched in 1989 and put into a near polar orbit around Venus in 1990, the Venusian surface was explored and a comprehensive mapping of its topographical features was accomplished. Since the atmosphere is opaque to visible light, the imaging was done mostly with synthetic aperture radar with a resolution down to 100 m (see Chap. 15). Most of the surface is covered with volcanic flows with an average temperature at the surface estimated to be in the neighborhood of 500°C, that is, comparable to Mercury, which is much nearer to the sun. This fact has prompted the speculation that this is a severe result of the *greenhouse effect*. It will be recalled that this effect results from the property of many gases, typically carbon dioxide, to readily allow sunlight to be transmitted through them, but will strongly absorb infrared (heat) radiation. In the case of an actual green house, the glass used must have, to some degree, the same property of admitting sunlight and absorbing infrared radiation.

Other space missions devoted to Venus are the European *Venus Express* put into orbit in 2006 and the Japanese *Akatsuki* (*Dawn*) launched in 2010. They were both designed to study the lower and upper atmosphere and climate on Venus using multiple wavelength imaging. Unfortunately the Japanese attempt to put their spacecraft into orbit failed and it flew past the planet. It is planned to try again in the future.

4.4 The Superior Planets

Mars, "the red planet," is the most readily observable and bears the closest kinship to our own planet, although its atmosphere of carbon dioxide makes it uninhabitable. There has been an assiduous effort to find water on the planet, and very probably water does exist as ice. When it snows in the polar regions of Mars it comes down in the form of frozen carbon dioxide, "dry ice"! Mars has about half the diameter of Earth and only about one-tenth its mass and has therefore slightly less mean density. In comparison with the Earth it rotates at about the same rate with a period of 24 h 37 m and revolves around a more eccentric elliptical orbit in about 687 days with a semimajor axis about one and a half times as large. Its equator is inclined to its orbital plane by about 24° that is slightly more than the Earth's 23° 27′ and therefore should have a similar phenomenon of seasons, which last twice as long. Unlike Venus, there is generally little cloud cover on Mars to obstruct the view of its surface. Mars has two moons named *Phobos* and *Deimos*, which have diameters of only about 25 km and 13 km, respectively. They were first observed in 1877 by Asaph Hall, a largely self-taught American astronomer at the U.S. Naval Observatory. They are essentially much cratered, irregular orbiting rocks, as revealed by photographs taken by NASA's Mars orbiter, Mariner 9, launched in 1971. The USSR also made a successful orbiter mission to Mars in the same year. NASA has launched a series of orbiters/landers over the years, including *Viking2* in 1975, the *Mars Global Surveyor*, and the *Mars Pathfinder* in 1996. Missions continued to be launched through 2007, including notably the *Mars Reconnaisance Orbiter* in 2005 which carried a sounder to detect subsurface water and analytic equipment to identify surface minerals and look for the transport of dust and water in the Martian atmosphere. The *Phoenix Mars Lander* was successfully launched in 2007. The surface temperature of Mars reaches a maximum of about 30°C at the equator, falling during the night to around −75°C. At the South Pole the temperature can fall to −150°C. Telescopic observation of the Martian surface reveals a landscape tinted various shades of orange and red; of some interest (less so now) are the so-called *canals* first noted by the Italian astronomer Schiaparelli in 1877. These canals, or *canali*, as the Italian astronomer called them, were extensively mapped by the American Astronomer Percival Lowell, believing them to be too organized to be natural. The Mariner missions put an end to the speculation about these canals: they don't exist.

The largest by far among the planets, appropriately called Jupiter, king of the Roman gods, has been known since ancient times, but holds a special place as the planet first studied using a telescope by Galileo in 1610 and found to have satellites circling around it. He observed the four largest of these moons named *Io*, *Europa*, *Ganymede*, and *Callisto*. The two largest have a diameter 1.5 times Earth's moon. There are in addition nine other, much smaller satellites and a ring discovered later by the NASA Voyager mission in 1979. Jupiter has a radius about 11 times that of Earth, but a mass only 318 times as large. The semimajor axis of its somewhat eccentric orbit ($\varepsilon = 0.048$) is about five times larger than the Earth, with a (sidereal) orbital period of almost 12 Earth years. Its equator is inclined to the orbital plane by only 3.1°, compared with Earth's 23.5°, and its sidereal rotational period is only about 10 h.

In appearance the surface of Jupiter appears multicolored with striated dark belts and light zones across its surface, assumed to be created by strong E–W winds in its upper atmosphere. A surface feature called *Great Red Spot* has been noted since the 1,800 s and thought to be a cyclonic storm in the atmosphere. The composition of the outer atmosphere is similar to the sun, consisting of hydrogen and helium. The fact that such light molecules can remain bound to the planet is due to the strength of its gravitational pull. It is predicted that deep in the atmosphere the temperature and pressure conditions would favor the condensation of hydrogen into liquid form and that at greater depths it may convert into a solid metallic form. At the center it is possible that there is a hard core the size of Earth. The very rapid rotation of the planet is expected to produce a much stronger magnetic field than the Earth's, giving rise to a significant *magnetosphere* in which ions and electrons are trapped in plasma form.

The NASA Galileo missions 1995–2003 have yielded a wealth of data on Jupiter's atmosphere and magnetosphere. They have also recorded a more complete picture of the satellite system, and for the first time, the *rings* around the planet. A mission called *Juno* is planned for 2011 to further study Jupiter from a polar orbit.

Saturn is the second largest planet in the solar system, and its distinctive rings mark it as a unique symbol of a planet in the night sky. It is the farthest planet that was known prior to the advent of the telescope. Galileo noticed the rings in 1610 with his primitive telescope, but was unable to resolve their structure. It was left to Huygens, presumably with a better telescope, to see their true form in 1655. There are three prominent rings beyond the very faint innermost one, now designated as the *D-ring*. The three are now designated as the *C-ring*, the *B-ring*, and, after a gap called the *Cassini division*, are followed by another prominent ring the *A-ring*.

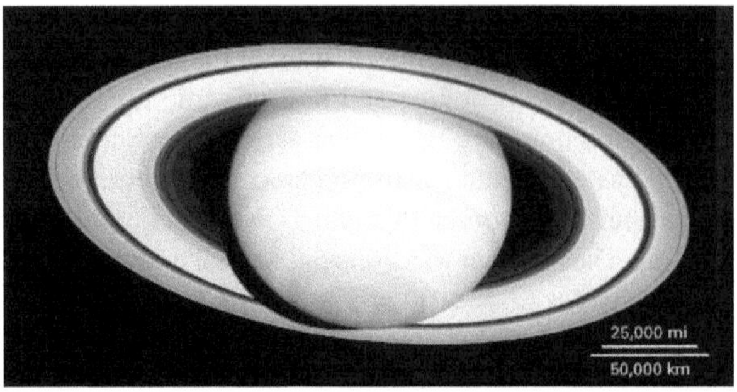

Fig. 4.3 The rings of Saturn taken by Voyager 2 (1981) (NASA/JPL photograph)

There are at least 23 named narrow rings around the planet of different widths and brightness. The outermost ring reaches a diameter more than twice the diameter of the parent planet. The rings are very thin, no more than perhaps 1 km across; seen edgewise they virtually disappear. They are believed to be composed of billions of small particles, although some radar data indicate that some particles must be over 1 m in diameter. They are believed to be fragments of different size resulting from the breakup of an asteroid or comet. The planet's invariant equatorial plane, that is, the plane perpendicular to its axis of spin, is inclined to its orbital plane by 28°. As the planet follows its orbit around the sun, to an observer on the Earth this means that there are times when Saturn's rings are seen edge on and other times when they are tilted one way or the other. Saturn is a little smaller in diameter than Jupiter, almost 1,000 times the volume of Earth but with a mass only 95 times the mass of the Earth, making it unique in having a mean density less than water (0.7 g/cm^3). The semimajor axis of its orbit is about 9.5 AU and eccentricity $\varepsilon = 0.056$. It rotates on its axis approximately once every 10 h and completes its orbit in about 29.5 sidereal years. This means that its position varies slowly relative to the stars and therefore useful in celestial navigation. The escape velocity on its surface, that is, the minimum velocity any body projected vertically must have in order never to fall back under gravity, is high enough that the planet can retain the lighter gases, hydrogen and helium. We recall that for a gas at a given temperature, the lighter its molecules are, the higher their average velocity.

The NASA/JPL planetary missions such as Voyager 1 (1980) and Voyager 2 (1981) flyby missions have contributed extensively to our knowledge of the structure of planets and the atmospheres around them, and in particular in the case of Saturn, its familiar rings shown in Fig. 4.3.

Saturn's outer layers are much like Jupiter, consisting mostly of hydrogen in different forms and helium, with trace quantities of other gases. The Cassini mission, launched in 1997, was designed to study Saturn's ring system and satellites and carried Europe's *Huygens probe* which was released in 2005 to parachute down to the surface of Titan, Saturn's largest satellite. This was the first probe landing in the outer solar system that yielded detailed information about Titan's atmosphere.

The planets beyond Saturn are Uranus, Neptune, and until recent times Pluto. The planet Uranus was not known as such prior to the eighteenth century undoubtedly because, with the naked eye, it is an undistinguished point of light much like any other distant star. It was discovered telescopically in 1781 by the astronomer William Herschel. This German-born Englishman, Court Astronomer to King George III, a fellow German, was, without exaggeration, the most prolific observational astronomer of all time. After an early career in music he developed a strong interest in astronomy and became an exceptional builder of optical telescopes. He excelled in the making of superior mirrors made of *speculum* alloy and ultimately built a telescope with an aperture of 1.2 m, a record that stood for 50 years. With his telescopes he embarked on a career of systematically scanning the night sky. His first major discovery was the *Orion nebula*, but more relevant to our present subject is his discovery of a new planet that he wanted to name for King George, but in keeping with the use of the names of Greek gods for the other planets, the name Uranus was accepted. He went on to catalog no fewer than 2,500 objects including nebulae and star clusters.

Uranus is about 3.7 times the size of Earth with a mass almost 14.5 times larger; thus it has a mean density 3.5 times *less* than Earth. The semimajor axis of its elliptical orbit is about 19 AU with an eccentricity almost three times larger than Earth. It has the unique property that its equatorial plane is tilted with respect to its orbital plane by 82°! That is, its axis of rotation is within 8° of lying *in the plane of the orbit*. This means that at certain times during its journey along its orbit, an observer on Earth may be confronted with the planet's Polar Regions. The sense of rotation is also odd: it rotates in the opposite direction to all other planets except Venus. It takes Uranus 84 *years* to complete its orbit and takes almost 11 h to complete its rotation about its axis. Uranus is surrounded by an atmosphere of hydrogen and helium with some methane gas. Viewed through a telescope it has a bluish-green hue, resulting from the absorption by methane in the red end of the spectrum. Its internal structure is expected to be similar to Saturn, although its mean density is more than twice as large: 1.6 versus 0.7 g/cm^3.

The most exceptional fact about Neptune, the next planet in the solar system, is that it represents a major triumph of theoretical astronomy [7]. It was mathematically predicted on the basis of Newtonian theory. In attempts to account theoretically for the observed irregularities in the orbit of Uranus, the known perturbations due to the gravitational pull of the neighboring planets, Jupiter and Saturn, proved inadequate to provide a satisfactory answer. Thus around 1843 a young English astronomer, John Adams, and independently the mathematician Urbain LeVerrier set themselves the task to analyze the irregularities in the observed motion of Uranus and to investigate whether the presence of another unknown planet could account for them. Their calculations indicated that indeed the existence of another planet more distant than Uranus could account for the observed behavior. In 1845 Adams sent his prediction of the coordinates of this putative planet to the Astronomer Royal, Airy, with an estimated error of only 2°! The following year, quite unaware of Adams' work, Leverrier published his result, which Airy noted agreed with Adams to within 1°. This convinced Airy to suggest that the Cambridge Observatory should search the indicated region of the sky. Unfortunately the Cambridge Observatory working with inadequate star charts in the region of *Aquarius* where the planet was predicted to be failed to recognize the planet among the stars. In the meantime LeVerrier sought to have the Berlin observatory look for the planet, a fortunate development since subsequent observations showed that there would have been significant movement in the position of the planet over time. The very same night that the Berlin Observatory received the request to search for the planet, it was found less than 1° from LeVerrier's prediction!

Neptune is so far from Earth that it subtends an angle of only 2 s of arc; its diameter is almost 4 times that of the Earth, with a mass 17 times that of the Earth. Its (sidereal) rotation period is about 16 h, and revolves in a nearly circular ($\varepsilon = 0.0086$) orbit with semimajor axis of about 30 AU, making it the farthest of the *gas giants*. Its orbital plane is inclined by 1.8° to the ecliptic and 28° to its equatorial plane, compared with 23.5° for the Earth. It completes its orbit in 164.8 years, which means it has had time to complete its orbit only once since its discovery! Because of the inclination of the spin axis relative to the orbital plane it is expected that there should be seasons on Neptune similar to Earth, but because of the nearly circular orbit their duration would be nearly equal in length. The planet also has a magnetic field whose strength is a little weaker than that of the Earth, about 1.5×10^{-5} T at the equator, with the oddity however that the magnetic poles are on an axis tilted with respect to the body axis by about 47°. Voyager 2, the only flyby mission to study Neptune, was able to determine the period the planet's rotation as being 16 h through the periodicity of high-frequency signals associated with the magnetic field.

Like the other gas giants Neptune's atmosphere is about 75 % hydrogen and about 25 % helium with some methane. The presence of the latter again accounts for absorption at the red end of the spectrum giving the planet a bluish-green hue. The temperature of the atmosphere varies with altitude; a minimum of −223° occurs at an altitude where the atmospheric pressure is about 10^4 Pa (1 Earth atmosphere = 10^5 Pa). At the 10^5 Pa level the mean temperature is −199°C. The temperature increases rapidly at depths below this level reaching astronomically high (!) temperatures and pressures at the center. Like the other giants there are high winds confined to zones along the parallels of latitude that circulate around the planet. Images of the planet also show a *great dark* spot indicative of a storm center driven by the high winds. The presumed constituents of the interior of the planet are again mostly frozen crystalline or amorphous compounds grouped under the term *ices* and rock.

Thanks to Voyager 2 Neptune is known to have 13 moons and 6 rings. The largest of the moons is *Triton*, a satellite of special interest since it oddly revolves around Neptune in the opposite direction to its parent's rotation. Like the moon–Earth system, Triton's rotation is locked in synchronism with its orbital motion causing it to always present the same side to the planet. Triton is enveloped by an atmosphere of mainly nitrogen with a thin layer of frozen nitrogen covering the surface. Another unique observation made by Voyager 2 beyond surface temperature (−235°C) and composition of the surface is the presence of *volcanic activity*.

In 2006, Pluto, once classified as the ninth planet in the solar system, was reclassified as a *dwarf planet*. The history of its discovery is a remarkable example of an accident of fortune. It began when astronomers, including Percival Lowell, obsessed with what appeared to be residual deviations from the predicted orbit of Neptune, began to try to account for these based on the possible presence of a ninth planet, not realizing that those residuals could be accounted for as mostly errors of observation! At the time Lowell started pursuing this, Neptune had not been observed long enough to have an accurate knowledge of perturbation on its orbit, and so the calculations mainly involved Uranus. His calculations indicated two possible coordinates where a perturbing planet may be located; the more likely one was in the constellation *Gemini*. He tragically spent the rest of his life vainly seeking it. The search was far worse than "finding a needle in a haystack". One is looking for that one star-like object in the middle of literally millions of identical points of light! The task was made more manageable by the invention of what is called the *blink camera*. In this the observer's field of vision is automatically switched between precisely the same field of stars taken at different times, so that if one point of light has shifted, it is immediately noticeable. In 1930 14 years after Lowell's death, Clyde Tombaugh found an object within 6° of Lowell's prediction. The irony is that the residuals that Lowell was so worried about were based on observations of Uranus' position made in the eighteenth century. The errors in those observations were clearly so great as to render the "residuals" irrelevant! The success in finding a planet more or less where Lowell predicted is purely fortuitous.

Pluto has about 1/5 the radius of Earth and a mass 1/500 Earth's mass. It rotates once every 6.5 days in a direction opposite to its orbital motion, which it completes in 248 years. It has a very eccentric orbit ($\varepsilon = 0.25$) with a semimajor axis of 39.5 AU. The plane of the orbit is inclined to the equatorial plane by −58° (retrograde revolution). Its surface temperature is in the range −233°C to −223°C.

Beyond Pluto extending to 55 AU from the sun are hundreds of thousands of icy bodies in a wide range of sizes, one even as large as Pluto, in a belt beyond Pluto called the *Kuiper Belt*. Since 1992 when the first object was seen 42 AU from the sun, just beyond Pluto, about 1,300 *KBOs* (Kuiper Belt Objects) have been identified. An unusually large KBO was discovered in 2005 that was actually *larger* than Pluto. At that point the question was raised as to whether this new object that was named *Eris* should be classified as the tenth planet, or reclassify Pluto and Eris as belonging to another category. In 2006 a new category of *Dwarf Planet* was inaugurated with Pluto and Eris its first members. Since then, by 2009 four new dwarf planets had been added.

4.5 The Stars

Beyond our solar system are the myriad "fixed" stars that are scattered throughout the celestial globe, some bright, others less so, in patterns and groupings (constellations) that through the ages have been given names to serve as mnemonic aids. The appearance that members of a constellation belong together as an outline of a goat (Capricorn) or a ram (Aries) is of course pure imagination; they only appear to line up the way they do to an observer on Earth, but are otherwise totally unrelated.

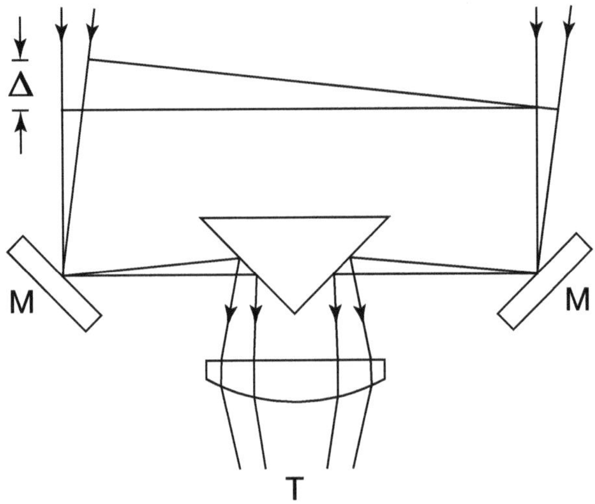

Fig. 4.4 Michelson's stellar interferometer

All stars are so far from our solar system that their shapes are not resolved by the largest optical telescopes. There are a few, however, whose diameters are almost resolvable; these may be within the capability of a technique originally suggested by H. Fizeau in 1868 involving the use of light interference. It was successfully realized in 1920 by A. A. Michelson of the "Michelson–Morley experiment" fame, which confirmed the absence of an "ether drift." It is a method that makes the *effective* aperture of a telescope very much larger without actually grinding a larger mirror. Michelson's original experiment used a system of four widely separated plane mirrors rigidly mounted in front of the telescope, as shown in Fig. 4.4.

The principle is to observe the intensity maxima and minima formed by the wave interference of star light entering the telescope by way of the two mirrors, whose separation can be varied. If the star light reaching the telescope indeed consisted of parallel rays coming from a point source, and the separation of the mirrors is slowly scanned, then because of interference between light intercepted by the two mirrors, a 100 % modulation between dark and light would be seen at the eyepiece. If however the star has a finite diameter and the rays falling on the mirrors have a *range* of directions, then for each direction there is a different mirror separation for maximum or minimum intensity at the eyepiece. If the mirror separation is such that when light from one side of the star produces a maximum and the light from the other side of the star produces a minimum, then the contrast between maxima and minima is lost. The procedure is to vary the distance between the two plane mirrors until the intensity modulation disappears. From the geometry of the apparatus and the wavelength of the light the angular size of the star can be calculated. Needless to say the success of the experiment hinges on keeping any mechanical vibration of the mirrors extremely low. Furthermore since stars emit a continuous spectrum, a color filter is necessary to select a narrow band of wavelength. Little progress was made with this technique after Michelson, a testament to his skill in working with light interference experiments. However in the 1970s the work took a new direction thanks to A. Labeyrie and C.H. Townes and others, using interferometry with separated telescopes. In more recent times the synthesis of multi-mirror arrays to achieve higher optical resolution has been implemented in ground installations, and no doubt will in the future be implemented in space.

The distances from the solar system to even the nearest stars is so large that the preferred unit of distance is the *light-year*, a unit of distance sufficiently large to accommodate with simple numbers the vast scale of astronomical distances. It is defined as the distance that light travels in 1 year. Since light travels in vacuo approximately 300,000,000 m in 1 s, one light-year amounts to 9.4×10^{12} km. The vast distances to the constellations also mean that their apparent positions change very little on a human timescale; they are noticeably unchanged since Babylonian times. However the stars *do move* relative to the sun at possibly high speed, but imperceptibly to the human eye because of their great distance. The motion of a star relative to the Earth is specified in terms of its *radial* velocity and *proper motion.* The radial velocity, as the term suggests, is the speed along the line-of-sight, which is close to the direction away or towards the sun. This is determined from the shift in the frequency (color) of the light received from the star due to the *Doppler effect*, the change in the observed frequency of a wave, whether sound or light, due to the relative motion between the source and observer. In the case of light there is a shift towards the blue end of the spectrum if the source and observer are moving towards each other and red if they are moving apart. The effect was postulated by the Austrian Christian Doppler in the context of starlight, verified acoustically by Buys Ballot, and independently arrived at experimentally for the much more difficult case of light waves by the French physicist Armand H. L. Fizeau. In France it is sometimes referred to as the Doppler–Fizeau effect. Using the Doppler shift, the radial component of the relative velocity of any star can be determined if enough radiation is received to analyze its spectrum. Of course that may not be the total velocity vector; there may be a perpendicular component that would give rise to a change in the direction of the star. The rate at which this occurs expressed as the number of seconds of arc per year is curiously called the *proper motion.* In this case the accuracy of measurement critically depends on the resolution with which the star is observed relative to adjacent background stars. The angular shift is usually determined by comparing high-resolution photographs taken of the given region of the sky at times separated by at least decades. Of course the proper motion, as defined, does not give the actual *transverse velocity* of the star: that requires also knowledge of its distance. When combined vectorially with the radial velocity, the transverse velocity yields the total *space velocity.*

The orbital motion of the Earth affects the accuracy of star observations in several ways. The first we shall consider is called *aberration.* The term aberration is most commonly used in the context of optics, but here we refer to the error in sighting a star by reason of the *apparent* change in its direction due to the observer's own motion. To illustrate the phenomenon of aberration an analogous example is the experience of someone running through vertically falling rain and getting wet in front! The reason is that motion is relative, and in the person's frame of reference the rain *does* fall with a component of velocity equal and opposite to his velocity.

In the case of star observations, since the average velocity of the Earth in its orbit is around 30 km per sec the corresponding aberration angle (in radians) $v/c = 30/3 \times 10^5 = 10^{-4}$ or about 20.5 s of arc. Detection of this aberration and its variation throughout the year is conclusive evidence that the Earth is in motion with respect to the stars. This is a very small effect, beyond the ability of early instruments to detect, a fact that explains the reluctance of early astronomers to accept the heliocentric model.

Another phenomenon affecting star observations is parallax. It is simply the apparent change in the *relative* directions to distant objects resulting from a change in the position of the observer. This phenomenon provides a method of determining the distance to some of the nearer stars, by measuring the apparent angular displacement of the star against a background of stars known to be very much more distant. As the Earth moves around in its nearly circular orbit, a near star, in general, appears to move relative to distant ones around a small ellipse called the *parallactic ellipse*. Near stars that happen to be in a direction perpendicular to the Earth's orbit will appear to describe the same figure as the Earth. On the other hand if the star is in the plane of the Earth's orbit, it will appear to simply oscillate along a straight line. The semimajor axis of the parallactic ellipse is called simply the stellar parallax. Since distances to the stars are generally given in units of *AU* (*astronomical unit*) defined as the mean distance from the Earth to the sun, it follows that the *stellar parallax* is equal to the angle subtended at the distance of the star by 1 AU measured at right angles to the direction of the star. Another unit of stellar distance is the *parsec (pc)*, defined as the distance of a hypothetical star that has a parallax of 1 s of arc. It takes light 3.26 years to travel 1 pc, that is, 1 pc = 3.26 light-years. For example the nearest star and brightest visible in the northern hemisphere is *Sirius* at a distance of 2.6 pc or about 8 LY, that is, it takes 8 light-years to reach us from that star; this should be compared with the 8 min it takes light to reach us from the sun.

Finally we mention *atmospheric refraction* caused by the presence of our atmosphere, principally composed of oxygen and nitrogen covering the Earth, for which we can be very grateful! It happens that in the temperature range that exists on the Earth's surface, the thermal velocities of the molecules of these gases are fortunately below the gravitational escape velocities. Such is not the case for helium gas. When light travels obliquely across an interface between two layers of the atmosphere that have different optical densities the light rays are refracted, or change directions. This is attributed to the fact that the velocity of light is smaller in the denser medium, as first explained by Huygens on the basis of the wave theory of light. The law governing the degree to which the light bends is called *Snell's Law*, which states

$$n_1 \sin\theta_1 = n_2 \sin\theta_2, \tag{4.2}$$

where θ_1 and θ_2 are the angles of the incident and refracted rays relative to the perpendicular to the interface between the two media, and $n = c/V$ is defined as the *refractive index* where V and c are the velocity of light in the medium and in

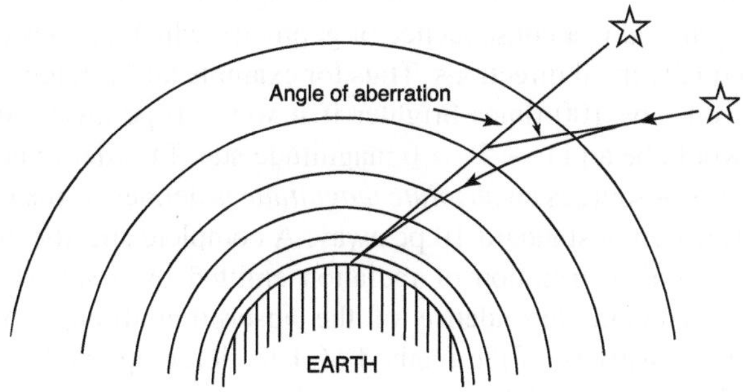

Fig. 4.5 Atmospheric aberration

vacuo, respectively. It follows from this law that if the light passes through a transparent medium consisting of plane parallel layers of different refractive indices n, the total refraction depends on only the values of θ and n at the top and bottom strata and *not in between*. Atmospheric refraction refers to the phenomenon in which the increasing density of the atmosphere near the Earth's surface causes star light entering it obliquely to curve towards the vertical. If the atmosphere is modeled as a refracting medium consisting of plane parallel strata, the aberration angle R in the sighting of a star is simple to calculate; in fact it would be given by

$$R = (n_0 - 1)\cot(A), \tag{4.3}$$

where A is the altitude of the star and n_0 is the refractive index of the atmosphere at the Earth's *surface*. A more realistic model would of course consist of an atmosphere that is in concentric spherical layers, but the calculation in that case would be very much more complicated. The aberration is negligible for stars near the zenith but becomes increasingly serious lower in the sky, as shown in Fig. 4.5. Thus events such as sunsets are very much affected and explain the observed flattening distortions seen in a sinking sun's disc.

The brightness of stars is specified using the *magnitude scale*, which has its origin going back to the second century B.C. when Hipparchus devised a subjective scale of brightness extending from the brightest to the barely visible in steps numbered from 1 to 6. A more quantitative scale was proposed in 1856 by N. R. Pogson based on Fechner's Law in physiology, which states roughly that the response of the eye to a stimulus is proportional to the *logarithm* of the stimulus. Noting that a magnitude 6 on the old scale corresponded to about 100 *times lower* intensity than a magnitude 1, he proposed that the scale be defined as a *logarithmic* scale in which a unit increase on that scale represents *division* by a factor of about 2.512 ($=100^{1/5}$) in brightness. Thus, for example, a magnitude 11 is 10 steps numerically above 1 and therefore $(2.512)^{-10} = 1/10{,}000$th of a magnitude 1 star. In modern times luminous intensity can be measured photometrically, making the scale purely objective. The relative brightness of a celestial object as observed on the Earth does not determine the brightness at the source. At distances from a star large compared with its size, the light intensity falls off according to the *inverse square law*, a consequence of geometry which applies to anything that spreads out equally in all directions. Thus for example an 11 magnitude star that is 100 pc away would be 100 times brighter if it were 10 pc away, and since $(2.512)^5 = 100$ it would be a $(11 - 5) = 6$ magnitude star. In order to compare the *intrinsic* brightness of sources an *absolute magnitude* is defined as the magnitude a star *would* have if it were a standard 10 pc away. A complete specification of the total amount and spectral distribution of radiation emitted by a star should take into account the wavelength dependence of the absorption through our atmosphere as well as the sensitivity of the method of detection employed, whether it be the naked eye, photographic plate, or photomultiplier. A magnitude scale based on the entire spectral range emitted by a star is called the *bolometric magnitude*, and the value it would have at the standard distance of 10 pc is the *absolute* bolometric magnitude. It is quoted in units of *ergs per second* and may be called the *luminosity* of the star. For the *sun* the radiant energy falling on unit area (1 m^2) outside the atmosphere is about 1.37×10^6 W. This quantity is called the *solar constant.* If the spectrum of the light received from the sun is analyzed with a spectrometer, for example, one based on a glass prism or a diffraction grating, a remarkable feature is observed, almost as startling as the famous "black spots" on the face of the sun. Several distinct *black* spectral lines are seen superimposed on the usual bright colors of the rainbow. They were first observed by Wollaston, but were systematically studied by Joseph von Frauenhofer who by around 1814 had listed about 500 lines and by 1823 had observed these lines in the continuous spectrum of stars. They are called *Frauenhofer lines* in his honor. The origin of these lines is the selective absorption of radiation at certain wavelengths characteristic of ions or atoms in the cooler outer atmospheres of the stars. The presence in the atmospheres of some nuclei that have recaptured electrons to form bound systems, that is, ions or atoms, will result in their absorbing certain wavelengths in making quantum jumps between their energy levels, thereby removing these wavelengths from the spectrum. The conditions in the atmosphere must be such that after a quantum absorption the ion does not reradiate the photon, but loses the energy through collisions with other ions The presence or absence of dark lines and their location in the spectrum yield valuable information about the temperature and composition of the stellar atmosphere. The spectra of different stars differ in significant ways and to bring order to a bewildering array of spectra, they have been assigned to a set of seven *spectral classes*. They are designated by the seven letters O, B, A, F, G, K, and M. They roughly correspond to increasing temperature of the atmosphere from O to M, as indicated by the presence or absence of certain atoms or ions. For example, class O includes the highest temperatures (over 25,000°K) at which no neutral atoms are present, but multiply ionized elements may be present since they require extreme temperatures to strip them of all the electrons. At the other end we have the lowest temperature class M where some lines may exist belonging to neutral atoms and molecules.

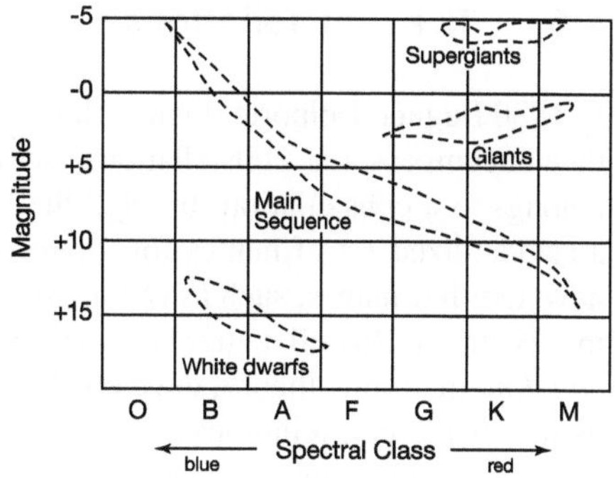

Fig. 4.6 The Hertzsprung–Russell diagram

To the extent that a star can be approximated as a spherical *black body*, in the sense of a perfect absorber/emitter, the luminosity can be related to its temperature through the Stefan–Boltzmann law which states

$$L = 4\pi r^2 \sigma T^4, \tag{4.4}$$

where L is the luminosity, r the radius, σ the Stephan–Boltzmann constant, and T is the absolute temperature. A plot made of the absolute star magnitude versus the spectral (temperature) class is called the Hertzsprung–Russell diagram, as shown in Fig. 4.6. For a very large class of stars the plotted points lie close to a simple S-curve, while some other stars give plots that cluster in a different region of the plane. The stars belonging to the S-curve are called *main sequence* stars.

The European Space Agency (*ESA*) launched in 1989 the *Hipparcos Space Astrometry Mission* to make a comprehensive survey of all stars with *magnitude* less than 11, to precisely fix their positions, and compile a reference catalog. It continued to send down data for three and a half years yielding an unprecedented body of highly precise stellar data embodied in the *Hipparcos General Catalog* comprising 118,218 stars. An auxiliary star mapper provided additional stars with somewhat less accuracy. With the *Tycho catalog* of more than 10^6 stars and *Tycho II* completed in the year 2000 a total of 2,500,000 stars are cataloged, comprising about 99 % of all stars brighter than magnitude 11. By analyzing the proper motions and velocities of a large body of stars surrounding the solar system it is possible to form a picture of the movements of the sun (a star like any other) in relationship to its neighbors as well as the overall motion with respect the totality of stars in our own galaxy the *Milky Way*. Recent calculations on the motion of the sun through the galaxy, based on a sample of one million stars from the Hipparcos catalog, suggest that the stars in the galaxy are revolving around the galactic center with a constant velocity. Also, on the basis of one model of the galaxy the sun has traversed its spiral arms more than once during the past 5×10^8 years. What is remarkable is that geological studies would seem to show that the Earth experienced long periods of unusually low temperatures during periods that coincide with these traversals.

4.6 The Constellations

In 1930 Eugene Delporte divided the entire celestial globe into contiguous regions that encompass the constellations, so that every point on the celestial sphere belongs to a constellation. In 1922 the International Astronomical Union adopted a standardized set of names for the recognized 88 constellations, many of which have familiar names, such as *Orion, Leo, Virgo, Cygnus, Gemini, etc.* The astronomer Henry N. Russell, better known to students of physics for the *Russell–Saunders coupling* in atomic theory, developed a system of three letter abbreviations for the names of the constellations which was adopted by the IAU. The designations of stars *within* a constellation were given various forms to accommodate the increasing number of observable stars. An early method in popular use was initiated by Johann Bayer in 1603. According to this, a star is specified first by using a letter of the Greek alphabet to place it in the order of its relative magnitude in the constellation followed by the name of the constellation in Latin (in the genitive case). For example the brightest star in the constellation *Centaur* (the Bull) would be *α Centauri*. One anomalous case is *β Geminorum (Pollux)* which is the brightest in *Gemini* (*the twins*), whereas α *Geminorum* (*Castor)* is in fact the second brightest. One difficulty is that with the growing ability to detect weak stars, their number can exceed the number of letters in the Greek alphabet! In Table 4.1 are listed some of the more familiar constellations and the brightest stars in them. It will be recalled that *right ascension* corresponds to "longitude" and is the angle along the celestial equator from the vernal equinox to the hour circle of the star in units of time, while the *declination* corresponds to the "latitude" and is the angle in degrees along the hour circle from the celestial equator to the given star. In Table 4.1 are listed the brightest stars, having magnitude <1 [8].

Table 4.1 The brightest stars: magnitude <1

Star	Constellation	APP.MAG	Distance	Color
Sirius	α-Canis Majoris	−1.46	8.6 LY	Blue
Canopus	α-Carinae	−0.72	74 LY	Blue-white
Arcturus	α-Bootes	−0.04	34 LY	Blue
Rigil Ken	α-Centauri	−0.01	4.3 LY	White-yellow
Vega	α-Lyrae	+0.03	25.3 LY	Blue
Capella	α-Aurigae	+0.08	41 LY	White-yellow
Rigel	β-Orionis	+0.12	815 LY	Blue
Procyon	α-Canis Minoris	+0.38	11.4 LY	Blue-white
Archena	α-Eridani	+0.46	69 LY	Blue
Betelguese	α-Orionis	+0.5	650 LY	Red
Hadar	β-Centauri	+0.61	320 LY	
Altair	α-Aquilae	+0.77	16.8 LY	Blue
Aldebaran	α-Tauri	+0.85	60 LY	Orange-red
Spica	α-Virgo	+0.98	220 LY	Blue

References

1. H. Jeffreys, *The Earth, its Origin, History and Constitution*, 2nd edn. (Harvard University Press, 1962)
2. G.O. Abell, *Realm of the Universe*, 5th edn. (Saunders College, Philadelphia, PA, 1994)
3. A. Einstein et al., *The Principle of Relativity* (Dover, Mineola, NY, 1952), p. 163
4. L. Combrink, in *Sciences of Geodesy-1*, ed. by Guochang Xu (Springer, New York, 2010)
5. A. Villiers, *Captain James Cook* (Charles Scribner's Sons, New York, NY, 1967)
6. G.O. Abell, Op.Cit. p.210
7. G.O. Abell, Op. Cit. p. 183
8. G.O. Abell, Op. Cit. p. 404

Navigation at Sea

5

5.1 Terrestrial Coordinates

In all but the most exacting applications, such as global oceanic or satellite navigation, the earth may be assumed to be a sphere. Therefore the most appropriate coordinate system to use to define positions on its surface is what would be called in coordinate geometry spherical polar coordinates. The axis of spin is taken to be the polar axis and the plane perpendicular to the axis that passes through the center of the sphere is the equatorial plane. The position of any point on the surface of the sphere is specified with reference to an orthogonal set of lines of latitude and longitude. The lines of longitude, or equivalently, meridians, are great circles that pass through both poles, formed by the intersection of planes passing through the axis and the spherical surface. The lines of equal latitude are formed at the intersection of the spherical surface with circular cones having vertices at the center and drawn around the axis. The angle of a cone at its apex is called the *colatitude*, that is, 90° minus the latitude, so that if the angle of the cone is 90°, corresponding to the *geographical equator,* the latitude is 0°. To distinguish between the northern and southern hemispheres the latitude must be specified as N (north) or S (south). Since the longitude lacks any natural zero from which to begin measurement, an arbitrary point was chosen and agreed to by international convention in 1884, namely the meridian passing through the observatory at Greenwich, England. That meridian is called the *prime meridian,* and a plane through that meridian and the axis divides the globe into two hemispheres the east (E) and west (W). The geographic coordinates are illustrated in Fig. 5.1.

A representation of the earth as a globe is ill-suited for charting courses and preserving navigational information. These functions clearly require a metrical representation in a plane. Unfortunately a sphere is topologically incompatible with a plane; it is impossible to project a spherical surface onto a plane without distortion. Nevertheless many types of useful projections have been devised, each with its own advantages and disadvantages. The most important for navigators is the *Mercator Projection*, created by a Flemish map maker named Gerardus

F.G. Major, *Quo Vadis: Evolution of Modern Navigation: The Rise of Quantum Techniques*, DOI 10.1007/978-1-4614-8672-5_5,

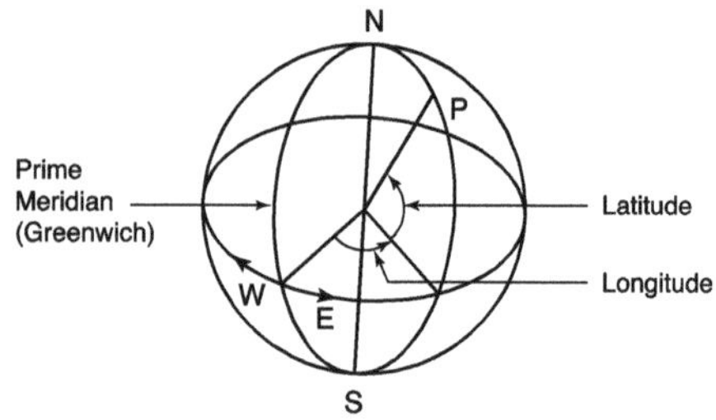

Fig. 5.1 Geographic coordinates: latitude and longitude

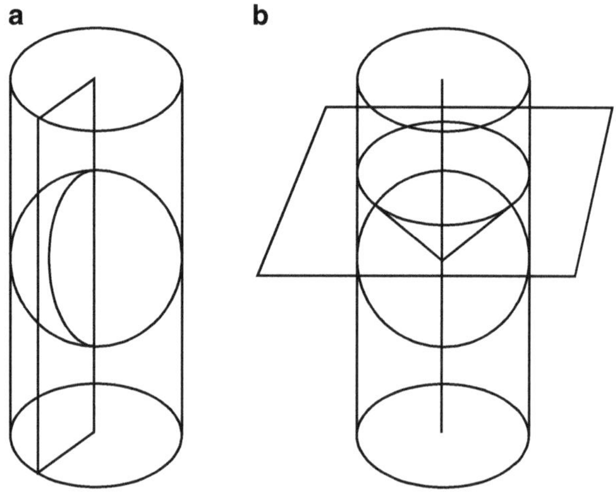

Fig. 5.2 The Mercator projection

Mercator, around 1569. Its distinctive features are that the parallels of longitude appear as *equally* spaced parallel lines and the parallels of latitude are also parallel lines at right angles to those of longitude but unfortunately become rapidly more widely spaced at higher latitudes. The Mercator projection is produced simply by circumscribing the spherical earth with a cylinder of the same radius parallel to its axis, as shown in Fig. 5.2. The planes through the axis that define the meridians will intersect the cylinder in equally spaced parallel lines; while the cones with equal increments of apex angle will intersect the cylinder in parallel circles, whose spacing is not constant but increases rapidly toward the poles. The cylinder can of course be unrolled without distortion to produce the desired flat representation.

The Mercator projection as described is *not conformal*: it does not preserve the shape of contours on the surface of the earth. This can be seen by considering the polar regions where distances along parallels of latitude converge to zero at the poles, whereas the projection leads to constant spacing! To reduce this distortion, the scale along the meridians must be correspondingly increased towards the poles; this is done mathematically. The result is that features such as coastlines in the polar regions are magnified, but at least their *shape* is nearly preserved, that is, nearly *conformal*. The Mercator projection is widely used for marine charts because it has the great virtue of enabling navigators to plot a *constant bearing* course by a simple straight line. Such a course however does *not* represent the shortest distance between the end points on the sphere. The curve on the sphere that projects into a straight line is called a *rhumb line* which, if continued, would curve around one of the poles! The shortest path between two points on a sphere is along the great circle joining the two points called the *geodesic,* defined as the intersection of a plane passing through the two points and the center of the earth. A Mercator chart representing a large area of the earth's surface will show serious distortions at the higher latitudes and geodesics project as lines with a curvature that increases in significance over large distances. Nevertheless, over relatively small areas of the earth's surface, which may extend over several 100 miles, it is by far the most commonly used projection. Other projections onto cylinders are of course possible in which the cylinder is oriented at different angles to the earth's axis, for example, at right angles to it. This would yield a truer representation of the polar regions, but of little use for general purposes.

Another type of projection of a spherical surface is called a *gnomonic projection.* It is simply constructed by projecting lines radiating from the center of the sphere onto a *plane* touching the sphere at some chosen point. There are three choices of the point of contact: (1) at the equator, called an *equatorial gnomonic*; (2) at a pole, called a *polar gnomonic*; or (3) at an arbitrary point on the sphere, called an *oblique gnomonic.* This type of projection has the interesting property that geodesics (that is, great circles) are projected as straight lines, while rhumb lines are curves. Thus great circle routes can be drawn as straight lines, and coordinates read along the route to be plotted later on a Mercator chart.

5.2 Astronomical Coordinate Systems

To erect a reference framework for the description of celestial objects and to account for their observed motions, we begin with the concept of the celestial sphere. This is a conceptual sphere of infinite radius centered on the earth, one that the ancients regarded as real, on which the stars appear to be fixed, and which appears to an observer on earth to rotate about the points called the celestial poles where the earth's spin axis cuts the sphere. The plane perpendicular to the rotation axis of the earth and passing through its center is called the equatorial plane. The corresponding *celestial equator* is the circle where the equatorial plane intersects the celestial sphere. The *ecliptic* is defined as the circle of intersection of the earth's *orbital* plane with the celestial sphere. The points of intersection between the ecliptic and equator are called the *equinoctial points,* since at those points the earth's axis is perpendicular to the earth's orbital minor axis and the periods of night and day are of equal length. The great circles passing through the celestial poles and intersecting the equator at right angles in the way that lines of longitude do on earth are called *hour circles.*

The equatorial and orbital planes are both very nearly fixed in space, apart from a very slow precession of the spin axis. As the earth revolves around the sun in its orbit, to an observer on earth the sun appears to move in the opposite direction along the orbit, crossing at the equinoctial points alternately from north to south of the equator. At the summer solstice (longest day) around June 22 in the northern hemisphere, the sun appears at noon about 23.5° north latitude. Hence points on the earth that are at 23.5°N will have the sun pass through the zenith at noon, that is, vertically overhead. This latitude on earth is called the Tropic of Cancer and the similarly defined one for the southern hemisphere is the Tropic of Capricorn. We recall that Eratosthenes observed that the sun reached the zenith at noon in a certain location in Egypt, a fact he used to measure the radius of the earth.

Since the earth is very nearly a uniformly rotating platform of observation, a description of the observed positions of celestial objects begins with a suitable coordinate system that can describe their apparent motions in a simple manner. There are two principal systems, among others, that are most commonly used: briefly described as the *horizon* and *celestial* coordinates. Astronomers also use the *galactic* coordinate system, but that is outside our present concern. The multiplicity of coordinate systems reflects the different aspects of the celestial system. The three coordinate systems mentioned are all basically spherical polar systems of the type familiar in coordinate geometry in which a z-axis is defined and the angular coordinates θ, the colatitude, and φ the azimuthal (longitude) angle are defined with reference to a sphere.

The horizon system of coordinates is defined as follows: it is a spherical polar coordinate system in which what would be the z-axis in coordinate geometry is taken as the *vertical* direction as determined by a plumb line, or perpendicular to a free liquid surface at the location of the *observer*. Where this vertical line pierces the celestial sphere above is called the *zenith* and the point at the other end is called the *nadir*. If the earth were a uniform sphere, the vertical line would pass through its center. The circle of intersection between the celestial sphere and a plane perpendicular to the vertical axis and passing through the center of the earth is called the *celestial horizon*. The plane that passes through the vertical axis and the celestial poles intersects the celestial sphere in what is called the *local celestial meridian* or just *meridian* at that point. The points of intersection of the meridian with the horizon are called the north and south points. To specify the position of a celestial body in this system requires the *azimuth* and *altitude*. The (true) azimuth is the angle (0°–360°) measured clockwise on the horizon from the north point to the intersection point on the horizon of the circle through celestial body and the poles, as shown in Fig. 5.3. The altitude is defined as the angle above the horizon measured along the vertical circle passing through the body on the celestial sphere. The distinguishing feature of this system is that it refers to the position of a particular observer.

The celestial system is defined in terms of the celestial equator which is the circle of intersection between the celestial sphere and the plane perpendicular to the earth's axis of spin, the equatorial plane of the earth. The great circles intersecting at the poles that are perpendicular to the equator are called *hour circles*. The hour circle that passes through the zenith of an observer is, as already mentioned, called his meridian; in other words, the meridian passes through both the celestial pole and the zenith of the observer. In this system the coordinates of a celestial body are the

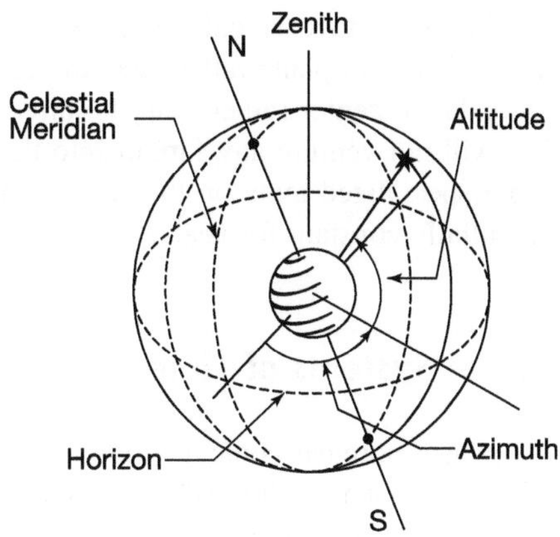

Fig. 5.3 Horizon system of coordinates

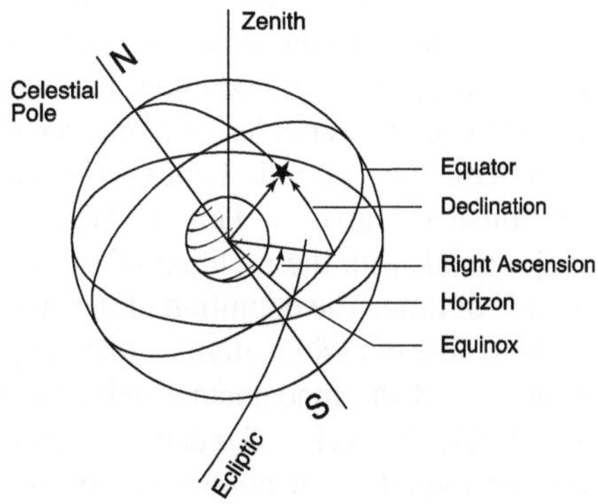

Fig. 5.4 Celestial system of coordinates

hour angle and declination. The hour angle is the angle measured westward along the equator from one of three possible reference meridians to the hour circle through the body. The three reference meridians are the meridian passing through a point fixed on the celestial globe called *Aries*, giving the *sidereal hour angle* (SHA), the meridian of Greenwich (GHA), or the observer's meridian, called the local hour angle (LHA). The hour angle may be expressed either in units of angle, degrees, minutes, and seconds, or of time, hours, minutes, and seconds. The sidereal hour angle is specified independently of the observer and measured in a westerly direction from the vernal equinox which is called the *First Point of Aries(ram)* because in ancient times the vernal equinox was in fact located in that constellation; however now the vernal equinox is in the constellation *Pisces (fish)*. The shift in the position of the equinox, it will be recalled, is due to the slow precession of the earth's axis of spin. On a human time scale the sidereal hour angles of stars changes very little. The other coordinate in this system is the *declination,* defined as the angle along the hour circle of the body from the equator to the body. It is specified as north or south (or positive/negative) according as the body is north or south of the equator, as shown in Fig. 5.4.

Finally we have the galactic system which uses the mean plane of the Milky Way as the principal plane rather than the equatorial plane, and it uses the direction to the galactic center rather than the direction of the vernal equinox as a reference. We will not venture any farther into the galactic system; the reader is referred to more specialized texts on the subject. These coordinates for a given star remain essentially constant for years.

5.3 Systems of Time

In an astronomical context time plays a role much like it does for players in an orchestra. In a metaphorical sense it coordinates the relative positions and motions of a vast array of celestial bodies. Since the observer is on a rotating platform, the earth, time is an integral part of the coordinates defining the position of a celestial body. Astronomers' use of time is inextricably linked with the concept of angle in the context of the periodic motion of the earth; in fact the angular position of the meridian is interchangeably quoted either in units of angle or units of time. There are many defined cyclical variables that generate different timescales, the most important are the International Atomic Time (TAI), Sidereal Time (ST), Universal Time (UT), and Coordinated Universal Time (UTC). The International Atomic Time represents a major break with the tradition of using the motions of celestial bodies to establish a scale of time. The definition of the atomic time unit is based on an average among an ensemble of cesium atomic clocks at principal laboratories around the world. The definition of the fundamental unit of time, the (atomic) second, is the duration of 9,129,631,770 cycles of radiation corresponding to the quantum transition between the two hyperfine quantum levels of the ground electronic state of Cs 133. Atomic clocks play such a critical role in the realization of satellite navigation systems that there will be an extended discussion of them in a later chapter.

The unit of sidereal time is defined as the mean interval between successive transits of a fixed point in space across the same meridian due to the earth's daily rotation. That fixed point is taken to be the vernal equinoctial point. Universal time is based on the mean solar day which is the length of the solar day averaged over the entire orbit of the earth. Because the speed at which the earth revolves around the sun varies according to Kepler's second law, the apparent *solar* 24-h day varies in actual duration throughout the year. Therefore a mean solar day is defined with reference to a fictitious mean sun which moves at constant speed throughout the year. The correction that must be applied to the apparent solar time to obtain the universal time is carried out using what is quaintly called the equation of time [1]. The correction (apparent solar time—mean solar time) varies throughout the year due to the eccentricity of the earth's orbit and obliquity of its axis, as already explained. The correction as shown graphically in Fig. 5.5 is fitted to within 1 % by the following formula:

$$\Delta = 9.87\sin 2\theta 7.67\sin(\theta + 78.7), \tag{5.1}$$

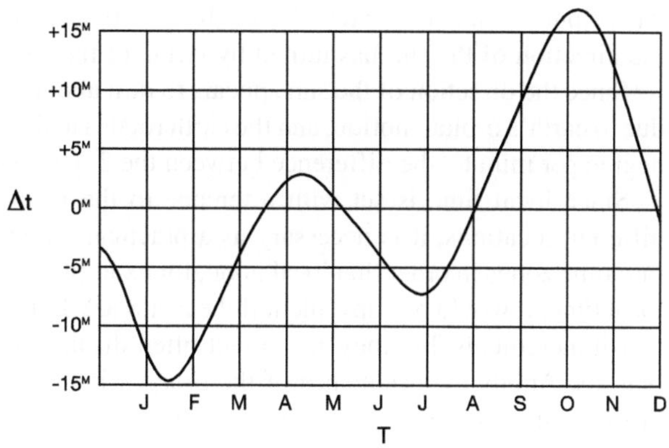

Fig. 5.5 Equation of time

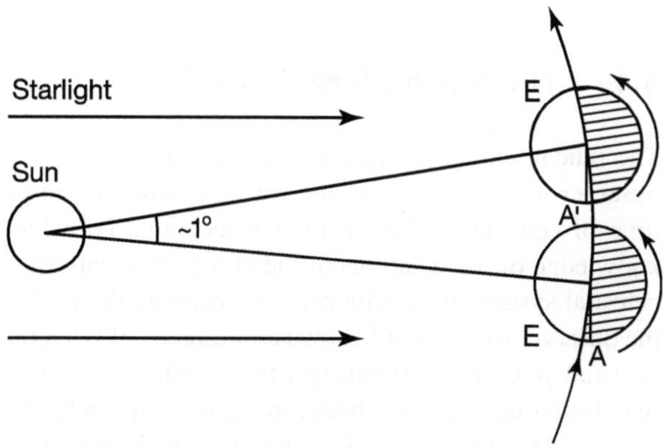

Fig. 5.6 Difference between the solar and sidereal day

where Δ is the equation of time in minutes, and $\theta = 360\ (n - 81)/365$ for the nth day of the year, that is, $n = 1,2,3\ldots$ for January $1,2,3\ldots$ etc.

The universal time at the prime meridian (Greenwich) is, not surprisingly, called Greenwich Time (GT). The Coordinated universal time differs from atomic time by a number of intercalary ("leap") seconds introduced to keep atomic time within 0.9 s of a corrected form of universal time (UT1).

We note that there is an important distinction between a time based on the mean solar day and the sidereal day, defined with respect to the transit of the vernal equinox, or more accurately with respect to the fixed stars. Since the earth revolves around the sun in a nearly circular orbit while it is spinning on its axis, the time between successive transits of a given meridian under the sun differs from what it would be in the absence of the orbital motion. This is made clear by noting that if the earth had no spin at all, the sun would still cross a given meridian every time the earth completed a revolution around the sun. The difference between the lengths of the solar and sidereal days may easily be calculated if we assume the earth's orbit is a circle. Since the sense of rotation (whether clockwise or counterclockwise) of the earth's spin and revolution is the same, we see by reference to Fig. 5.6 that the solar day is longer than the sidereal day by the time it takes the earth to turn the angle that the direction of the sun has turned by virtue of the earth's revolution around it.

Since the direction of the sun appears to turn about $360/365.25 = 0.985°$ per day due to earth's orbital motion, and the earth rotates at the rate of $360/24 \times 60 = 0.25$ degree per minute, the difference between the 2 days is about 3.95 (sidereal) min.

Since local time is set with reference to the local meridian, which differs at different locations, it is necessary as a practical matter to have the globe divided into time zones in order that local noon, for example, occurs approximately at 12:00 local time. It would be convenient if the zones are $360/24 = 15°$ wide to correspond to 1 h increments, but they are in fact often displaced for totally nonastronomical reasons. Another consequence of the progression of time as we travel around the globe is the need for a "cut", that is, an international date line where the date is increased or decreased by one depending on whether it is crossed in a westerly or easterly direction.

5.4 The Navigational Triangle

The fundamental objective in celestial navigation is to find the coordinates of any observer on the surface of the earth in terms of the positions observed at a given time of celestial bodies of known location. To achieve that goal requires firstly knowledge of the positions of the chosen navigational bodies with reference to the celestial system of coordinates and secondly the positions of the bodies as seen by the observer in terms of his own coordinates. If two great circles are drawn from the celestial pole, the first through the zenith of the observer, and a second through the chosen navigational body, the two great circles form two sides of a spherical triangle. The third side is the vertical circle drawn through the zenith of the observer and the body. Naturally to specify the angle between the two great circles at the pole requires knowledge of the exact hour angle of the observer and the body. The problem of celestial navigation is finding the sides of the celestial triangle shown in Fig. 5.7; thus it is reduced to solving a problem in spherical trigonometry [2].

Since spherical trigonometry is obviously more challenging than the plane variety studied in school, the solving of the spherical triangle is sufficiently complex as to generate historically several methods of solution. The interested reader may consult the standard reference in these matters, namely the volume on *Navigation and Nautical Astronomy* by Dutton, commander U.S. Navy, first published in 1926. However, as a practical matter, an algorithmic approach is generally adopted for general application by navigators.

The first step is to relate the celestial triangle to the terrestrial coordinates that will define the observer's position on the earth's surface. This is done by constructing what is called the *navigational triangle*. In this, the position of the celestial body is in effect projected radially onto the surface of the earth where its position is called the *geographic position* (GP) of the body. Every celestial body, including the sun and moon as well as the stars, is represented by a GP point on the earth's surface. The GP of the given celestial body forms one vertex of the

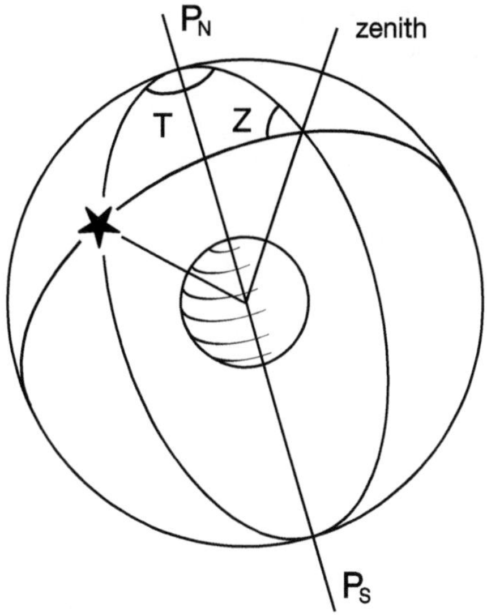

Fig. 5.7 The celestial triangle

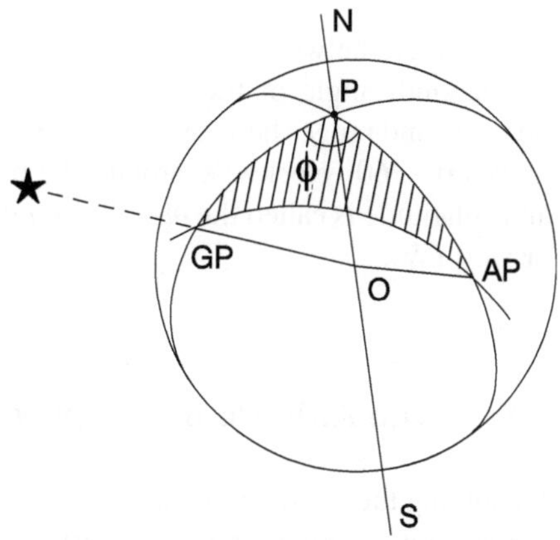

Fig. 5.8 The navigational triangle

navigational triangle, while the geographical position of the observer forms another vertex. The latter is, of course, the unknown in the navigational problem and conventionally called the *assumed position* (AP). The third vertex of the navigational triangle corresponds to the celestial pole in the observer's hemisphere, and is called the *elevated pole*, which may be designated as P_n or P_s depending on whether the observer is in the northern or southern hemisphere.

The sides of the navigational triangle are called the *colatitude*, the *coaltitude*, and the *polar distance*. The prefix *co-* in the first two simply comes from the fact they are "complementary" angles in the trigonometric sense: that is, 90° minus the angle. Thus the "length" of the colatitude is 90° minus the latitude of AP, and the coaltitude is 90° minus the altitude of the celestial body. The polar distance is simply the angle from the pole to the GP of the body, equal to 90° $\pm$ the declination of the GP. These are illustrated in Fig. 5.8.

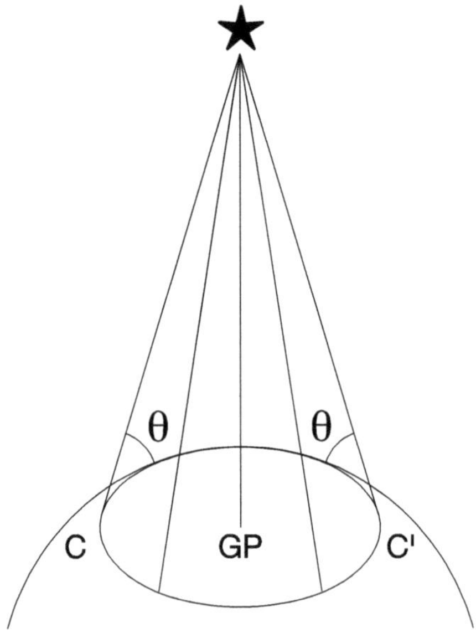

Fig. 5.9 Circle of equal altitude

The angle between the colatitude and coaltitude, marked φ in Fig. 5.8, is called the azimuth angle of the star with respect to the meridian passing through the observer and ranges between 0° and 180° E or W. This is not the *true azimuth* which ranges 0°–360° clockwise around the pole. The angle between the polar distance and colatitude is called the *meridian angle;* like longitude it ranges 0°–180° east (E) or west (W).

5.5 The Altitude Intercept Method [3]

By solving the navigational triangle, the *altitude* of any celestial body can be found for any given observer on the surface of the earth. This requires obtaining the declination and Greenwich hour angle (GHA) of the body from the *Nautical Almanac* for the given time. But the altitude alone does not define a unique point on the earth's surface; in fact all points on a circle with center at the point directly below the celestial body, its *geographic position* (GP), will lead to the same altitude. This locus of points where the observed altitude is a constant is called the *circle of equal altitude*, and is illustrated in Fig. 5.9. In principle, if the observer determines the circle of equal altitude for a second body, where they intersect marks his position. To resolve the ambiguity of having two points of intersection requires an independent estimate of position, obtained in practice by some other means of navigation, such as *dead reckoning*. Unfortunately, to achieve sufficient accuracy would require the circles to be drawn on an impractically large scale. Thus an altitude less than 80° would correspond to a circle of radius greater than 600 miles on the earth's surface!

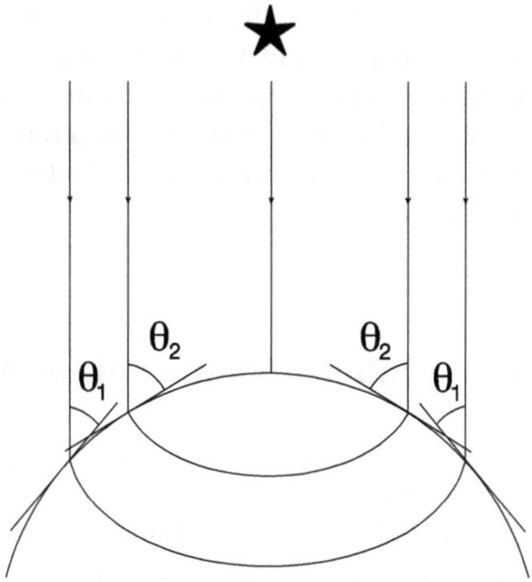

Fig. 5.10 Circles of equal stellar altitude

The altitude intercept method is designed to obviate the need to plot entire circles of equal altitude in order to establish a celestial *line of position* (LOP). The starting point is to *assume* a hypothetical "assumed position" (AP) for the observer. Naturally the closer the AP is to the actual location of the observer the better, but it can be shown that the choice is not material to the outcome. Using that AP, the altitude of the navigational body and its azimuth for the particular time of observation are obtained from an almanac and what is called a *sight reduction table,* about which more will be said later. A short arc of a circle, approximated by a straight line, representing the locus of equal altitude based on the assumed position, is drawn on the chart intersecting at right angles the azimuth line from the GP. Then the actual altitude of the body is *observed* and the result used to plot a second line of equal altitude parallel to the first. This second line is drawn nearer to the GP of the body, or farther from it according to whether the observed altitude is larger or smaller than the value based on the assumed position (see Fig. 5.10).

In summary, the celestial LOP is a straight line drawn perpendicular to the true azimuth line at a point on that line shifted from the AP by the amount of the difference between the calculated and observed altitudes. If the observed altitude is larger than the calculated one, then the LOP must be moved nearer the GP of the body. The size of the shift is called the *intercept* difference.

In essence, by assuming an approximate position of the observer and computing the corresponding true azimuth of the navigational body, the observer makes it possible to plot only a very short, approximately linear portion of the circles of equal altitude. Since the line of position, as constructed, is the locus of points having altitudes equal to the value actually observed, it follows that the position of the observer must indeed be somewhere on that line.

In order to fix the actual position of the observer, which is after all the fundamental objective, it would require that the same procedure be followed for another celestial body. The position of the observer would obviously be at the point of intersection of the two lines of position. To ensure accuracy in finding the point of intersection, the two chosen bodies should lead to LOP's that intersect at a substantial angle.

5.6 Tabular Solution of the Navigational Triangle [4]

As previously stated, the altitude intercept method of finding a celestial LOP begins by assuming an approximate position of the navigator, generally arrived at by dead reckoning (DR), and calculating the altitude and true azimuth of the chosen body for that assumed position. Since practicing navigators would simply not have time (nor presumably the inclination) to solve the problem in spherical trigonometry, the usual precomputer practice would have been to use comprehensive tables of the computed coordinates of heavenly bodies at regular intervals of time called *ephemerides.* Even these, the *Nautical Almanac* and the *Air Almanac,* are complicated enough that special training is required to use them. These almanacs are used to carry out the first step of obtaining the Greenwich hour angle (GHA) and declination, and hence the GP, of the chosen navigational body at the time of observation. These coordinates together with the assumed position of the observer are inputs for the *Sight Reduction Tables,* which yield the solution to the navigational triangle and the values for the altitude and true azimuth required for the altitude intercept method and the plotting of the line of position. Rather than relying on printed tables a modern navigator would retrieve the information automatically from a memory as part of a navigational computer program.

The American Nautical Almanac is published annually by the U.S. Naval Observatory and available from the U.S. Government Printing Office. It is essentially a tabulation, with respect to GMT and date, of the declination and either the sidereal hour angle (for the stars) or the Greenwich hour angle of the celestial bodies of interest to navigation: this includes the sun, moon, and four planets. A total of 57 navigational stars are tabulated alphabetically for each hour of GMT. The actual use of the almanac requires converting the time of observation to GMT, and conversely the GHA value read from the almanac has to be converted to LHA. The entries in the tables must of necessity have finite increments in the time (1 h); this means to increase time resolution requires (linear) interpolation. This is justified for stars since their GHA and declination do not abruptly change in time. For example if the GHA of Aries on December 17 at 17:00 h is 340° 54.4′ and at 18:00 h it is 355° 56.9′, then for example at 17:22 h the GHA is about 340° 54.4 + (22/60)(355° 56.9–340° 54.4). In the case of bodies other than the stars a simple linear interpolation may not be sufficiently accurate since they are closer and appear to move faster. At the bottom of the columns for planets are recorded corrections that must be applied to GHA readings to account for disturbances in motion. These corrections are available separately in the *Increments and Corrections* table at the back of the almanac.

As already indicated the declination and LHA of the chosen body derived from the almanac form two sides and the included angle of the navigational triangle which must be solved to determine the altitude and azimuth angle of the body in terms of the assumed position of the observer. Again in practice the method entails using prepared solutions in the form of minutely detailed comprehensive tables. One such compendium that runs to six volumes is the *Sight Reduction Tables for Marine Navigation No. 229.* It is a tabulation of computed altitude H_c and azimuth angle for navigational bodies. The input arguments for the tables are degrees of LHA, assumed latitude of the observer, and declination of the body. Values of the calculated altitude and zenith angle are tabulated in increments of whole degrees with interpolation tables.

5.7 Lunar Distance Method

In the discussion of celestial navigation so far, it has been assumed that the precise time is known at the point of observation and simultaneously at the prime meridian (Greenwich). But prior to the eighteenth century, keeping sufficiently precise time at sea was out of the question. The common timepieces available, sundials and sand glasses, were barely good enough to keep the time of day, let alone the time at Greenwich. Reliable marine chronometers, the subject of the next chapter, were not available until the late eighteenth century.

However navigators, for about a century prior to the development of mechanical chronometers, were able to use a technique called the *lunar distance method* to find their longitude. It is a method by which a navigator, through celestial observation of the moon's position in relation to the fixed stars, was able to *deduce* the time at Greenwich. The idea originated much earlier in the sixteenth century, long before the time when it could be applied because of the lack of precise ephemerides giving the positions of celestial bodies. It was a German astronomer by the name of Mayer who in 1753 published lunar tables of sufficient accuracy to prove the validity of the method. Later the English Astronomer Royal, Maskelyne, included these tables in the *British Mariner's Guide.* Maskelyne was to become embroiled later in the controversy surrounding the award of a prize for solving the problem of finding longitude at sea. The next chapter will be devoted to the longitude problem.

Through the urgings of astronomers in England in the seventeenth century, King Charles II established the Royal Observatory at Greenwich. By the eighteenth century ephemerides of the moon, planets, and several bright stars had been published, including, presumably from the work of Mayer, the (angular) distances of the moon from certain stars during its 27.3 (sidereal) day orbital period. Henceforth navigators with a sextant were able to measure the angular distance between the moon and a bright star along its path, while noting the local time at which the observation was made. From the published lunar distance tables, the Greenwich time at which that distance occurs is noted. The difference between Greenwich time and local time gives directly the longitude. To determine local time may require a clock that is accurate for at most one day, and not the duration of the journey. The method is made practicable by the fact that the moon travels among the fixed stars at a relatively rapid pace amounting to ½ minute of arc per minute of time.

The principle of the method is quite simple. The angular distance between the moon and a specified bright star along its path is the same observed from any point on earth, after correcting for parallax caused by the disparity in distance between the moon and stars. The accuracy of the method therefore depends on the degree to which the significant variation of the parallax across the globe can be taken into account. Nevertheless since the moon and stars are quite far from the observer, the parallax correction, though significant, need only be approximated. To determine it precisely would require knowledge of the position of the observer, which is unknown. In addition to the correction for parallax, there is also atmospheric refraction which is less accurately predictable and can significantly affect the observations made at different locations across the earth.

It is not surprising that Christopher Columbus, like other explorers, carried with him on his journeys into the unknown all the charts and ephemerides of the moon and principal stars that were available. One widely used almanac, carried by Columbus, was published by a German astronomer/astrologer known by his Latin pseudonym Regiomontanus (1436–1476). On his fourth and final voyage Columbus was to endure a harrowing experience in Jamaica [5]. While exploring the coast of Central America he had to abandon two of his ships because of an infestation of wood worm and was forced finally to beach the last two caravels on the island of Jamaica in June, 1503. At first the natives were not unfriendly and he and his crew were able to secure food and water; however as time passed relations soured and he felt the situation called for desperate measures. What happened next is worthy of a Hollywood production. It began while poring over the ephemerides of Regiomontanus when he noticed that a total eclipse of the moon was predicted to occur soon, on February 29, 1504. This gave him an idea: he would use the eclipse to prove to the natives that he had magical powers. He knew that the tables were prepared for Nuremberg, Germany, but decided it was worth trying to use them for a bit of innocent deception that played on the superstition of the natives to win their respect. So the day before the predicted eclipse, Columbus let it be known to the natives that he was very angry with them and that if they did not mend their ways he would work his magical power and darken the moon! The natives were unimpressed and told him in effect to "do thy worst!" So on the day of the eclipse he watched nervously as the time he calculated approached. Soon, there for everyone to see, the shadow of the earth fell across the moon and it was a total eclipse. The natives, terrified, pleaded with Columbus to bring back the moon. Columbus retired to his quarters and using his timer waited until the eclipse was half over and announced that if they behaved he would relent and return the moon to its former glory, which of course it did. From that time on the natives saw Columbus in a different light! By measuring the local time of the eclipse, Columbus could have estimated his longitude, at least relative to Nuremberg, but he was otherwise occupied at the time.

5.8 The Sextant

The marine sextant is a handheld instrument for the measurement of the angle between the directions of two distant objects; it is the modern culmination of devices intended to perform the same function that date back centuries. It is an icon, in the modern sense, of celestial navigation; its principal application is the measurement of the altitude of celestial bodies above the horizon. Its name derives from the Latin meaning one-sixth, a reference to the fact that its angular scale, called the *arc*, is about one-sixth of a circle. The principle of the sextant is to use plane mirrors to allow an observer to receive parallel light through a telescope from two sources, one directly, and the other by reflection, and by rotating one mirror a measured angle, bring the images into alignment and thereby determine the angle between the directions of the two sources.

Figure 5.11 shows the geometrical optics of a marine sextant. In the triangle M_SNM_H we have the angle $d + j = i$ and in the triangle M_SM_HT we have $2i = x + 2j$. By multiplying the first equation by 2 and subtracting one equation from the other we easily find that $d = x/2$. Thus for example to determine the elevation of the sun above the horizon, the directions of the sun's rim and horizon are brought into coincidence as observed in the telescope; this requires the relative angle of the two mirrors to be turned *half the angle* between sun and horizon. It follows that since the range of the arc is nominally 60°, the instrument is capable of measuring angles between objects as much as 120° apart.

Figure 5.12 is a photograph of a modern sextant the Astra III manufactured by the Celestaire company. It is a precision instrument with a rigid metal frame equipped with a grip and called, not surprisingly, the *frame*. On the frame are mounted essentially two mirrors, the arc, and a telescope. The structure below the telescope called the *limb* is designed to measure precisely the angle between the mirrors. It consists of a 60° arc of a precise worm gear that engages a corresponding gear on a micrometer drum; rotation of the worm gear advances the index arm along the arc. One complete rotation of the micrometer drum advances the index arm 1°. The gradations inscribed on the scale and micrometer drum are actually at half degree intervals, but are marked as full degrees in order to give directly the desired angle between the directions of the objects being sighted. The *index arm* is mounted so as to pivot precisely about the center of the arc. The plane of the index *mirror* M_S (Fig. 5.11) is perpendicular to the plane of the arc and rigidly mounted at the top of

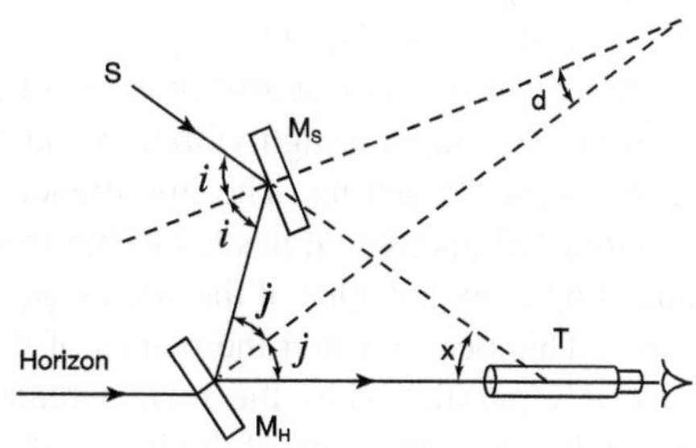

Fig. 5.11 Ray diagram illustrating the principle of the sextant

Fig. 5.12 The marine sextant (Courtesy of Celestaire Inc.)

the index arm. The *horizon glass* H has the bottom half mirrored, while the upper half is clear glass. It is mounted so that at *zero reading* on the arc it is parallel to the index mirror. The telescope T is inserted in a collar joined to the frame to direct the line of sight perpendicular to the horizon glass. To illustrate the principle of the sextant, suppose we wish to measure the altitude of the sun above the horizon. When the telescope is directed at the horizon, light passes through the clear part of the horizon glass to form an image in the focal plane of the telescope. The sextant is now rotated about the axis of the *telescope* until it is vertical and the image of the horizon thus appears horizontal in the field of the telescope. Rotate the index arm carefully until the image of the sun (dimmed by a dark filter) in the focal plane as seen through the mirrored part of the horizon glass just touches the line of the horizon as seen through the clear part of the horizon glass. The index arm is then clamped in position and the final adjustment is made using micrometer drum and vernier. When the altitude of the moon or stars is to be observed some procedure is necessary to acquire their image. One possibility is to have prior knowledge of the approximate azimuth and altitude for the time of observation. Failing that, the sextant is set to zero angle and bodily rotated until the star is seen through the telescope. Then unclamp the index arm and slowly rotate it up as the body of the sextant is turned down keeping the star in the field of view.

As a precision optical instrument it is inevitable that over time some of the critical adjustments may, through vibration and other environmental effects, "walk away" from its proper setting. The importance of not getting erroneous fixes through a maladjusted instrument has led to the recommendation of elaborate procedures for adjusting a sextant. One of the crucial adjustments is to reduce what is termed *index error*. This occurs when the planes of the index mirror and horizon glass are not precisely parallel when the sextant reads 0° 0.0′ of arc. This can be checked by observing the continuity of the line of the horizon in the horizon glass.

5.9 The Magnetic Compass

The magnetic compass is a direction finding instrument that remains in use even today. It has a rich history which we touched on in Chap. 2. Its basic simplicity, particularly when used in a nonmagnetic environment (aside from the earth's field of course), makes it universally relied upon, at least as a backup. In modern times it has been superseded by the mechanical gyrocompass on large ships and aircraft, and even more recently by laser-based gyros; these will be touched on in later chapters.

If the compass is to be used in an environment where magnetic materials are present such as steel structures, as on a modern ship for example, then the compass will in general not point along the local magnetic meridian but deviate by an angle called the *compass deviation.* A marine compass must of course provide for a means of annulling the magnetic deviation by using compensating magnetic elements: this is called *compass adjustment*, and is obviously critical aboard a steel ship. Elaborate procedures to annul perturbing magnetic fields around the compass have evolved over the years since steel-hulled ships first came into use. The object of compass adjustment is simply to annul all the magnetic field effects of the steel in the ship, so that the compass reading is as though it were in a nonmagnetic ship. It consists generally of positioning magnetic elements as part of the compass housing *(binnacle)* to cancel the disturbing field. In addressing this, we must recognize an essential difference between two types of magnetic materials that may contribute to the field at the compass: so-called permanent magnetic materials and soft magnetic materials. Permanent magnetic materials such as hard (high carbon) steel are said to have a large coercive force, that is, it takes a very intense magnetic field applied to them to change their magnetization, and therefore the relatively weak magnetic field of the earth has negligible effect on the field they produce. In contrast, so-called soft magnetic materials such as iron or low carbon steel will alter their magnetization in response to even a weak applied field, such as the earth's. Thus unlike hard steel, their magnetization is determined by the earth's field, and the field distribution around the compass can vary *with respect to the ship* if the latter moves in the earth's field. Let us consider the easier case of the hard magnetic materials first. The field they produce at the compass can be analyzed into three mutually perpendicular components along the three coordinate axes fixed in the ship, with the z-axis along the vertical, x- axis along the fore-aft line, and y-axis abeam. Since the compass needle is on the deck (x–y) plane, a vertical component of the disturbing field would tend to only tilt the needle in the z-direction. The components of the perturbing field do not change with respect to the ship as it moves, assuming the ship is a rigid structure. However if, for example, the ship is on a N–S bearing, the magnetic field at the compass will have two components: the earth's field along the x-axis, and the z-component of the perturbing field, but the latter has no effect on the compass needle as long as the ship is on even keel. However if the ship rolls, the perturbing field now has also a component in the y-direction causing the needle to deflect from true north. As the ship rolls back and forth the needle will oscillate. If the ship's heading had been E–W the oscillation

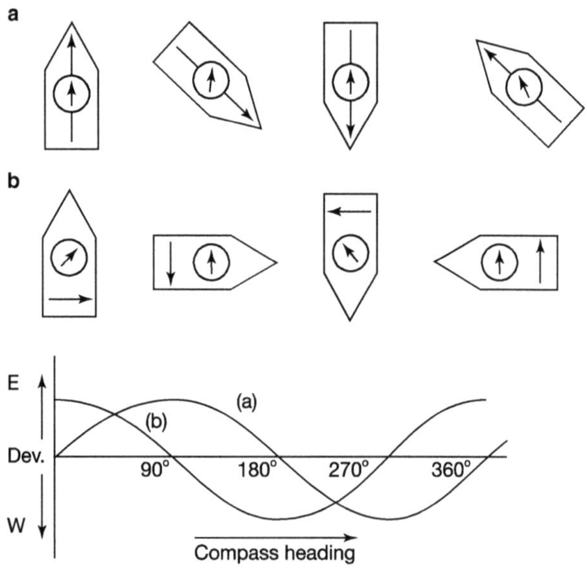

Fig. 5.13 Marine compass deviation due to ship's permanent magnetization: (**a**) fore-aft direction, (**b**) abeam direction, and (**c**) graph of deviation as a function of compass heading

would occur if the ship pitches. The x- and y-components of the perturbing field will produce a varying deviation as the ship yaws or swings in heading on an even keel. In Fig. 5.13 are shown effects of a permanent disturbing field component along the three axes of the ship.

In the case of induced magnetism, not only does it create a perturbing magnetic field, but the direction of that field with respect to the ship itself will depend on the heading and roll of the ship, as well as the magnetic latitude. Since most ships (except aircraft carriers) are symmetrical about the fore-aft line, in the case that it is on an even keel, the vertically induced magnetism will not vary with heading, and will cause a deviation similar to the hard vertical magnetization. An x–y horizontal distribution of soft iron that is symmetric about the fore-aft line produces a compass deviation that reverses at all four cardinal points giving a curve called *quadrantal* because it is different in each of the four quadrants. If the horizontal induced magnetization is not symmetric, another quadrantal curve is obtained but shifted 45°. In addition to the sources of deviation already cited the presence of both vertical and horizontal soft magnetization can lead to deviations resulting from rolling motion of the ship.

The procedures to annul compass errors are an attempt to design a set of appropriate placements of permanent magnets and soft iron masses at different points around the compass in such a way that the corrections are as independent as possible. Otherwise if the adjustment of one item changes the settings of the others, then a satisfactory configuration can never be reached. The task is made a little easier by the fact that permanent magnetism and induced magnetism are mostly independent and may be treated separately. Thus the standard components that are commonly used to correct for the deviation of the compass include some permanent

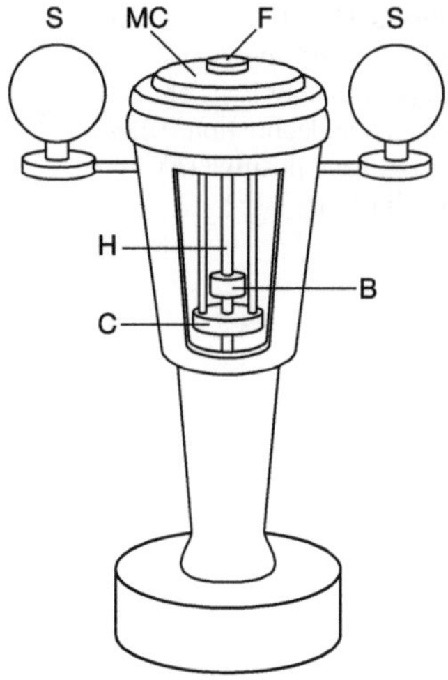

Fig. 5.14 Classical marine compass binnacle. *MC* magnetic compass, *S* soft iron spheres, *F* Flinders bar, *H* heeling magnet tube, *B* fore-aft magnet trays, *C* abeam magnet trays

magnets to annul the field due to hard magnetic materials and soft iron elements to cancel the field due to induced magnetization of the ship. Figure 5.14 shows an outline of the compass binnacle showing the location of the different correctors; in particular these may be listed as follows:

1. Vertical permanent magnet inside the *heeling magnet tube*, mounted below the compass
2. Permanent magnets set in fore-aft direction in their trays
3. Permanent magnets set abeam in their trays
4. Vertical soft iron cylinders (Flinders bar) set in an external tube
5. Two hollow soft iron spheres, one on each side of the compass

Not only must the magnetic compass be carefully adjusted for magnetic deviations arising from the ship's structure, but ultimately correction must also be made for any deviation in the magnetic field of the earth itself from the assumed theoretical model of a dipole field lined up with the magnetic N–S axis. In the past, accurate maps showing contours of equal variation of the magnetic compass from true north, also called *declination of the compass*, were essential to navigators; however with the later development of gyroscopes and particularly now with the advent of global satellite navigation systems, the magnetic compass is all but obsolete. Figure 5.15 is an approximate world map showing some lines of equal magnetic declination, called isogonic lines, particularly the isogonic lines corresponding to zero deviation, called agonic lines.

The actual distribution of the earth's magnetic field is not only space variable but time variable as well. There are many factors that can affect the magnetic field vector on the earth's surface, some of which were mentioned in Chap. 2. Since this does not pretend to be a manual or a history of marine navigation, we will resist the temptation to pursue the intricate procedures that have evolved to correct the readings of a magnetic compass aboard a steel ship.

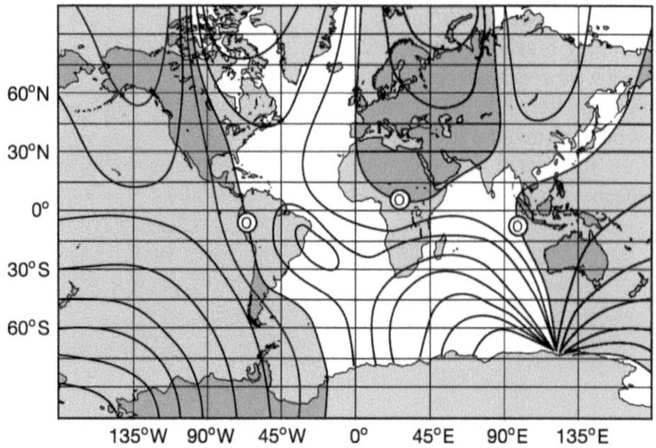

Fig. 5.15 Illustrative map of the world showing approximately lines of equal magnetic declination. *Shaded areas* have positive (*E*) declination

References

1. R.R. Hobbs, *Marine Navigation*, vol. 318, 3rd edn. (Naval Institute, Annapolis, MD, 1990)
2. B. Dutton, *Navigation and Nautical Astronomy* (United States Naval Institute, Annapolis, MD, 1957). Rev. ed
3. R.R. Hobbs, op. cit. p. 306
4. B. Dutton, op. cit. p. 373
5. B. Landström, *Columbus*, vol. 180 (McMillan, New York, 1966)

6 The Longitude Problem

6.1 Early Clocks

In our outline of the methods of celestial navigation in the last chapter, one essential input that is often glossed over is the precise time when the observations are made. The fundamental fact is that to navigate on a global scale, the only recourse up till very recently is to deduce positions on the earth by reference to observations of the celestial bodies. The modern global navigation satellite systems will be discussed in later chapters. On a human scale we cannot span the globe just by dead reckoning or some form of geodetic triangulation; we have to rely on celestial observations. This would not be so bad, were we not on a spinning observation platform and the celestial globe appears to be in constant rotation. Since an observer's angular position about the earth's axis is his longitude, a consequence is that a determination of longitude is inherently a matter of precisely tracking that apparent rotation in time. It is ironic that the very cyclical motion of the sun and stars which have historically been the means of keeping time is the same motion that requires us to have the ability to transfer time in order to navigate; that is, we need a precise clock. If the local time at some fiducial point on the earth (e.g., Greenwich) is available at any point on the earth's surface as carried by a stable clock, then a measurement of the local time at that point will yield directly the longitude of that point. Since the earth turns through 360° in 24 h, it follows that every hour difference in local time corresponds to a difference of 15° in longitude.

The history of ways to measure time goes back to ancient civilizations. A rudimentary example of a water clock has been found among ancient Egyptian artifacts dating back to 2000 B.C. It was a stone vessel filled with water that escaped slowly through a small hole at the bottom provided for that purpose. Later versions of the water clock had a conical shape, presumably arrived at by noting that it results in a more constant flow. The hydrodynamic problem of achieving a uniform rate of flow is not trivial since it depends on the size and shape of the opening at the bottom. The hydrostatic pressure at the mouth of the opening will fall as the level of the water falls, but how the rate of flow of water depends on the pressure and therefore what

F.G. Major, *Quo Vadis: Evolution of Modern Navigation: The Rise of Quantum Techniques*, DOI 10.1007/978-1-4614-8672-5_6,

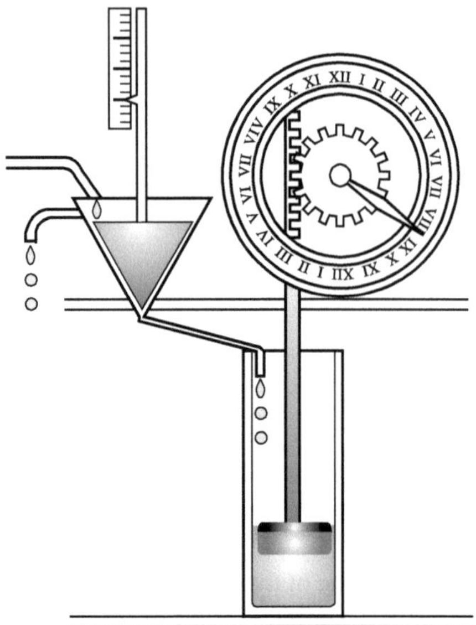

Fig. 6.1 The clepsydra Greek water clock

shape to make the vessel were presumably arrived at through experience. In a later development, rather than attempt to shape the vessel, a constant pressure was maintained at the outlet by introducing a continuous supply of water and having an overflow outlet to maintain a fixed level. The time was displayed by a mechanism actuated by the collected water level. That briefly is the principle of the water clock, the *clepsydra*, shown schematically in Fig. 6.1, whose invention is attributed to an Alexandrian Greek by the name of Ktesibios around 250 B.C. [1].

The accomplishments of Ktesibios include other hydraulic inventions such as a water pump. The clepsydra was in common use throughout classical times. It is written that these water clocks were used to allot time to speakers in a debate. When the water ran out it was time to stop talking! It is interesting that as late as the seventeenth century, Galileo used in his experiments on gravity a timer which worked on the same principle as a clepsydra. He measured the total time by collecting the water in a jar!

Water clocks of a different kind, built in China as early as the sixth century A.D. developed along a more elaborate mechanical path. They were very large water wheels whose rate of rotation was controlled in discrete steps by the flow of water. Around the rim of the wheel were regularly spaced buckets, and by an elaborate mechanism the wheel could not turn until a measured amount of water had been collected in each bucket, at which point it turned until the next empty bucket was in place to be filled.

The hour glass, or sand glass, that we introduced in Chap. 2, is of course the solid state analog of the clepsydra. It operates on the same principle; however, the flow of solid particles has characteristics that are quite different from liquid flow. One striking difference is that unlike a liquid, solid particles filling a container do not build up a hydrostatic pressure increasing linearly with depth below the surface [2]. The sand glass was still in use aboard ships in the nineteenth century to time the watch. Its limited range and poor accuracy severely limited its usefulness beyond that. It was not until the eighteenth century that a satisfactory mechanical marine timekeeper was developed.

6.2 Mechanical Clocks

Much has been written on the history of mechanical clocks and clock making, a testament to the enduring fascination of the subject. We will limit ourselves to a brief review of the mechanical design of clocks from the vantage point of modern design criteria, leading up to the marine chronometers of John Harrison.

To assess the merits of a clock design, it is useful to separate the mechanism into three functional parts: first, a constant source of energy which has taken the form either of falling weights or a coiled spring, second, a mechanical element capable of stable periodic motion of constant frequency to serve as time regulator, and third, a means of displaying the elapsed time in the desired units, achieved with a suitable gear train and a dial. Of these the most critical and challenging is the regulator; the history of advancement in clock making is the history of the development of this part.

The two principal requirements in regulator design are first, an oscillatory system whose time to complete an oscillation (period) is constant and insensitive to operating and environmental conditions, and second, a mechanical design must be found that allows the regulator to control the transmission of power from the source to the clock mechanism with minimum reaction on the regulator affecting its period. However, paradoxically, *some* reaction on the regulator is necessary to maintain its periodic motion; however, this reaction must be small compared to the energy in the regulator oscillation. As with any mechanical system there will always be frictional forces present which will cause the motion of the regulator to come to rest in the absence of an adequate source of excitation. With improper design the reaction on the regulator would perturb its supposedly free oscillation, compromising its function as a regulator. Ideally, the regulator should be free to oscillate without disturbance except for a transient "kick" at the appropriate phase in its oscillation.

To achieve these seemingly contradictory conditions requires the following: first, the regulator must have very low inherent friction (in the language of electrical engineers, a high Q) in order that a very weak periodic impulse is sufficient to maintain its oscillation. Second, it must exert its control in a "trigger" fashion, that is, a weak interaction of the regulator acting for a small fraction of the period of oscillation must control a much larger force transmitted from the energy source to the gear train driving the clockwork. The mechanical means of achieving this is called the *escapement*.

An early design of an escapement that was widely used for tower clocks in public squares, churches, etc. in the fourteenth century and continued in use for at

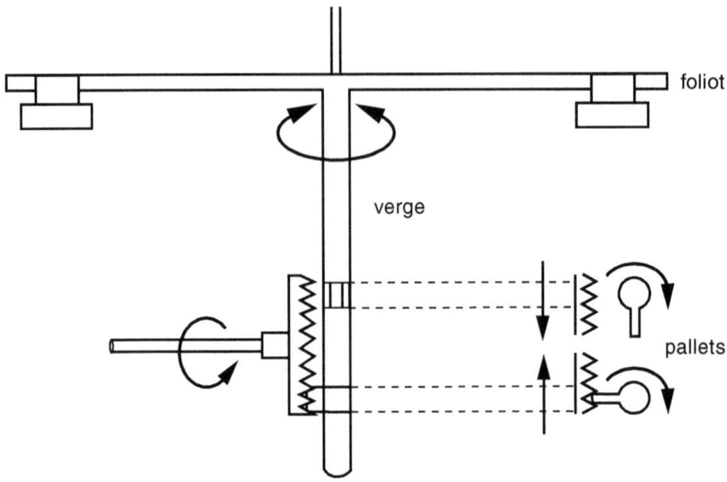

Fig. 6.2 The verge and foliot escapement

least three centuries is called the *verge* and *foliot* [3]. It is shown in its basic form in Fig. 6.2. The foliot is a balanced horizontal beam suspended at its middle with equal weights attached to its ends. The verge is a vertical spindle rigidly attached at right angles to the foliot at its suspension point. Along the length of the verge are rigidly attached two flat vanes parallel to the verge called *pallets* projecting at right angles to the verge and to each other. The pallets engage at opposite ends of a diameter the horizontal cogs on a vertical wheel, called the *scape* or *crown wheel*. The planes of the pallets are parallel to the axis of the verge, and the two are oriented 90° apart. This balanced foliot-verge system is capable of free simple periodic oscillation about a vertical axis, the torsion in the suspension providing the restoring force, whose period can be adjusted by varying the weights suspended at the ends of the foliot. As the latter rotates back and forth, the crown wheel is alternately blocked by one pallet then the other. The rate of rotation of the wheel, which is driven by the power source, is thus regulated by the oscillation of the foliot. The reaction back on the foliot during contact between the pallets and the crown wheel cog is relatively strong, and it is only by making the foliot massive enough, that the clock rate is made stable and less sensitive to environmental conditions. Thus, the tower clocks suffered from at least two major deficiencies: the first is in its design of the escapement; that is, the force of reaction on the pallets is too large and acts for too large a fraction of the cycle, and the second is the large angle of swing of the foliot, which makes the period significantly dependent on the width of the swing. It is only for small angle of torsion of the suspension that the motion is *simple harmonic* with a period independent of the amplitude. We now briefly describe the classic piece of time keeping, the pendulum, still to be found typically as a long case grandfather clock. The discovery of the *isochronism* of a pendulum, that is, the property of taking the same time to complete an oscillation, no matter how widely it swings, dates from 1583 and is attributed to Galileo Galilei of Pisa. The often cited story of his observing the swinging of a chandelier at the cathedral in Pisa may be true but it is not at all clear that he was thinking of the possible use of a pendulum to regulate a clock. As a medical student, he was more interested in measuring the rate of the heart beat in a person [4]. However, it is recorded that a few months before his death in 1642, Galileo did in fact suggest such an application, and since by then his sight had failed him; he dictated a design to his son Vincento, who made drawings but never actually built a working model.

The credit for first use of a pendulum as a regulator in a mechanical clock went to Christaan Huygens [5], perhaps better known for his wave theory of light, 14 years later. His announcement was immediately challenged by the compatriots of Galileo, even though the latter had not produced a working model. Galileo hardly needs the pendulum to secure his place in the history of science! Huygens used the traditional verge escapement, but after analyzing the oscillatory motion theoretically, realized that it was necessary to limit the angle of rotation; this he did by introducing gears between the verge and the pendulum. The rate of the clock could be adjusted by varying the length of the pendulum. In 1673 Huygens had a pendulum clock built in which a flexible pendulum swung against *curved cheeks* causing the end of the pendulum to sweep a curve called a *cycloid*, the isochronous curve demanded by the theory for constant period at all angles of swing. However, this added complexity was found in practice not to produce commensurate improvement in performance.

The properties of a pendulum are well known to any student of physics since its motion is representative of a very common form of dynamic behavior. The essential pendulum is a body that is pivoted or suspended so that it swings freely. The essential characteristic, as Galileo noted, is that for *small* oscillations the time to complete a swing does not depend on the width of the swing provided it remains small. This runs counter to intuition since it is natural to assume that the wider the swing, the longer it should take to complete it. This isochronism extends also to the torsion pendulum in which the oscillatory motion is a rotation about a vertical axis, with the same restriction to small oscillation. An essential difference between the two types of oscillators is that the period of a torsion pendulum depends on the elastic properties of the suspension, whereas a swinging pendulum has a period that depends only on its length and the acceleration of gravity. The susceptibility of a pendulum to temperature variations, arising from the expansion of the length, can be mitigated by several methods: the first is to choose a material that has a very small thermal expansion coefficient; such an alloy was discovered by Charles Guillaume called *Elinvar* for which he was awarded the Nobel Prize in 1920. Elinvar, a nickel-steel alloy, is ideal in that not only does it have a very small expansion coefficient, it is also nonmagnetic and rust proof. Prior to the discovery of Elinvar, one technique to overcome thermal expansion was to use a combination of two metals of different expansion coefficients assembled in such a way that the expansion of one compensates the other. The other factors affecting the period of oscillation are atmospheric pressure and humidity. The period of oscillation depends on the atmospheric pressure since it has a damping effect on the motion. The most serious drawback, however, particularly as regards its possible use to

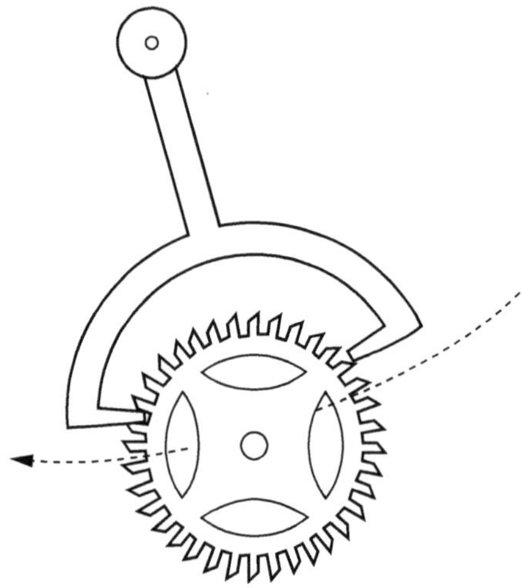

Fig. 6.3 The anchor escapement

regulate a marine clock is its sensitivity to inertial forces; it is difficult to see how a freely swinging pendulum could be made immune to the rolling and pitching of a ship. Another disadvantage is its dependence on the acceleration of gravity. In fact a pendulum clock has been used to make a global survey of the strength of gravity in order to determine the exact shape of the earth. However, in principle charts showing lines of equal gravity could be prepared to correct a marine pendulum clock at different point on the globe. This would be no more onerous than having to correct the reading of a magnetic compass for variation. In fact Huygens did have a pendulum clock tested at sea, but it failed to meet accuracy standards because it was inadequately resistant to the motion of the ship.

Returning to the crucial part of the design of a clock, namely the escapement, major advances in design were made between about 1670 and 1715. First came what is called the *anchor escapement* followed by Graham's *deadbeat anchor* [6]. The essential design of the anchor escapement is shown in Fig. 6.3.

The central part of the anchor escapement is an *escape wheel* having specially shaped radial teeth that are engaged by the pallets of the oscillating anchor thereby regulating the rotation of the escape wheel. There are several important improvements over the older foliot-verge design: first and most important, in making contact with the escape wheel the anchor pallet moves at *right angles* to the motion of the wheel. This means that the force of interaction between the anchor and the escape teeth produces minimum torque back on the anchor. The second improvement is the smaller angle that the anchor has to rotate to engage the escape wheel. This is an advantage for a mechanical oscillator since the oscillation period tends to be more constant with respect to amplitude for small oscillations, and lastly, the anchor is in contact with the escape wheel for only a fraction of the period of oscillation, further isolating the period-determining function from the power train.

6.3 The Spring-Balance Wheel Clock

The use of the pendulum clock for fixed installations, such as in observatories, cathedrals, and other public places, had become widespread by the eighteenth century. However, it is ill suited for any mobile application where it may be subjected to variable inertial forces. This includes, as Huygens found out, one environment of particular relevance to us, namely aboard a ship. Furthermore, any attempt to scale down the size of the pendulum would incur difficulties on many other fronts: increased importance of air resistance, friction, and the curvature of the knife edge on which the pendulum is pivoted. Furthermore, there is the variability of the acceleration of gravity across the globe, although as mentioned earlier, that should be no more unacceptable than the variation of the magnetic compass.

The challenge is to design a clock that is totally insensitive to inertial forces arising from external motions, as well as insensitive to the environment in general, including temperature, atmospheric pressure, and humidity. It occurred to several physicists and clockmakers that the oscillation of a metallic spring could form the basis of such a clock. After all, the oscillation of a mass on a spring is the most common example of simple harmonic oscillation in physics. Little wonder that it would occur to Robert Hooke, best known for his "Hooke's Law" pertaining to the elastic deformation of matter, that the oscillations of some form of spring could be used as a clock regulator. However, the period of oscillation of a simple helical coil is constant only if the amount of distortion of the wire is small. The crucial point is that a large coil of very thin resilient metal must be used. The end of such a coil, whether spiral or helical, can move a significant distance with only a *small* deformation of the wire along its length. The most compact form of coil meeting that requirement is a flat spiral coil. Hooke had apparently included a spiral coil among the many other forms and innovations he had considered, but only after Huygens' announcement of the success in using a spiral spring-balance as regulator did he hastily order a model of a watch to be made using some form of spring, with less than positive results. It has been conjectured that Hooke had ambitions to exploit his ideas in an entrepreneurial spirit (not unheard of among scientists), so he held off perfecting his invention while attempting to raise funds to launch his enterprise. He never successfully "reduced to practice" his ideas, in the words of patent attorneys.

Thus, it was Christaan Huygens again who deserves the credit for introducing the spiral coil-balance wheel into clock making in 1675. His precedence was immediately and indignantly challenged by Hooke. Whenever a scientific or technical breakthrough is made, there is a propensity to attach one person's name as sole innovator, when in reality often these advances are bubbling under the surface when one of the bubbles suddenly pops! We know that almost always there are other thinkers and inventors who set the stage for the advance. In a letter to Robert Hooke, Newton is quoted as saying "If I have seen further, it is by standing on the shoulders of giants." When the contest for recognition is between two prominent scientists, the controversy is resolved, if at all, along national lines; so the spiral coil-balance wheel is an English invention to the English, and a Dutch invention to the Dutch.

Huygens went further and in 1675 also designed the first experimental marine chronometer, which he ordered to be made by a distinguished clockmaker in France. Unfortunately in sea trials the experimental model proved unequal to the harsh buffeting of rough seas and atmospheric fluctuations of temperature and humidity; it became clear that more development work was needed.

From the latter part of the seventeenth and well into the eighteenth century, the maritime nations particularly were consumed with the problem of finding an accurate, practical way of measuring longitude at sea. Certainly the European countries, particularly Britain, France, Spain, and the Netherlands, were all vying to find a solution. There were really only two competing methods: we have already discussed the astronomical lunar distance method, which had adherents among some astronomers including Newton, but only because they were not convinced that a clock could be built with sufficient accuracy to compete with it. The other method, using an accurate clock to determine longitude at sea, is generally attributed to the Flemish mathematician and mapmaker Regnier Gemma Frisius, who proposed it around 1530. Of course at the time it could only have been a conjecture; clocks then were far too crude for such an idea to be taken seriously. However, with the advances in horology that occurred in the next 150 years, the idea not only became feasible, but represented the best hope of achieving the goal of determining longitude at sea.

The quest for that goal was deemed most urgent by the British, who pride themselves, with reason, in their prowess at sea. As evidence of this sense of urgency, we have the proclamation of King Charles II himself that an astronomical observatory be built at his park at Greenwich for the expressed purpose of solving "the longitude problem." Thus, the Royal Observatory at Greenwich was born in 1675. The original method envisaged was still the method favored by many astronomers, namely the lunar distance method. The observatory would provide the almanac necessary to implement this method. By act of parliament in 1714, the British Government offered a prize of £20,000 (a very considerable amount in those days) to anyone who could measure longitude to within one half of a degree on an ocean voyage to the West Indies. To administer and judge the prize, a special Board of Longitude was established.

The directness and simplicity of measuring longitude by time was understood and seen as an irresistible challenge among clockmakers everywhere. The £20,000 prize offered by the British government as well as valuable prizes offered by France and other countries no doubt contributed to the vigor with which they set about perfecting their clocks. Among the most prominent to continue work on developing precise chronometers was Pierre Le Roy, who had the distinguished title of *Horloger de Roi*, or king's clockmaker, and worked at the Académie des Sciences in Paris. He was winner of the first longitude prize, the one awarded by the French academy in 1769. The case can be made that, not withstanding John Harrison's skill and inspiring story, Le Roy was in fact the most influential in the development of the modern marine chronometer.

In England the challenge was taken up by the aforementioned John Harrison [1], a woodcraftsman from Lincolnshire. He was born in Yorkshire in 1693 but his family moved to Lincolnshire when he was young. He followed his father's trade as a carpenter (this should be contrasted with Pierre Le Roy who followed his father as the king's clockmaker), but clearly young Harrison was more than the description as woodworker would connote: he was not only a craftsman but also was endowed with an extraordinary creative drive and the mechanical ability to implement his ideas. He built his first *long case (grandfather) clock* completely out of wood in 1713 when he was only 20 years old. His meticulous joining and fitting of the intricate clockwork in wood was masterful. Then during the 1720s John Harrison assisted by his younger brother James designed and built a series of long case clocks incorporating new refinements to improve performance . Notable among these was the so-called *grid iron pendulum* [7]. This reduced the sensitivity of the oscillation of a pendulum clock to ambient temperature due to the expansion/contraction of its pendulum. We recall that the period of a pendulum depends on it *radius of gyration* and increases with thermal expansion of that radius. The Harrisons' solution was simple and effective: they used two kinds of metal, brass and iron, which have different coefficients of expansion, the brass having the larger coefficient. A simplified design to illustrate the principle is shown in Fig. 6.4. If we assume the temperature rises, the three parallel iron rods expand, lowering the lower bridge, while the upper bridge rises just the right amount to keep the position of the bob unchanged. But of course the overall length of the rods has increased; therefore, they must have negligible mass compared with the bob. It follows that the rods must in fact be stiff wires. Since the difference in expansion coefficients between iron and brass is small, the Harrisons had to elaborate on the same idea and use many more parallel wires to achieve the desired result. The method proved to be inadequate for the highest precision clocks and was ultimately superseded by the mercury compensated pendulum. This type of pendulum uses essentially columns of mercury in place of the classic bob. Mercury has a relatively large coefficient of expansion so that if the temperature rises, for example, the mercury level *rises* and with it the center of mass; this compensates for the increased length of the suspension and keeps the radius of gyration constant.

At first Harrison made long case clocks with precisely contoured wooden gears, but used a metal escape wheel and anchor escapement, a common exception in wooden clocks of the time. He was acquainted with Huygen's work on pendulum design and incorporated Huygens' idea of an isochronous cycloidal swing. In designing a turret clock for Brocklesby Park in Lincolnshire, a clock that was exposed to the elements, he faced the serious problem of avoiding the buildup of dirt and corrosion at the escapement surfaces causing friction and loss of accuracy. His solution may have begun with the principle of the anchor escapement, but ultimately developed into a complicated system involving *articulated* pallet arms to engage the escape wheel. Its complex motion is strongly reminiscent of a *grasshopper* in motion, no doubt a fascinating thing to watch; but for all its complexity it worked beautifully. It avoided sliding friction at the points of contact between the

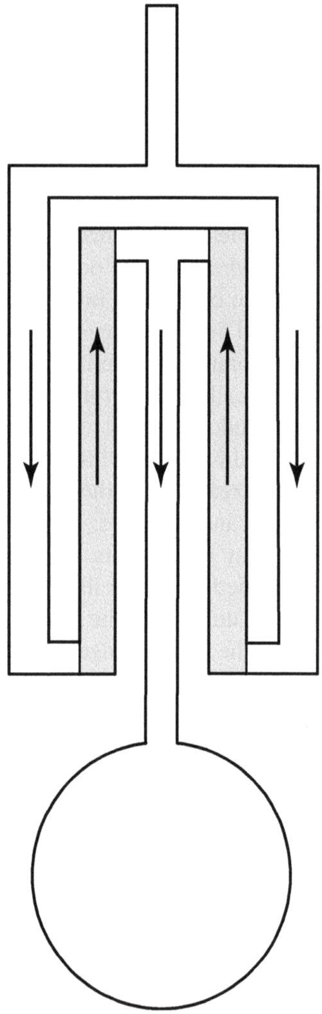

Fig. 6.4 The gridiron pendulum of Harrison

pallets and the teeth of the escape wheel, and therefore avoided the constant need for cleaning and oiling the mechanism. It is depicted schematically in Fig. 6.5.

The idea did not become popular; however, presumably because few people had Harrison's mechanical talent to make it work; nevertheless, it represented a revolutionary idea, testament to the inventive genius of Harrison.

Harrison's first attempt at constructing a marine chronometer to meet the specifications of the Board of Longitude was begun in 1730 and presented in 1736. His design of this clock, later dubbed H1, was a large, heavy, complex instrument in which he had made many ad hoc changes to his wooden clocks in an effort to meet the new environmental conditions. According to Landes [8] it is a "wondrous device" to observe, with its intricate motions keeping a regular quiet beat. It was spring-driven and ran for one day without rewinding. Of particular

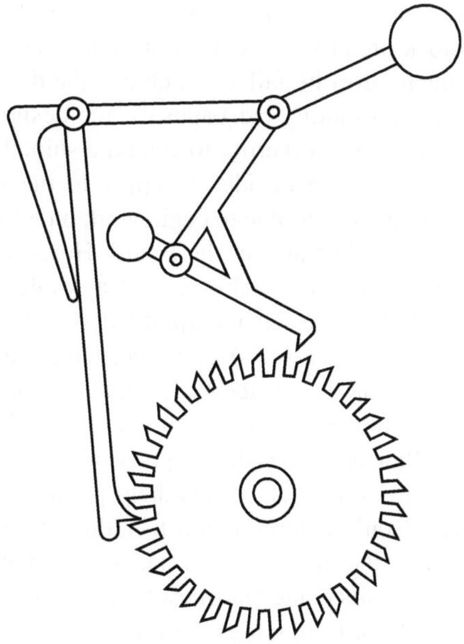

Fig. 6.5 The grasshopper escapement of Harrison

interest is that it was designed to be insensitive to its orientation in space and less affected by inertial forces, a critical property on a rolling ship. This is achieved by designing a regulator in which *two* counter-rotating balances are coupled in such a way that any change in the oscillation of one is compensated by the other. Other features include the nearly frictionless grasshopper escapement, a temperature *compensating curb* that in effect adjusts the length and therefore the period of the balance spring, and a torque equalizing *fusee* (pronounced "foozay") to deliver constant torque from the mainspring. This last device, the *fusee* (French from the Latin meaning *spindle*, an obvious reference to its shape), is an approximately *conical* (actually hyperboloid) pulley with spiral grooves wound with a chord or chain attached to the mainspring drum. As the mainspring is wound, the connecting chain begins wrapping around the fusee at its large end and as the winding proceeds the chain winds around smaller and smaller radii of the fusee. It is effectively a way of having a continuously variable gear ratio. With proper design the torque at the axle of the fusee will remain constant as the torque from the mainspring winds down. It is clearly evident that he implemented all possible design features he could imagine that would make his clock immune to sources of instability aboard ship.

Harrison put his creation on public display in London in 1735 [8], and by all accounts was a sensation among not only clockmakers but the public as well, who undoubtedly took it to be a form of kinetic art. H1 preceded the work of the French kinetic artist, Jean Tingley, by over 200 years! In 1736 H1 was taken on a test run aboard the ship *Centurion* to Lisbon and back on the *Oxford*. The clock performed well; however, Harrison felt that there was so much more to do to perfect his instrument that, instead of pursing the prize with H1, he requested financial support from the Board of Longitude to continue his development work.

The next 3 years (1737–1740) Harrison spent working on H2, also a large, heavy clock broadly of the same design as H1 but with some innovations. Even after so much effort he failed to achieve the desired performance in his revolutionary two counter-rotating bar balances; the design did not make the clock rate sufficiently immune from inertial forces on a ship. To continue the work he would need more money, so he decided to apply to the Board of Longitude for additional funds. In 1740 he was able to begin work on what became H3, his third attempt to meet the rigorous demands of the Board. However, long before H3 was completed, he was coming to the conclusion that a whole new approach would be required to meet those demands. He incorporated into H3 two novel devices to improve temperature stability and lower friction. For temperature stabilization he used a bimetallic strip to control the balance wheel against temperature changes. A bimetallic strip is made by bonding together two parallel strips of metal having different expansion coefficients. It has the property of bending according to the temperature, a mechanical response that makes them popular as sensors in temperature control devices. Harrison's determination to reduce friction to the maximum degree possible led him to the invention of what is known as the *caged roller bearing* [8]. It had been known since ancient times that rolling friction is far less than sliding friction, and the use of rolling logs to move heavy weights was known to the ancient Egyptians. However, the design of a bearing that uses caged rollers, rotating without touching each other around a circular race was due to Harrison. The crucial feature is that the rollers are prevented from touching each other, for otherwise sliding friction would come into play between the rollers at the line of contact, defeating the whole idea.

In spite of these innovations and the many years of effort to perfect them, Harrison came to the conclusion that he needed to make a new beginning. While still working on H3, in 1753 Harrison commissioned a London watchmaker, by the name of John Jefferys, to construct for his personal use a *portable* watch according to Harrison's own design. With a background of working on the scale of tower clocks, outside the realm of delicate watches, Harrison was able to see with a new eye innovative design changes that could transform the humble pocket watch into a chronometer. Indeed by 1757 he had seen the superior performance of the watch he had commissioned Jefferys to make. In that year, Harrison sought additional funds from the Board of Longitude to continue work on H3, but more significantly, to work as well on the development of two *portable* watches, one that would be a pocket watch and the other somewhat larger, presumably destined to become a ship's chronometer. And thus work on his fourth attempt H4 was begun.

H4 is a portable timepiece, encased in a double silver box measuring only about 15 cm in diameter and weighing about 1.5 kg. It looked very much like an oversized pocket watch. The three noteworthy features of H4 are: first, what Harrison called a *going fusee*, second, a *compensation curb*, and third a *remontoire train*. We have already had a brief encounter with the part that has the fancy French name, the fusee; this time the reference is to one of Harrisons many innovations the "going fusee," which means it is a version of a fusee in which there is provision to maintain power to the clockwork while the mainspring is being rewound. In very broad terms it is a mechanism mounted on the base of the fusee body consisting of a secondary short spring kept constantly under tension by the mainspring, so that when the clock is being rewound and the mainspring is no longer driving the clockwork, the secondary spring continues the motion without interruption. The secondary spring acts though a ratchet wheel which is stopped from reverse motion by what horologists call a *click*. Readers who would like to learn more about going fusees, and clicks are encouraged to consult the many books published on the subject, one or two of which are listed at the end of the book.

The compensation curb has already been alluded to as a temperature compensation device applied to the balance spring. The remontoire (from the French *remonter*, to wind) is simply a small secondary source of torque that drives just the regulator assembly; it is periodically rewound by the mainspring. As for the critical escapement, he obviously could not use his grasshopper escapement this time; therefore, he perfected a high precision almost *deadbeat* escapement with jeweled pallets. The introduction of the remontoire was clearly motivated by a desire to achieve the utmost stability in the new design against any variation in the power of the mainspring. Harrison demonstrated again his drive to achieve perfection and to try to solve mechanical problems no matter how daunting, earning him high praise for his ingenuity and workmanship from his peers.

Harrison entrusted his son William to accompany the H4 on board the ship *Deptford*, which set sail for the West Indies on November 18, 1761. On arrival in Jamaica on January 19 1762 there was cause to celebrate: the watch had lost only 5.1 s over a passage that took 2 months. That is a drift of only one part in 10^6 under shipboard conditions. But the Board of Longitude was still not satisfied: it decided that further tests should be conducted. On August 17 of the same year the Board asked Harrison if he would be willing to let H4 be transferred to the Greenwich Observatory for further tests under the supervision of the Astronomer Royal, but Harrison would not hear of it. Like the creative craftsman he was, he must have felt an attachment to his creation that transcended the possession of a piece of machinery. Besides, there *was* a great deal of money at stake, and he did not want to let it out of his hands. Finally, a second trial was agreed to: this time William would accompany H4 to Barbados aboard the ship *Tartar*, leaving on 28 March 1764. Also on board to observe the daily performance of H4 was Nevil Maskelyne, who was to become Astronomer Royal the following year. The passage took 47 days and in that time the clock's error was calculated to be 39.2 s, much worse than the previous trip, but still three times better than the terms originally set by the Board for the award of the £20,000 prize.

Harrison was treated pretty shabbily by the Board of Longitude; there is no other way to describe it. It implied that the performance of H4 may very well be pure chance, and that it cannot be assumed that it will perform as well in the future. Reluctantly, if not grudgingly, the board offered to pay him *half* the amount of the prize, £10,000, as soon as he surrendered all his drawings and mechanisms to the Astronomer Royal, and the remaining half would be given after other clocks of the exact same design had been built and tested by someone else. Harrison was hardly in a position to argue with the Board since he apparently did not have any supporters in positions of power. Except for Maskelyne, who was biased in favor of the astronomical approach, he was most likely dealing with an entrenched, probably hereditary, bureaucracy no more knowledgeable about horology than Gilbert and Sullivan's Major General was about the square of the hypotenuse! Harrison hesitated for several weeks, but finally gave in. In August of 1765 Harrison met with six expert representatives of the Board to examine H4 and all the working drawings pertaining to it. After a delay of 1 week during which the experts were presumably puzzling over the intricate designs, they were satisfied that a full disclosure had been made and signed a certificate to that effect. The Board still insisted that all four of his clocks be handed over for closer examination. One further condition remained to be met: the construction of an exact duplicate of H4 strictly following the working drawings to the same tolerances. The Board asked Harrison if he could suggest anyone who might be entrusted with carrying out that demanding task. He suggested Larcum Kendall, an expert watchmaker who had probably assisted him in the past. For the price of £400, to be paid in advance, Kendall consented just to make exact copies of the parts and assemble them, without being held responsible in any way for the performance of the final product. Finally, the Board ran out of objections and Harrison received *half* the prize money.

On May 5, 1766 the Astronomer Royal, Nevil Maskelyne, received custody of H4 from the Lords of the Admiralty in order to carry out a series of tests that took a period of 10 months, until March 4, 1767. Its performance confirmed the excellent impression it had made on the voyages to the West Indies. It kept constant time, particularly when the ambient temperature was not extreme. Unfortunately on one occasion when the room temperature was allowed to fall to 26 °F (that is 6 °F below freezing) its rate, contrary to expectations, did *not* change appreciably. This was taken to be a *flaw* by the learned astronomers, because Harrison had predicted its rate may change at extreme temperatures. Of course it demonstrated how extraordinarily effective was Harrison's temperature compensation. Maskelyne reported that H4 "cannot be depended upon to keep the longitude to within 1° in a voyage of 6 weeks, or within one half of one degree for more than 14 days, and then only if the temperature is kept some degrees above freezing, etc." He goes on to say that "nevertheless it is a useful invention, in conjunction with observations of the distance of the moon from the sun and fixed stars, may be of considerable advantage to navigation". We should note that a 1° error in 6 weeks is equivalent to a time error of 240 s in 3.6×10^6 s, that is 66 times the drift of H4 on the first journey.

To qualify for the second half of the prize money, the Board required Harrison to personally build two more copies of H4 in order to carry out further tests, while requiring the original H4 to be kept at the observatory. It only added more luster to Harrison's meticulous workmanship that he would keep such exact records of his design that he could reconstruct such an intricate machine without benefit of seeing it before him. Now in his 70s, old and frail without the benefit of modern medicine, he and his son William worked on what was to become H5. Kendall's watch, now designated as K1, was completed in 1769 and submitted to the Board in early 1970. At the completion of H5 Harrison pleaded with the Board to allow him to submit H5 and K1 as the two copies of H4 that they had demanded. The response was unequivocal that *both* units must be made by the Harrisons. Now 79 years old, John Harrison decided to appeal to King George III himself. He had previously been commissioned by the King to make a watch for him. It had been tested at his private observatory at Richmond and showed remarkable stability, which could not have failed to impress the King. An appeal was made to the King by a letter to his private astronomer Stephen Demainbray. In reply William was asked to appear in person before the King to plead his case. After hearing William describe the conduct of the Board, the King is said to have remarked [8]:

> *….. these people have been cruelly wronged…….. by God, Harrison, I will see you righted….*

The same year H5 was subjected to tests under the King's own orders and patronage and its performance met all expectations. However, the Board of Longitude apparently convinced of its own superior judgment *refused* to recognize the assessment of the King's astronomer. Why the members of the Board were not convicted of high treason is not recorded. However, undaunted to the end, the Harrisons petitioned parliament "to be righted" as the King put it. By an Act of Parliament in 1773, the Harrisons were finally awarded £8,750. Presumably the advances they had already received were subtracted from half the prize money. Of far greater importance to John Harrison was the long sought recognition that he had indeed solved the longitude problem. At least the members of Parliament representing the people voted that he had, even if the Board and in particular the astronomer Maskelyne had reservations.

It is interesting, in passing, that the clock which Larcum Kendall's made, K1, as a simplified version of H4, was completed in 1772, just in time for Captain Cook to take it on his second voyage of exploration in the south Pacific Ocean on the *Resolution.* Cook also took on board three other clocks made by another English expert watchmaker, John Arnold. Kendall's chronometer performed extremely well on that daunting 3 year expedition, taking the ship from tropical waters through rough polar waters, and crossing twice the Antarctic Circle in a vain search for a "southern continent" that some believed "must" exist. The subcontinent of Australia had previously been discovered by a European in 1606 and identified with the mythical *Terra Australis Incognita* by the Dutch explorer Willem Janszoonin, but apparently that was not enough to put to rest the mistaken belief that more was needed to "counterbalance" the large land mass in the northern hemisphere. The success of Kendall's K1 watch left little doubt that longitude can be determined adequately at sea for long periods of time and under extreme climactic conditions by a well-made watch, such as one based on Harrison's design.

6.4 Modern Chronometers

It was inevitable that we should be drawn to the human story of John Harrison and his son, but of course there were many other clockmakers working in Britain and on the continent who have contributed to the ultimate development of exquisitely precise timekeepers. Culturally, the eighteenth century was an age enthralled by mechanical devices. Thus a model of planets orbiting around the sun, called an orrery, or an intricate music box might adorn a parlor, or even automata appear in operas by Mozart and Offenbach. The Franklin Institute in Philadelphia came into possession of a remarkable antique automaton that it restored and discovered the maker to have been a Frenchman by the name of Henri Maillardet. It is programmed to perform complicated human-like functions, such as sitting at a desk, for example, and writing English and French verse! So it was in the spirit of the times that there was so much creative energy devoted to perfecting clocks.

Among the notable watchmakers that followed Harrison and continued the development of timepieces, we should mention John Arnold and Thomas Earnshaw who developed the detached *détente escapement* (often anglicized to detent) popular in marine chronometers. In this, the balance wheel is free to oscillate more than once before receiving the impulse from the escape wheel. It has the special advantage that it does not require constant cleaning and lubrication: critical parts are immune from temperature fluctuations, and mounted by gimbals, it is unaffected by bodily motion.

The escapement invented by Thomas Mudge was later perfected by Abraham-Louis Breguet, a name synonymous with upper class watches (as well as prices), to become the *Swiss lever escapement*. It is found in some form in most modern wrist watches. Breguet is a very prominent figure in the history of horology, a prodigiously creative man, who lived through trying times, to put it mildly. He was born in Switzerland but founded a watchmaking establishment in Paris, where he built up a brand of high quality luxury watches bearing his name to this day. When the bloody French revolution erupted, he took refuge in his native Switzerland. He was highly respected in the trade and was a close friend of the Englishman John Arnold. He is given credit for many innovations; among them are automatic winding, improved lever escapement, antishock mounting, a spiral hair spring with an over-coil named for him, a double escape wheel chronometer, etc. His company's marine chronometers set the standard, and their basic design widely duplicated by the navies of the USA, Germany, and Japan. At the outbreak of World War II, the US Navy, which had previously relied too heavily on marine chronometers imported from Switzerland and Germany, had to quickly build up a domestic supply. Thus it was that in 1939–1940 the US Naval Observatory put out requests for proposals to build large numbers of marine chronometers to tight military specifications. This implied mass producing these precision instruments: no mean feat. A watchmaking company named Hamilton in Lancaster, Pennsylvania, expressed interest but required an existing chronometer to study its design and judge whether it would be within their capability to produce it on a mass scale. The Naval Observatory agreed to provide such a chronometer, which may well have been a Breguet unit. After several months' study by Hamilton of the problems attending the mass production of fine instruments that until then had been the products of skilled craftsmen, and the manner in which a production line might be set up, finally agreed to the contract. Almost exactly one year after signing the contract, Hamilton delivered two prototypes to the Naval Observatory, where they were subjected to acceptance evaluation. The results were very positive and everyone presumably gave a big sigh of relief. To mass produce something of that precision was unprecedented, and before the end of the war Hamilton had delivered thousands of units: a magnificent achievement.

World War II marked the end of the mechanical chronometer as the ultimate timekeeper and the emergence of the quartz crystal-controlled clock and the transmission of time by radio. Radio broadcasting preceded the quartz revolution, but it took the impetus of the war to spur development at microwave as well as radio frequencies. We take up quartz oscillators at some length in the next chapter because of their present importance in frequency/time measurement.

References

1. L. Sprague de Camp, *The Ancient Engineers* (Balantyne Books, New York, NY, 1990), p. 142
2. A.A. Mills, S. Day, S. Parkes, Eur. J. Phys. **17**, 97 (1996)
3. A.V. Roup et al., in *Proceedings of the American Control Conference*, Arlington, VA, 2001, p. 3245
4. G. Galilei, *Dialogue Concerning the Two Chief World Systems* (University of California Press, Berkeley, CA, 1967), p. 230
5. D.S. Landes, *Revolution in Time* (Harvard University Press, Cambridge, MA, 1983), p. 116
6. W.J. Gazeley, *Clock and Watch Escapements* (Robert Hale, London, 1992)
7. D.S. Landes, *Revolution in Time* (Harvard University Press, Cambridge, MA, 1983), p. 146
8. D.S. Landes, *Revolution in Time* (Harvard University Press, Cambridge, MA, 1983), p. 156

7 The Quartz Revolution

7.1 Historical Background

The drive to improve the performance of clocks inevitably led on a path from large lumbering mechanical clock mechanisms with long periods of oscillation to increasingly smaller mechanisms that run at much higher frequency. The hair spring-balance wheel clocks evolved over the eighteenth and nineteenth centuries into mechanical instruments of such perfection that further improvement could only come painstakingly in small increments at the cost of higher complexity. There is a limit to the precision that can be attained with a relatively low frequency reference oscillator; in a fundamental sense, there is a limit to how finely one can split a period of oscillation. It was therefore inevitable when the electronic revolution arrived in the 1950s with the development of the transistor and smaller batteries that a higher frequency electrical watch would soon emerge; as indeed it did in the form of the Bulova Accutron. This watch uses a *tuning fork* as the reference vibrator, so called not so much because it resembles a fork, although it does have two tines, but rather it is the name given to a two pronged device that produces a pure audible tone when struck, an instrument that has a long history as a standard of acoustic frequency, or musical tone. The tuning fork in the Accutron was designed to vibrate at the audible frequency of 360 Hz, hence the watch hums rather than ticks; another immediately noticeable difference about the new watch is that the second hand rotates continuously, and not in discrete steps. The principal initiator of this development in 1960 was an electrical engineer by the name of Max Hetzel [1]. Initial tests showed that the new watch was more stable than the classical balance-wheel watches by at least an order of magnitude. It was found, however, that the initial tuning fork design made the watch perceptibly sensitive to its orientation in space, with deviations of ±5 s per day. This is still below the limit required for chronometer rating, but nevertheless the design was modified to minimize this effect. The tuning fork movement was licensed also to other quality watchmakers, such as Omega and Tissot. These and other fine chronometers designed by Piaget

F.G. Major, *Quo Vadis: Evolution of Modern Navigation: The Rise of Quantum Techniques*, DOI 10.1007/978-1-4614-8672-5_7,

were the ultimate mechanical timepieces of the day; timepieces that were carried by the Apollo astronauts.

However the age of the tuning fork watch was short. As already mentioned, the electronic revolution, which began in the 1950s with the invention of the transistor, made possible not only tuning fork watches but also the far superior quartz-controlled watches. The microelectronic revolution that followed in the 1970s completely changed the face of electronics with integrated circuit boards replacing individual components connected together with wires The quartz-controlled movement in watches has since become so undistinguished that we forget what revolutionary advance in technology it represented: it made it possible for cheap imported quartz *wrist* watches to be almost as accurate as the most sophisticated mechanical chronometers!

The use of vibrating quartz crystals in electronics can be traced to the development at the beginning of the twentieth century of radio communication, which requires the generation, transmission, and detection of radio waves in a narrow *frequency band*. A quartz crystal can perform the critical function of stabilizing the base *frequency* of the transmitted radio waves. By restricting the band of frequencies that are transmitted, it is possible to have a large number of independent transmitters of radio waves without interference between them. A further critical reason for wanting to restrict the frequency bandwidth is concern for the ratio of signal to noise. There is an ever present *random* electrical fluctuation (*noise*) in any electrical system, noise that, if analyzed in terms of its frequency components, has a spectrum that extends over a wide range that is characteristic of the system. Therefore, the wider the band width required to accommodate fluctuations in the frequency of the transmitted or received radio waves, the greater the amount of noise accompanying the signal. There are two kinds of noise: *amplitude* noise and *frequency* (or *phase*) noise. In a radio signal, amplitude noise is generally much more serious than frequency noise; hence, a signal sent in the form of amplitude modulation (AM) of the carrier radio wave is noisier than one sent as a frequency modulation (FM). But whatever type of modulation is used, the base frequency that is broadcast must be very stable for the reasons just described. The allocation of broadcast frequencies in the radio spectrum, and the frequency stability required of transmitters, is the responsibility of the Comité Consultatif International Radioéléctrique (CCIR).

Historically, the impetus to develop quartz oscillators and detectors came also from another direction, namely submarine warfare in World War I. There was an urgent need for sensitive acoustic detectors to track enemy submarines under water. There was also a drive to develop powerful *ultrasonic* sources and detectors to use in an echo-location method for submarine detection, an early form of sonar and precursor of radar. To achieve a measure of directionality in the transmission and reception of sound waves under water, it is necessary that their wavelength in water not exceed around 1 m; otherwise the sound waves would spread out and lose directionality. A wavelength of 10 cm in water corresponds to a frequency of about 15 kHz. Vibrations are said to be ultrasonic if their frequency in air is beyond the audible range for humans, which for most individuals extends to around 15 kHz.

The early implementation of the idea of sonar is associated with the names of the Russian Constantin Chilowski and the noted French physicist Paul Langevin [2], better known for his work in magnetism. At first they tried a *singing condenser* as a source of ultrasonic waves in water. The term *condenser* was formerly used to mean what is now called a *capacitor*. The singing condenser is simply two plane parallel metal plates between which an alternating voltage is applied causing a periodic force between them which makes them vibrate. Conversely, if sound waves in the medium surrounding the plates cause them to vibrate, then a voltage applied between the plates will cause an alternating current to flow in the circuit supplying the voltage. That is the principle of the sound detector called the *condenser microphone*. The physics of the condenser microphone is elementary: we recall that a capacitor having a voltage V applied between its plates, and a capacitance $C(t)$ that is fluctuating in time due to a pressure wave in the surrounding medium, will have an amount of electrical charge $\pm Q(t)$ on its plates given by $Q(t) = C(t)V$. The charge $Q(t)$ can fluctuate only through the flow of current from one plate to the other, by the law of conservation of charge. To measure that current, a known resistor is placed in the circuit which applies the voltage to the plates; the voltage across this resistor is the output of the microphone. This line of development of underwater devices for submarine detection was ultimately superseded by the use of pure quartz crystals for both ultrasonic source and receiver. By 1917 a quartz-based form of sonar was developed with a range of about 6 km.

Returning to the stabilization of the frequency of radio broadcasts, what is required is a device that has an exceptionally sharp resonant response to a radiofrequency electric/magnetic field in the desired frequency range. The sharpness of resonance of a system is specified by what is called its Q-factor. There are a number of equivalent ways of defining Q: one is in terms of the sharpness (on a frequency scale) of the resonant response to a periodic driving force, thus $Q = \omega_0/\gamma$, where ω_0 is the center frequency and γ is the frequency width of the resonance curve. If the oscillating element suffers from dissipative forces such as internal friction that converts oscillatory energy into heat, its resonance will be broadened, and its *Q-factor* diminished. Among a select class of crystals, quartz was known to have excellent elastic properties, that is, it has extremely small internal friction so that as it vibrates there is an extraordinarily small amount of energy converted irreversibly to heat. As a consequence, it has an extremely high intrinsic Q. Of equal importance is a property that only crystals of a certain symmetry class have, namely it exhibits a phenomenon called *piezoelectricity*. This means that if pressure is exerted along a particular direction with respect to the crystal axes, an electrical polarization is created, as if the crystal were in an electric field. This enables a quartz crystal subjected to an ultrasonic pressure wave to generate a high frequency electrical output, thus acting as a detector. Conversely if an electric voltage is applied across a crystalline quartz plate, a mechanical stress is produced along the direction of the electric field as if a pressure or tension were being applied to the crystal. This effect provides a way of exciting vibrations in the crystal by applying an alternating voltage to metal electrodes on opposite faces of the crystal. A detailed study of piezoelectricity, and the restriction on the class of symmetry the crystal must have, was first published in 1880 by Jacques and Pierre Curie (the latter was to become famous also as the husband of Marie Sklodowska, the discoverer of radium). Although it was recognized even then that piezoelectric crystals were potential sources and detectors of ultrasonic waves, it was not until the later invention of the triode vacuum tube amplifier in 1906 that the use of quartz transducers became possible.

The invention of the vacuum tube provides another example of the ambiguity that often beclouds what should be celebrated as an intellectual achievement. In this case, John Fleming, working for Marconi in Britain, invented in 1904 the vacuum diode rectifier, inspired by the *Edison Effect*. This effect was discovered by Edison through his search for a suitable filament material for an incandescent lamp. He noticed that a measurable electric current flowed from the heated filament to other conducting surfaces in the glass tube. Within 2 years of Marconi's patent application for Fleming's diode, Lee de Forest in the USA came out with a tube much like Fleming's except it had a third open wire electrode near the cathode to improve its performance as a radio detector, which he called an audion. Since de Forest claimed that his device owes nothing to Fleming's discovery although it clearly was an outgrowth of it, patent infringement litigation between Marconi and de Forest was inevitable, and continued for many years. De Forest's audion tube was not evacuated of air to a high vacuum; rather, de Forest believed that some air was needed for its proper operation. He failed to realize that by evacuating the tube to high vacuum, the third electrode, that came to be called the *grid*, could much more usefully *control* the flow of current, resulting in an *amplifier*. De Forest may have simply elaborated on an existent invention, the diode; however, his work did lead eventually to the development of the triode amplifier by GE and others, enabling continuous radiofrequency oscillations to be produced and amplified. Previously, even during World War I, radio transmitters used high voltage sparks to generate radio waves, not unlike Heinrich Hertz's original experiments in Germany to test Maxwell's theory.

With the development of the triode vacuum tube, it was not long before oscillators were built in which a continuous radiofrequency output was generated at a frequency tuned by an inductance–capacitance (LC) "tank" circuit. It was again World War I that provided the stimulus to experiment with piezoelectric crystal resonators to control the frequency of radio transmitters. It is nice to think that something good came out of World War I in addition to the evils of battle field slaughter and World War II. Alexander Nicholson working at the Bell Telephone Laboratory had been actively exploring piezoelectric crystal applications in electronics, including microphones, loudspeakers, and phonograph pickups. He was the first to be granted a patent in 1918 for incorporating a Rochelle salt crystal in an electronic oscillator. Walter Cady of Wesleyan University was also active in this field and had studied the resonances of *quartz crystals* in electronic oscillators during the same period. Initially, Cady was interested in the crystal resonances more from the point of view of *measuring* the frequencies of resonance. It was only after the end of the war that it occurred to him in 1921 to *control* the oscillator frequency by the crystal vibrational resonance. He had noticed an unexpected reaction on the behavior of the oscillator at the crystal mechanical resonance frequencies tending to lock the frequency of the oscillation to the crystal resonance. Cady challenged Nicholson's prior claim to having observed the same thing; again the lawyers were busy. It was about this time that G.W. Pierce developed his widely popular design of a circuit for a crystal-controlled oscillator, used to this day.

7.2 Crystalline Quartz

The elastic properties of quartz, that is its properties that relate to stresses and strains, have been known to be extraordinary for a long time, with extremely small internal friction, high tensile strength, and low thermal expansion coefficient. *Fused* quartz fibers have long been used for suspension in torsion balances since they have the same breaking strength as a steel wire of the same diameter, but with a much smaller modulus of rigidity under torsional (twisting) stress. This latter property is of course important for a sensitive torsion balance. However, its most important property is its nearly perfect elasticity, that is, whatever work is done when it is deformed, that energy is stored without loss as elastic energy and will be almost totally regained when the stress is relieved. A low intrinsic rate of elastic energy loss means that when subjected to an oscillating force that coincides with one of its natural modes of vibration, it will have a response that varies very sharply with the frequency; that is, it will have a very high Q-value. This is extremely important for a resonator since it means that it can be minimally coupled to the amplifier and therefore less affected by the latter's noise and instabilities in gain. The power level of the excitation of the crystal must be stabilized at a low value to ensure good frequency stability. The frequency stability of the oscillator will ultimately be limited by the presence of electrical noise in the system. Long-term instability causes a drift in the frequency described as "aging" in the crystal, which in the intermediate term is determinate and can be taken into account. The stability of quartz oscillators is discussed at greater length later in this chapter.

In addition to its superior elastic properties, crystalline quartz is far less sensitive to environmental conditions such as temperature and humidity than other solids, and further has an extremely small thermal expansion coefficient. This explains its extraordinary resilience under extreme thermal stress; thus for example, a quartz rod can be heated to red heat and plunged into icy water without cracking. Naturally, the quality of any material must be judged with reference to the application to which it is to be put. In fact, it turns out that the qualities that set quartz apart from other materials are the very ones that leave room for improvement: thermal expansion and long-term stability of elastic properties.

The chemical composition of quartz is silicon dioxide (SiO_2), otherwise called *silica*, and is abundant as the principal component of sand and is an indispensable constituent in the formation of different types of glass. In its crystalline form, it is a three-dimensional lattice held together by what chemists call *covalent bonds*. Each Si atom is surrounded by *four* O atoms at the vertices of a regular tetrahedron, and each O atom is joined to two Si atoms. Chemical bonds are basically electrical in

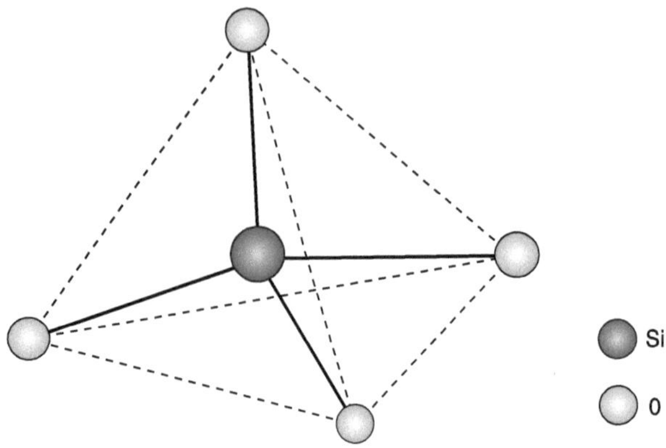

Fig. 7.1 Tetrahedral symmetry of silicon bonds

nature; however, classical electrostatics cannot predict a stable configuration of electrical charges; to explain their existence, quantum theory is required. Major contributions to that theory were made by Linus Pauling, for which he was awarded the Nobel Prize in 1954. We distinguish between two types of chemical bonds: the *ionic bond* and the just mentioned *covalent bond*. In the case of the ionic bond, the outer (*valence*) electron(s) of one atom is (are) transferred to the other atom to form a negative ion, leaving the first atom as a positive ion with a net positive electrical charge. The other common bond is the *covalent bond* in which the valence electrons of the two atoms are "shared" between them. In the case of crystalline quartz, we are dealing with covalent bonds, but not with a stable combination of just three atoms, as the chemical formula SiO_2 might suggest, but with the entire aggregate of atoms. That is, the problem does not separate into small groups of atoms we call molecules; the compound can form into a crystalline lattice or be in an amorphous glassy state called *fused quartz*. It is in this latter form that it is most commonly found, as ultraviolet-transmitting optical windows, lenses, and fibers. One should compare SiO_2 with carbon dioxide CO_2 which of course does consist of separate CO_2 molecules as a gas under normal conditions. It would be inappropriate to say much more about the theory of molecular structure; however, it is necessary to briefly describe the crystallographic properties of quartz, in order to describe the crystalline orientation of the various quartz plates used to control oscillators.

The four valence (binding) electrons in the Si atom form four covalent bonds along four lines having equal angles to each other leading to tetrahedral symmetry, as shown in Fig. 7.1.

The O atom has two unsatisfied covalent bonds (technically known as *orbitals*) one of which links up with a Si atom in a single bond leaving the other orbital to link up with another Si atom. In the case of crystalline quartz, this continues throughout the solid in one interlocking three-dimensional lattice. The bonds between the atoms in the three-dimensional lattice are very strong, resulting in the crystal having a very high melting point at 1,710 °C. In the case of CO_2, the four covalent bonds of C are satisfied by two *double* bonds with two O atoms, thus satisfying all bonds within each separate CO_2 molecule, and the result is a gas at normal temperatures.

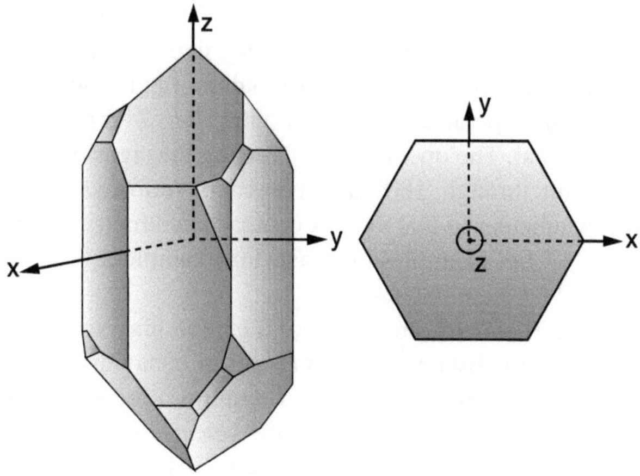

Fig. 7.2 Quartz crystal axes

Since chemical bonds between the constituent atoms of a crystal generally have specific relative orientation and spatial symmetry, these are reflected in the macroscopic surface planes of a crystal. It was a French crystallographer named René-Just Haüy (pronounced Hawee) who in the late eighteenth century discovered the crystalline property of *cleavage planes*; planes along which a crystal fractures easily because the interatomic bonds perpendicular to them are relatively weak. Moreover, he discovered that a crystal can continue to be subdivided along these planes to produce smaller and smaller crystals that preserve the same relative angles between planes. He was led to postulate that a crystal is made up by stacking the same unit he called *molecule intégrante*. Haüy was concerned that the edges where planes intersect in a crystal did not appear to be stepped as expected, even under a microscope. We now know that they were simply beyond the resolving power of the microscopes available to him. There is indeed a *unit cell* such that when translated along parallel lines generates the entire crystal structure; however, the cleavage planes that determine the way the crystal easily breaks up may not lead to the unit cell as Haüy's idea implies.

Crystalline quartz, like all crystals, has certain characteristic symmetry properties. The tetrahedral arrangement of the Si–O bonds suggests that the crystal has a threefold axis of symmetry, that is, if the crystal is rotated about an axis along a Si–O bond it passes through three positions where it is indistinguishable from the first. It has another symmetry that is difficult to see without a physical model, namely, *three twofold* symmetry axes perpendicular to the threefold axis, and because of the latter's threefold symmetry, the three twofold axes must be separated by 120°. The threefold axis is conventionally taken to be the *z-axis* of a coordinate system, and one of the twofold axes is taken to be the *x-axis*, while the *y-axis* is defined as perpendicular to the *x–z* plane to produce an orthogonal system. Figure 7.2 illustrates the typical cleavage planes of a quartz crystal, and the assignment of coordinate axes.

A type of symmetry whose *absence* is critical for a crystal to display a piezoelectric effect, and which is absent in quartz, is a center of symmetry. A center of symmetry would exist in a crystal if, when every atom is moved to a diametrically opposite position with respect to any fixed point in the crystal as center, the result is indistinguishable from the original configuration. Mathematically, it is symmetry under the operation of reversing the *signs* of all the coordinates of all the atoms in the crystal. The operation of reversing the signs of all coordinates is equivalent to a rotation through 180° about an axis followed by a reflection through a plane perpendicular to that axis. A center of symmetry is a rare property among common objects, but examples do exist, for example a brick or football; however, among crystals it is quite common.

The symmetry properties of a crystal play an important role in determining the presence or absence of certain types of behavior, such as, in the present context, the piezoelectric effect. This effect requires a pressure applied to the crystal to cause a separation of electric charge, called electrical *polarization*. The crystal as a whole is neutral, and remains so; however, the balance of charge in the atoms is distorted by the outer electrons being displaced relative to the positive inner cores. The result is that electrical charges of opposite sign appear on opposite surfaces of the crystal. Now assume the crystal is subjected to the inversion transformation just described, namely change the sign of all the coordinates in the system. If the crystal has a center of symmetry and is therefore the same after as before the transformation, it would have the effect of reversing the direction of the polarization vector, with the same pressure, since pressure acts in both directions equally. This is possible only if the polarization is zero. The conclusion is that if a crystal is to exhibit piezoelectricity, it must *not* have a center of symmetry. In passing we may note that another extremely important application of crystals that requires the same restriction on symmetry is in nonlinear optics, such as second harmonic generation of light waves. In this case, the crystal is required to display a dependence of its polarization on the power in the light field E^2 which, like pressure, is unchanged by a reversal of the signs of the coordinates. The first successful experiment to produce the second harmonic of a light beam was done by focusing a helium–neon laser beam on a quartz crystal.

7.3 X-Ray Crystallography

An essential property of crystalline media is their anisotropy. Therefore, knowledge of the orientation of the crystalline lattice with respect to some external reference, such as its surface geometry, is a prerequisite to their application. This is a problem for which crystallographers in the past have developed techniques based on the angles between natural cleavage planes and anisotropy in physical and optical properties. But since the early years of the twentieth century the most powerful technique has been the use of *X-rays*. We recall that X-rays, so named by their discoverer Wilhelm Roentgen because they were a complete mystery, were discovered accidentally when Roentgen, running a high-voltage gas discharge tube on the top of a cabinet in his laboratory discovered that photographic plates stored in

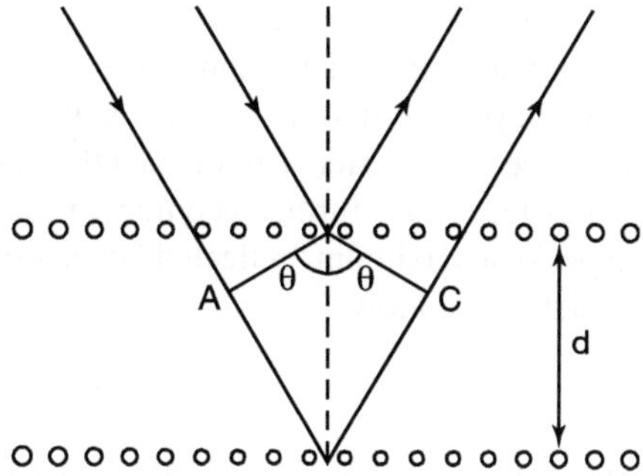

Fig. 7.3 Bragg reflection of X-rays from atomic planes

the cabinet underneath the discharge became fogged. His exact exclamation on seeing this is not recorded, but it can be assumed that he indulged in some form of German profanity. X-rays, we now know, are electromagnetic waves, as are radio waves and light waves; the only fundamental difference between them is their short wavelength; X-rays have a wavelength that is on the order of 1/3,500 of a light wave and the energy of their quanta therefore 3,500 times larger. They will of course pass through materials that are totally opaque to visible light.

In 1912 at the suggestion of Max von Laue, two of his young assistants performed an experiment analogous to one well known at the time, namely the diffraction of light waves by a fine *grating*. Here, a grating is a reflecting or transparent polished surface with closely spaced parallel grooves precisely ruled on it. When monochromatic light falls on such a grating, the intensity of the light reflected (or transmitted) will have maxima along directions where the diffracted light waves from adjacent stripes are in phase and reinforce each other, that is, differ by a whole number of wavelengths. To apply the same technique to X-rays would be impractical since it would require a grating with the spacing of the grooves scaled down to the wavelength of the X-rays. It occurred to von Laue that the regularly spaced atoms in a crystal may act as a three-dimensional "grating" and scatter X-rays along directions dependent on the phase relationship of waves scattered from different atoms in the lattice. This indeed turned out to be the case, and a powerful tool for analyzing the structure of crystals was quickly developed. Major players in this development were W.H. Bragg and W.L. Bragg, father and son, who established X-ray diffraction as a powerful tool for crystal structure analysis.

It is to W.H. Bragg that we owe a powerful way to analyze X-ray diffraction patterns. It begins by assigning all the atoms in the crystal to a set of parallel planes, called *atomic planes*. The regular geometrical order of atoms in the crystal ensures that this is always possible, as illustrated in Fig. 7.3.

Imagine a monochromatic (single wavelength) X-ray wave falling obliquely on the atomic planes in the lattice, and that secondary scattered waves radiate out from each atom, waves that are phase coherent with the incident wave. The secondary waves originating from one atomic plane will differ in phase from ones originating from an adjacent atomic plane. The phase difference depends on the perpendicular distance between the atomic planes, the wavelength of the X-rays, and significantly, the direction with respect to the atomic planes. If there had been atoms in just a single plane, then the secondary waves would all be in phase provided only that they obeyed the first law of optics, namely that the angle of incidence must equal the angle of reflection. But the fact that the X-rays penetrate deep into many parallel planes leads to a further geometrical condition limiting the value of the incident angle for a maximum "reflected" intensity. This condition for a *Bragg reflection* can be stated as follows:

$$2d \sin \theta = n\lambda, \qquad (7.1)$$

where d is the perpendicular distance between consecutive atomic planes, θ is the angle between the incoming parallel X-ray beam and the atomic planes, n is an integer, and λ is the X-ray wavelength.

As already mentioned, it was von Laue, or more accurately his assistants Friedrich and Knipping, who first succeeded in observing the diffraction of X-rays in 1912 at the University of Munich. This was at a time when X-ray sources consisted of high-voltage electrical discharge tubes to produce "cathode rays" (electrons) that bombarded a cooled metal target. Friedrich had another X-ray project setup and so he with the help of Knipping could quickly test von Laue's idea of using a crystal as a three-dimensional diffraction grating. It is perhaps interesting that it was only a very short time before that von Laue happened to have learned from Ewald of the lattice dimensions in crystals. The experiment seen through modern eyes would appear as a simple one: collimate an X-ray beam with defining apertures, place a crystal in its path, and surround the crystal with photographic plates. In fact, with the X-ray sources available at the time, several hours exposure times were required, aggravating the problem of eliminating stray radiation. At first they tried backscattering, putting the photographic plate on the *same* side as the source with negative results. Undaunted, they tried forward scattering and after hours of exposure obtained a rather blurry pattern of spots arranged in circles around the point of incidence of the beam on the photographic plate. The X-ray tube probably had a copper or nickel target, and therefore the emitted X-ray spectrum would have had several intense lines on a broad base of *bremsstrahlung* (German for breaking rays) produced by the sudden slowing down of the electron beam in the target. Since the atoms in a crystal can be grouped into parallel atomic planes oriented in many different directions with respect to the incident X-ray beam, those obeying Bragg's condition will give rise to reflected beams as if by plane mirrors. The result recorded on a cylindrical photographic film with the crystal on its axis will be a symmetrical pattern of dots called a Laue pattern produced by the "reflections" from the atomic planes. This pattern reveals the orientation and symmetry properties of the crystal. A quantitative determination of the absolute distance between atomic planes requires knowledge of the X-ray wavelength; however, a great deal can be learned by the *relative* distance which can be deduced from angle measurements of the reflections. This requires an instrument called a *goniometer*, which consists essentially of a rotatable crystal mount at the center of a circular scale graduated in degrees, and a radiation detector on a rotatable arm whose angle relative to the direction of the incident X-ray beam can be accurately measured. Generally, it is not possible to arrive at the detailed structure of an unknown crystal without some prior knowledge or assumption as to its structure; the consequences of the assumption can then be verified by X-ray diffraction.

In the context of quartz crystals as resonators for the control of frequency standards, X-rays are an invaluable diagnostic tool; they make it possible to see the distribution of the amplitude of vibration in a crystal plate. This ability relies on the dependence of the intensity of X-ray "reflection" on the interplane separation; even a small change in that separation leads to a departure from the Bragg condition and an observable change in reflected intensity. This is particularly important in the development of high-Q quartz resonators since the effect of special contouring on its mode of vibration can be tested.

7.4 Artificial Quartz Crystals

Crystalline quartz occurs fairly commonly in nature, generally formed like many other crystalline deposits from water solutions of minerals such as silica under conditions of high temperature and pressure. The first attempts to grow quartz crystals were made in 1845 by C.E. Schafhautl in an effort to understand the conditions under which they are formed in nature. He succeeded in growing tiny crystals from silica dissolved in alkaline water at high pressures and temperature gradient. With the advent of the use of quartz crystals to control the frequency of radio transmitters and time standards, the demand for crystalline quartz jumped to a level that only artificially grown crystals can satisfy. The major industrial nations became involved in developing processes for mass producing artificial quartz, particularly in Japan.

The industrial method currently used attempts to duplicate the natural process. A.C. Walker and E.C. Buchler are given credit for making a systematic study of the crystal growing process and ultimately succeeding in growing 100–300 g crystals at a rate of 2.2 mm per day. The process is described as *hydrothermal* [3], that is, it involves water solutions and high temperatures. The alkaline water solution of silica is subjected to high temperature and pressure in a large chamber called an autoclave. The upper half of the autoclave is kept at a lower temperature than the bottom half and material entering the top half from the bottom is supersaturated and will crystallize on a seed crystal suspended there. On an industrial scale, autoclaves have been built that produce up to 5 mm of crystal per day extending over 200 days.

As resonators, quartz crystals have been subjected to extensive study to improve their quality factor, Q [4]. Studies of the temperature dependence of the intrinsic Q-value of different samples of quartz revealed that a strong peak in the vibrational energy loss, and a consequent lowering of the Q occurs at −223 °C. This was ultimately traced to the presence in the lattice of *sodium* impurity ions which it was discovered could be removed electrolytically by applying a voltage between electrodes in the molten silica. Infrared absorption spectroscopy has been used effectively in the detection of impurities, as part of quality control in the manufacturing process. The recognition that the presence of impurity ions can cause vibrational energy loss has contributed greatly to improving the purity and therefore *Q*-factor of quartz resonators.

7.5 Quartz Resonator

By their very crystalline nature, quartz plates exhibit the piezoelectric effect in a way which very much depends on the directions that the mechanical stress and/or electrical polarization have in relation to the crystallographic axes. For example, an electric field applied along the direction we earlier specified as the *x*-axis is coupled to a longitudinal strain along the *same* *x*-axis together with a strain of the same magnitude but *opposite* sign along the *y*-axis. A change in sign of a strain simply means changing, for example, a compression into an extension, and vice versa. A quartz plate may exhibit several types of strain; four are shown in Fig. 7.4.

Other types of strain include *torsional* (twist), *flexural* (bend), *shear,* and *longitudinal*, both extension and compression. The different modes are used to cover different ranges in the frequency spectrum. Thus, a flexural mode in a bar may be used at the lower end of the frequency spectrum up to 100 kHz, whereas an extension mode extends the range to 300 kHz. A face shear mode covers the range of 300 kHz to 1 MHz, and a thickness shear mode may reach 30 MHz. For the higher frequencies, it is preferable to use overtones (harmonics) than the fundamental resonance of an extremely thin fragile plate. The best precision crystal plates typically vibrate in the fifth overtone of a thickness shear mode at a frequency of 5 MHz.

The selection of exactly where the quartz plate is cut from the crystal lattice is important since it determines how efficiently the desired mode of vibration is excited and even more importantly, what the temperature sensitivity of the resonant frequency will be. Since the electrical coupling to the vibrator is through a pair of parallel plane electrodes, it is often cut in the shape of a flat circular or rectangular wafer from the body of the crystal with the desired orientation with respect to the crystallographic axes. Figure 7.5 shows the orientation of the three most common cuts: the *x*-cut, and the two rotated *y*-cuts.

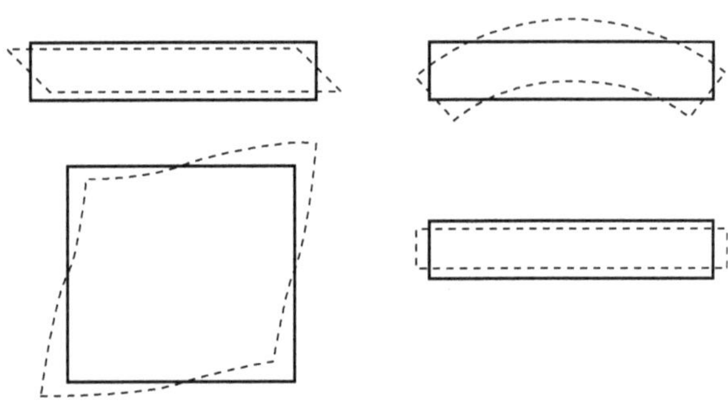

Fig. 7.4 Modes of strain in a quartz plate

Fig. 7.5 Common cuts of quartz plates

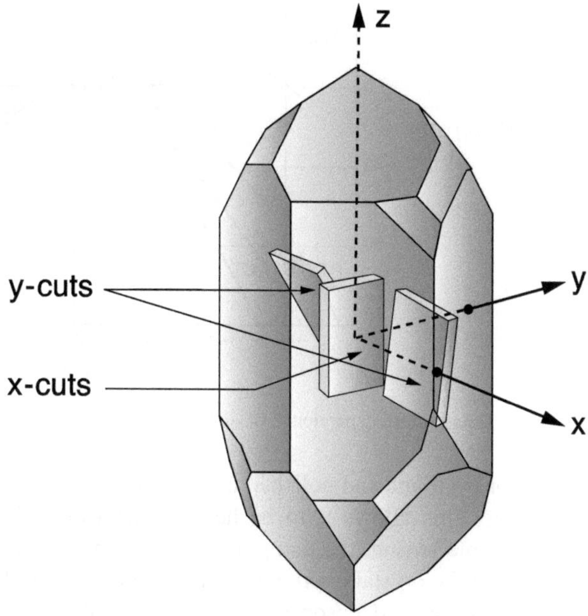

The x-cut has faces perpendicular to the x-axis, and the electrodes being parallel to the crystal faces, the electric field is therefore along the x-axis. The field produces a stain in the same direction; hence, it gives rise to what is described as a pure *longitudinal* (extensional) standing wave along the x-axis. There are two cuts designated as y-cuts which are in planes rotated in opposite directions about the x-axis, such that the perpendiculars to the plates, and hence the electric fields, make angles of $-59°$ and $+31°$ with respect to the z-axis in the y–z plane. These directions are pure mode directions for *shear* waves; no other modes are ideally excited.

In discussing the frequency stability of quartz oscillators, we distinguish between short-term stability (phase noise) and long-term stability (drift). Let us first consider long-term stability, also called *aging*. The resonance frequency of a quartz resonator typically exhibits a long-term smooth drift in frequency, which can be plotted and extrapolated with fair confidence, at least in the intermediate term, say over a period of months, if not years. The aging rate of high frequency thickness-shear mode resonators has been reduced to impressively low values by improved production and mounting methods. The factors that can contribute to drift are changes involving imperfections in the crystal lattice, easing of thermal or mechanical stress caused during production or present in the mount, and the adsorption and desorption of gasses at the surface. Aging rates vary from a few parts per million per month for solder-sealed metal envelopes, to as low as a few parts per million for the first 2 *years* for cold welded metal envelopes.

The mounting of the quartz wafer is extremely important in achieving the highest possible Q-factor. For a high-performance frequency standard, the crystal

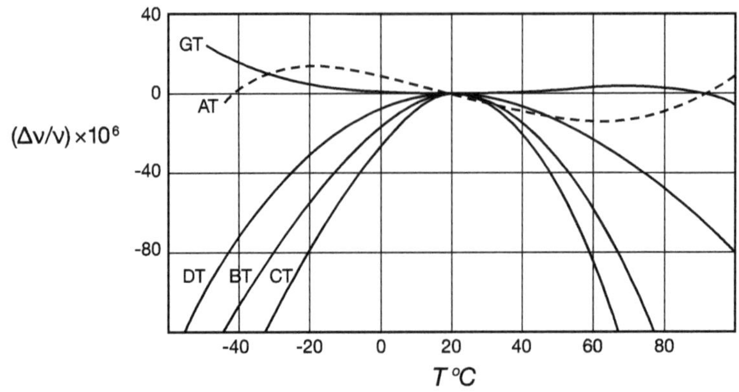

Fig. 7.6 Temperature dependence of different cuts of quartz [5]

resonator is ground and polished into a plano-convex disc that vibrates with the rim quiescent. This allows it to be held by the rim without degrading its Q-factor. For the same reason, the damping effect of an atmosphere must be eliminated by evacuating the crystal oven; a vacuum will also have the effect of reducing aging. The adsorption and desorption of gasses at the surface of the crystal have an observable effect, a fact that is remarkable when we consider that it involves a surface gas layer probably no more than one molecule thick. This is indicative of the otherwise extreme stability of the quartz resonance frequency. Thanks to the advanced state of vacuum knowledge honed over a half century when electronics depended on the vacuum tube, we know that the surfaces of all materials, particularly metals, will "outgas" in a vacuum. Even after the system has been "outgassed" by baking it under vacuum and raising the temperature of metal parts to red heat, they will continue to outgas slowly, with molecules diffusing from deeper and deeper layers to the surface of the metal. In the electronic tube industry, the preferred method of maintaining the vacuum in a sealed off tube was to use a *getter*. This typically consisted of a bright silvery surface of metallic barium deposited by evaporation of a small quantity of barium metal using induction heating by a coil outside the tube. The barium reacts chemically with all but the inert gases, helium, neon, etc., a reaction that produces nonvolatile compounds with low vapor pressure.

The temperature is the most important physical parameter affecting the frequency of a quartz resonator [5]. This is ultimately due to the thermal expansion of the crystal, which like other physical characteristics is anisotropic in a crystal, with the coefficient a function of the orientation of the cut with respect to the crystallographic axes. By correctly choosing that orientation the temperature dependence of the frequency can be minimized at the operating temperature. In Fig. 7.6 are reproduced graphs showing the dependence of the fractional frequency variation $\Delta\nu/\nu$ as a function of temperature for several common cuts of quartz crystal. It is no accident that the curves all have a stationary behavior around a temperature of 20 °C, the usual operating temperature. We note that two cuts, AT and GT, have inflection points at that temperature, which means that they are stationary to second order and therefore even less sensitive to a temperature change.

Even when the operating temperature is set at the stationary point on the curves, there is still room for improving the frequency stability by placing the quartz unit in a constant temperature oven. The use of the term "oven" might suggest a high temperature, whereas in fact it has to operate only moderately above the highest ambient room temperature in order to be able to depend on cooling through contact with the environment when the oven tries to correct a rise in oven temperature. To stabilize the oven temperature proportional control is used in which the *rate of increase* of heating power delivered to the oven is proportional to the temperature "error," that is the difference between the actual temperature and the set temperature. To implement this type of servo control involves considerably more sophisticated electronics than the on–off control using a mercury switch or bimetallic strip.

Another possible source of frequency instability is the electrical field intensity driving the crystal, and the electrical circuits associated with it. At the level of stability demanded of a frequency standard, the actual output frequency may conceivably deviate from the natural resonance frequency due to "pulling" by circuit resonances or overdriving the crystal into the nonlinear region. Experience has confirmed that like classical oscillators, the crystal resonator should be driven at a *constant*, *low* power level. In order to implement this, a control circuit is required of a type known from the early days of radio as *automatic gain control* (AGC). This is a *negative* feedback control loop, called a *servo-loop*, in which the amplitude of the electrical oscillation at the crystal is detected and compared with a reference voltage to obtain an *error signal*; this is amplified and passed through an integrator (for stability) and applied to a point in the crystal circuit that controls the amplitude of oscillation to bring the error to zero.

7.6 Quartz Resonator as a Circuit Element

There is a parallelism between oscillatory mechanical systems and electrical ones reflected in the fact that they both involve oscillations between potential and kinetic forms of energy and their variables obey similar second-order differential equations. In order to analyze the behavior of the quartz oscillator and its interaction with the electronic circuit driving it, the ability to model the quartz crystal as an electronic circuit radically simplifies the analysis. Such an electrical model is called an *equivalent circuit*, made up of the basic elements of an electronic circuit namely resistance, inductance, and capacitance. A basic equivalent circuit for a quartz plate is shown in Fig. 7.7.

Referring to the figure, the inductance L and capacitance C_2 represent the inertia and elastic constant that determine the resonance frequency of the crystal. The capacitance C_1 represents the capacitance of the crystal with its metal electrodes. The resistance R accounts for the energy loss in the crystal and its mount. For a quartz crystal, R is extremely small, $C_1 \gg C_2$, and L is extremely large. If L were the inductance of an actual coil, it would be a meter in diameter with a winding of

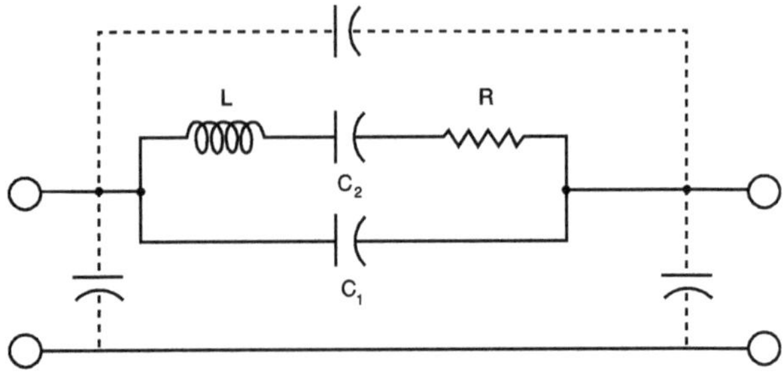

Fig. 7.7 The equivalent circuit of a quartz plate

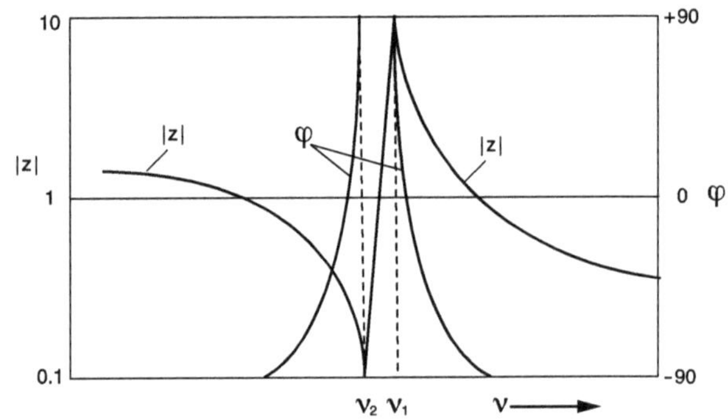

Fig. 7.8 Response of quartz crystal near resonance

several thousand turns. The diagram shows the circuit as a four-terminal network even though the crystal unit itself has only two; the other two are the common ground. When the crystal is incorporated into an electronic circuit to form an oscillator, the same criterion of minimal reaction on the frequency determining element from the power source applies, as in a mechanical clock. The condition on the capacitances $C_1 \gg C_2$ is imposed to meet that requirement. The higher the value of Q, and values reaching 10^6 are now not uncommon, the greater the tolerance for fluctuation in the operating conditions of the amplifier. Thus, any small variation in the amplifier phase shift can be compensated in a high Q crystal by a correspondingly small shift in oscillation frequency.

To obtain the frequency response curve of the crystal assembly, imagine a variable frequency sinusoidal voltage is applied between one electrode of the crystal and ground, while the voltage and relative phase of the voltage appearing on the opposite electrode are measured. We would find that in the neighborhood of a resonant vibration in the crystal, the relative phase would jump from $-90°$ to $0°$ at the center of resonance to +90° on the other side, as shown in Fig. 7.8.

The frequency width of the resonance at half the maximum amplitude is determined by the value of Q; thus, for $Q = 10^6$ at a frequency of 1 MHz the width is only 1 Hz.

7.7 Stability of Oscillators

The determination of the instability of an oscillator presumes the existence of a reference oscillator whose stability is unquestioned. This raises a broader question regarding primary standards of frequency: an evaluation of their stability must be deduced from agreement between members in an ensemble of oscillators. We could go even further and insist that the oscillators not have the same physical design that may make them drift the same way! All frequency/time standards are, to varying degrees, subject to variations in their output frequency, and will over time inevitably drift. It is essential from the point of view both of manufacturers and users that there be an agreed upon way of specifying that instability. Since the instability is in part random, a statistical analysis is required which generally begins by assuming the property of *stationarity*. This means roughly that the frequency (or phase) as a function of time does not change if the instant from which time is measured is displaced. It is obvious from our description of the various ways in which quartz, for example, may change deterministically in time that stationarity would not really comport with the actual behavior of a real system. The way around this is to separate the behavior into random fluctuation and deterministic variation, or drift. Of course the "long-term drift," if followed for a sufficiently long time may reverse and appear more random; in other words, perhaps the distinction is only a matter of time scale; nevertheless, in practice the distinction is useful. In the theoretical discussion of frequency stability it is assumed that the deterministic contribution is subtracted out, and that it is legitimate to use statistical methods on the random part.

There are two complementary ways of specifying the random fluctuations in the output frequency or phase of any oscillator. The first is a *time domain* description in which the deviations in (fractional) frequency with respect to a frequency standard are measured at regular intervals, usually denoted by τ to generate a time series. The measure of instability is now conventionally taken to be the Allan variance [6], a statistical two-sample variance which solves the convergence problems in the case of certain types of noise, such as flicker noise and random walk, about which more will be said later. It is defined as follows:

$$\sigma^2 y(\tau) = \frac{1}{2}\left\langle (y_{i+1} - y_i)^2 \right\rangle, \tag{7.2}$$

where the average $< >$ is taken over a set of relative frequency fluctuations y_i measured at equal intervals τ. A complete analysis requires a large number of determinations of frequency during time intervals extending say from 1 to 10^4 s. To magnify the relative deviation of the frequency of an unknown oscillator, a standard is chosen having close to the same frequency, and a crystal mixer used to obtain a signal at the low *difference* frequency. This difference is measured with a frequency counter using the frequency standard for the time base. By recording the count at regular intervals, a set of numerical values for the frequency *deviation* is obtained from which the Allan variance can be computed.

The time domain method is unsuitable for short-term instability studies [7]. For that the *frequency domain* description is used, which is in terms of the (*Fourier*) *spectrum* of the relative fluctuations in frequency.

In addition to common man-made or *artificial* sources of noise, electronic devices in general are subject to two fundamental types of noise that can be traced to the atomic nature of matter and electric charge. These are the *Nyquist–Johnson noise* due to *thermal* agitation of electrons and lattice ions in a conductor, and noise due to the discrete particle nature of the charge carriers, the electrons, named metaphorically *shot noise* by Schottky in 1918. In a simplified model, we visualize a conductor as consisting of positive ions arranged in a lattice imbedded in a sea of electrons. However, it would be wrong to imagine a picture of the electrons as small classical particles bouncing around in the conductor: the behavior of electrons in a metal is a strictly quantum mechanical problem. At all temperatures above absolute zero (−273 °C), the ions are in a state of thermal agitation. In 1928, Nyquist published a theoretical treatment of thermal noise [8], which is based on a model involving the modes of electrical oscillation on a transmission line. He showed that for frequencies sufficiently low that the quantum of energy $h\nu$ is much less than the thermal energy kT, the mean square voltage $<\Delta V^2>$ across a resistance R in a frequency interval $\Delta\nu$ is given by:

$$\left\langle V^2 \right\rangle = 4RkT\Delta v. \tag{7.3}$$

From this result, we see that the spectral density of thermal noise is given by $4RkT$, independent of frequency: so-called *white noise*. Unlike shot noise, this thermal noise is universal and applies to all conductors; only the validity of the quoted formula is restricted to the given physical condition. In fact, the presence of thermal noise is so universal that a figure of merit for amplifier sensitivity is *noise temperature*. On the other hand, shot noise is manifested only where conduction involves a stream of independent charge carriers as from the cathode of a vacuum diode operating at low cathode temperature, or in devices such as tunnel junctions, where electrons are transmitted randomly and *independently*. In this case, it can be shown that the spectral density is again independent of frequency and simply proportional to the average current, I. If the motion of the electrons is completely uncorrelated then the shot noise assumes a maximum value of $2eI$, where e is the electron charge. The same cannot be said of electrons in a large metallic conductor, where the electrons interact with lattice vibrations; in this case shot noise is not observed.

Another type of noise that is evident in electronic devices is called *flicker noise*. The term originated in the age of electronic vacuum tubes in reference to the so-called *flicker effect*, which described fluctuations in tube current resulting from variations in the electron emission from the hot cathode. It is *frequency* noise characterized by a spectral density with a $1/\nu$ dependence. For that reason, it is often referred to as *1/f noise* (*f* for frequency). This type of noise also occurs in

Schottky barrier diodes. This is a particularly unwelcome type of noise in an oscillator since it diverges at lower frequencies. This means that slow variations get larger and increase without limit for longer observation times. This would seem to present a discouraging prospect: it would seem to suggest we can never build a clock that does not eventually drift without limit! Thankfully, it turns out that flicker noise is not fundamental in the sense that shot noise or thermal noise are fundamental; in fact, its presence in electronic tubes was significantly reduced by proper design and manufacture.

Another statistically distinct type of noise that is also thought to be a result of instrumental effects is called *random walk* [9]. The term is familiar in the context of a classical problem in statistics whose solution and ramifications have historically occupied many well-known mathematicians. The problem is easily stated and is of such importance that we sketch it as a little diversion. Given p *white* objects and q *black* objects enclosed in a black box, what is the probability of drawing n white objects in N tries. If the actual number of successes is tabulated for a very large number of sets of N tries, it is expected that the *average* ratio of whites to blacks would be p/q. But of course the ratio for individual trials is scattered above and below the average. A measure of the dispersion around the average is called the *standard deviation*, symbol σ, defined as the root mean square of the deviations from the average. We state without proof that for the present problem $\sigma^2 = Nr(1 - r)$, where $r = p/q$. To apply this to the random walk problem, assume a person (usually assumed to be a drunk) takes equal steps of length L along a straight line with succeeding steps taken with equal probability in one direction or the other. The question is how far after N steps does he get. If m of the total steps are in (say) the positive direction, then he advanced a distance of mL and stepped back a distance of $(N - m)L$ for a net advance of ΔL given by:

$$\Delta L = mL - (N - m)L = 2\left(m - \frac{N}{2}\right)L. \tag{7.4}$$

It follows that $\langle \Delta L^2 \rangle = 4\langle m - N/2 \rangle^2 L^2$. But in this case $r = 1/2$, and therefore $\langle m - N/2 \rangle^2 = \sigma^2 = N/4$. Hence finally we have:

$$\left\langle \Delta L^2 \right\rangle = NL^2. \tag{7.5}$$

It is fortunate that the different common sources of frequency fluctuation in an oscillator produce a simple characteristic power-law Fourier spectrum. Hence, by plotting the noise spectrum using log–log scales on the two axes, the type of noise dominating in different regions of the frequency scale become evident. Thus, if the plot is a horizontal line, it is called white noise and must be due to shot or thermal noise, while flicker noise would give a line with negative slope. Figure 7.9 shows the Allan variance for some identifiable kinds of noise in oscillators [10].

Fig. 7.9 The Allan deviation of frequency fluctuations arising from **A** white and flicker phase noise, **B** white frequency and thermal noise, **C** flicker frequency noise, and **D** random walk noise

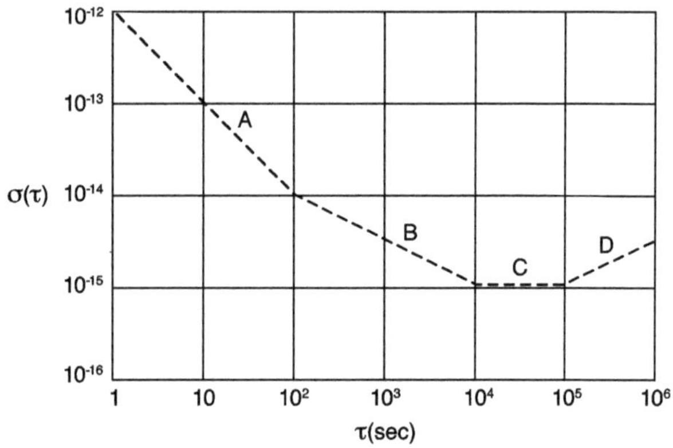

References

1. D.S. Landes, *Revolution in Time* (Harvard University Press, Cambridge, MA, 1983), p. 344
2. Christopher McGahey, PhD Thesis, *A History of Piezoelectric Quartz Crystal Tech. Community*, Georgia Institute of Technology, 2009
3. A.C. Walker, J. Am. Ceram. Soc. **36**, 250 (1953)
4. A.W. Warner, Bell Syst. Tech. J. **39**, 1193 (1960)
5. E.A. Gerber, R.A. Sykes, Proc. IEEE **54**, 103 (1966)
6. D.W. Allen, J.A. Barnes, in *Proceedings of the 35th Annual Frequency Control Symposium*, Ft. Monmouth, NJ May (1981)
7. CCIR Characterization of Freq. and Phase Noise, Report 580, p. 142 (1986)
8. H. Nyquist, Phys. Rev. **32**, 110 (1928)
9. R.M. Mazo, Phys. Rev. **A34**, 2364 (1986)
10. C. Audoin, B. Guinot, *The Measurement of Time* (Cambridge University Press, Cambridge, UK, 2001)

8 The Classical Atomic Clocks

8.1 Language of Quantum Mechanics

In trying to describe the workings of an atomic clock we are inevitably led into a world where a different language is spoken: the language of quantum mechanics; we shall encounter terms such as quantum states, quantum transitions, and even quantum degeneracy. Fortunately for us it is possible to derive useful results concerning the motion and radiative behavior of atoms without adhering to the strict quantum formulation of modern theory. In fact the early development of quantum theory went through a stage where certain quantum ideas were grafted onto the classical theory of mechanics and electromagnetism.

Prior to the introduction of the concept of a quantum of energy by Max Plank early in the twentieth century, by which he meant "packet" of energy, the reigning theory for the interaction of radiation with matter was the *electron theory* of H.A. Lorentz [1]. In this, the response of atoms to an electromagnetic wave (light) was expressed in terms of *atomic oscillators* pictured as electrons elastically bound to the atoms, being driven by the electrical component of the electromagnetic wave. Lorentz showed that if the oscillation of the electrons was repeatedly interrupted by collisions causing a random change in the phase of their oscillation, that the atoms would continuously absorb energy from the radiation. And conversely, if the elastically bound electrons were caused to vibrate by virtue of repeated collisions through thermal agitation, radiation would be emitted at some fundamental frequency *and its harmonics*. Unfortunately for classical theory this is *not* what is observed. The atoms in general *do* emit a spectrum that includes discrete frequencies, but they are not related as harmonics, as classical theory demands.

In the decade before World War I there was a wealth of experimental data on the wavelengths of lines in the spectra of many substances, and there was a general assumption that their frequencies should accord somehow with the classical *normal modes of oscillation* of some complicated structure. The problem as it was conceived then was: exactly what *is* that structure. The fact that none of the spectral line frequencies seemed to bear a harmonic relationship (1:2:3:. . ..) to any other

F.G. Major, *Quo Vadis: Evolution of Modern Navigation: The Rise of Quantum Techniques*, DOI 10.1007/978-1-4614-8672-5_8,

was a mystery. Nevertheless a Swiss physicist, a contemporary of A. Einstein, named Walter Ritz did publish in 1908 a formula which related the frequencies, now called the *Ritz combination principle* [2]. It states that the frequency $\nu_{n,m}$ of any line in the spectrum is characterized by two integers n and m such that the following holds:

$$\nu_{n,m} = \frac{a}{m^2} - \frac{a}{n^2}(m, n = 1, 2, 3, \ldots.). \tag{8.1}$$

At the end of the nineteenth century another mystery was occupying the minds of those who cared about such things, namely, the spectrum of what is called *black body radiation*. This is radiation in thermal equilibrium at some temperature T in a system in which the radiation is constantly being absorbed and reradiated by surrounding matter. Paradoxically light from the sun is approximately "black body" radiation. It is radiation characterized by a *continuous spectrum*, that is, the intensity as a function of the frequency ν is continuous, but more significantly, it does *not* depend on the chemical nature of the matter, only on the temperature. A source of this radiation for spectral analysis is typically a high temperature oven provided with a small viewing hole; such a hole would appear black at moderate temperatures since all radiation falling on it from the outside is not reradiated through it. All attempts using classical theory by eminent physicists of the day failed to explain the way the intensity of this thermal radiation varies with wavelength; the various approaches mostly led to what came to be called the *Rayleigh–Jeans formula* which states that the intensity should increase as ν^2 without limit. This was totally wrong; in fact the spectral intensity goes through a maximum and tends towards zero at the blue end of the spectrum. However a theoretical result obtained by Wilhelm Wien in 1896, known as *Wien' radiation law*, did in fact have the right asymptotic form as $\nu \to \infty$.

The final solution to the black body radiation problem was to become a pivotal point in the history of physics and our ability to understand the physical world. In one classical derivation of the formula for the spectrum it was first assumed as a mathematical device that the exchange of energy between radiation and matter occurs in finite increments and then proceeded to the limit of a continuous exchange. It was Max Plank who, in 1901, showed that a result can be obtained that agreed with experiment if one did *not* proceed to the limit of continuous exchange but retained discrete increments proportional to the frequency of the radiation. He called these increments *quanta* with an energy given by $E = h\nu$, where h is a universal constant with a value in SI of $h \sim 6.6261 \times 10^{-34}$ J s.

The understanding of the processes of emission and absorption of radiation by matter was significantly advanced by A. Einstein who postulated that electromagnetic radiation *itself* behaved as if it were made up of particles called *photons*, each having an amount of radiant energy equal to $h\nu$. On the basis of this idea Einstein was able to explain the details of the *photoelectric effect*, for which he was awarded the Nobel Prize in physics. The photoelectric effect, which is the emission of electrons (in vacuo) from metal surfaces irradiated with light, provides another example where the concept of quanta is necessary to explain the dependence of the energy of the emitted electrons on the intensity of the light falling on the metal. Classically, increasing the intensity of a light beam means increasing the amplitude of the electromagnetic field and would be expected to give rise to the emission of more energetic electrons. But it does not; it simply increases the *number* of emitted electrons. An equally significant result due to Einstein came out of his attempt to use photons to explain the spectrum of black body radiation; he found that the processes of absorption and spontaneous emission are not sufficient, one had to postulate *induced emission*, that is, an emission rate dependent on the number of photons already present, a process which is at the heart of laser operation.

On another front, just prior to World War I, profound advances were being made in our understanding of the atomic structure of matter by a team at Cambridge University headed by a New Zealander named (to give him his full title) Sir Earnest Rutherford of Nelson [3]. Nelson is a little town in the South Island of New Zealand where he was born. This was a time in the history of nuclear physics when particle accelerators and Wilson cloud chambers were decades in the future, and probing the structure of atoms consisted in observing the passage through gases or metal foils of particles emitted by naturally radioactive sources. The method of detection of high speed particles was by counting scintillations on a phosphor screen. This required graduate students with exceptional peripheral vision. At the time, physicists knew that atoms consisted of positive ions and electrons, but there was only speculation as to how these particles were arranged in an atom. An even deeper mystery is how these charges could form a stable entity considering that classical electromagnetic theory, which works extremely well on an everyday scale, says it is impossible to have a stable arrangement of electrical charges held together by electrical forces alone. It occurred to Rutherford that if you want to learn something about the distribution of charge in an atom, the best way is to shoot nuclear projectiles at it, and see what happens! He chose as projectiles α-particles (helium nuclei) that are emitted spontaneously from radioactive polonium; the target was gold foil. The angles of scattering were measured using the scintillations on a screen as already mentioned. He was amazed to see that some particles were deflected through what seemed unreasonably large angles; the expectation was that the positive and negative charges in the atom are somehow distributed throughout the volume of the atom. He applied classical mechanics to work out the angular distribution of the scattered particles on the assumption that the classical Coulomb inverse square law applied to the force between the projectile ion and the charged core of a target atom. That such a core may exist was made feasible by the large scattering angles observed. The experimental results agreed with his calculations and the nuclear model of the atom was born: namely, an atom consists of a positive nucleus where nearly all the mass resides surrounded by a cloud of negatively charged particles. Not to take anything away from Rutherford's accomplishment, but it should be pointed out how fortuitous his success was, on two counts: the nucleus of atoms is so small that it approximates a point charge, which he assumed, and secondly, it happens that the Coulomb inverse square law for the force between charges is the *only* known force law that gives the same result classically as a proper quantum treatment!

Building on Rutherford's nuclear model of atomic structure, around 1913 Niels Bohr proposed a model of the simplest atom, hydrogen, which brought together all the relevant information known at the time. He superimposed quantum notions on a planetary model that accords with Rutherford's conclusion that the mass resides in a central positive nucleus surrounded by negative electrons. He introduced two radically new postulates: the first and most radical postulate addresses the difficulty mentioned earlier of the classical instability of a system of electrical charges. He posited that the electrons form a planetary system about the nucleus and that there exist certain orbits in which an electron could circulate without radiating energy, contrary to the classical prediction that the electron would radiate energy and spiral into the nucleus. These orbits he called *stationary orbits*, and of all the orbits that classical theory might allow, only orbits that satisfy the following *quantum condition* on their *angular momentum* L are stationary:

$$L = \frac{nh}{2\pi}, \tag{8.2}$$

where n is an integer, and h is the same Planck's constant used to define the quantum of energy $h\nu$. For a simple circular orbit of radius r the angular momentum is given by

$$L = mVr, \tag{8.3}$$

where V is the velocity along the orbit and m the mass of the electron. Bohr further postulated that the frequency of the radiation emitted by an atom is *not* the frequency of rotation of the electron in its orbit as would be the case classically, but rather the atom radiates a quantum of energy hv when an electron makes a *transition* from one orbit to another; thus,

$$h\nu = E_2 - E_1. \tag{8.4}$$

The concept of stationary states postulated by Bohr became much more plausible thanks to the work of Louis de Broglie published 10 years later [4]. In this de Broglie argued that the particle–wave duality of light that had been the subject of much debate and speculation ever since Young's interference experiments is also a property of particles of matter. The success of Bohr's theory to predict the wavelengths of lines in the hydrogen optical spectrum seemed to suggest a wave behavior on the part of the electron, since the only known context in nature where integers were involved was in specifying resonant modes of vibration. De Broglie was able to associate the "corpuscular" and wave properties of matter in a theory he called *wave mechanics*, which he showed was consistent with Einstein's theory of *Special Relativity*. According to de Broglie, a particle of mass m moving with the velocity V has a wave associated with it "guiding" its motion whose wavelength λ, now called the *de Broglie wavelength*, is given by the following:

$$\lambda = \frac{h}{mV}, \tag{8.5}$$

where h is Planck's constant. If we apply this formula to Bohr's stationary orbit we find

$$2\pi r = n\lambda, \tag{8.6}$$

a familiar resonance condition which states that the circumference must be a whole number of wavelengths. The wave nature of matter was explicitly demonstrated in the laboratory in later years, for example, by the Davisson–Germer experiments on electron diffraction. The theory of Bohr and de Broglie as elaborated by Sommerfeld and now referred to as the "old quantum theory" dealt only with stationary states; other problems such as transitions between states, and collisions, required later developments in the theory.

The ideas of de Broglie stimulated further development towards a consistent theory free from ad hoc assumptions. Bohr's condition for a stationary orbit was so reminiscent of a condition for resonant modes of vibration in wave phenomena that it was inevitable that physicists sought to construct a differential *wave equation* analogous to the equations of sound and electromagnetic waves. Erwin Schrödinger published just such an equation in 1925. Following the spirit of de Broglie's ideas, Schrödinger sought an equation that determines a function of coordinates, called the *wave function*, represented by the Greek letter Ψ whose magnitude at any point in space gives the *probability density* that the particle occupies that position. This interpretation requires that the integral of that function over all space must converge to a finite value in order that the total probability that the particle is *somewhere* in space can be set equal to one, a process called *normalization*. The equation, which is a second-order differential equation analogous to the classical wave equation, is now called the *Schrödinger equation* [5]. It replaces the equations of motion of a particle in classical mechanics; the motion of a particle is no longer given by its spatial coordinates as a function of time, but rather by its wavefunction. Not every mathematical solution of the Schrödinger equation is admissible as representing a solution to the physical problem; only those qualify that satisfy certain mathematical conditions, among them the condition already cited. Acceptable solutions are called *eigenfunctions*, German for proper functions. Further discussion of quantum theory would take us too far afield; the interested reader can find a superabundance of published books on the subject. However an intelligent discussion of atomic clocks will require an acquaintance with the language of quantum mechanics and in particular the concepts of quantum *states*, *transitions*, and quantum *numbers*.

8.2 The Schrödinger Equation

We will illustrate these concepts with a one-dimensional idealized example of a particle confined between impenetrable walls, but is otherwise free (readers not mathematically inclined may skip this section without any great loss). Its motion is described by its wavefunction which is the proper solution of the appropriate Schrödinger equation:

$$\frac{d^2\Psi}{dx^2}+\frac{8\pi^2 mE}{h^2}\Psi=0. \tag{8.7}$$

This has the same form as for simple harmonic motion, with the well-known solution of the form:

$$\Psi=A\ \sin(kx), \tag{8.8}$$

where A is an arbitrary constant, and k must satisfy

$$k^2=\frac{8\pi^2 mE}{h^2}, \tag{8.9}$$

where m is the particle mass and E its energy. We must now impose the *boundary conditions* that $\Psi(x)=0$ at the walls, say at $x=0$ and $x=D$; thus, for Ψ to equal zero at $x=D$, we require that k be restricted to the following discrete values:

$$k_{\mathrm{n}}=\frac{n\pi}{D}, \tag{8.10}$$

where n is any integer. Hence the energy spectrum consists of discrete values given by

$$E_{\mathrm{n}}=n^2\frac{h^2}{8D^2m}. \tag{8.11}$$

The eigenfunctions of the problem are the following:

$$\Psi_{\mathrm{n}}=A\ \sin(k_{\mathrm{n}}x), \tag{8.12}$$

where the constant A must be chosen to normalize Ψ_{n}, that is, the total probability that the particle is *somewhere* between the two walls must equal unity; a simple integration yields $A=\sqrt{2/D}$.

A more interesting example from the point of view of atomic physics is a particle moving under the influence of a *central force*, that is, one directed radially towards a point, as we have in the hydrogen atom, where a single electron moves in the Coulomb electric field of a proton. As a three-dimensional problem with three independent coordinates, the solution is expected to involve three quantum numbers.

Fig. 8.1 Example of the probability distribution of an electrons in the state $n = 3$, $l = 1, m = 0$

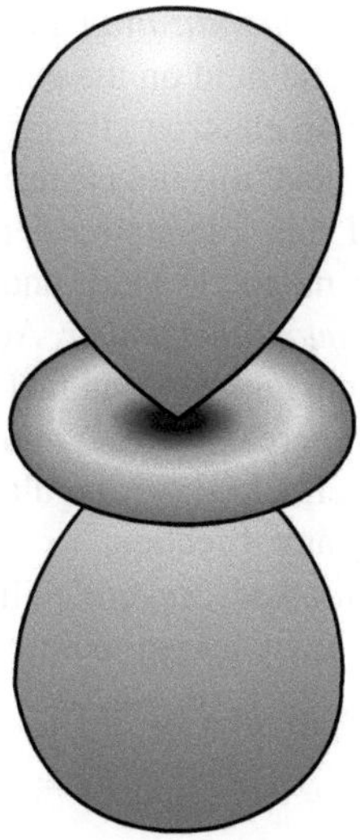

Because of the spherical symmetry of the problem it is most naturally solved using spherical polar coordinates (r, θ, φ), where as usual r is the radial distance, θ is the colatitude, and φ the azimuthal angle. We will not venture further into the mathematics of solving the Schrödinger equation beyond a description of the eigenstates and the quantum numbers specifying them. The equation is solved by a mathematical technique called *separation of variables*. A solution is sought in the form of a product of three functions each involving only one of the variables r, θ, φ. The result is a product of functions involving quantum numbers originating as separation constants that are conventionally designated as n, l, m:

$$\Psi = R_{n,l}(r)\Theta_{l,m}(\theta)\Phi_m(\varphi). \tag{8.13}$$

The values of the quantum numbers n, l, m are restricted to integers as follows: $n = 1, 2, 3\ldots$,while $l \leq (n - 1)$ and $m = -l, -(l - 1), -(l - 2) \ldots +(l - 2), +(l - 1), +l$. The radial function $R_{n,l}(r)$ has $(n - l - 1)$ zeros (nodes), and the function of the colatitude $\Theta_{l,m}(\theta)$ has $(l - m)$ zeros. Figure 8.1 illustrates the quantum hydrogenic probability function for the state $n = 3$, $l = 2$, and $m = 0$.

Following the conventional notation used by spectroscopists, electrons in an atom having $l = 0, 1, 2, 3\ldots$ are referred to as s-, p-, d-, f- ....electrons, respectively. The quantum number l that appears in the angular parts of the wavefunction is an expression of the quantization of the *orbital angular momentum*, and m its component along the arbitrarily chosen z-axis. The requirement that the function of the azimuthal angle φ around the z-axis, namely, $\Phi_m(\varphi)$, be single-valued as a function of φ imposes the restriction that m the component of the angular momentum along the z-axis be quantized. This is a radical departure from classical mechanics where no such restriction exists. It implies that angular momentum can assume only certain discrete directions with respect to a given axis. This is called *space quantization* and is of profound importance in the theory of the effect of a magnetic field on the energy levels of an atom: the *Zeeman effect*. The orbital motion of an electron gives rise to a magnetic interaction with an applied external magnetic field; this field is usually assumed to define the direction of the z-axis. The energy of this interaction depends on the component of the angular momentum along the magnetic field, and therefore along the z-axis, whence m is called the *magnetic quantum number*. According to quantum theory a state that has nominally the quantum number, l, in fact has angular momentum given by $[l(l+1)]^{1/2}$ in units of $h/2\pi$, whereas the maximum projection along the z-axis is l. This is a strictly quantum effect, since angular momentum is a vector and classically can point exactly in any direction.

The "allowed" *negative* eigenvalues (or proper values) of energy of the *bound states* of the hydrogen atom resulting from the equation for $R_{n,l}(r)$ is as follows:

$$E_{\mathrm{n}} = -\frac{1}{n^2}\frac{1}{(4\pi\varepsilon_0)^2}\frac{2\pi^2\mu e^4}{h^2} = -\frac{R}{n^2}, \tag{8.14}$$

where R is called the *Rydberg* constant in units of energy. For positive energy, all values are allowed in a continuum, and would correspond physically to the scattering of an electron with positive energy by a central charge. That is essentially the same problem that Rutherford solved classically with the good fortune that his result agrees with the quantum result. The expression for the bound energies happens to coincide precisely with the formula derived by Bohr using his planetary model; another example of a fortunate coincidence that ensured that his postulates would not be dismissed. It is significant that the energy predicted by the classical Schrödinger equation does *not* depend on the quantum numbers l and m; in cases where this is so, the states are said to be *degenerate* with respect to these quantum numbers.

A consequence of the electron having orbital angular momentum $lh/2\pi$, which classically would be pictured as arising from the electron's motion around the proton, is that it has a *magnetic moment*, that is, it has a magnetic field as of a small magnet. Hence in an external magnetic field applied along the z-axis it would have magnetic potential energy determined by the z-component of that magnetic moment and therefore proportional to the quantum number m. This is reflected in the atom's energy level structure, and therefore the optical line spectrum due to transitions involving these levels would show a splitting of the lines called the *Zeeman effect*. The experimental study of the Zeeman effect had a very significant impact on the development of atomic and elementary particle physics. In analyzing the structure of the spectra emitted by atoms in strong magnetic fields, Uhlenbeck and Goudsmit discovered in 1925 that they could explain complex atomic line spectra by assuming that in addition to orbital angular momentum an electron must be assumed to have an inherent angular momentum of ½($h/2\pi$) which Pauli called *spin*. This is one-half the quantum of angular momentum associated with orbital motion. Spin was postulated to be another invariable attribute of the electron like its rest mass or electric charge. The way this new phenomenon revealed itself is by "anomalies" in the observed Zeeman spectral patterns: an ad hoc explanation was that the electron has a different ratio of magnetic moment to angular momentum, a quantity termed the *g-factor*; its spin has an angular momentum of ½($h/2\pi$) but the *same* magnetic moment as orbital angular momentum of $h/2\pi$; the electron *g-factor* is said therefore to have an "anomalous" value of 2. The dependence of the energy on the quantum numbers l and s results in the splitting of spectral lines called *fine structure*. This anomalous g-factor of the electron was derived as a logically consistent consequence of a relativistic theory of the electron by Paul A.M. Dirac. However in the case of hydrogen, the relativistic quantum theory of Dirac (and by curious coincidence even Sommerfeld's relativistic version of Bohr's theory) predicts a degeneracy of states with equal j values, where $j = l + s$. To complicate matters further, Lamb and Retherford showed experimentally that this degeneracy does not agree with experiment; in fact they found what is now called the *Lamb Shift* of about 1.057 GHz. Modern high-resolution experiments on stored electrons have measured the g-factor to be 2(1.001159...), slightly at variance with Dirac's theory

8.3 Atomic Structure

The quantum description of multi-electron atoms is based upon the *central field approximation*. This assumes that each electron moves in a spherically symmetric average electric potential field $V(r)$ due to the nucleus and all the other electrons. The reason one might expect this to be a reasonably good approximation is that for the outer electron(s) responsible for the optical spectrum, the nucleus has Z times the charge of individual electrons distributed around the nucleus, where Z is the *atomic number*; the inner electrons form in effect a spherical cloud around the nucleus. To the extent that the electrostatic field is central, the angular parts of the atomic eigenfunction are the same as for hydrogen, so that in the absence of a magnetic field the m-states have the same energy and are said to be *degenerate*. However for the radial part of the wavefunction, the potential will not have the Coulomb $-1/r$ dependence and states with different orbital quantum number l will now have different energies. The restrictions on the values of n, L, m_L of the hydrogen atom still apply. The spin of the electron s = ½ (in units of $h/2\pi$) can have only two directions with respect to a given axis: either parallel or antiparallel, and thus requires an additional quantum number $m_s = \pm$½ to specify its direction. If an electron has orbital motion in the central electric field, so that l is not zero, then there is an interaction called the *spin–orbit* interaction, causing a *coupling* between l and s. This means that the quantum state of the electron must be expressed in terms of the total angular momentum ($l + s$) whose quantum number is conventionally represented by j. The symbols l, s, j, we recall, represent angular momentum *quantum* vectors of magnitudes $[l(l + 1)]^{1/2}$,...etc. in units of $h/2\pi$. This raises the question of how the operations of addition and subtraction of angular momenta are handled in a multi-electron atom. A rigorous development of the theory of

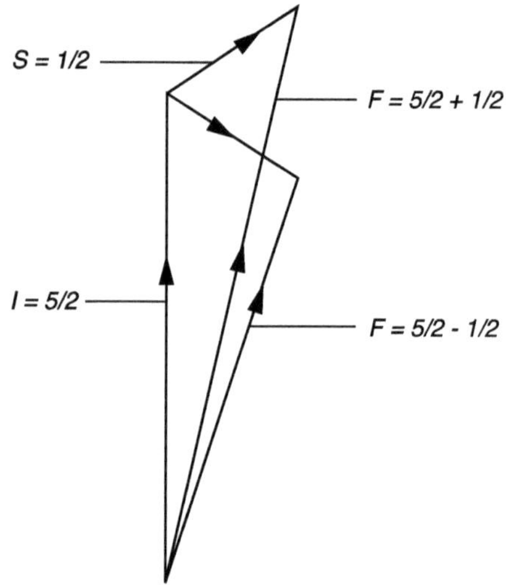

Fig. 8.2 Addition of quantum vectors

angular momentum in quantum mechanics would certainly take us too far afield. Luckily it is possible to obtain valid results if we treat them like classical vectors provided we superimpose some quantum conditions on them according to what is called the *vector model*. Classically two vectors $\boldsymbol{J}_1$ and $\boldsymbol{J}_2$ when added together can result in any vector between $\boldsymbol{J}_1 - \boldsymbol{J}_2$ and $\boldsymbol{J}_1 + \boldsymbol{J}_2$ depending on the angle between the two vectors. However in quantum theory according to the vector model the resultant can be only one of a *discrete* set of vectors that starts at $\boldsymbol{J}_1 - \boldsymbol{J}_2$ and by increments of *one* unit reaches $\boldsymbol{J}_1 + \boldsymbol{J}_2$. This is illustrated in Fig. 8.2 for the case $\boldsymbol{J}_1 = 2$ and $\boldsymbol{J}_2 = 3$.

Symmetry plays a central role in the theory of quantum mechanics. Of immediate concern to us is symmetry of wavefunctions that represent a group of electrons, which in this context are referred to as fermions. As first pointed out by Pauli, a wavefunction representing a group of electrons must be *antisymmetric* with respect to the exchange of the spatial and spin coordinates of any two electrons. An immediate consequence of this symmetry requirement is what is called the *Pauli Exclusion Principle*, which states that in a wavefunction representing many electrons no more than *one* electron can occupy the same state, that is, have the same set of quantum numbers. When such a state has an electron occupying it, it is said to be "filled." This principle has far-reaching consequences in atomic theory; in fact atoms as we know them would not exist otherwise! On the basis of this principle we are able to not only construct the periodic table of the chemical elements but also to understand their chemical valence, or "combining power." Let us enumerate all the electron states available around a nucleus with increasing number of protons. The states fall into groups called *shells* each corresponding to a particular value of n the principal quantum number. The number of states in a shell can be shown to be $2n^2$. The lowest or *ground* state of an atom is obtained by assigning electrons to states in such a way as to result in the lowest *total* energy.

As an example consider an atom of silicon, Si. It has a nuclear charge of $+14e$ and therefore the neutral atom will have a total of 14 electrons in all its shells, distributed as follows: two in $n = 1$ shell, eight in $n = 2$ shell, and four in $n = 3$ shell. We note that the $n = 3$ shell can accommodate 18 electrons, and so that shell is not filled. In the case of the noble gas argon, it has a nucleus with a charge of $+18e$ which means it would have eight electrons in the $n = 3$ shell, filling it. It is the closed shell outer structure that makes argon chemically inert. The reactive alkali elements: Li, Na, K, Rb, and Cs have only one electron in the outermost shell, and the alkaline earths Mg, Ca, Sr, and Ba have two electrons outside closed shells. However in the lowest *ground state* of an atom not all subshells corresponding to different l values are necessarily filled before the next shell begins to fill up. The distribution of electrons among the shells and subshells for the ground state is determined by requiring the total energy to be a minimum. Thus, we have what are called transition elements, which occur where it is energetically lower to go to a higher shell than to complete the previous one.

8.4 Atomic Spectra

The normal state of atoms is the lowest energy state called the *ground state*. They may be excited to higher energy states by thermal agitation at very high temperatures or by interactions with energetic electrons or photons. An atom in an upper excited energy state will generally *spontaneously* return to the ground state by emitting one photon (in rare cases two photons), or may be de-excited by collisions with other atoms. The frequency of the emitted photon is determined by the Bohr condition. Because of the discreteness of the energy levels of free atoms the emitted photon spectrum consists of series of lines. The mean time that an atom spends in an excited state before making a transition to a lower state is the *radiative lifetime*. The lifetime generally depends on the quantum numbers J, L, S of the excited state and the presence of suitable lower energy states to which it is "allowed" to make a transition according to certain quantum *selection rules*. The most probable and common types of transitions are called *electric dipole transitions*. The selection rules for these are the following:

$$\Delta J = 0, \pm 1; \Delta L \pm 1; \Delta S = 0, \tag{8.15}$$

where the Δ notation means the difference in the quantum number between the initial and final states. Transitions involving simultaneously more than one electron are rare and will not be considered.

Of crucial importance to the analysis of the spectrum emitted by a multi-electron atom is the way the orbital angular momenta $\boldsymbol{l}_i$ of the several electrons couple to $\boldsymbol{s}_i$ the spin angular momenta. In one scheme called the *Russell–Saunders* coupling, the orbital momenta $\boldsymbol{l}_i$ are coupled first to yield a resultant $\boldsymbol{L}$, and the spins are coupled to yield a resultant $\boldsymbol{S}$. Then the total angular momentum $\boldsymbol{J}$ is given by $\boldsymbol{J} = \boldsymbol{L} + \boldsymbol{S}$ and the component along a given axis is $m_J = m_L + m_S$. The other principal scheme of dealing with the angular momentum is the so-called *j–j coupling* in which the l_i and s_i values are added first to yield j_i and then these are combined to yield the total angular momentum.

Long before the arrival of quantum mechanics, spectroscopists had developed a nomenclature to describe spectra. Thus, for atoms with one electron outside closed shells, that is, the alkali elements, it was noticed that the lines fell into groups called *series* in which the lines are arranged with continuously decreasing spacing towards a limit at the high-frequency end. It was found that the frequencies of members of a series obey a formula that is a modified hydrogen formula displaced along the frequency scale; thus,

$$\nu = \nu_\infty - \frac{R}{n^{*2}}, \tag{8.16}$$

where ν_∞ is the series limit, n^* is the effective principal quantum number, and R is called the Rydberg constant (in frequency units). The effective number n^* for a series has the form of a running integer plus a correction that depends on the orbital quantum number l, so that for each value of l we have a different series. Historically the series are labeled S-, P-, D-, F-, G-series for $L = 0, 1, 2, 3, 4$. It should not be surprising that the series corresponding to different L values are displaced along the frequency scale, since states with low values of *angular* momentum are more stretched out and penetrate deeper into the closed inner shells. When the electron spin is included with its attendant spin–orbit coupling giving rise to *fine structure*, the labels of the states must reflect this. The format used is called *spectroscopic notation*. As an example, if only one electron is involved, as in hydrogen or the alkali elements, then the electron spin leads to a splitting into *doublets*; for example a P-state with $L = 1$ splits into $J = 1/2$ or $3/2$ and would be labeled $^2P_{1/2}$ or $^2P_{3/2}$. An S-state with $L = 0$ has no fine structure. In the case of the He atom and the alkaline earths Mg, Ca, Sr,... the two outer electrons couple to form two resultant spins $S = 0$, and $S = 1$. In this case since electric dipole transitions (the only kind that concerns us here) are not allowed between states of different S, it follows that we have in effect two totally independent sets of quantum levels as if they were different atoms. The prototype of this phenomenon is the helium atom: there are *ortho*-helium and *para*-helium atoms, having $S = 1$ and $S = 0$, respectively. The energy of an *ortho*-state having $L = 1$ and $S = 1$ will through spin–orbit coupling split into a *triplet* having $J = 0, 1, 2$, while a *para*-state with $S = 0$ remains a singlet. These levels (or *terms* in the language of spectroscopists) are designated as $^3P_0$, $^3P_1$, $^3P_2$, and $^1P_0$.

8.5 Hyperfine Interaction

When individual spectral lines are analyzed under conditions allowing the highest possible resolution, many are observed to be really groups of lines very close together, a feature called *hyperfine structure*. It was Wolfgang Pauli who first pointed out that this is not caused by some additional degree of freedom of the electrons but rather some interaction involving the nucleus. There are two ways in which the nucleus plays a role in the optical spectrum: first the element may have several naturally occurring stable isotopes, and second the nucleus may have a magnetic moment and spin.

Isotopes have identical nuclear charge and therefore outer electron structures, but different nuclear masses due to the presence of different numbers of neutrons in the nucleus. Although the nuclear mass is thousands of times larger than the electrons, nevertheless the motion of the nucleus about the common center of mass of the electron–nucleus system cannot be ignored. In fact it was the discovery by Urey and his coworkers in 1932 that in the spectrum of hydrogen each line had a close companion corresponding to a mass number of 2 that proved the existence of an isotope of hydrogen, now called deuterium. Little did Urey know that someday his discovery would be essential to building an H-bomb. His was a remarkable achievement considering that in nature deuterium constitutes only one part in 5,000 parts of hydrogen. In the case of the heavier elements the isotope shift is not entirely due to the mass; the nuclear radius and charge distribution within the nucleus can differ for different isotopes.

The constituents of atomic nuclei are protons and neutrons, both of which have half integral spins with accompanying magnetic moments. The simplest nucleus, the proton, has a mass 1,836 times the mass of the electron, yet has the same spin angular momentum as the electron: $\frac{1}{2}(h/2\pi)$. The fact that the neutron, an electrically neutral particle overall, has a magnetic moment indicates that it is not a fundamental particle in the sense that an electron is fundament; it presumably has some internal structure. The spin of the nucleus is quantized and given as $I(h/2\pi)$ where I is a quantum number which can be integral as well as half integral, as determined by the internal structure of the nucleus. The nucleus also has a magnetic moment expressed in *nuclear magnetons*, a unit defined as $\mu_N = eh/4\pi m_p$ where m_p is the proton mass. We recall that a similar expression in which the proton mass is replaced by the electron mass defines the Bohr magneton. The magnetic moments of the proton and neutron are often expressed in terms of the appropriate g-factor; thus, for example, $\mu_n = (1/2)\, g_n\, \mu_N$ is the magnetic moment of the neutron.

The question of the stable values of spin and magnetic moment for a given nucleus is a difficult one of nuclear structure, which has seen different models perfected, including the shell model, to explain the stability of nuclei. Fortunately we need not trouble ourselves with that; we will assume given values of the nuclear data. Since the magnetic moment of the nucleus is orders of magnitude smaller than the electron moment, it is expected that the different orientations of the nuclear spin will lead to very narrow splittings of the electronic states. For this reason the magnetic interaction between the nucleus and the outer electrons is called the *hyperfine interaction*. It is precisely this interaction that leads to the hyperfine spectrum and the extraordinarily narrow microwave transitions that are at the heart of atomic clocks. The hyperfine interaction is expected to be the strongest between the nuclear magnetic moment and that of an electron in an electronic S-state with $L = 0$. The reason is a distinctly quantum reality that the S-state is actually nonzero

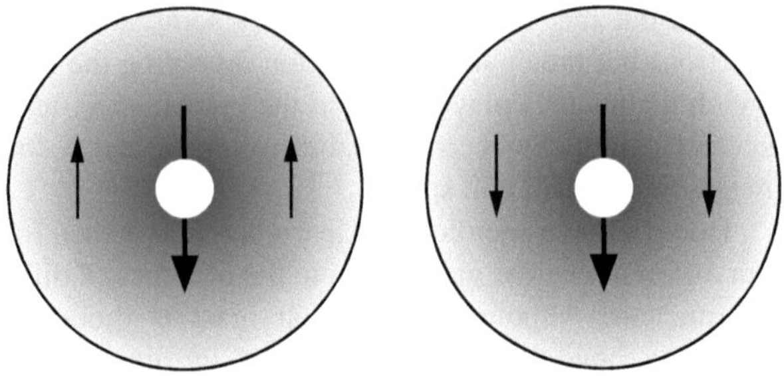

Fig. 8.3 Magnetic moment of nucleus interacts with electron cloud

at the center of the nucleus! Even the original Bohr theory had an orbit with zero angular momentum go right through the origin. Since the nucleus is of finite size it means that the electron has a finite probability of being inside it: a very unclassical behavior! It might be thought that the electron should be captured by the nucleus; in fact the phenomenon of electron capture does occur in some nuclei, but mostly of electrons in the innermost so-called K-shell of the atom. In the S-state the electron distribution is spherically symmetric about the nucleus and so its interaction with the nucleus is not like two magnets side by side, but rather one magnet, the nucleus, *imbedded* in a spherically symmetric magnetized medium, the electron, as illustrated in Fig. 8.3.

The following is the expression for the energy of interaction between an electron in an S-state and a structureless nucleus with nuclear spin I and magnetic moment μ_n, as first published by Fermi in the early days of the new quantum mechanics [6]:

$$\begin{aligned} E\left(\frac{I+1}{2}\right) &= \frac{2}{3}\mu_e\mu_n|\Psi(0)|^2 I, \\ E\left(\frac{I-1}{2}\right) &= -\frac{2}{3}\mu_e\mu_n|\Psi(0)^2|(I+1), \end{aligned} \tag{8.17}$$

where $|\Psi(0)|^2$ represents the electron density in the nucleus. This result is based on a very simple model of the atom which ignores the structure of the nucleus and the presence of other electrons. It can be rationalized by a simple argument based on elementary electromagnetic theory; for the interested reader we briefly sketch the classical argument; others may skip it. The derivation begins with the expression for the magnetic field inside a magnetized medium: $\mathbf{B} = (2/3)\mathbf{M}$ where $\mathbf{B}$ is the intensity of the magnetic field, and $\mathbf{M}$ is the magnetization defined as the magnetic dipole moment per unit volume. In this case $\mathbf{M}$ is given by the following:

$$\mathbf{M} = g_e\mu_B|\Psi(0)|^2\mathbf{s}, \tag{8.18}$$

where μ_B is the Bohr magneton, $|\Psi|^2$ is the probability density of the electron, and $\mathbf{s} = 1/2$. The electron is viewed as a continuous magnetized medium distributed around the nucleus according to the probability function. The magnetic field acting on the nucleus according to classical theory is given by

$$\mathbf{B} = \left(\frac{2}{3}\right) g_e \mu_B |\Psi(0)|^2 \mathbf{s}. \tag{8.19}$$

It remains to find the classical energy of the nuclear magnetic dipole in the field **B**, namely,

$$E = -g_n \mu_N \mathrm{I}\,\mathrm{B}\ \cos\theta. \tag{8.20}$$

Since the total angular momentum is the vector sum $\mathbf{F} = \mathbf{I} + \mathbf{s}$, we have $\cos\theta = (F^2 - I^2 - S^2)/2IS$. This does not quite accord with the quantum result, but shows in classical terms the origin of the hyperfine splitting. It provides a concrete picture, but is otherwise hard to justify. The quantum treatment is not expected to give a very accurate result for the heavier atoms; even for the simplest atom, hydrogen, a precise result requires the application of quantum electrodynamics. Because the hyperfine frequency in hydrogen can be measured with extreme accuracy, as we shall see in the next chapter, it has stimulated a great deal of theoretical activity to carry the calculations to higher and higher order corrections in order to test the limits of the theory. As of the time of this writing the hyperfine frequency of hydrogen is undoubtedly the most precisely measured physical number: it is known to 12 significant figures.

8.6 The Rubidium Standard

Among atomic frequency/time standards (to call them "clocks" hardly conveys their technological standing, not to mention their cost) the rubidium standard is the most rugged, compact, and therefore most portable. Ruggedized versions have been developed for shipboard and military missile-borne applications.

It is based on the resonant absorption of microwaves by free rubidium atoms as they make transitions between quantum states separated by virtue of the electron–nuclear hyperfine interaction. When a free rubidium atom is irradiated with microwaves, it will strongly absorb that radiation provided its frequency is such that the energy of its quanta exactly equals the energy difference between two quantum states in the atom; this is spoken of as a *resonant* response. In the isotope Rb^{87}, used in the rubidium standard, the resonant frequency is 6.83468... GHz, a number now known to over ten significant figures. In order to observe the sharpest resonance, that is, with the highest spectral resolution, the atom must be otherwise as free as possible, for as long as possible, while it is responding to the probing microwave field. If the atoms of rubidium were confined in a glass bulb under high vacuum, they would bounce around at high speed and would spend a very short time of undisturbed response to the microwave field. This would broaden the spectrum of the resonance and because of the electrical noise always present in a system would increase the uncertainty of fixing on the center of the resonance. What is required is a way to increase the atom's time of flight to reach the walls, but to do it in a way that does not disturb the atom as it is undergoing the transition from one quantum state to another. It was Hans Dehmelt in the early 1950s who first demonstrated that the atoms could be allowed to *diffuse* slowly through an inert gas such as argon, without broadening the resonance. Indeed the lengthened time of undisturbed interaction between the rubidium atoms and microwave field *sharpene*d the resonance. It was reasoned that since the inert gases have a closed shell structure (which is *why* they are inert) they have no unpaired electron spins and hence no magnetic interaction with the rubidium's outer electron. Hence it is safe to presume that the hyperfine interaction in the rubidium atom would not be affected in a collision. While the use of a gas-filled rubidium *absorption cell* certainly meant that the standard can be very compact and relatively rugged, it unfortunately does have two drawbacks; the first is that the collisions with the inert gas atoms do produce a *pressure shift* in the resonance frequency, and second, an optical method is required to detect the resonant transitions with an attendant *light shift*. These shifts disqualify it as an absolute standard; nevertheless it is widely used as an extremely stable clock.

Rubidium has two naturally occurring *isotopes*, that is, atoms having the same number of protons in the nucleus, and therefore outer electronic structure which identifies them as rubidium, but a different number of neutrons in the nucleus. Natural rubidium consists of about 72 % of the stable mass 85 isotope with a nuclear spin $I = 5/2$ and 28 % of the very weakly radioactive (life time 4.8×10^{10} years) mass 87 with nuclear spin $I = 3/2$. Following the quantum rules of adding vectors we find that the ground state of Rb^{87} which has $L = 0$ and $S = 1/2$ and represented as an $^2S_{1/2}$ state, after the addition of the nuclear spin $I = 3/2$, has the total angular momentum $F = 3/2 + ½ = 2$ or $F = 3/2 - ½ = 1$. The first two excited P-states with $L = 1$ form a so-called *fine structure doublet* with $J = 1 + ½$ and $J = 1 - ½$ obtained by combining the $L = 1$ orbital momentum with the spin ½ of the electron. These in turn will, when the nuclear spin $I = 3/2$ is added in, yield two possible states with $F = 0$ and $F = 3$ for one member of the doublet and states with $F = 1$ and $F = 2$ for the other. Similar arguments apply to the other mass 85 isotope. The quantum level scheme soon becomes a little complicated; they are illustrated in Fig. 8.4 which also shows their dependence on an external magnetic field.

Since the frequency of the standard is determined by the resonant absorption of microwave energy in a transition between a pair of the hyperfine states the dependence of these energy levels on the ambient magnetic field is obviously very important. It is inescapable that a magnetic field be present; we are immersed in the earth's magnetic field and are surrounded by man-made magnetic fields from steel structures, etc. Therefore it is important to examine the extent to which a magnetic field may displace the chosen transition frequency, and how it may be minimized. It happens that there is a pair of states between which the microwave transition frequency is far less sensitive to the external magnetic field than all the others. It helps to see this by plotting the energies of the different hyperfine states as a function of the intensity of an external magnetic field. Let us focus on the Rb^{87} isotope since it will prove to be the most suitable for a standard. It has two hyperfine components in the ground $^2S_{1/2}$ state with $F = 2$ and $F = 1$; in a magnetic field the states with the various values of m_F, the components of F along the direction of

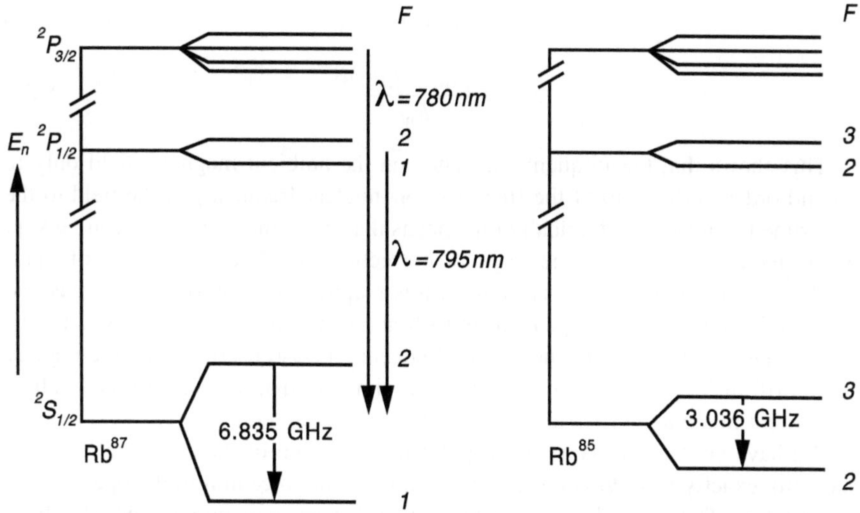

Fig. 8.4 Hyperfine structure of the ground state of the isotopes mass 85 and 87 of rubidium

the field, will have different energies, analogous to the energies required to turn a compass needle to different angles away from north.

In zero magnetic field the hyperfine states with $F = 1$ and $F = 2$ have energies that differ by the strength of the hyperfine interaction, as discussed in the last section. It is the frequency corresponding to *this* energy through Planck's formula $E_2 - E_1 = h\nu$ that is chosen as the reference in the rubidium standard. As the magnetic field intensity is increased slightly above zero, the states with different values of m_F, that is, components along the magnetic field direction, will split into different energies proportional to the values of m_F and the intensity of the magnetic field; hence the graphs are straight lines with slopes dependent on the values of m_F. This description in terms of (F,m_F), the total angular momentum, is valid provided the magnetic field is not so strong that its energy of coupling to the individual nuclear and electron spins is comparable to the hyperfine coupling. This is illustrated in the figure by the change that occurs in the graphs as the magnetic field becomes more intense. In the limit of large values of the magnetic field we see that the graph splits into *two groups* of graphs representing the two orientations of the electron spin with respect to the magnetic field, and *within* each group we have the different independent orientations of the nuclear spin.

The solution to the problem of finding, in the present system, the energies of all the m_F-states as a function of the external magnetic field is given by the *Breit–Rabi formula* [7]. It applies to the case of $I = 3/2$ and $S = ½$ and an interaction that depends on the scalar product **I.S**. If $E_{\rm hf}$ represents the hyperfine interaction energy, then the formula gives for the transition frequency between $(F = 2, m_F = 0)$ and $(F = 1, m_F = 0)$, the so-called (0-0) transition, the following:

$$h\nu = E_{\text{hs}}\sqrt{1+x^2}, \tag{8.21}$$

$$x = \frac{(g_s\mu_B - g_I\mu_N)B}{E_{\text{hf}}}. \tag{8.22}$$

This shows that the frequency depends on the ambient magnetic field only to second order in the ratio of the free electron–nuclear frequency in the field to the hyperfine frequency. Numerically this means that if the magnetic field can be kept below about 2.5×10^{-7} T, the magnetic correction would be on the order of 1 part in 10^{12}. Furthermore the system allows this magnetic correction to be measured using the magnetic field-dependent transitions in which m_F changes by ± 1. The weak magnetic field dependence of the frequency is also useful for correcting any small drift in frequency that may occur over time; all that is necessary is to adjust the current in a magnetic field coil around the rubidium cell.

We have yet to address a crucial point in the design of the pre-laser rubidium standard: exactly how do you *detect* those transitions so central to the operation of the standard. The method used is called *optical hyperfine pumping* in which a light beam resonant with the transition from the ground S-state to the first excited P-state is used to probe the populations of internal quantum states and detect the onset of transitions between them. We recall that in the interaction of atoms with radiation we have three fundament processes: absorption, spontaneous emission, and stimulated emission, and these can occur with photons of any wavelength, including microwaves. However there is a distinction that must be made between what are called *electric dipole* transitions and *magnetic dipole* transitions. In the visible region of the spectrum the transitions are strong electric dipole transitions from states whose energies are much higher than the mean thermal energy of the atoms at ordinary temperatures, so that in thermal equilibrium the number of atoms in the upper state remains low, and transitions are readily observable both in fluorescence and absorption. In contrast the microwave frequency standard uses the magnetic dipole type of transitions which is characterized by a vanishingly small rate of *spontaneous* emission. Since the probability of absorption of microwave energy is exactly the same as the probability of *stimulated* emission, it follows that if the upper and lower states are equally populated, there is no observable effect caused by transitions between them. We emphasize that the difference in energy between the hyperfine states is so small compared with thermal energies that in thermal equilibrium at normal temperatures the number of atoms in the upper state is very nearly equal to that in the lower; thus, transitions induced by resonant microwaves will cause no observable effect on either the atoms or the microwaves. In order to detect resonant transitions we need a way of creating a difference in populations in the two hyperfine sublevels and be able to monitor that difference as the frequency of the microwaves is swept through resonance.

This was achieved using *optical* resonance absorption of light by the atoms, following methods originally developed in the early 1950s by the French spectroscopist Alfred Kastler for magnetic resonance studies in free atoms. He used the absorption of *polarized* resonance light to produce spin orientation in atoms, that is,

Fig. 8.5 (**a**) Optical pumping. (**b**) Hyperfine structure of $\lambda = 780$ nm line in isotopes of Rb

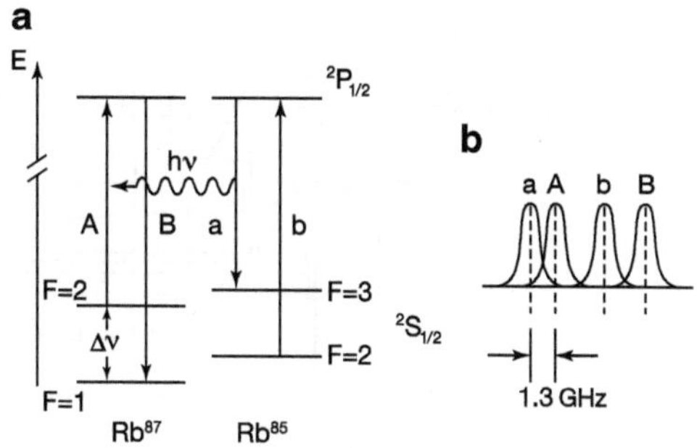

differences in populations of the magnetic *m*-substates. For the rubidium standard what is needed is a difference in population between hyperfine states. If an analogous optical pumping method is to be used for achieving this, it would require a source of light that is selectively absorbed by atoms in one hyperfine sublevel but not the other. Lasers were still decades in the future at the time that the first rubidium standard was successfully demonstrated by Bender, Beaty, and Chi in 1958. A pre-laser light source consisting of a classic rubidium vapor lamp was used, one in which a small glass bulb containing a small quantity of Rb and an inert gas is excited using a radiofrequency coil around it. In order to ensure that the light so produced would be selectively absorbed by one hyperfine but not the other, they noticed a fortuitous matching of wavelengths in the optical hyperfine spectra of the two isotopes of rubidium Rb^{85} and Rb^{87}. There are two strong "resonance" lines attributable to each isotope that arise from transitions between the ground state and the first two excited P-states, forming a fine structure *doublet* at $\lambda = 780$ nm and $\lambda = 795$ nm (1 nm $= 10^{-9}$ m) in the red part of the optical spectrum. Figure 8.5 shows the hyperfine structure of the resonance line at $\lambda = 780$ nm of the two isotopes, neglecting the hyperfine splittings of the upper P-states since they are much smaller than for the ground S-state. As evident from the figure, the spectrum emitted by an ordinary Rb vapor lamp consists of four resolved lines due to the hyperfine splitting of the ground states of the two isotopes, but most significantly there is an appreciable *overlap* between only the longer wavelength lines of the two isotopes. It occurred to the researchers at NIST that if enriched isotopes of Rb were available, one could choose (say) Rb^{87}as the standard "working substance" in an optical absorption cell, with the output of an enriched Rb^{87}lamp with a separate *Rb^{85} filter* to produce light that resonates with only one hyperfine line. This would cause the population of that resonant ground state sublevel to be depleted because after each excitation to the upper P-state, the atom reradiates to *both* ground state sublevels. After repeated absorption and reemission, atoms of Rb^{87} in one hyperfine sublevel are "pumped" into the other. This process is similar to the Kastler optical pumping technique for magnetic sublevels.

The beauty of this optical pumping technique is that the monitoring of sublevel populations, and therefore the detection of transitions, is an integral part of it. Thus, as the atoms are pumped from one hyperfine sublevel to the other, fewer atoms are left to absorb the pumping light, and therefore the Rb^{87} vapor in the bulb becomes

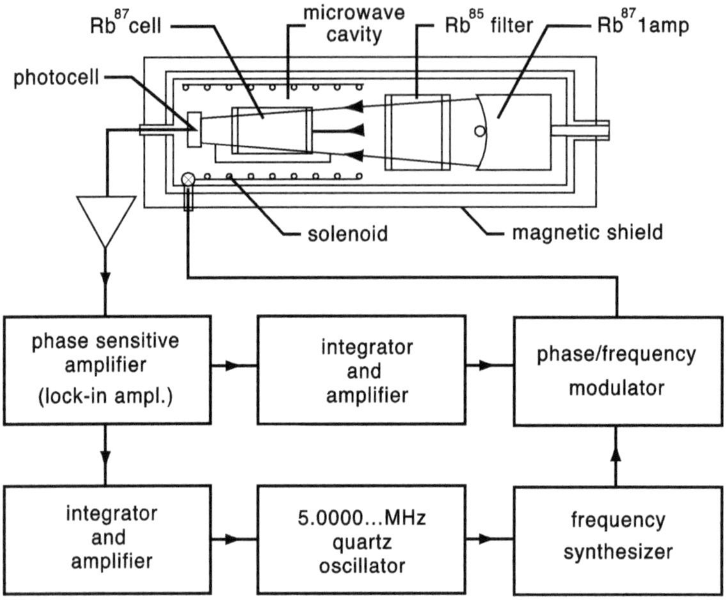

Fig. 8.6 The classical design of Rubidium frequency standard

more transparent. In order to display the microwave resonance, it only remains to irradiate the Rb absorption cell with microwaves whose frequency, and therefore energy $h\nu$, can be swept through resonance with the ground state hyperfine splitting. The transitions caused by the microwaves will tend to equalize the populations of the hyperfine states leading to a drop in the transmitted intensity of the pumping light. A photocell and amplifier are then used to measure the transmitted light intensity and display the resonance curve as the frequency of the microwaves is swept. A simplified schematic drawing of a rubidium standard is shown in Fig. 8.6.

As the heart of a frequency standard, the rubidium absorption cell has been exhaustively studied, particularly as regards long-term stability. Possibly the most pernicious source of long-term drift is the desorption of gases from the walls of the glass cell. Another source of instability is fluctuation in temperature; however, that can be mitigated by choosing the proper mixture of gases; for example, it has been found that a mixture of 12 % neon and 88 % argon reduces the sensitivity of the frequency to temperature. The presence of the buffer gas has one beneficial effect in addition to lengthening the atoms' interaction time with the microwave field, and, that is, it can narrow the resonance frequency width by reducing the Doppler width through the *Dicke effect*, as briefly described in the next section.

Central to the accuracy and stability of atomic frequency standards is the question of the width and shape of the spectral resonance curve whose "center" constitutes the reference for the standard. The presence of any asymmetry and noise fluctuation contributes to errors in the standard. The universal presence of noise in the system dictates that the sharper the resonance, the more precisely the center frequency can be determined. A universal source of spectral line broadening in optical spectra is the Doppler effect, which we recall is the observed shift in frequency of a source whenever there is any relative motion between it and the receptor. Its universal presence is familiar to optical spectroscopists as setting the limit to spectral resolution, that is, the ability to separate close lines. For the *optical* lines of atoms in a gas where the mean distance the atoms travel between collisions is very much longer than the wavelength of the radiation, the Doppler shift is computed simply on the basis of the motion of the absorber through the undulations of the wave: the rate at which the crests are crossed, and therefore the frequency, depends on the velocity of the observer. This leads to the familiar classical formula for the Doppler shift in frequency. The particles in a gas in thermal equilibrium have velocities distributed according to a law attributed to Maxwell. This distribution law of velocities is converted to a distribution of frequencies through the Doppler effect. We mention this only by way of background.

If we apply the first-order Doppler formula,

$$\nu = \left(1 \pm \frac{V}{c}\right)\nu_0, \tag{8.23}$$

to the case of Rb atoms diffusing in a buffer gas with an average velocity of about 10^4 m/s, we find that the Doppler shift in the Rb hyperfine frequency is about 230 kHz, which is over 1,000 times larger than the width of the resonance actually observed. Clearly in the microwave region of the spectrum the situation is quite different from the optical region: the atoms do not continue to travel for large distances many times the wavelength of the radiation but typically distances that are much shorter. This important distinction and the consequences of it were pointed out by Robert Dicke in 1953 [8]. To understand how the theory of the Doppler effect must be modified, consider the simple case of a single atom constrained to move back and forth along a straight line with simple harmonic motion, and suppose it is irradiated with a microwave with some definite frequency ν_0. Because of the Doppler effect the atom "sees" a microwave that is frequency modulated (FM). The spectrum analysis of such frequency modulated radiation is well known through its use in radio broadcasting. There are three quantities, aside from its amplitude that are required to specify a frequency modulated oscillation: the center frequency, the frequency deviation, and the frequency of the modulation. We will not venture into the theory of frequency modulation beyond pointing out some properties that may be counterintuitive. The most striking result is that the spectrum consists of individual *lines* with a central frequency and *sideband frequencies* at equal intervals with diminishing amplitudes extending to *infinity*. We emphasize that the frequency does not go "through" all the intermediate values as it swings from one extreme to the other. Another counterintuitive observation is that the sidebands are *not* zero beyond the "instantaneous" values of the frequency at the end of its oscillation. Let us define the *modulation index* (m) as follows:

$$m = \frac{\Delta\nu}{\nu_m}, \tag{8.24}$$

where $\Delta\nu$ is the maximum deviation and ν_m the modulation frequency. If the deviation is small compared with the frequency of modulation, that is, $m \ll 1$, the sidebands will diminish rapidly to zero and the central frequency will dominate. If we apply this to our oscillating atom we find the following for the maximum deviation using the Doppler formula:

$$\Delta\nu = \frac{V_m}{c}\nu_0, \tag{8.25}$$

where V_m is the atom's maximum velocity and ν_0 is the frequency of the radiation, but $V_m = 2\pi a\nu_m$ where a is the amplitude of the oscillation; therefore substituting back into the previous equation and simplifying we finally get

$$m = \frac{2\pi a}{\lambda}. \tag{8.26}$$

This suggests that if the atom's movement is confined to a distance small compared with the wavelength, the modulation index is small and the sidebands rapidly decrease; thus, the spectrum it "sees" is very much narrower than the full Doppler shift of a freely moving atom. Dicke worked out the full quantum derivation of the effect and the theoretical line shapes under these conditions. In cases where the particle motion is in fact simple harmonic, the Dicke effect has profound ramifications in their spectrum, as we shall see when we consider the spectrum of ions oscillating in a trap. In the case of the rubidium standard the mean free path of the rubidium atoms in the buffer gas is in the range of a few millimeters whereas the resonant microwave wavelength λ is very much larger at about 44 mm; the observed linewidth is very much consistent with Dicke's theory.

The use of the classic rf-excited spectral lamp and hyperfine pumping in the Rb standard continued until suitable solid-state lasers became available in the early 1980s. We will take up the laser-based design of Rb standards for satellite-borne applications, and in particular GPS, in a later chapter.

8.7 The Cesium Standard

The cesium standard has been elevated to the unique role of *defining* the unit of time, displacing the motions of the earth and heavenly bodies from their privileged position as regulating and even defining the very concept of time, a position that has endured since early man. The General Conference of Weights and Measures in 1967 adopted a new definition of the international unit of time. The wording of the new definition is as follows: "The second is the duration of 9,192,631,770 periods of the radiation corresponding to the transition between the two hyperfine levels of the fundamental state of the atom of Cs-133." This is of course an expression of the belief that all factors that may affect the frequency of the cesium spectrum are well understood and, when all deviations are accounted for, the same constant frequency is obtained in all embodiments of the standard. It is interesting to compare this definition with that of the kilogram, which is still defined with reference to a material artifact. The kilogram is the mass of a cylinder of pure platinum–iridium kept with six other identical prototypes at the *Bureau International des Poids et Mesures* in France. The essential difference is that one relies on atomic processes at a fundamental level while the other involves matter on a complex macroscopic level. The ten-digit number in the definition of the second was chosen, of course, in order that the atomic second agree with the then existing ephemeris second. The drawback to the astronomical second was the practical one that it was available only through the intermediary of stable clocks. With the advancements in atomic clocks it soon became evident that they were becoming the de facto standards against which the astronomical data were being evaluated.

The cesium standard uses precisely the same kind of quantum transitions that are used in the rubidium standard. The essential difference between the rubidium and cesium standards is really only in the different ways of achieving a long, free interaction time between the atoms and the resonant microwave field. As we have learned, the rubidium standard attempts to achieve this by diffusion through an inert gas; in the cesium standard, on the other hand, atoms collimated in a beam are in free flight in high vacuum. These two configurations dictate different methods of detecting resonant transitions between states. With a rubidium absorption cell we learned that hyperfine optical pumping is used; such optical pumping would be difficult with conventional (non-laser) light sources in a long beam apparatus; instead, in the classical cesium standard, the atoms pass through what amount to a magnetic state separator and analyzer. There is, of course, no reason in principle why cesium vapor cannot be optically pumped in a cell or a rubidium beam magnetically state selected. In fact with the availability of laser pumping sources able to be tuned to the wavelength of individual lines in the hyperfine spectrum, it is no longer necessary to depend on the fortuitous matching of the spectra of isotopes as in Rb, and optically pumped Cs units have been developed. But the feature that sets a beam standard apart from all other types of standards is the almost collision-free environment of the atoms as they make transitions. That is, unlike the rubidium standard and the hydrogen maser to be discussed later, atoms in a beam are not subjected to perturbations that have unpredictable effects on the frequency of resonance. Since Cs had long been studied as an atomic frequency standard and has properties that enable, in principle, a higher accuracy and long-term stability than Rb, it was uniquely chosen to define the new international second.

The formation of atomic/molecular beams draws on a technique developed in the early years of the twentieth century when spectroscopists were engaged in trying to achieve the sharpest possible spectral lines by reducing the broadening due to the omnipresent Doppler effect. The idea was to do spectroscopic observation on a collimated stream of atoms that can be viewed at right angles, thereby eliminating the first-order Doppler effect. The principal challenge at the time would have been to produce a sufficiently high vacuum in the experimental apparatus; the atmospheric pressure must be reduced at least until the free path of the atoms is as long as the required beam. High vacuum technology was still in its early stages; the mercury diffusion pump was still under development when Dunoyer demonstrated in 1911 the formation of an atomic beam of sodium in a glass vacuum system. The most significant early experiment involving magnetic deflection of an atomic beam was the 1923 experiment of Otto Stern and Walther Gerlach. They carried out one of those pivotal experiments in physics that introduced a new world view of how nature operates. It demonstrated unequivocally that space quantization is real and that atoms do take up orientations at discrete angles, as quantum theory required. To test this essentially quantum phenomenon they passed a beam of well-collimated silver atoms between the poles of a powerful magnet shaped to produce the steepest possible gradient in field intensity across the beam. The silver atom has a ground state with one unpaired electron; thus, quantum theory requires that the beam should split into two, since there are two possible projections of the electron moment along the magnetic field: $\pm 1/2 g_e \mu_B$. The beam was received on a glass plate, and after a lengthy exposure *two* spots could clearly be seen, contradicting classical expectations.

Cesium is a heavy alkali element like rubidium with one s-electron outside closed shells in the ground state. Its atomic number is 55 and has only one stable isotope with mass number 133 and a nuclear spin of $I = 7/2$. Also like Rb the nuclear moment I of the nucleus couples with that of the electron spin $S = 1/2$ of the s-electron to produce a hyperfine splitting into two levels with quantum numbers $F = 4$ and $F = 3$. In zero external magnetic field, the energy separation between these sublevels is given by the frequency cited in the definition of the international second. In a weak external magnetic field B, that is, one that satisfies

$$\nu_{hfs} \gg (S g_e \mu_B - I g_n \mu_N) B, \tag{8.27}$$

where S and I are the electron and nuclear spin, μ_B and μ_N the Bohr and nuclear magnetons, and g_e and g_N their g-factors, the atoms in the $F = 4$ and $F = 3$ states behave like bar magnets with components along the field obeying space quantization rules; the projections along the magnetic field are given by $m_F = -4, -3, -2, -1, 0, +1, +2, +3, +4$ for $F = 4$, and $-3, -2, -1, 0, +1\ +2, +3$ for $F = 3$. In the limit of very intense external magnetic fields, the nuclear and spin moments are decoupled and the quantization is in terms $m_I = -3/2, -1/2, +1/2, +3/2$, with $m_S = +1/2$ and $m_I = -3/2, -1/2, +1/2, +3/2$, with $m_S = -1/2$. Figure 8.5 shows the energies of these states as a function of the intensity of an external magnetic field. Since the electron magnetic moment is much larger than the nuclear magnetic moment, the first group of four magnetic sublevels with $m_S = +1/2$ have graphs that tend to parallel lines with a positive slope and the set with $m_S = -1/2$ tend to lines with a negative slope (Fig. 8.7).

Thus, in an external magnetic field strong enough to decouple I and S, thus requiring the states to be described in the (m_S, m_I) representation, the atoms having positive values of m_S will have their magnetic moment pointing *opposite* to the magnetic field because of the negative electron charge and therefore have *positive* potential energy that *increases* with the strength of the field. The reason is that the potential energy of a magnetic dipole is positive when it is pointing in the

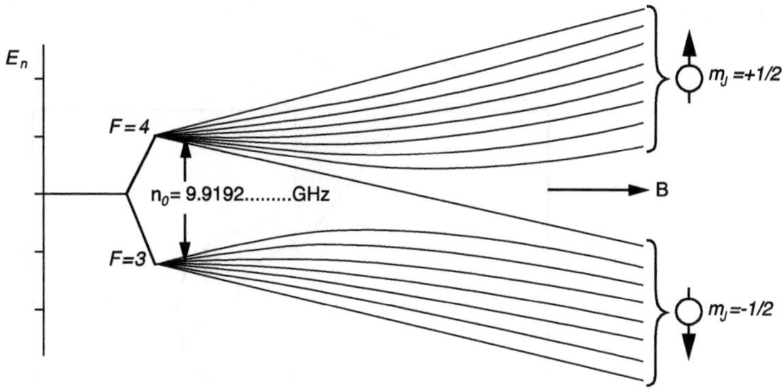

Fig. 8.7 The energy of magnetic sublevels in Cs^{133} as a function of external magnetic field

opposite direction to the field because it can do work in turning into the *same* direction as the field. The significance of this is that if atoms of Cs atoms collimated in a beam pass through a strong magnetic field with a steep *transverse* gradient in intensity, they will experience a *force* that will deflect the beam in the direction of lower intensity. The result is that atoms in states with opposite signs of m_S are deflected in opposite directions, as in the Stern–Gerlach experiment.

The classical cesium standard uses techniques derived from the classical work of I. Rabi and his coworkers who made a critical expansion of the Stern–Gerlach type of experiment by introducing a combination of *two* Stern–Gerlach magnets with a uniform field magnet between them, where quantum transitions are induced by a high-frequency field. The original Stern–Gerlach design is of a non-focusing two-pole magnet with one pole having a rectangular groove and the other a sharp wedge shape. Design studies of magnetic field distribution ultimately led to a design with a constant field gradient over a larger aperture. Figure 8.8 is a schematic drawing of the Cs state-selecting magnet.

In the quantum transition region between the state-selecting magnets the magnetic field must be as uniform as possible and very weak, but not zero. It must be weak to take advantage of the absence of the first-order magnetic field dependence of the (0-0) hyperfine transition frequency. When the difference in energy between the several hyperfine states (F, m_F) is expressed as a power series in the magnetic field B, the first term in the series for the $m_F = 0$ states is the quadratic term, which reduces to the form [7]:

$$\Delta E_{0-0} = \frac{(g_e\mu_B - g_n\mu_N)^2}{h\nu_0} B^2, \tag{8.28}$$

where as before μ_B and μ_N are respectively the Bohr and nuclear magnetons. Typically B may be on the order of 10^{-5} T (0.1 G). Since the energy levels inevitably have a finite width it is necessary to have the magnetic field strong enough to prevent overlap. The value of the hyperfine frequency is deduced by applying the magnetic field correction to the observed frequency. One way to measure the field correction

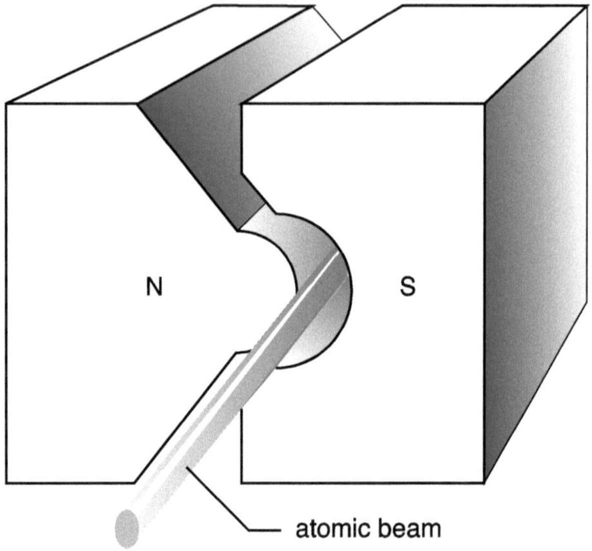

Fig. 8.8 Constant gradient magnet

would be to observe some field-dependent transitions, for example, $(F = 3, m_F = 3) \rightarrow (F = 4, m_F = 4)$. But to induce these $\Delta m_F = \pm 1$ transitions would require the magnetic component of the microwave field to be perpendicular rather than parallel to the static field. There are several transitions in which $\Delta F = \pm 1$, $\Delta m_F = 0$. The theoretical frequencies for all transitions as a function of external magnetic field strength are calculable using the Breit–Rabi formula [7], from which the magnetic correction can be deduced. Needless to say the magnetic field must be as free from environmental disturbances as possible; to achieve this, the transition region is magnetically shielded by multiple layers of high permeability alloys. In high-quality standards the magnetic field would be held at a constant value also through a servo feedback system.

To obtain a sharp resonance signal in the extended distribution of atoms in a beam, it is critical to find a way to reduce the Doppler effect. We recall the argument made by Dicke on the dependence of the Doppler effect on the number of wavelengths traversed by an atom during its interaction with the radiation field. In the case of resonance observed on an atomic beam, the problem is to have an interaction time as long as possible without the atom traversing many wavelengths, an apparently contradictory requirement. A brilliant solution was demonstrated in 1949 by Norman Ramsey [9]: it consists of using two separate, short, *phase-coherent* transition regions at the extreme ends of the transition region, which, for microwave fields, requires a special cavity design now called a *Ramsey cavity*. It is the microwave analog of an echo chamber consisting of a long section of rectangular section waveguide with 90° bends at its ends where apertures are provided for the Cs beam to pass through the standing microwave field pattern. The microwave power is fed symmetrically at the center of the waveguide. The polarization of the microwave magnetic field, that is, the direction of the magnetic field vector,

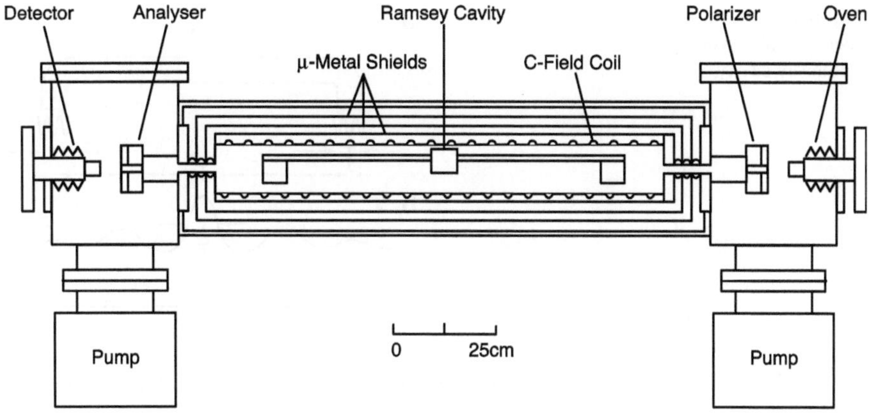

Fig. 8.9 Elements of magnetic deflection Cs beam frequency standard

must be parallel to the external uniform magnetic field in order to induce transitions between the $F = 4$, $m_F = 0$ and $F = 3$, $m_F = 0$ states. The Ramsey cavity is shown schematically in Fig. 8.9 as part of the classical magnetic deflection Cs beam frequency standard.

The source of the Cs beam is a constant temperature enclosure called the *oven* which is provided with a carefully designed *effuser*, that is, a multichannel opening typically a bundle of hypodermic needles that serves to make the Cs atoms emerge in a roughly collimated beam. The metallic Cs charge in the oven is electrically heated to the operating temperature of about 100°C, at which temperature the vapor density is such that few collisions between atoms occur within the effuser. The assumption is that atoms whose velocity in the oven is within a narrow cone parallel to the effuser will continue in a straight line through it as part of a nearly parallel beam. Of course things are never that simple: a great deal of attention has been paid to optimizing the rate of atoms issuing from the oven, and avoiding clogging of the channels. The need to be able to operate the standard over long periods of time without interruption has motivated the development of recirculating ovens. The atoms issuing from the oven pass through beam-shaping stops and enter the strong inhomogeneous A-magnet where I and S are decoupled and atoms with opposite directions of electron spin are deflected in opposite directions because of the steep transverse gradient in the field intensity. The atoms then pass through beam stops to remove the group of atoms in the $m_S = -1/2$ states, after which they return to their coupled (F, m_F) states as the magnetic field continuously drops till they enter the uniform C-field region and the Ramsey cavity. If the microwave frequency in the C-field is *off* resonance and no transitions are induced, the Cs beam will be deflected in the B-magnet in the same direction as the A-magnet, doubling its deviation from the axis of the apparatus and thereby missing the detector. On the other hand if the microwave frequency resonates with the desired (0-0) hyperfine transition, the atoms making the transition to the $F = 3$ hyperfine state will on entering the high field region of magnet B have I and S decoupled again, now occupying the $s = -1/2$ state, and will be deflected in

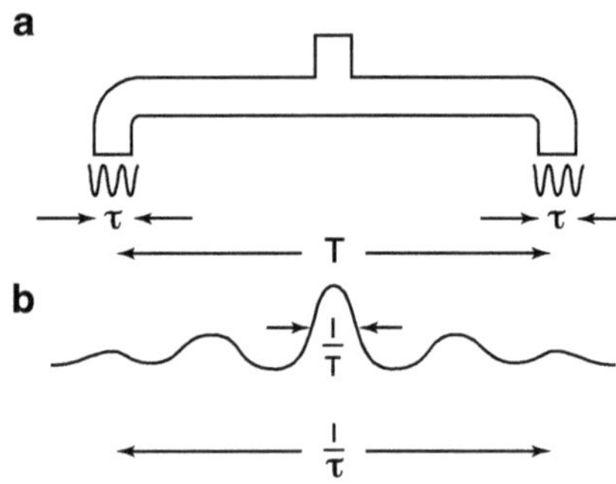

Fig. 8.10 Resonance spectrum observed by a Ramsey cavity

the opposite direction and onto the detector. This is called a "flop in" way of operating the experiment.

The availability of an efficient low noise detector of Cs atoms played an important part in the early success of the Cs beam machine as a frequency standard. The method of detection was based on the phenomenon of *surface ionization:* when cesium atoms come into contact with the surfaces of certain metals such as tungsten and niobium, surfaces heated to a high temperature to remove surface adsorbed layers of gas, they lose the outer electron to the metal. This is possible since the binding energy of electrons in these metals (called the *work function*) is larger than the binding energy of the outer electron in a Cs atom. The process is remarkable not only that it happens at all (the electron has to go through a potential barrier that is classically forbidden), but that it happens with such high probability. The main drawback is that other species of impurity ions are also emitted from the hot metal; hence a mass analyzer must be used.

The major factors that determine the spectral resolution of the Cs beam standard, that is, the precision with which the desired frequency can be determined, are the time of flight in the Ramsey cavity and the number of cesium atoms contributing to the signal. For atoms with a velocity distribution characteristic of thermal equilibrium at absolute temperature T passing through a Ramsey cavity of length L the resonance spectrum resulting from the finite observation time is computed to be as shown in Fig. 8.10.

A measure of the spectral resolution is the width of the central peak as given by the separation of the two minima on either side of the central maximum which Ramsey has shown [9] equals $\Delta\nu = 2\alpha/L$, where α is the mean thermal velocity along the axis. This will be recognized as an example of a general result, namely, $\Delta\nu_{1/2}\ \Delta T = 1$, where T is the observation time. It follows that for the sharpest resonance, the beam should be as long as possible. Of course there is a practical limit on how long the apparatus can be made, but more fundamentally, as the beam is made longer, the beam intensity will fall since the deflection magnets are non-focusing; this loss of intensity tends to offset the advantage of a narrower resonance. Moreover gravity will cause the path of the atoms to pursue a parabola, rather complicating the design. We will learn in a later chapter how gravity is exploited to make what has been called an *atomic fountain* standard.

The other fundamental factor that determines the accuracy with which the center of the resonance curve can be determined is the intensity and stability of the atomic beam, that is, the average number of Cs atoms that are counted per second by the detector and the stability of that number. The ultimate noise is due to the fact that we are dealing with discrete atoms which arrive randomly at the detector in an uncorrelated fashion. This type of noise arising from the discrete particle nature of the atoms is called *shot noise*. It is the type of electrical noise first observed by Schottky in the electron current of a diode tube operating at low cathode temperature. It can be shown [10] that if the *average rate* of atoms arriving per second at the detector is defined as $(dn/dt)_0$ then assuming a Poison distribution in the number of atoms arriving at the detector, the standard deviation from the mean, σ, a measure of the noise, is given by

$$\sigma^2 = \left(\frac{dn}{dt}\right)_0 \tau, \tag{8.29}$$

where τ is the length of time during which the atoms were counted. It follows that since the signal is proportional to $(dn/dt)_0$ to the first power, the ratio of signal to noise is improved by increasing the flow of atoms in the beam, provided, of course, that the source does not have some systematic instability. This shows the importance of a well-designed Cs oven with a clean effuser.

As a standard, the Cs beam system has been subjected to an extraordinary degree of scrutiny to determine any *systematic* errors in the observed frequency. Units built in different laboratories and different countries are routinely tested and compared, particularly the units maintained by the standards laboratories around the world. There are many fundamental sources of systematic error that have to be accounted for in addition to the magnetic field, for example, the second-order Doppler effect, the thermal radiation in the environment, and collisions between Cs atoms causing spin exchange shifts in frequency. In addition to all these there are always instrumental errors; an important one has to do with the phase of the microwave field in the Ramsey cavity, and another with the velocity distribution of the atoms in the beam. However it would be out of place to dwell on these here, particularly since the magnetic deflection standards have been superseded by the atomic fountain standards, which we will take up in a later chapter.

References

1. H.A. Lorentz, *The Theory of Electrons*, 2nd edn. (Dover, New York, NY, 1952)
2. G. Herzberg, *Atomic Spectra and Atomic Structure*, 2nd edn. (Dover, New York, NY, 2010)
3. M.S. Longair, *Theoretical Concepts in Physics* (Cambridge University Press, Cambridge, 1984), p. 32
4. Louis de Broglie, *Matter and Light*, Dover English edition 1939
5. R.P. Feynman, R.B. Leighton, M. Sands, *Lectures on Physics*, vol. III (Addison-Wesley, Reading, MA, 1965)
6. N.F. Ramsey, *Molecular Beams* (Oxford University Press, Oxford, 1990). new edn

7. N.F. Ramsey, loc cit. p. 86
8. R.H. Dicke, The Effect of Collisions Upon the Doppler Width. Phys. Rev. **89**, 472–73 (1953)
9. N.F. Ramsey, op. cit
10. H. Margenau, G.M. Murphy, *The Mathematics of Physics and Chemistry*, 2nd edn. (Young Press, 2009)

Atomic and Molecular Oscillators 9

9.1 The Ammonia Maser

Studies of radiation in the microwave region of the spectrum really became possible through what amounted to a crash program during World War II to develop the technology necessary to support the new invention: RADAR. Radiation laboratories sprang up at a number of universities in the USA, notably at the Massachusetts Institute of Technology, and elsewhere. Before the war, a microwave generator called a *magnetron* had been invented which later evolved into the *resonator cavity magnetron*, a powerful microwave oscillator used in RADAR during the war. Of course now magnetrons are in every home in the industrialized world; they power the microwave ovens. Just prior to the start of the war in Europe in 1939, two brothers by the name of Varian invented at Stanford University a microwave tube called a *klystron* which was *tunable* and produced a stable output of microwaves with wavelengths ranging from millimeters to several centimeters. Incidentally the same Varian brothers developed the sputter-ion pump that made the achievement of ultrahigh vacuum universally available on a large scale, for example, in space simulators at NASA, or large particle accelerators.

The availability of microwave sources and detectors opened up the vast field of spectroscopy in that region of the spectrum. This brought with it a drive to improve the sensitivity of detecting absorption lines in *molecular* spectra, which because of the rotational motion of atoms in a molecule display a wealth of lines in the microwave region of the spectrum. What was lacking was an efficient means of amplifying weak signals at microwave frequencies; vacuum tube amplifiers could reach only ultrahigh frequencies (UHF). The idea of using molecular transitions, and in particular s*timulated emission* to amplify microwave radiation was first described by Gordon, Zeiger, and Townes of Columbia University in 1954 and independently proposed by Basov and Prokhorov of the Lebedev Institute for Physics. The device is known by the acronym *MASER*: *m*icrowave *a*mplification by *s*timulated *e*mission of *r*adiation. The device was conceived as a high-resolution spectrometer or a stable microwave oscillator. It combines the techniques of

F.G. Major, *Quo Vadis: Evolution of Modern Navigation: The Rise of Quantum Techniques*, DOI 10.1007/978-1-4614-8672-5_9,

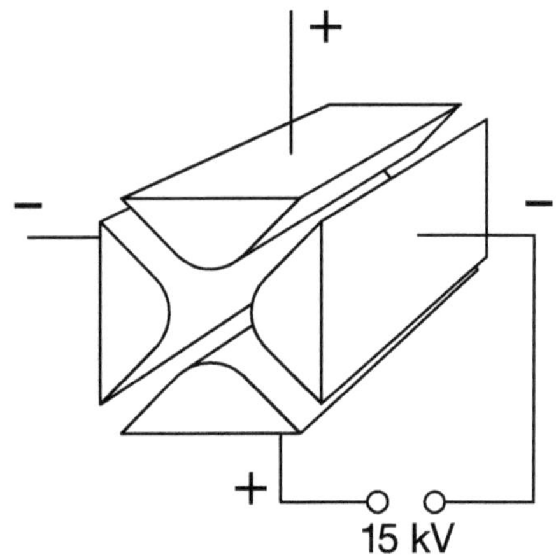

Fig. 9.1 Electric quadrupole NH_3 state selector

microwave absorption spectroscopy and molecular beam deflection techniques. The initial attempt to test the idea was on the spectrum of ammonia (NH_3), a molecule that, because of the strength and abundance of its lines in the microwave region, played an important role in the development of microwave spectroscopy. Even before the maser, an absorption line in the spectrum of ammonia was used by Lyons and his associates at what was then the U.S. National Bureau of Standards to stabilize the frequency of a quartz crystal oscillator. This was one of the earliest attempts to stabilize a clock using a quantum transition.

To describe the principle of the maser we can do no better than follow the original description by Townes [1]. Ammonia gas (boiling point $-33°C$) fills the source chamber at a reduced pressure of around 100 Pa, and issues from a number of small holes into a region of sufficiently high vacuum to form a beam. This requires high-speed pumps to keep the background pressure of gas sufficiently low; otherwise it would be meaningless to speak of a beam. Since the freezing point is a readily achievable $-78°C$, the use of cold "traps" can reduce the required speed of the vacuum pumps. The beam is made to pass through a quadrupole electric field which acts as a state selector analogous to the A-magnet in the cesium beam standard. In this case the *electric dipole* moments of the ammonia molecules are acted on by the electric field produced between parallel rods carrying the voltages shown in Fig. 9.1 for a simple quadrupole. The state selector is designed to focus molecules towards the axis if they are in the upper quantum state whose energy has a positive increasing value as the electric field increases, and to diverge away from the axis if they are in the lower state and their energy becomes more negative with increasing field. The state-selected beam consisting now of mostly upper state molecules enters a microwave resonant cavity tuned to the transition frequency between the upper and lower states. If a *weak* microwave signal is introduced into the cavity, the power level will rapidly increase if the frequency of the signal resonates with the transition frequency of the molecules since they are *stimulated* to emit microwave energy and return to a lower state. By sweeping the frequency of the signal it may resonate with more than one molecular transition, thereby producing a spectrum.

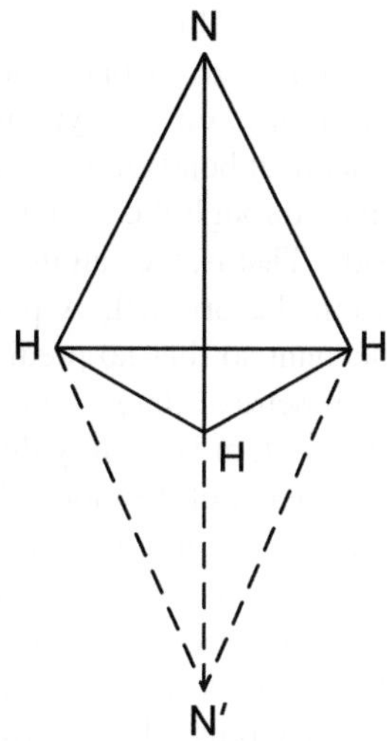

Fig. 9.2 The structure of ammonia

The focusing electrostatic field may be a multipole field produced by an even number of parallel cylinders arranged around the surface of a cylinder carrying voltages that are alternately positive and negative. The simplest field is formed by four conductors called a quadrupole field whose components may be written in the form:

$$E_x = -\frac{V_0}{{r_0}^2}x, \quad E_y = +\frac{V_0}{{r_0}^2}y \quad E_z = 0. \tag{9.1}$$

The electrostatic potential function of this field is given by

$$\phi = \frac{V_0}{2{r_0}^2}\left(x^2 - y^2\right). \tag{9.2}$$

Hence the surfaces of equal potential are sheets parallel to the z-axis with hyperbolic sections. In practice these surfaces may be approximated by cylinders with a circular cross section. The two opposing pairs of sheets have potentials of opposite sign. Since $E^2 = {E_x}^2 + {E_y}^2 = {E_0}^2\ (x^2 + y^2) = {E_0}^2\ r^2$ we see that the electric field is radial and cylindrically symmetric with zero on the axis. The focusing action of this field whether convergent or divergent depends, as we have already stated, on the way the energy of the quantum state of the molecule varies with the electric field. States whose energy increases with increasing intensity of the electric field would be deflected toward the axis where the field intensity is lower, whereas the opposite is true for ones whose energy decreases.

As already mentioned, the first experiments were conducted on the ammonia molecule, a molecule that exhibits a feature in its spectrum that is attributed to a mode of oscillation that has no classical analogue. To understand this we must first describe the molecular structure of ammonia, NH_3. The geometrical arrangement of atoms in ammonia is shown in Fig. 9.2.

The N–H chemical bonds lie along the edges of a triangular pyramid; the three H-atoms are in one plane at the vertices of an equilateral triangle, and the one N-atom is on the symmetry axis at the apex of the pyramid. If we imagine the chemical bonds to be like elastic bands, we could pull the nitrogen atom along the axis, through the center of the H-plane, to the mirror image position on the other side. That is, we can *invert* the molecule, and the result would be indistinguishable from the original. A proper quantum description must take that symmetry into account so that no prediction based on that theory will distinguish between these symmetrical states. It happens that the minimum energy of the system occurs when the N-atom is slightly displaced (in either direction) from the H-plane. This means the energy of the molecule as a function of the position of the N-atom along the axis has *two* minimum positions, symmetrically located on either side of the H-plane, and a local *maximum* between them at the center. This way of describing the dynamics of the molecule treats the motion of the electrons as if the motion of the nuclei is negligible, an approximation justified by the much higher velocities of the electrons. The electronic energy is then assumed to determine the slower motion of the N and H nuclei. Because of the symmetry, a solution in which the N-atom is at just one potential minimum cannot represent a stationary state, that is, it cannot remain in that state indefinitely. In fact in quantum theory the molecular wavefunctions are "symmetrized" combinations of wavefunctions representing the N-atom at both positions at once! We will not attempt to pursue this any further than to say we can picture the molecule as having the N-atom periodically oscillating between the two positions. The spectrum that results from this motion is called an *inversion* spectrum. In ammonia the inversion frequency is about 23.9 GHz.

In order to convert the ammonia maser into a self-sustaining oscillator all that is necessary is to ensure that the stimulated emission of microwave power in the cavity exceeds all the ohmic and other losses of power in the cavity. When that condition is satisfied the system will break into oscillation because the stimulated emission is in phase with any radiation already present. In summary, there are at least three important conditions that must be met for oscillation to occur: First the number of state-selected molecules in the upper state entering the cavity per second must sufficiently exceed the number in the lower state, second, the mode pattern of the field in the cavity must be as long as possible without phase reversal in order to lengthen the time of interaction with the beam, and third, the losses in the cavity must be low, that is, its Q must be sufficiently high. Quantitatively, the formula for the threshold condition for oscillation given by Townes is as follows:

$$N_2 - N_1 \geq \frac{3hV\Delta\nu}{16\pi^2 Q\mu^2}, \tag{9.3}$$

where N_2 and N_1 are the numbers of molecules in the upper and lower states, $\Delta\nu$ is the width of the molecular resonance, μ is the dipole matrix element for the molecular transition that determines the transition probability, and Q and V are the quality factor and volume of the cavity. To form an idea of the magnitudes of the physical parameters involved we cite some early results: With a cavity $Q = 12{,}000$ the minimum state selector voltage required to start oscillation was 11 kV with a source pressure of 800 Pa.

9.2 The Rubidium Maser

The initial efforts to construct a rubidium standard in the form of a maser oscillator (or simply maser) are associated with the names of Hartman at the Ecole Normale Supérieure in Paris and Davidovitts and Novick at Columbia University, who announced their success in 1966. Later Jacques Vanier, then at Laval University, did a thorough experimental study of the rubidium maser. The essential difference between the passive rubidium standard and the maser is that in the latter the bulb containing the Rb vapor is placed in a high-Q microwave cavity. The atoms are optically pumped into the upper hyperfine state, from which they are stimulated to emit radiation and return to the lower state. If the total rate of radiation emission by the atoms is greater than the losses in the cavity, oscillations at the hyperfine transition frequency will be maintained. This differs from the ammonia maser only in that the *population inversion*, as it is has been called, is not produced by electric fields acting on molecules in a beam but rather by optical pumping, as in the rubidium frequency standard.

There is, however, a fundamental difference between an ammonia maser and the rubidium maser in the nature of the quantum transitions involved: whereas the ammonia molecules emit and absorb radiation through an oscillating *electric* dipole, the rubidium transitions are much weaker *magnetic dipole* transitions. This raises considerably the threshold for oscillation; it requires optimal design of the high-Q cavity, the Rb absorption cell, and the optical pumping source. The high-Q cavity usually has the form of a right circular cylinder, operated in a mode designated as TE_{011} which has the electromagnetic field distribution shown in Fig. 9.3. We note that the microwave field has the same phase throughout the space around the center of the cavity where the bulb containing the Rb vapor is placed. The Q-factor of an isolated unloaded cavity is determined by the power loss on the inner surfaces of the cavity and to a lesser extent loss in the material of the bulb, chosen with that property in mind. Since microwaves do not penetrate deeply

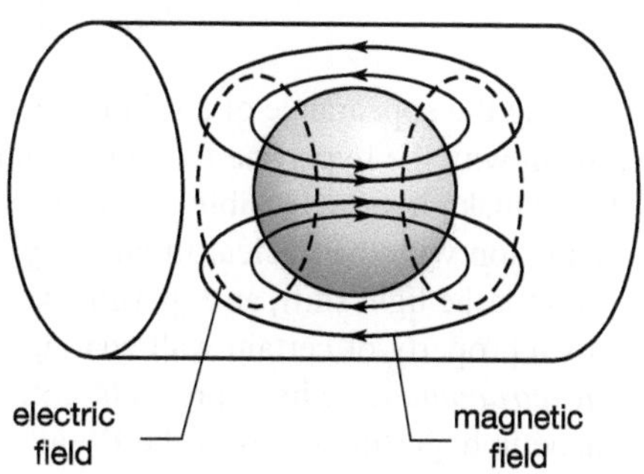

Fig. 9.3 The TE_{011} resonant mode of the electromagnetic field in a cylindrical cavity

into a conductor, the Q-factor can be increased by plating the interior surfaces with a good conductor, such as silver.

In the early experimental designs, oscillation was possible only when the optical pumping source was intense and the vapor density in the absorption cell relatively high. These two requirements aggravated two problems with optically pumped standards, namely, the light shift, and another process we have not previously mentioned, the *spin-exchange* shift in frequency. The spin-exchange collision process can be illustrated as follows: A(↑) + B (↓) → A(↓) + B(↑) where A and B are two atoms of Rb and the arrows indicate the two possible directions of the outer electron spin with respect to a given axis. This process can be shown to cause a shift in the hyperfine frequency by an amount that increases with the collision frequency between the atoms and therefore the density of the vapor. The process can be explained only in terms of quantum mechanics: In the collision between atoms A and B in which A and B form a transient "molecule," the part of the wavefunction that represents the spin state must be properly "symmetrized" as follows:

$$\uparrow\downarrow\rightarrow\frac{1}{2}(\uparrow\downarrow+\downarrow\uparrow)+\frac{1}{2}(\uparrow\downarrow-\downarrow\uparrow). \tag{9.4}$$

The first term on the right is clearly symmetric and the second antisymmetric with respect an exchange of the electrons. But the total electron wavefunction consisting of the product of the spatial and spin parts must be antisymmetric according to the Pauli Principle. Therefore the spatial factors which determine the scattering have opposite symmetry to the spin factors with respect to an exchange of coordinates. Thus during a collision between the two alkali atoms the scattering amplitudes of the two (symmetrized) spin states will be different, resulting in the appearance of atoms that have exchanged spins. To show this explicitly, suppose that after the collision the colliding atoms emerge in the following spin state:

$$\frac{a}{2}(\uparrow\downarrow+\downarrow\uparrow)+\frac{b}{2}(\uparrow\downarrow-\downarrow\uparrow). \tag{9.5}$$

Then if we rearrange the terms we get

$$\frac{a+b}{2}(\uparrow\downarrow)+\frac{a-b}{2}(\downarrow\uparrow) \tag{9.6}$$

showing the appearance of a spin flipped amplitude of $(a-b)/2$. So far this does not explain why the hyperfine frequency should be shifted by this process, although it should make it more feasible, since the collisions do affect the electron spins whose interaction with the nuclear moments gives the hyperfine structure.

As for the light shift, an experimental solution has been developed exploiting the known property of certain wall coatings, one of which is the long-chain hydrocarbon *tetracontane*, which prevents collisions with the walls from perturbing the magnetic hyperfine states of the Rb atoms This allows the use of *two* bulbs joined by a tube; the optical pumping is done in one bulb, and the atoms, now mostly in the upper hyperfine state, diffuse through the tube to the bulb inside the cavity. Unfortunately collisions of the Rb atoms with the wax-coated walls will still suffer a shift in hyperfine frequency, a shift that now very much depends on the condition of the coating and its long-term stability. However the strong signal-to-noise ratio that has been achieved and the narrow spectrum of the output give it exceptional short-term stability.

In a later chapter we will discuss more recent developments in Rb standards design making use of laser sources of light.

9.3 The Hydrogen Maser

Conceptually the H-maser was a natural outgrowth of the continuing drive to improve the spectral resolution of atomic beam resonance signals by increasing the interaction time between the atoms and the probing microwave field. The theoretical basis for this can be traced to the Heisenberg uncertainty principle that the uncertainty in the frequency of a transition is inversely proportional to the length of time it is observed. In the type of quantum transitions involved between hyperfine levels, called *magnetic dipole transitions*, the mean lifetime before spontaneous transitions occur is extremely long, calculated to be years. Therefore it is not the natural lifetime of the upper state but the finite length of time during which the measurement is made that determines the spectral width of the resonance. This contrasts with the *electric* dipole transitions in the optical part of the spectrum, where the radiative lifetime is so short that it usually determines the spectral line width. There are three ways in which the interaction time can conceivably be lengthened: First, reduce the velocity of the atoms; this unfortunately would reduce the atomic flux causing a loss in the signal-to-noise ratio. Moreover a reduction in velocity by a substantial factor requires hydrogen at an extremely low temperature, complicating the source design. Second, make the Ramsey cavity longer, but there is an obvious practical limit, not to mention the curvature of path of the atoms under gravity. Third, keep the atoms interacting with the resonant field by confining them within inert walls that simply deflect their motion without affecting their internal state. This was the approach pursued in Ramsey's laboratory at Harvard University. A number of different configurations and inner surface coatings of bulbs were tried, culminating in the discovery that a certain fluorocarbon polymer called *Teflon* allowed an extraordinary length of perturbation-free confinement time. In fact Teflon proved to have such a high degree of inertness with respect to hyperfine states of hydrogen that it is doubtful that anyone would have predicted it. The ability to store hydrogen atoms unperturbed in a finite volume for a considerable time with an inverted population between hyperfine states means that a maser oscillator analogous to the ammonia maser is possible, despite the much weaker magnetic coupling to the radiation field.

The nomenclature for the specification of the quantum states in the hydrogen atom was described in the last chapter. Hydrogen is of course the simplest atom

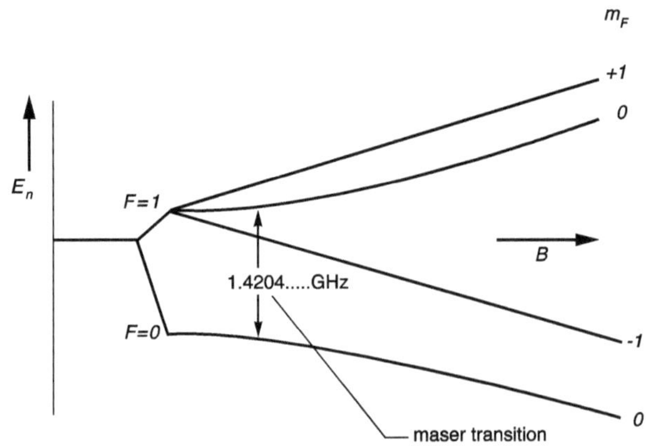

Fig. 9.4 The energy of states in H as a function of magnetic field

consisting of a single proton surrounded by an electron. It has another stable isotope with an additional neutron in the nucleus called deuterium, and a third radioactive isotope, tritium, but neither heavy isotope concerns us here. Hydrogen occurs naturally in the form of diatomic molecules and must therefore be dissociated into individual atoms in the maser source. Unlike the heavier elements, as a two-body problem the quantum theory of the hydrogen atom can be solved exactly, but as mentioned in the last chapter, the experimental results of Lamb and Retherford showed that the Dirac theory fails to explain what is now called the *Lamb shift* between the two $n = 2$ states with $j = ½$: one state with $l = 0$ and the other $l = 1$. This difference in frequency amounts to about 1.057 GHz and can mostly be explained by the successor to Dirac's theory: Quantum Electrodynamics (QED). The spectrum of hydrogen, including the hyperfine structure, has led to exhaustive theoretical calculations and serves to confirm the validity of theory to an extraordinary degree.

We have already treated the quantum description of hydrogenic systems, that is, ones with a single electron in a central field. The ground state of the electron is $^2S_{1/2}$ and the proton nucleus has $I = ½$ and therefore the coupling between the electron and proton magnetic moments requires a representation in terms of the vector F where $F = I + S$, so that the hyperfine structure states in the ground electronic state are labeled ($F = 1$, $m_F = +1, 0, -1$) and ($F = 0$ $m_F = 0$). The hyperfine energy difference between the $F = 1$ and $F = 0$ states is approximately 1,420 GHz with a wavelength of about 21 cm; radiation of that wavelength is familiar to radio astronomers. The plot of the energies of the hyperfine states as a function of an external magnetic field is much simpler than for Cs, but serves the same purpose; it is shown in Fig. 9.4.

As with Cs we note that the curves for ($F = 0$, $m_F = 0$) and ($F = 1$, $m_F = 0$) start horizontally from $B = 0$ and curve in opposite directions with increasing field intensity. Again this means that the lowest order variation of these two states with respect to the magnetic field is the second order, critically reducing the magnetic shift or broadening of the so-called (0–0) transition between those states. For very large magnetic field intensity the four curves split into two pairs of curves: one pair

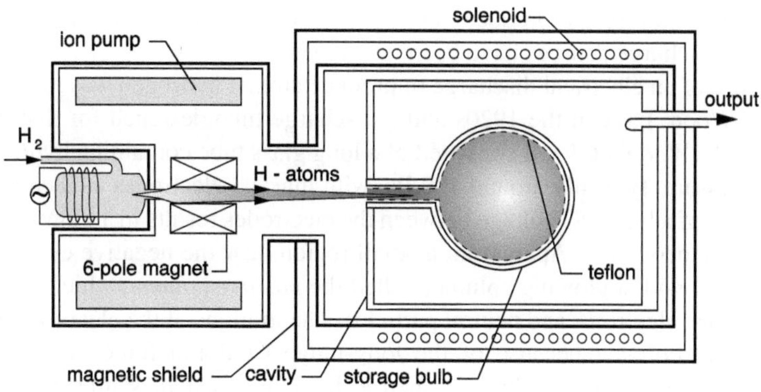

Fig. 9.5 The hydrogen maser

corresponding to $m_S = +½$ and the other $m_S = -½$. Again as with Cs, the curves with positive slope correspond to increasing energy with increasing field intensity, and those with negative slope correspond to decreasing energy. It follows that in a nonuniform magnetic field the atoms in the two pairs of states experience forces in opposite directions.

A description of the hydrogen maser may be divided into four essential parts: the source of atomic hydrogen, the quantum state selector, the hydrogen storage bulb, and the microwave cavity. The assembled system is shown schematically in Fig. 9.5.

As already mentioned hydrogen gas naturally occurs as diatomic molecules and must be efficiently dissociated into atoms. This requires each molecule be given at least 4.4 electron-volts of energy to break the bond between the two atoms. There are two practical ways in which this may be accomplished: either by collisions with energetic electrons or by sufficiently high-temperature collisions involving the molecules. The thermal dissociation method is simply the heating of the molecular gas to as high a temperature as the material of the container will allow. Tungsten has the highest melting point at about 3,370°C, but even at this temperature the average kinetic energy of a hydrogen molecule is only about 0.5 electron-volt. However because the kinetic energy is statistically distributed according to Maxwell's law, there will be a significant number of molecules dissociated. In the work of Lamb and Retherford on hydrogen, a degree of dissociation of 64 % was reported in a tungsten oven operating at 2,200°C. Its main drawback from the point of view of its application in a maser is the relatively high velocity with which the atoms emerge; this, as we shall see, reduces the solid angle of acceptance of the focusing state selector. The performance of the high-temperature source is predictable and stable but not efficient and is now only of historical interest in having been used in some classical work on atomic hydrogen. For a time during the early development of the maser, some research was done on developing this type of source when it seemed the electrical discharge source suffered from long-term aging, and sometimes inexplicably developed a condition called "the whites", indicating a failure to produce atomic hydrogen. Since then highly reliable electrical discharge sources have been developed.

The use of an electrical discharge to produce atomic hydrogen was first reported by R.W.Wood, back in the 1920s and a discharge tube designed for that purpose was called a *Wood's tube*. It consisted of a long glass tube containing hydrogen gas at low pressure, typically around 100 Pa, with internal electrodes at each end. The application of a high DC voltage between the electrodes results in a glow discharge, much like a neon sign. Apart from a small region near the negative electrode, the tube is filled with a glowing column, called the *positive column*, which consists of positive ions, electrons, and neutral particles. The motions of the electrons and ions are strongly coupled because of the long-range Coulomb force between them, causing them to act like a fluid and hence the name *plasma*. While the ions tend to have a temperature only slightly above the neutral gas, the electrons have a much higher kinetic energy and are able to ionize enough neutral gas particles to replenish the number neutralized on the walls and thus maintain the discharge. Electron collisions with neutral gas particles will mostly excite them to higher quantum levels from which they may make transitions to lower states by emitting optical radiation and hence the glow. Some electrons will have a high enough kinetic energy to cause dissociation of the molecule according to the process:

$$\mathrm{H_2 + e \rightarrow H + H + e} + h\nu. \tag{9.7}$$

Under conditions where a preponderance of atomic hydrogen is being produced, the glow has a clear pink color. Hydrogen atoms will readily recombine to form molecular hydrogen if they come into contact with a solid surface, particularly a metal. Two freely moving atoms coming into contact in space cannot form a stable molecule with negative energy; there must be a third body to satisfy the conservation laws. A single hydrogen atom hitting a metal surface on the other hand has a good chance of forming a molecule with an atom adsorbed on the metal surface.

The modern incarnation of the Woods tube is a more efficient and compact electrodeless high-frequency discharge in a pyrex glass bulb, typically about 2.5 cm in diameter. It is evacuated to high vacuum and then back filled with pure hydrogen through a palladium–silver "leak" consisting of a palladium–silver barrier that, when heated, allows only hydrogen to go through; it is either a freestanding thin-walled tube or supported by a porous substrate. It is closed at one end and sealed at the other to a glass tube leading to the source chamber. Around the leak is wound a heater wire to raise the temperature and thereby control the rate at which pure hydrogen enters the source chamber. A glass jacket is sealed around the leak and connected to a hydrogen source. The hydrogen in the source bulb is dissociated by an electrodeless discharge using the high-frequency field of a resonant coil driven by a 100–200 MHz power oscillator. The critical part of the source is the beam-forming collimator which may consist of one or more short capillary tubes. The atoms issuing from the collimator form an atomic beam in the high vacuum of the main body of the maser. The collimator does not have the problem of material adhering to the surfaces of the capillaries, as can happen with the Cs source, but

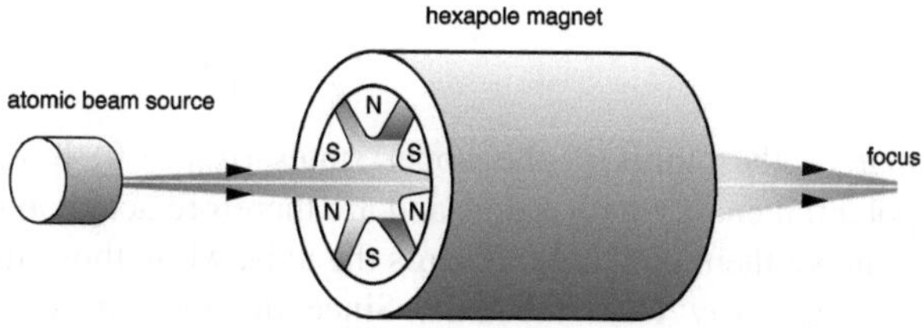

Fig. 9.6 The hexapole focusing magnetic field

there is the possibility of the atoms recombining to form molecules at those surfaces. The angular spread of the emergent atomic beam is considerably bigger than the acceptance angle of the focusing magnet that follows, limiting the utilization factor of the hydrogen to little more than 0.01 %. As a consequence, large capacity vacuum pumps are required to keep the background pressure in the main body low enough that scattering of the beam is minimized. Fortunately the background gas will be predominantly molecular hydrogen and will therefore not take part in the microwave emission–absorption processes in the cavity. It may however cause a small shift in the maser frequency. The requirement of fast differential pumping is met by using ion pumps since they are very efficient in pumping hydrogen, but unfortunately their powerful magnets add to the shielding problems.

The required population inversion between the populations of the upper ($F = 1$, $m_F = 0$) and lower ($F = 0$, $m_F = 0$) states is achieved using a hexapole focusing magnet, of the type originally proposed in 1951 by W. Paul [2] for applications in magnetic resonance studies. The field of this powerful magnet focuses towards the axis atoms in states that have a positive magnetic energy that increases in the radial direction and deflects away from the axis, the ones in states with a negative energy. The result is that more atoms in the upper ($F = 1$, $m_F = 0$) state enter the microwave cavity than the lower ($F = 0$, $m_F = 0$), as required for sustained oscillation. The hexapole field has a threefold axis of symmetry and, since there are no electric currents, is derivable from the (magnetic) harmonic potential function:

$$\Phi(r) = \Phi_0 \left(\frac{r}{r_0}\right)^3 \cos(3\theta). \tag{9.8}$$

The surfaces of constant potential that ideally determine the shape of the pole pieces are straight parallel cylinders with hyperbolic cross sections, as shown in Fig. 9.6. In practice the pole tips must be designed to concentrate the field.

From the potential function we obtain the components of the magnetic field:

$$B_r = 3\frac{\Phi_0}{r_0{}^3} r^2 \cos(3\theta); \quad B_\theta = 3\frac{\Phi_0}{r_0{}^3} r^2 \sin(3\theta). \tag{9.9}$$

It follows that the energy of the atoms in the limit of large B is given approximately by

$$E = 3\frac{\Phi_0}{{r_0}^3}\mu_{\text{eff}}r^2, \tag{9.10}$$

that is, the atoms in the upper states ($\mu_{\text{eff}} > 0$) have an increasing (magnetic) potential energy with radius and are therefore acted on by a restoring force tending to make them converge towards the axis, while those in the lower states ($\mu_{\text{eff}} < 0$) diverge away from the axis. Since the motion is simple harmonic inside the magnetic field, finding the trajectory of the converging atoms is straight forward. Clearly given the thermal energy of the atoms, only those having a transverse kinetic energy less than the maximum potential energy at the pole pieces of the magnet can emerge as a focused beam. The maximum acceptance solid angle of the magnet is given by [3]:

$$\frac{\Omega}{4\pi} = \frac{1}{4}\frac{{r_0}^2}{L^2 + \left(\frac{V}{\omega}\right)^2}, \tag{9.11}$$

where L is the distance from the source to the magnet, V is the mean thermal velocity of the atoms, r_0 is the radius of the bore of the magnet, and ω is defined as

$$\omega^2 = \left(\frac{2\mu_{\text{eff}}B_0}{M{r_0}^2}\right). \tag{9.12}$$

Early studies [3] of shaping the magnet, that is, varying the inner radius r_0 along the length of the magnet, in order to better utilize the magnetic deflection and increase the acceptance angle, have not generally been judged to be justified by the improvement realized in practice.

The atomic beam is directed at the small entrance aperture of the storage bulb inside the microwave cavity. In the space occupied by the cavity the external magnetic field must be very weak, in the range of 10^{-7} T (1 mG) and as uniform as possible. Since the atoms must traverse the space between the strong field of the state selector and the weak uniform field of the cavity, it is important that the magnetic field make the transition continuously without passing through zero or making abrupt changes. This applies to any Rabi-type magnetic resonance beam machines; it avoids what are called *Majorana* transitions that can occur when a spin passes through such conditions of the field. The microwave cavity is surrounded coaxially by layers of high permeability magnetic shielding material. These materials can develop “hard spots” as a result of spot welding and shaping and therefore must be annealed. The uniform field is produced by a solenoid with its axis coincident with the axis of the microwave cavity in order that the microwave field have the correct polarization to induce the (0–0) transition.

The material of the bulb is usually fused quartz, chosen because of its particularly low loss of microwave power through heat dissipation, which would lower the Q of the cavity. It is usually a spherical bulb on the order of 15 cm in diameter with a collimator similar to the one used in the source to admit the beam while reducing the escape of atoms from the bulb. As already mentioned the inner surface is coated with a polymer that Dupont designated by the trade name of Teflon. The process by which a uniform coherent coating is prepared is more art than science. The basic procedure consists of first thoroughly cleaning the inner surface, then wetting the surface with a liquid suspension of Teflon, and thoroughly drying it. Finally the bulb is baked at 360–380°C to fuse the particles of Teflon together into a coherent coating, while circulating through it air or oxygen to help oxidize and remove impurities. From this description of the preparation of the wall coating of the bulb, it is clear that the surface cannot be ultimately defined in physical terms and therefore the maser is unacceptable as an absolute standard.

The bulb in which the population-inverted atoms are stored is placed at the center of a cylindrical microwave cavity, whose TE_{011} stationary mode of oscillation resonates with the hydrogen frequency of 1.420 GHz. In order to avoid spectral broadening, it is important that the microwave field acting on the atoms in the cavity be of constant phase. A cylindrical cavity that approximately conforms with the shape of a spherical bulb has equal length and diameter, each about 27.6 cm. The magnetic field pattern of the axially symmetric TE_{011} mode was shown in Fig. 9.3. The Q-factor of an isolated unloaded cavity is determined by the loss of microwave power due to resistive heating at the inner surfaces of the cavity walls. A silver-plated cavity as described above would have a theoretical Q-factor of about 87,000. The presence of the quartz bulb may slightly reduce this number, but will significantly detune the cavity, lowering its resonant frequency. A hairpin loop mounted on one of the end plates of the cavity, with its plane perpendicular to the magnetic component of the microwave field at that point, serves to couple power to a 50 Ω cable.

The crucial practical question of course is: At what minimum rate must population-inverted hydrogen atoms enter the microwave cavity in order to have self-sustained oscillation; in other words what is the *threshold* value of the atomic flux? This threshold is determined by the requirement that the rate at which the atoms are induced to radiate by the microwave field already present in the cavity should be at least large enough to make up all the losses in the cavity. This latter quantity is specified by the Q-factor, which must therefore be high enough to achieve oscillation for a given influx of atoms into the cavity. Thus for a total electromagnetic energy U of the microwave field in the cavity, the total power dissipated P may be expressed in terms of Q, as follows:

$$P = \omega \frac{U}{Q}, \tag{9.13}$$

where ω is the (angular) frequency. The total energy U must be expressed in terms of the magnetic component of the microwave field in the cavity. We have in SI units:

$$U = \frac{1}{2\mu_0} \langle B^2 \rangle_c V_c, \tag{9.14}$$

where $< B^2 >_c$ is the mean square amplitude of the microwave magnetic field averaged over a cavity of volume V_c. The radiated power P at the hyperfine frequency from an atomic flux of I atoms/s in the upper state entering the bulb in which the mean coherence time is $1/\gamma$ leads to the following result :

$$P = \frac{1}{2} Ih\nu \frac{x^2}{\gamma^2 + x^2}, \tag{9.15}$$

where $x = 2\pi\mu_H B_z/h$. Since the magnetic field in the cavity is not uniform across the bulb, it is necessary to take the appropriate average over the volume of the bulb. It can be shown that we may write $<x^2> = (2\pi\mu_0)^2 <B_z>_B^2$; readers interested in the details are referred to the classic paper by Kleppner et al. [4]. In order to find the threshold condition we must relate the power emitted by the atoms to the rate of loss in the cavity; for this we need to find the relationship between the quantity that determines the radiation rate, $<B_z>_B^2$, and the quantity involved in the Q-factor of the cavity, $<B_z^2>_c$. The ratio of these two quantities is called the *filling factor*, η. Assuming this quantity can be computed for the given microwave mode of the cavity, the threshold condition can be stated in SI units as follows [4]:

$$I_{th} = \frac{hV_c\gamma^2}{2\pi\mu_0 {\mu_H}^2 Q\eta}. \tag{9.16}$$

So far we have assumed the phenomenological coherence time $1/\gamma$ to avoid a considerably more complicated picture involving so-called *relaxation* collisions between H atoms that affect the radiation process. Let us first recall that the microwave transitions are between two magnetic states and that the intrinsic probability of making a transition in one direction is the same as in the other. Hence a net emission requires more atoms in the upper state than the lower. In describing the motion of the magnetic moments of a collection of hydrogen atoms in response to an oscillating microwave field as we have in the maser bulb, it helps to think of the atoms as having a magnetic moment acted on by the magnetic component of the microwave field. There are two characteristic times used to describe the random processes that may affect their response to the microwave field: The first, called T_1, is the average time for the atoms as a whole to reach thermal equilibrium, in which the states are nearly equally populated. The other, called T_2, is the average time for the decay of *phase coherence* among the atoms in their response to the oscillating microwave field. Such coherence can be lost if for example the atoms move through an inhomogeneous magnetic field, causing the transition frequency of different atoms to be different. The inclusion of these relaxation processes in computing the transition probability leads to a modified result in which γ^2 is replaced by $(T_1T_2)^{-1}$, where

$$\begin{aligned} 1/T_1 &= 1/T_x + 1/T_b, \\ 1/T_2 &= 1/(2T_x) + 1/T_b, \end{aligned} \tag{9.17}$$

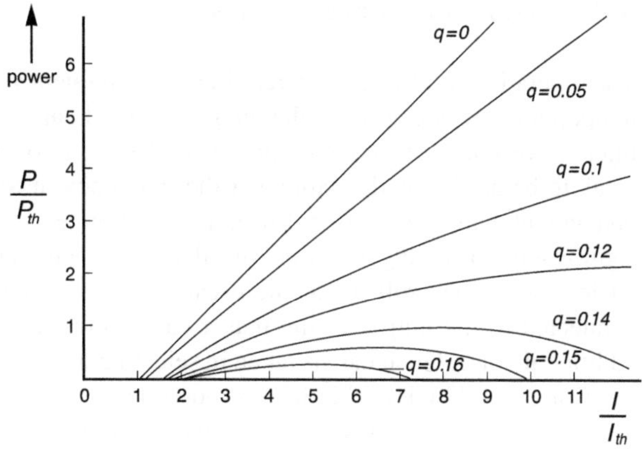

Fig. 9.7 H-maser output power as a function of atomic flux for different q [5]

where the subscripts x and b refer to spin exchange and escape from the bulb, respectively; the first causes a loss of coherence and the latter a loss of radiating atoms from the bulb.

We have already encountered the *spin-exchange* collision process in our discussion of the Rb maser. The same process occurs in collisions between H atoms with similar effects, namely, an interruption of the radiation process, which broadens the resonance spectrum as well as shifts the center frequency. The spectral broadening of the atomic transition from the upper to lower state reduces the probability of induced radiation, with the immediate effect of raising the threshold for the onset of oscillation. In fact when spin exchange and loss of atoms through the input aperture are taken into account the power radiated P is given by the following [5]:

$$\frac{2P/h\nu}{I_{\mathrm{th}}} = \frac{I}{I_{\mathrm{th}}} - \left[1 + 3q\frac{I}{I_{\mathrm{th}}} + 2q^2\left(\frac{I}{I_{\mathrm{th}}}\right)^2\right], \tag{9.18}$$

where the parameter q is given by

$$q = \frac{\sigma V_{\mathrm{r}} h V_{\mathrm{c}}}{2\mu_0 {\mu_{\mathrm{H}}}^2 \eta V_{\mathrm{b}} Q}\left(\frac{I_{\mathrm{tot}}}{I}\right). \tag{9.19}$$

The curious thing about this result is that unless $q < (3\text{-}2\sqrt{2})$ there is no real solution to the equation $P = 0$. Now q is proportional to the hydrogen beam flux, that is, the number of atoms that enter the cavity per second; thus we have the counterintuitive result that you can have too *many* atoms entering the cavity to produce oscillation! In Fig. 9.8 is reproduced a set of graphs showing the microwave output power as a function of the atomic flux for different values of the parameter q. We note that as q increases and approaches the limiting value of about 0.171, the range of flux over which oscillation occurs tends to zero (Fig. 9.7).

9.4 Associated Electronics

The operation of the maser requires low-frequency electronics to control the temperature of the cavity, the magnetic field, and the atomic source. At the microwave end, the maser output at 1.42 GHz is on the order of 10^{-13} W, too weak to be used directly. Moreover the frequency must be coherently divided to a convenient value such as 5 MHz; the challenge is to accomplish this without significantly affecting the free oscillation of the maser, a situation obviously analogous to the problem of the escapement in mechanical clocks. Whatever is connected to the output of the maser cavity must cause a minimum detuning or additional noise. Great care is taken in selecting a low-noise solid-state preamplifier operating at 1.42 GHz with the best possible *noise figure* to isolate the maser from subsequent circuits. An ultrastable quartz oscillator is used in conjunction with frequency synthesizer techniques involving phase lock loops to produce a phase-coherent signal at the desired standard frequency, typically 5 MHz. The detailed design of such electronics can of course vary widely with varying claims of performance.

We can illustrate the general principles by briefly describing the pioneering system designed by H. Peters in the late 1960s for the NASA tracking stations. The essential functional elements of the system are shown schematically in Fig. 9.8. Starting with a stable 5 MHz quartz oscillator, its output is multiplied by 280 to produce a 1,400 MHz signal which is mixed with the pre-amplified output of the maser to give a heterodyne frequency (IF) of about 20,405 kHz. This is amplified by a tuned IF amplifier, followed by two further heterodyne IF stages with frequencies at 405 and 5 kHz. The 5 kHz signal is amplified and connected to a phase detector (comparator) whose reference is derived from the 5 MHz quartz oscillator through a digital synthesizer. The frequency of the latter, which can be advanced in steps of 0.0001 Hz, determines the precise frequency to which the quartz oscillator will be stabilized; its actual setting is determined by the corrections for the magnetic field

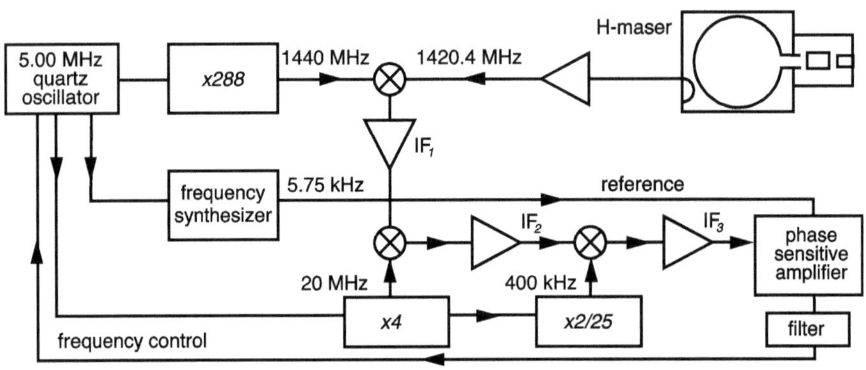

Fig. 9.8 A block diagram of the basic electronics of a hydrogen maser

and wall shift. The output of the phase comparator constitutes the error signal which, after passing through a filter to ensure stability of the servo loop, is connected to the frequency control element of the quartz oscillator.

9.5 Performance of the Hydrogen Maser

There are at least two important aspects to the performance of a time/frequency standard: absolute accuracy/reproducibility and stability. We begin with accuracy: we can safely assume that a hydrogen atom is identical to any other hydrogen atom no matter what its provenance. However the manner in which we observe its hyperfine frequency may affect the outcome. There are several major sources of uncertainty in the observed hyperfine frequency of hydrogen that must be accounted for in arriving at the standard frequency. These are, among others, the *wall shift*, the *spin-exchange shift*, the *frequency pulling* by the cavity, the *magnetic shift*, and to a lesser extent the *second-order Doppler shift*; we will consider these briefly in order.

The very property of Teflon that made the hydrogen maser possible, namely,s its ability to deflect the motion of hydrogen atoms without causing transitions between magnetic quantum states, unfortunately is accompanied by a residual distortion of the atoms during the collision that causes a small shift in the hyperfine frequency, called the *wall shift*. As we already saw in the case of Rb standards, a similar shift in frequency occurs as a result of collisions between the Rb and buffer gas atoms. In the hydrogen maser the atoms bounce around randomly within the coated spherical bulb until they happen to escape through the entrance aperture or associate with other hydrogen atoms on the wall to form molecules. To estimate the dependence of the wall shift on the size of the bulb we need to estimate the mean length of an atom's free flight between wall collisions. For a bulb of radius R that mean distance is $4R/3$, and thus the mean free time is $4R/3V_{\text{ave.}}$. When a hydrogen atom approaches the *Teflon* surface, one of several types of collisions can ensue: the most likely outcome is that the atom dwells on the surface for a finite time trapped in a potential well. The well is formed by an initial attraction as the atom nears the surface followed by a sharp repulsion as the atom overlaps the atomic structure of the wall. The length of time an atom remains trapped depends strongly on the depth of the well and the thermal energy of the atom, that is, the temperature. If the atom forms a chemical bond with other atoms or impurities on the surface it is lost from the radiating ensemble, thereby limiting both relaxation times T_1 and T_2 equally. If the atom is not trapped but goes through the strong electric field at the surface, the electron distribution around the nucleus is distorted, changing the electron density at the nucleus and hence the hyperfine interval, that is, the maser frequency. It is known that the frequency is reduced during that part of the atom's trajectory when it is approaching the wall with an attractive *van der Waals force* acting on it. On the other hand the subsequent repulsive field forcing the atom to rebound causes a rise in the frequency, with a net shift that depends on the details of the collision with the surface. The duration of the collision is extremely short compared with the period of the microwave field and strongly depends on the temperature and the amount of energy lost by the atom to the surface. The disturbance of the hydrogen state is a sharp impulse which may cause non-radiative hyperfine transitions, leading to a shortening of T_1, or more seriously, cause phase jumps which shorten the coherence time T_2, and, since they tend to be in one direction, a shift in frequency. To bring out the dependence of the wall shift on the diameter of the bulb we make use of the mean free time between wall collisions already mentioned, namely, $<T_c> = 4R/3V_{ave}$. Thus if we assume an average increment $<\Delta\varphi>$ in phase occurring per collision, the mean frequency would change by the amount $<\Delta\nu> = (1/2\pi)<\Delta\varphi> 3V_{ave}/4R$. Experimental measurements have verified the essential correctness of this result. The quantity $2R<\Delta\nu>$ has been found experimentally to be on the order of -0.4 Hz.cm. at 40°C for a particular Teflon (FEP 120 made by Dupont). The negative sign indicates of course that the frequency is shifted down; for a bulb having a diameter of 15 cm this corresponds to a fractional shift of -1.88×10^{-11}. This of course is a very small shift, but not in the context of modern frequency standards.

The shift in the observed hyperfine frequency due to *spin exchange* was introduced in the discussion of the Rb maser. There it was suggested that the exchange of spin directions by colliding atoms was fundamentally explained in quantum terms as arising from the need to describe the collision in terms of wavefunctions properly symmetrized with respect to an exchange of the two electrons. During the collision, the two atoms may be thought of as forming a transient H_2 molecule, which, if the colliding atoms have electron spins in opposite directions, must be regarded quantum mechanically as a superposition of properly symmetrized wavefunctions. We recall that a system that has the spin function ↑↓ must be treated as a sum of a "singlet" spin function ½(↑↓ − ↓↑) and a member of a "triplet" ½(↑↓ + ↓↑). Because of Pauli's principle the former spin function must be associated with a symmetric space function and the latter with an antisymmetric function. These symmetrized functions determine not only the scattered amplitudes, and therefore the exchange of spin directions, but also the relative phase of the scattered wavefunctions, resulting in decoherence and a shift in the mean hyperfine frequency. Credit for identifying the spin-exchange frequency shift is given to S.B. Crampton [6] and P. Bender. The theoretical result for the frequency shift is expressed in terms of the phase shift Φ between the two transient molecular hydrogen wavefunction, designated as $^1\Sigma$ and $^3\Sigma$. The approximate frequency shift due to spin exchange may be written as follows [7]:

$$\frac{\Delta\omega_{se}}{\omega_0} = \frac{1}{8Q_l} f(\Phi)\lambda, \tag{9.20}$$

where $Q_l = \omega_0/2\gamma$ is the quality factor of the hyperfine resonance, and λ is defined (in SI units) as follows:

$$\lambda = \frac{V_c}{V_b} \frac{h}{\mu_0 \mu_H{}^2} \frac{V_{HH}\sigma_{HH}}{\eta} \frac{1}{Q_l}. \tag{9.21}$$

The frequency of oscillation of the maser is also "pulled" appreciably by the coupling of the atomic transitions to the resonant cavity, since the latter is required to have a high Q in order to achieve oscillation. It is a common phenomenon in any two coupled oscillating systems. The approximate result for the amount of cavity pulling of the frequency is as follows:

$$\frac{\Delta\omega_{cp}}{\omega_0} = \frac{(\omega_c - \omega_0)}{\omega_0} \frac{Q_c}{Q_l}. \tag{9.22}$$

Because of these shifts in frequency an important part of the design of the maser is an *automatic cavity tuner*. The first attempt at such a tuner takes advantage of a significant feature of the above results, namely that under the assumed approximations, both shifts are inversely proportional to the Q_l of the atomic resonance, that is, its frequency width. This width can be controlled by varying the atomic flux entering the bulb, since this affects the collision rate of the atoms. Moreover the cavity pulling can be in either direction, positive or negative; therefore it must be possible to vary the atomic flux while tuning the cavity and arrive at a point where a change in the atomic flux does not change the frequency of the maser; at that point the spin-exchange shift exactly cancels the cavity pulling. In practice modulating just a specified experimental parameter such as the atomic flux requires care to avoid affecting other parameters at the same time. For example, simply changing the excitation of the source or the pressure of hydrogen may also affect the velocity distribution of the atoms. Probably the "cleanest" way to modulate the beam would be to use some type of mechanical shutter. The main practical difficulty in implementing this method is the need for an auxiliary frequency standard of sufficient stability to monitor the change in the maser frequency. In the early days of development, a second hydrogen maser was used as a frequency reference. However, a number of effective electronic modulation methods have been developed over the years relying on ultrastable quartz frequency standards. An early electronic approach was reported by H. Peters [8] in which special pulse/logic circuitry was used to handle the error signal obtained with a crystal standard in a way that did not degrade the stability of the maser. Another approach, also developed by Peters, makes use of the fact that the atomic coherence time T_2 is longer than the corresponding time for the cavity, and therefore it was argued that the tuning of the cavity could be modulated relatively fast while the atomic system maintains its oscillation, causing the amplitude of oscillation of the cavity to be modulated at the same frequency if the cavity is mistuned. This would constitute an error signal for stabilizing the cavity using a servo feedback loop. Another approach described by C. Audoin [9] involves injecting into the cavity a frequency-modulated signal with the frequency switching at equal intervals (square wave modulation) and the carrier frequency suppressed by setting the frequency of modulation an even submultiple of the modulation depth; this again relies on the fact that the atomic coherence time T_2 of the cavity is shorter than that of the radiating atoms.

We now consider the sensitivity of the maser frequency to the strength of the ambient magnetic field. As in other standards, the quantum states between which the hydrogen maser transitions occur were chosen because there is no first-order (linear) dependence of their frequency on an external magnetic field. However there is a second-order shift requiring that the field intensity be measured accurately; this is usually accomplished using microwave transitions involving the states with $m_F = \pm 1$, whose frequency depends linearly on the field. The displacement of the frequency of the maser due to a magnetic field B is obtained from the Breit–Rabi solution for the energy levels of a system involving an electron spin ½ coupled to a nuclear spin I in a magnetic field. In our case of hydrogen, the difference in energy ΔE between $(F = 1, mF = 0)$ and $(F = 0, mF = 0)$ varies with the magnetic field as follows:

$$\Delta E = \Delta E_0 \sqrt{1 + x^2}, \tag{9.23}$$

where x is defined as $x = \frac{\left(-\mu_e,/S,+\mu_{\mathrm{p}},/,I\right)B}{h\nu_0}$. Substituting numerical values

$$\Delta\nu_0 = 2.75 \times 10^{11} B^2 - 2.68 \times 10^{13} B^4, \tag{9.24}$$

where the magnetic field intensity B is in teslas. For a typical field of 10^{-7} T the shift would be about two parts in 10^{12}. If the field is uniform to within 1 % over the bulb the consequent fractional field broadening would be on the order 10^{-14}.

Finally we consider the Doppler effect on the frequency by virtue of the motion of the atoms in the bulb. First we note that the first-order (linear) Doppler effect should be negligible since there is no *average* movement of the atoms, and moreover their motion is confined to a space less than the microwave wavelength, so that the Dicke line shape theory should apply (see Chap. 8). However the second-order Doppler effect is another matter. It is given by

$$\frac{\Delta\omega_0}{\omega_0} = \frac{3}{2}\frac{kT}{mc^2}, \tag{9.25}$$

where m is the mass of the hydrogen atom and T the absolute temperature.

Ultimately the accuracy of the hydrogen maser as a standard must be judged by its ability to deliver *independently* a constant frequency without calibration against any other device. This presumably will happen only if that frequency is the actual frequency of transition between hyperfine states in the hydrogen atom. The largest contributor to the uncertainty of the maser output frequency is the wall shift, which amounts to a fractional error of about 5×10^{-12}. To keep things in perspective, however, the hyperfine frequency of hydrogen is one of the most precisely known physical constants in all of physics; in terms of the international

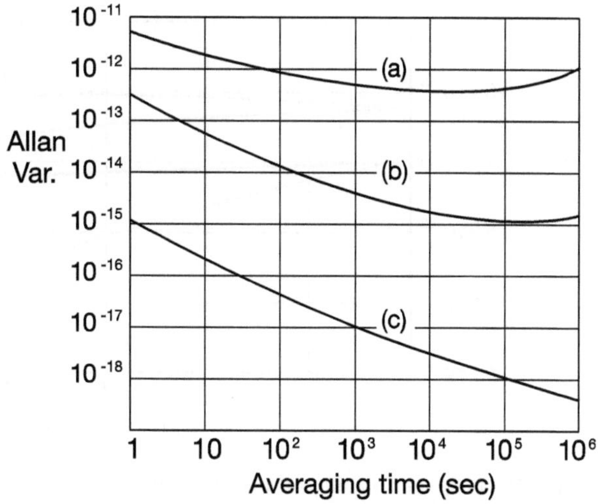

Fig. 9.9 Allan variance plotted against sampling time for (**a**) Rb passive standard, (**b**) H-maser at room temperature, and (**c**) theoretical thermal noise-limited H-maser at liq. He temperature

atomic timescale as defined by the *Bureau Internationale de l'Heure*, it has the value: $\nu_H = 1{,}420{,}405{,}751.778 \pm 0.003$ Hz.

Finally is the question of frequency stability, both medium term and long term; in this context medium term is taken to mean up to 3 months or so while long term is measured in years. We recall that in our discussion of the stability of the quartz standard it was pointed out that the accepted way of specifying the stability of an oscillator is by the Allan variance, defined as follows:

$$\sigma_y^2(\tau) = \S\left\langle \left(y_{i+1} - y_i\right)^2 \right\rangle, \tag{9.26}$$

where the average < > is taken over a large set of relative frequency fluctuations y_i measured at equal intervals τ. As pointed out in the discussion alluded to, different sources of statistical fluctuation, that is, *noise*, can be identified through the functional dependence of the Allan variance on the sampling time τ. By drawing a log–log plot of σ versus τ, power law dependence of σ on τ is quickly apparent and the physical process giving rise to it identified. A comparative plot of the Allan variance in frequency for an H-maser at room temperature, a passive Rb standard, and a cryogenic H-maser is shown in Fig. 9.9

We see that the fractional mean variance in frequency of the typical H-maser can reach as low as around 10^{-15} for a sampling period of about 1 h, while a maser with a cavity operated at the temperature of liquid helium has potentially a variance 10^2 smaller. We learn from the graphs that for times shorter than about 20 min, the graph has a slope of about $-1/2$, which indicates "white" thermal *frequency* noise, that is, the Fourier spectrum of the fluctuations in frequency is independent of frequency. On the other hand for sampling time longer than about 3 h the graph is consistent with a gradient of $+1/2$, which would indicate *random walk* in frequency.

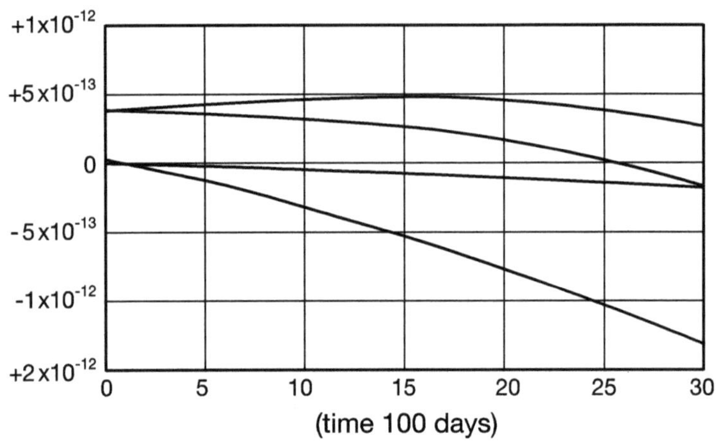

Fig. 9.10 Long-term stability of four NIST H-masers measured against a primary Cs atomic fountain standard

Recently [10] the Time and Frequency group at the National Institute of Standards and Technology (NIST) carried out a series of intermediate term frequency stability tests on six hydrogen masers using their Cs fountain standard (to be discussed in the next chapter) as reference. In Fig. 9.10 are reproduced the results plotted for four masers over a period of over 8 years. It is interesting that the results were actually plotted versus *mean Julian days* (MJD) as a measure of the time presumably to establish the period within a broader sweep of time in order to track how the masers might behave decades later. The results show remarkably smooth variation of the maser frequencies over the long term, some drifting more steeply than others.

References

1. H. Friedburg, W. Paul, Naturwissenschaften **38**, 159 (1951)
2. R.L. Christensen, D.R. Hamilton, Rev. Sci. Instrum. **30**, 356 (1959)
3. A. Lemonick, F.M. Pipkin, D.R. Hamilton, Rev. Sci. Instrum. **26**, 1112 (1955)
4. D. Kleppner, H.M. Goldenberg, N.F. Ramsey, Phys. Rev. **126**, 603 (1962)
5. D. Kleppner et al., Phys. Re. **138**, A972 (1965)
6. S.B. Crampton, Phys Re. **15**, 57 (1967)
7. N.F. Ramsey, Metrologia **1**, 7 (1965)
8. H.E. Peters et al, in *Proceedings of the 23rd Annual Frequency Control Symposium*, Atlantic City, 1969
9. C. Audoin, Rev. Phys Appl. (Paris) **16**, 125 (1981)
10. P.F. Kuhnle, 11th Annual PTTI, Nov 1979
11. T.E. Parker, S.R. Jefferts, T.P. Heavener, Frequency Control Symposium, IEEE International, New Port Beach, CA, June 2010

Field Confinement of Ions

10

10.1 Introduction

As already emphasized, the spectral line width of *magnetic* dipole transitions in atomic systems is determined in practice not by the natural lifetime of the upper state, but the available perturbation-free time for the observation. This has led to the development of devices that depend on diffusion through an inert gas in the case of the Rb standard and confinement in a bulb with inert wall coating in the case of the hydrogen maser. The short-term stability achieved in the hydrogen maser has been extraordinary and still surpasses any other device in the short term, but it has two important drawbacks: first and most serious is the shift in frequency due to collisions with the walls of the containment bulb, a shift which disqualifies it as an absolute standard, and second, its lack of portability due to the size of the cavity and the need for elaborate magnetic shielding. The experimental approach to overcome these limitations has evolved along two distinct directions: first the extension of the atomic beam observation time using a *vertical* beam, the *atomic fountain* standard, and second, the use of electrically charged atoms, that is ions, trapped by electromagnetic fields. We will treat ion confinement in fields in this chapter and deal with the atomic fountain technique after we have reviewed the subject of lasers in the next chapter.

In this chapter we will limit the discussion to the pre-laser confinement of atomic particles using relatively low-frequency (sub-optical) electromagnetic fields on charged particles at ordinary temperatures. Later we shall learn of the use of laser radiation to slow down the thermal motion of *neutral* atoms to the microkelvin range of temperatures and the formation of what is commonly called *optical molasses*. At normal temperatures, it is impractical to use electromagnetic fields to deflect the motion of neutral atoms sufficiently to confine them; however an electrically charged ion, formed when the parent atom loses an outer electron, can

F.G. Major, *Quo Vadis: Evolution of Modern Navigation: The Rise of Quantum Techniques*, DOI 10.1007/978-1-4614-8672-5_10,

readily be deflected by moderate electric and/or magnetic fields. This is radically different from the interaction of atoms with a wall coating which is also basically electrical in nature; the fields used to confine ions can be controlled and measured accurately. The question is how to achieve simultaneous confinement in all three dimensions. Here we are up against a theorem in electrostatics called *Earnshaw's theorem*, which states: a charged body placed in an electrostatic field cannot be maintained in stable equilibrium under the influence of electric forces alone [1]. This is a consequence of a fundamental property of the equations governing an electrostatic field in charge-free space: the electrostatic potential energy of a test charge cannot be a minimum at an isolated point in space. This means there is no point in the field from which a displacement in any direction would cause an increase in potential energy. At most we can have the potential increase along one direction, but decrease along another; thus the shape of a surface of equal potential in space can never be like a bowl, but like a saddle. This would seem pretty discouraging from the point of view of trapping an ion, but we are not limited to *static* electric fields, we simply have to look to either time-dependent fields, or a mixture of electric and magnetic fields.

10.2 The Penning Trap

We begin with a combination of an electrostatic field and a homogeneous magnetic field which has come to be called the *Penning trap* because of its resemblance to a vacuum gauge which uses a similar combination of fields. The gauge was first described by Penning in 1936 as one that extends the range of operation to much higher vacuum than the gauges then in use. These *ionization vacuum gauges* operated on the principle that the number of ions produced by a sufficiently energetic stream of electrons passing through a gas is proportional to the pressure (density) of that gas. The Penning design extended the range by using a strong magnet in conjunction with the high voltage electrodes to maintain a discharge even beyond the "blackout" point, which, as the name suggests, refers to the degree of vacuum when the discharge through a gas no longer emits a glow. The action of the magnetic field is to cause the electrons in the discharge to move in tight spirals, thereby increasing their path before reaching the walls of the tube, and hence increasing the probability of ionizing further atoms to maintain the discharge.

The pure quadrupole electric field geometry and the motion of charged particles subject to it were first described by J.R. Pierce of the Bell Telephone Laboratories in 1954 [2]. It has since been commonly used in various miniature forms in conjunction with a uniform axial magnetic field for electron and ion confinement; as such it is commonly called a *Penning trap*. It is illustrated in Fig. 10.1.

For the confinement of positive ions around the center, the hourglass-shaped cylinder must carry a negative electrical potential and the two end caps equal positive potential with respect to the center. The opposite would be true, of course,

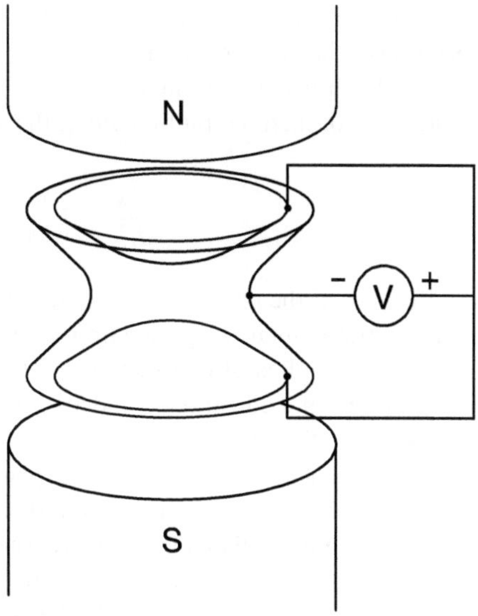

Fig. 10.1 The Penning trap: an electric quadrupole with a uniform axial magnetic field

for the confinement of negative ions. In terms of circular cylindrical coordinates (r, z) the electrostatic potential field V has the form:

$$V = \frac{V_0}{2{r_0}^2}\left(2z^2 - r^2\right). \tag{10.1}$$

The surfaces of equal potential having the values $\pm V_0/2$ are given by

$$\pm{r_0}^2 = 2z^2 - r^2, \tag{10.2}$$

which represent two hyperboloids of revolution about the z-axis, one corresponding to $+V_0/2$, and the other $-V_0/2$. The electrodes are conducting sheets coinciding with these equipotential surfaces. Note that the electrode potentials are chosen for the field to be symmetric about the origin, where the potential is zero. We note that along the z-axis the potential varies as z^2 increasing in either direction from the origin, and therefore a charged particle with energy $E < qV_0/2$ placed there would execute simple harmonic motion. Thus if the particle has a charge e and mass M the equation of motion for the z-coordinate is simply as follows:

$$\frac{d^2z}{dt^2} = -\omega_z^2 z; \quad \omega_z^2 = 2\left(\frac{e}{M}\right)\frac{V_0}{r_0^2}. \tag{10.3}$$

The presence of the axial magnetic field, to the extent that it is uniform and parallel to the z-axis, does not affect the particle motion in the axial direction, which remains as described.

However, the motion parallel to the x–y plane is considerably more complicated because of the crossed electric and magnetic fields, and is more simply described in terms of Cartesian coordinates x and y. To begin, let us assume that the magnetic field $B_z = 0$; then the motion along the x-axis, for example, would be

$$\frac{d^2x}{dt^2} = +\left(\frac{e}{M}\right)\frac{V_0}{r_0^2}x(B_z = 0). \tag{10.4}$$

Because of the plus sign on the right-hand side of the equation, the solution is an exponentially increasing function of time with a positive exponent; the same conclusion applies of course to the y-coordinate. Thus without the magnetic field the motion is not constrained and a particle in such a field would accelerate towards the cylinder. The effect of an axial magnetic field on the radial motion of an ion is to cause its path to curve around in a plane perpendicular to the field in a kind of cycloidal path. We recall that a cycloid is the path traced out by a point on a circle as it rolls around another circle, a favorite mathematical topic of Ptolemy and his epicycles; in this case the center of one circular motion describes another circle about the magnetic field axis. If the magnetic field intensity is large enough to dominate over the electrostatic field, the motion takes the aspect of a rapid rotation about a center which rotates more slowly around the field axis. The rapid rotation frequency is given by the following:

$$\nu_C = \frac{1}{2\pi}\frac{eB_z}{M}. \tag{10.5}$$

This is recognized as the *cyclotron frequency*, the frequency with which a charged particle goes around a uniform magnetic field. The name is derived from the cyclotron particle accelerator, the invention of Ernest Lawrence, in which the independence of this frequency with respect to the particle energy (at least at nonrelativistic energies) makes the accelerator possible. The drift velocity of the center of the cyclotron motion around the axis of the system can be shown to correspond to the velocity at which the Lorentz force on the particle is balanced by the electrostatic force. We recall that a charged particle moving through a magnetic field experiences a force in a direction perpendicular to both the magnetic field and the particle velocity; this is the force named for that prominent nineteenth century physicist. Since the radial electric field in the quadrupole is proportional to the radial distance of the particle, a balancing Lorentz force requires the particle velocity to increase at the same rate; the result is a constant *angular* velocity around the axis, called the *magnetron frequency*.

The magnetron, as mentioned in the last chapter, is a high power microwave generator used widely in radar transmitters and microwave ovens. Unlike the Penning trap the magnetron has a cylindrical electron-emitting cathode along the axis surrounded by a coaxial copper cylinder forming the anode. A radial electric field between the cathode and anode cylinder is produced by applying a positive potential to the anode. A series of equally spaced microwave cavities are machined out of the inner surface of the anode. A strong *alnico* permanent magnet provides an axial magnetic field which causes electrons emitted by the cathode to curve around and across the cavities in the anode, inducing oscillation in them at a microwave frequency. This is analogous to blowing across the end of a glass tube to make it whistle. Note that the magnetron motion does not depend on the size of the mass or charge of the particle, but only on the electric and magnetic field intensities.

The general solution to the radial equation of motion for the Penning trap yields the following two characteristic frequencies:

$$\nu^{\pm} = \frac{\nu_{\mathrm{C}}}{2} \pm \sqrt{\left(\frac{\nu_{\mathrm{C}}}{2}\right)^2 - \nu_{\mathrm{E}}^2}, \tag{10.6}$$

where ν_{E} is a function of the electrostatic potential, and ν_{C} a function of the magnetic field, given by

$$\nu_{\mathrm{E}}^2 = \frac{eV_0}{4\pi^2 M r_0^2} \nu_{\mathrm{C}} = \frac{eB}{2\pi M}. \tag{10.7}$$

If, as often is the case, $\nu_{\mathrm{C}} \gg \nu_{\mathrm{E}}$, the two frequencies may be written simply as

$$\nu^{+} = \nu_{\mathrm{C}} - \frac{\nu_{\mathrm{E}}^2}{\nu_{\mathrm{C}}} \quad \nu^{-} = \frac{\nu_{\mathrm{E}}^2}{\nu_{\mathrm{C}}}. \tag{10.8}$$

We note that the sum of the two frequencies is exactly the cyclotron frequency, a fact that has been used in highly precise g-factor measurements of multiply ionized atoms.

The reader may be excused for wondering at this point what all this has to do with navigation. The fact is that there are already applications of single ion storage techniques to fields as diverse as quantum computing; even now micro-gyroscope using a *single* ion trapped in a miniature ion trap and optically pumped by a suitable laser should not be many years in the future.

10.3 The Paul Trap

The Penning trap however is not suitable for the containment of reference ions whose magnetic hyperfine structure is used as a frequency standard because their spectrum would be strongly affected by the magnetic field. However for the hyperfine magnetic transitions contemplated here, an electric field would have a secondary effect. There is an ion trap that does not depend on any magnetic field: it gets around Earnshaw's theorem by using a *time-dependent* electric field. It is the Paul trap, named for Wolfgang Paul, who with his coworkers at Bonn, published in 1955 a description of the precursor of the ion trap, namely, the radiofrequency ion beam *mass filter*. It consisted of four parallel metal rods between which a radiofrequency electric field was applied. Paul et al. demonstrated that if the

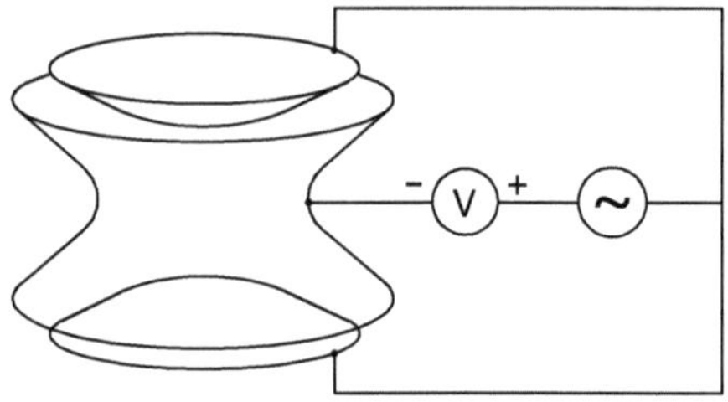

Fig. 10.2 The original Paul rf quadrupole ion trap

amplitude and frequency of the electric field were suitably chosen, ions of a certain mass would be brought to a focus, while others would diverge away from the beam and are lost. It was soon realized that this radial converging action of the radiofrequency field could be generalized to all three dimensions, that is, to confine selected ions in three dimensions; Paul gave it the name *ionenkäfig* which means ion cage [3]. The electrode geometry is identical to the Penning trap with the one essential difference: instead of applying a DC voltage between the electrodes with the polarity appropriate to the sign of the ion, the polarity is switched at radiofrequency, so the sign of the ion is now irrelevant. The behavior of the ion motion under such an alternating force, switching from convergent to divergent, critically depends on the spatial dependence of the amplitude of oscillation of the electric field. If an ion were acted on by an oscillatory field that was constant in amplitude from point to point, the ion motion would simply be an oscillation superimposed on its unchanged initial motion. In the case of the quadrupole field however, the field amplitude *increases* linearly both in the radial and axial directions. In this case although the ion will oscillate, it may under the right conditions do so asymmetrically with a tendency to return towards the center of the trap. This is an example of what is called the *strong focusing principle.* A simpler optical example of the application of this principle is the average *focusing* property of a sequence of alternating convergent and divergent lenses of equal power. Figure 10.2 illustrates the original electrode geometry used by Paul and his coworkers at the University of Bonn.

The ion motion in the Paul trap is amenable to exact analysis; let the electrical potential applied between the cylinder and end caps have the form:

$$V = \frac{(U_0 - V_0 \cos \Omega t)}{r_0^2}\left(z^2 - \frac{r^2}{2}\right), \tag{10.9}$$

then the equations of motion have the following form:

$$M\frac{d^2z}{dt^2} = -2\frac{e(U_0 - V_0 \cos \Omega t)}{r_0^2}z \quad M\frac{d^2r}{dt^2} = \frac{e(U_0 - V_0 \cos \Omega t)}{r_0^2}r. \tag{10.10}$$

These equations are of a form known as Mathieu equations after a French mathematician who studied them in connection with the problem of the vibrations of an elliptical membrane. In fact these equations crop up in many applications of

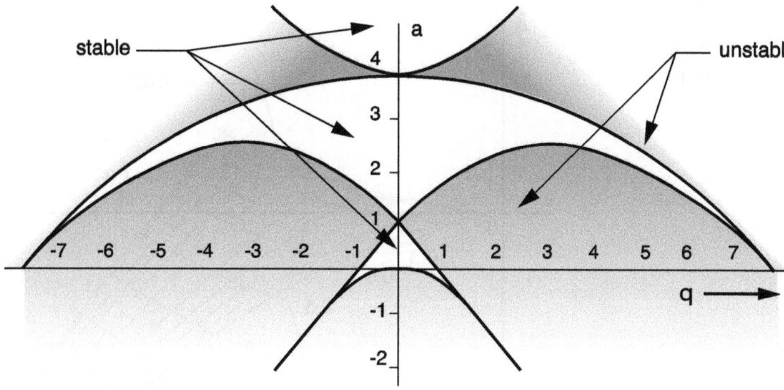

Fig. 10.3 The a–q stability diagram of the Mathieu equation

physical interest such as parametric oscillations. They are usually written in the standard form: where

$$\frac{d^2u}{d\theta^2} + (a - 2q\cos(2\theta))u = 0, \tag{10.11}$$

$$u = r, z \quad \theta = \frac{\Omega t}{2} \quad a_z = \frac{-8eU_0}{M\Omega^2 r_0^2} \quad q_z = \frac{4eV_0}{M\Omega^2 r_0^2},$$

with $a_r = -a_z/2$, $q_r = q_z/2$.

The most important property of the solutions of the Mathieu equation that concerns us is that they can be stable or unstable according to the values of a and q. That is, in our case the coordinates of the particle in the trap may be bounded and *oscillatory* or unbounded depending on the values of the coefficients a and q. Of course for particle confinement in three dimensions we require that the solutions have an upper bound simultaneously for both the r and z coordinates. To determine the conditions on the physical parameters of voltage amplitude V_0 and frequency Ω required to have stable solutions, it is useful to plot the boundaries separating regions of the a–q plane where the solutions are stable from those that are unstable, as shown in Fig. 10.3.

In order for a particular choice of voltage and frequency to lead to stable confinement in three dimensions, clearly the a and q values for both the radial and axial motion must simultaneously be in stable regions of the a–q plane. To aid in finding the experimental parameters, Paul et al. made a composite plot of U_0 against V_0 showing the stability boundaries for both r and z on the same set of axes, allowing for the difference in scale and sign of the a and q parameters for the two coordinates. Nearly all Paul traps have been designed to operate in the first stability region with the smallest values of a and q; the combined plot for that region is shown in Fig. 10.4.

As a concrete example suppose we wish to confine mercury ions (mass number 199) in a trap with $r_0 = 1$ cm operating at a frequency of 500 kHz, then a possible

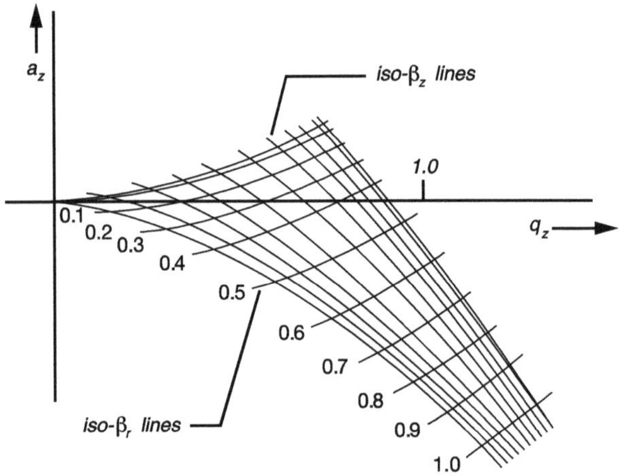

Fig. 10.4 Paul plot for simultaneous stability of motion in the r and z directions

choice of the operating point on the a–q stability diagram would be $a_z = 0.01$ and $q_z = 0.6$. Using these values, the corresponding values of U_0 and V_0 can be calculated.

On the basis of the general solution of the Mathieu differential equation for a and q in a stable region, the ion motion has a spectrum of discrete oscillation frequencies with amplitudes that are functions of a and q; the frequencies are as follows:

$$\omega_n = \left(\pm n + \frac{\beta}{2}\right)\Omega, \tag{10.12}$$

where n is an integer and the constant β is a function of a and q. If $a, q \ll 1$ it can be shown that

$$r(t) = A\left(1 + \frac{q_r}{2}\cos\Omega t\right)\cos\frac{\beta_r\Omega}{2}t \quad \beta^2 = \left(a + \frac{q^2}{2}\right), \tag{10.13}$$

and the motion can be regarded as an oscillation at the low *secular frequency* $\beta\Omega/2$ (corresponding to $n = 0$) with a small amplitude high-frequency jitter at the trapping frequency Ω. The dependence of the secular frequency on the DC voltage provides a convenient way to sweep the ion frequency for the purposes of detection, as we shall see.

There is an essential difference between the nature of the secular motion and the oscillation at the field frequency. In situations where collisions occur that tend to thermalize the motion, only the secular motion is affected, while the motion at the field frequency is driven by the field and cannot be thermalized. It is interesting to note that the combined kinetic energy of the high-frequency and secular motions is constant in the limit of $q \ll 1$; the energy merely swings periodically between the slow secular and high-frequency jitter motions. As the jitter amplitude increases with increasing distance from the center, the secular motion slows down. For this reason the motion is called *adiabatic* since there is no energy exchange with the outside surroundings. A more familiar example of such motions is that of a charged particle moving parallel to a magnetic field of gradually increasing intensity in which the particle will slow down as its cyclotron motion energy increases; this is the basis of the device called a *magnetic bottle*.

The main practical questions that must be addressed in designing a Paul ion trap are the initial formation of an ion population in the trap, the detection of the trapped ions, the maximum number of ions that can be accommodated, and last, the length of time of entrapment.

The buildup of an ion population in the trap is usually accomplished using an electron beam to ionize the desired atoms in the trap, although in one published exception, ultraviolet light was used to photo-dissociate iodine molecules into ion pairs [4]. Unlike the static electric field of the Penning trap in which the ion energy is a function only of its position in the electric field (the magnetic field makes no contribution to the energy), the high-frequency electric field of the Paul trap makes the ion entrapment, dependent on the phase of the field at the time of its creation in the trap or its injection from outside the trap. However it is possible to predict the maximum amplitude of the ion motion in terms of its initial position and velocity at a given phase of the field. Paul et al. have shown that for a given phase of the field at the time of the ion creation, a plot of the values of velocity and position corresponding to a given upper limit on the amplitude of the motion yields an ellipse. The parameters of this ellipse depend on the phase of the field oscillation at the time of the ion creation. Since ions produced by electron collisions have relatively small velocity, the computation of the fraction of ions produced by an electron beam that is confined within the boundary of the electrodes is relatively straightforward; it is represented merely by the area of a narrow strip along the x-axis.

The problem of quantitative detection of the trapped ions has received a great deal of attention in the context of the Paul trap as a mass spectrometer with emphasis on accuracy in determining the *number* of ions contributing to the signal. The simplest method of detection of the ions is the one originally used by Paul et al., namely, resonant excitation of the ion motion at the secular frequency by a weak high-frequency potential applied between the two end caps of the trap. This is achieved by connecting across the end caps a high-Q coil, forming a resonant parallel LC circuit tuned within the range of the low secular frequency ω_0; such a coil presents negligible impedance at both the high frequency Ω, and zero frequency (DC). The onset of excitation of the ion motion, as ω_0 is swept through resonance, causes a lowering of the Q-value of the coil and the voltage amplitude across it due to the absorption of energy by the ions. The same principle is used in nuclear magnetic resonance detection. We recall that the secular frequency is given by $\omega_0 = \beta\Omega/2$, where β is a function of the parameters a and q; therefore to aid in tuning the high-Q coil it is convenient to plot on the composite a–q graph, previously described, the iso-β lines, that is, lines of equal β, as shown in Fig. 10.4.

The signal-to-noise ratio achievable in detecting the trapped ions is ultimately determined by the presence of two fundamental types of noise: shot noise and thermal, or Johnson noise. In order that the shot noise exceed the Johnson noise at the detector input, we require $2e < i > \Delta\nu > (4kT/R)\,\Delta\nu$, that is, $R < i> > 2kT/e$. This shows the importance of having the detector circuit resistance as high as possible at resonance; this requires high inductance of the high-Q coil and low inter-winding capacitance. It is a remarkable experimental accomplishment that a coworker in Paul's laboratory was able to optimize the detection system to the point where he could see the discrete nature of the ions producing the signals by analyzing the Poisson distribution of the ion signal heights. This requires that the detection system be able to discern an increment in the ion population of *one ion*. Today, with the use of lasers to observe and cool ions, it is no longer unusual to confine and observe a *single ion* for extended periods of time; more will be said about that after the chapter on lasers.

The ion storage feature of the Penning and Paul traps coupled with their ability to display the mass spectrum of the trapped ions has seen these devices undergo intensive development as mass spectrometers. Later refinements in ion detection methods have been made aimed at increased sharpness of resonance and hence mass resolution; these include different methods of ion counting, achieved by the resonant excitation of the ion motion to the neighborhood of the limit of confinement followed by rapid extraction of the ions from the trap, accelerating them, and then counting them with a particle counter.

In any ion trap design, the maximum number of ions that can be contained is determined by the space charge field of the ions and thus by the depth of the potential well created by the high-frequency fields. But that number is attained by filling that potential well to the maximum, leading to a spread of kinetic energy among the ions and an increased Doppler broadening in their motional spectrum; thus there is a trade-off between signal-to-noise ratio and frequency deviation and broadening of the ion motional spectrum. Early in the development of ion containment an alternative approach was considered to improve the trapping time and reduce the spread in ion energy, namely to use higher multipole fields. Consider for example a *linear* multipole field produced by 2 N straight parallel metal strips along the z-axis, equally spaced around a circle in the x–y plane at angles $n\pi/N$, where $n = 1,2,3 \ldots..2\,N$. Suppose that an electric potential $\pm\, V_0 \cos(\Omega t)$ is applied to alternate strips, then it can be shown that the potential boundary value problem is solved using a harmonic function expansion to yield:

$$\Phi = V_0 \left(\frac{r}{r_0}\right)^N \cos\left(N\phi\right) \cos\left(\Omega t\right). \tag{10.14}$$

The simplest corresponds to $N = 2$ with four rods giving harmonic confinement in the radial direction. To achieve confinement along the axis, two approaches have been commonly used: in one, a pair of opposite rods is segmented and carries only a DC potential difference between the segments to produce a restoring DC field to

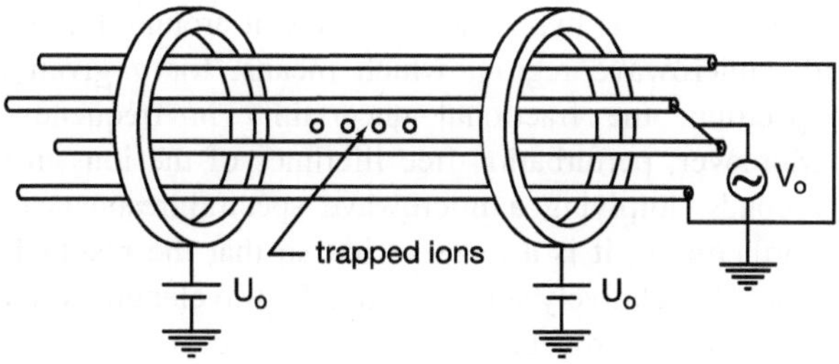

Fig. 10.5 Linear NIST radiofrequency ion trap

trap ions, and in the other, ring electrodes are mounted coaxially with the rods and carry a DC potential. The latter is exemplified by the design shown schematically in Fig. 10.5 which has been used by the NIST group [5] to do groundbreaking work on the coherence of quantum states involving individual ions.

10.4 Spectroscopy on Trapped Ions

The first proposal to use ions confined in isolation in a Paul trap as the basis for a new portable atomic frequency standard suitable for space applications was made by the author in 1969 at the NASA Goddard Space Flight Center and achieved with the collaboration of G. Werth in 1972 [6]. The new concept broke away from the practice of using magnets and atomic beams to observe the hyperfine frequency in neutral atoms, and proposed using atomic ions in a Paul trap interacting with an optical beam. It was seen as an approach capable of extraordinary frequency stability and accuracy in a light compact device suitable for mobile applications. This was well before laser technology had developed to the point of eventually opening up new technical frontiers allowing the observation of *individual* trapped ions. The application of the Paul trap in the field of radiofrequency spectroscopy had previously been demonstrated on helium ions [7] and molecular hydrogen ions [8]. These experiments were motivated by purely scientific interest and are mentioned here only as the first successful applications of ion traps in the field of spectroscopy rather than mass spectrometry for which they were originally intended. The helium ion experiment made use of spin-dependent collision processes between the ion and spin-polarized cesium atoms, while the molecular hydrogen ion was subjected to ionizing polarized ultraviolet radiation. The earliest experiments on the *optical pumping* of trapped ions were attempted by the author on strontium ions at Yale University in 1965 using the second resonance line in the rubidium spectrum as pumping source. This proved to be impractical at the time since it required the development of a new optical pumping source, but it became ultimately an eminently fruitful approach combining the techniques of optical pumping with ion confinement in electromagnetic traps.

The original arguments [9] for basing a spacecraft clock on the ground state hyperfine frequency of the mercury 199 ion have since been amply vindicated. The ion has one electron with $S = ½$ outside a closed shell structure, and a nuclear spin $I = ½$. Its hyperfine frequency at around 40.5 GHz is near the upper end of the microwave region, which means, for a given lifetime broadening of the spectrum, the fractional uncertainty in frequency is as small as possible. Moreover, perturbation-free lifetimes of the ions in the trap are typically many seconds, implying a microwave spectral resonance line-Q approaching 10^{11} ! Furthermore, it is a massive ion so that the residual second-order Doppler shift would be relatively small. Lastly, the wavelength is small at 7.4 mm compared with 21 cm for hydrogen, and therefore the dimensions of the microwave components are correspondingly small, an important consideration for a portable standard. But the property that was crucial before the advent of suitable compact laser systems is that among the mercury isotopes there are two stable ones: mass 199 and mass 202 such that the hyperfine spectral line emitted by one isotope can be used to pump the states of the other, as is done in the rubidium standard. The transition frequency between the two hyperfine states that was to serve as the standard was roughly deduced from optical data on the isotopic shifts in the optical spectral lines of mercury ions. After a prolonged search at NASA–Goddard, a microwave resonance at around 40.5 GHz was observed that had a linewidth that was clearly due to the microwave source rather than the ion resonance itself. It was not until a decade later that significant progress on the Hg^+ standard was published, particularly by the group at the Jet Propulsion Laboratory, drawing on the work at Goddard.

Since the mercury ion was to play a prominent role in the further development of portable ion frequency standards, we will outline here its atomic properties. The mercury atom, atomic number 80, has many isotopes, eight of which are stable; the mass 199 isotope has 16.9 % natural abundance and a nuclear spin $I = ½$ with a nuclear magnetic moment of +0.506 μ_n. The singly charged ion has 79 electrons around the nucleus with only a single electron in the outermost $n = 6$ shell in a 6 s state with zero orbital angular momentum, similar to the alkali atom, cesium. The first resonance wavelength of the ion is at 194.2 nm in the ultraviolet region of the spectrum. Since the ground state of the ion is an s-state, the electron will have a *contact* hyperfine interaction with the nuclear magnetic moment, just like the neutral cesium atom. The result is that in low magnetic fields the electron and nuclear moments are coupled and must be described in terms of the total angular moment F which may take on the values of $F = (I + ½)$ or $F = (I - ½)$. The fact that the nuclear spin is only ½ and yet has a strong hyperfine interaction energy with the outer electron is an added advantage of the chosen mercury isotope, since it leads to a simple quantum level structure, with fewer competing states in the attempt to place the ions in the one upper state of the desired transition. The graphs of the energies of the different $(F.m_F)$ states as functions of an external magnetic field B are shown in Fig. 10.6; they are similar to those of the hydrogen atom, except of course, for the much larger hyperfine splitting.

As with the other microwave standards, the magnetic field-independent $(F = 1, m_F = 0) \rightarrow (F = 0, m_F = 0)$ transition is used as the frequency reference, and we must therefore create a difference in the populations of these two states. As already mentioned the circumstance that made possible the achievement of the population difference in the hyperfine states by optical pumping, before the advent of lasers,

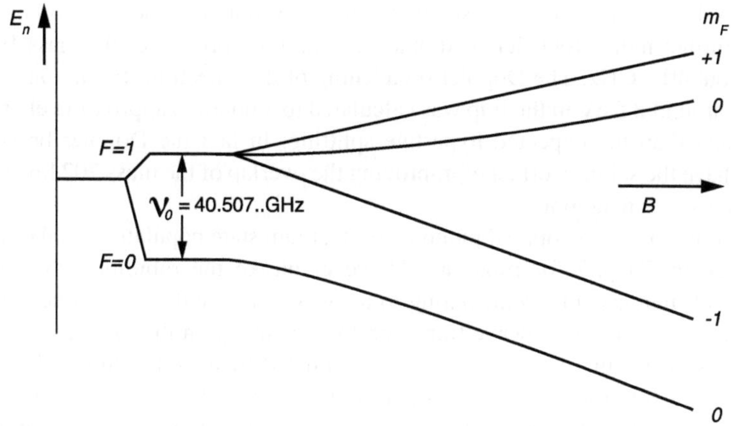

Fig. 10.6 The hyperfine state energies of the ground state of Hg^{199} as a function of magnetic field

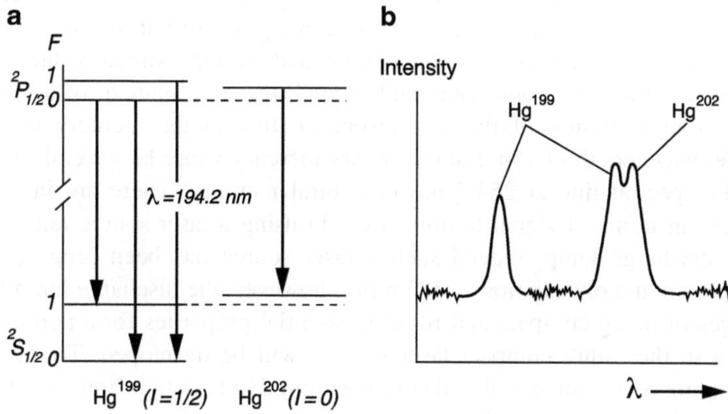

Fig. 10.7 (a)The ground quantum states of $^{199}Hg^+$ and $^{202}Hg^+$ and (b) their observed spectrum

was the discovery of the fortuitous proximity of the wavelength of the isotope 202 first resonance line with only one of the two hyperfine components in the isotope 199 absorption spectrum. The pumping radiation is thus supplied by a lamp filled with enriched mercury 202 which has $I = 0$, and therefore no hyperfine structure. The principle of the optical pumping cycle is identical to the rubidium case. The relevant quantum states of the ions of mass 199 and 202 isotopes of mercury are shown in Fig. 10.7.

The figure also shows the hyperfine components of the resonance emission line at 194.2 nm from an isotope-enriched mercury lamp using a high-resolution spectrograph; note the coincidence of the mass 202 line with just one of the mass 199 lines. This coincidental near overlap is explained by what spectroscopists call an *isptope shift* caused by the difference in nuclear structure of the two isotopes. The initial evidence for this is from the high-resolution spectroscopic work of

Mrozowski [10] on the isotopic shifts in mercury. From the same spectroscopic data it was further indirectly calculated that the hyperfine splitting of the mass 199 line was about 40.5 GHz. The Doppler broadening of that spectrum for an ion of mean energy around 2.5 eV in the trap was calculated to amount to approximately 6 GHz, much less than the expected hyperfine splitting. In fact the Doppler broadening would have the salutary effect of improving the overlap of the mass 202 line with its close mass 199 neighbor.

The principle of the optical pumping of quantum state populations is the same as described in Chap. 7 for rubidium. However unlike the rubidium standard, the number of mercury ions contributing to a signal is so small that it is necessary to observe the desired microwave transition by the change in the *fluorescence* of the ions. This means that the ions must be illuminated by a well-collimated pumping beam and the fluorescent light reemitted by the ions at *right angles* to the beam must be collected with wide-angle optics and focused on the cathode of a photomultiplier tube. Since the radiation is in the ultraviolet region of the spectrum, we are able to use a "solar blind" photomultiplier with a very low dark current for the detection of the fluorescence, a critical advantage in choosing the mercury ion. The practical realization of such a standard hinged on overcoming two serious obstacles: the first was ignorance of the hyperfine frequency beyond an approximate value deduced from ultraviolet spectroscopic data, and second, the development of a lamp that favored strong emission of the first resonance line of the mercury ion at the ultraviolet wavelength of 194.2 nm; ordinary mercury vapor lamps emit predominantly the spectral line at 253.7 nm of neutral mercury. There are inestimable advantages in terms of signal-to-noise ratio in using a laser source rather than a classical discharge lamp; indeed such a laser source has been proposed using the third harmonic of radiation at 582.6 nm; however, the discharge lamp has the advantages of being compact and rugged, essential properties for a portable unit. No doubt in the future compact laser sources will be developed. To gain some understanding of the numerical scale of the required light output from the lamp and the rate at which pumping radiation is absorbed and reradiated, we can assume a useful result concerning the resonance scattering of radiation by an atomic system. Given a beam of photons of resonant frequency having flux density j_ν photons crossing unit area per second, scattered by ions having a natural line width of $\Delta\nu_n$ and resonance linewidth $\Delta\nu_i$ in the trap, then the following approximate equation holds:

$$P = \frac{\lambda^2}{4}\left(\frac{\Delta\nu_n}{\Delta\nu_i}\right) j_\nu, \tag{10.15}$$

where P is the probability of photon absorption per second per ion. From the form of this equation we see that $\lambda^2/4$ is the cross section that the atom (or ion) presents to a beam of resonant radiation. In our case the numerical value of this cross section is about 9×10^{-11} cm^2, which it should be noted is more than five orders of magnitude larger than the typical atomic scattering cross section.

Nevertheless, on the basis of some reasonable assumptions about the line widths it is clear that the standard will need an extraordinarily bright conventional light source and special optics to minimize the stray scattering of light from different parts of the apparatus. Unfortunately as already mentioned, the usual high-frequency discharge lamp has the deficiency of emitting predominantly a line in the spectrum of the neutral atom at $\lambda = 253.7$ nm, the radiation responsible for the fluorescence in what will be the lamps for domestic lighting of the future. In order to enhance the emission further into the ultraviolet region it is necessary to raise the mean energy of the electrons in the discharge. One approach is to strike a high-frequency discharge in mercury vapor in vacuo without any buffer gas, which would normally be used to slow down the electrons and increase their diffusion time to the walls. This high-frequency vacuum arc mode of operation does indeed produce enhanced emission at the desired wavelength, but unfortunately is difficult to start, and shortens the life of the lamp by the discoloration of the quartz bulb. Therefore in practice some buffer gas such as helium is used with intensified effort to discriminate against undesirable radiation by filtering and using a photomultiplier detector that is insensitive to it.

The fundamental drawback to the classical discharge lamp, in sharp contrast to laser sources, is that its intrinsic *brightness* is limited, that is, the light output is from a relatively extended surface, and any optical system that focuses the light to a smaller area inevitably increases the angular divergence of the beam. This is a consequence of a general result called in this context the *brightness theorem.* It limits our ability to concentrate the output of the lamp on the ions and thereby limit the stray light that could reach the detector and swamp the weak fluorescence from the ions. In the usual configuration of pumping beam and detector placement, the beam passes through apertures in the cylindrical electrode along a diameter ending in a *Wood's horn,* a horn-like tube, tapered and curved, whose inner surface is coated with an absorbing material where the beam is totally absorbed. Careful attention must be paid to the edges of the apertures in the cylinder on a microscopic scale to avoid scattering of the pumping beam from points near the ions. The resonant light reradiated by the ions, as well as the stray background light, is of sufficiently low intensity that a photomultiplier is used to count the photons. The "solar-blind" photomultiplier, typically a 2 in. tube with a Cs–Te semitransparent end-on cathode, is mounted outside the vacuum envelope along the trap axis, perpendicular to the pumping beam. The ability to use a solar-blind photomultiplier is an enormous experimental advantage since it is not only insensitive to the radiation from the electron source and discriminates against the 253.7 nm radiation from the lamp, but even more critically, it has an extremely low dark current with its attendant noise. To further block the 253.7 nm light as well as other visible light from reaching the photomultiplier cathode narrow band interference filters are used centered on 194.2 nm. The general layout of the original NASA mercury 199 ion experiment is shown in Fig. 10.8.

In this design the inside surfaces of the solid quadrupole electrodes were machined to conform to the hyperbolic geometry with $r_0 = 1.13$ cm and $r_0/z_0 = \sqrt{2}$. Thin sheets of molybdenum spun into the proper hyperbolic shape would have been a better choice from the point of view of achieving ultrahigh

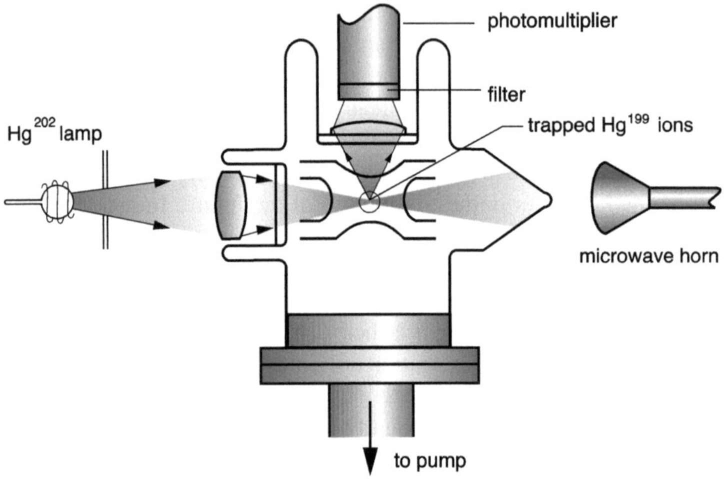

Fig. 10.8 Schematic drawing of the original NASA mercury 199 ion experiment

vacuum, which requires thorough outgassing of metal parts, typically by induction heating. As we shall see in a later chapter, present day micro-traps designed to trap a small number of ions, or even single ions, often take the form of a miniscule metal ring or a small circular hole in a thin metal sheet.

To observe the microwave resonance to the degree of spectral resolution expected of the mercury ion requires a source of microwave power of matching spectral purity. In view of the degradation in spectral purity that can occur in a straight high-order frequency multiplication from a quartz oscillator for example, it is clear that advanced frequency synthesizer techniques are required, possibly starting with a cavity-stabilized microwave oscillator such as a Gunn diode. This is a solid-state device based on what is called the *Gunn effect,* which consists of a fast pulsating current exhibited for example in a uniformly doped n-type GaAs crystal when a voltage applied to it exceeds a certain threshold value. The sharpness of the pulses is such that their Fourier spectrum extends into the microwave region of the spectrum. Used in conjunction with a very high-Q cavity it provides a relatively low-noise source of microwaves to observe the mercury resonance.

Unlike reincarnations of the NASA experiment more than a decade later, when laser sources became available, the low brightness of the *rf* discharge Hg lamp meant extremely weak fluorescent signals, swamped by the presence of scattered background radiation, even after extreme measures were taken to avoid it. The number of trapped ions decays in time and the number of ions produced by successive electron beam pulses are subject to fluctuation. There will also be fluctuations in the intensity of the pumping light with possibly long-term drift. Ideally the photon count data must be taken and reduced in such a way that yields the desired resonant frequency independent of the number of ions or intensity of the pumping light. The frequency synthesizer that supplies the resonant frequency is programmed to step the frequency between two values slightly above and below a center frequency which is swept and ultimately servo-controlled to lock on to the ion resonance frequency. Ideally the two frequencies should be set at values where the photon count is most sensitive to frequency on the two sides of the resonance curve. For an ideal Lorentzian line shape, such as would result from lifetime broadening, this occurs at $\nu = \nu_0 \pm \sqrt{2/3}\,\gamma$, where γ is the half-width of the resonance. Ideally the photon counts on the two sides should be identical only if the frequencies are symmetrical about the maximum of the resonance curve. Since we subtract one count from the other and average over many cycles we should have a number that would pass from a negative value through zero to a positive value as the microwave frequency is swept through the ion resonance. This is precisely the kind of error signal required for locking on to the resonance since it is zero when the frequency is right and distinguishes between the frequency being too high or too low. But things are never that simple; it is expected that the ion population in the trap slowly but continuously drops *during* each cycle in addition to fluctuations between cycles. Since it is expected that the duration of each cycle is a small fraction of the mean ion lifetime in the trap, we may assume that the ion number varies at a constant rate and an adequate correction consists of taking the second differences between photon counts. This requires that the frequency be switched back to its initial value to complete the cycle; thus if S is the reduced photon number for one cycle, and $n_1(\nu_1)$, $n_2(\nu_2)$, and $n_3(\nu_1)$ are the photon counts, then

$$S = [n_1(\nu_1) - n_2(\nu_2)] - [n_2(\nu_2)n_3(\nu_1)]. \tag{10.16}$$

It is assumed that the counts are accumulated over equal intervals of time; therefore any *linear* variation of the ion number or intensity of the light or even any aging of the photomultiplier, though this is very unlikely, cancels out. A more sophisticated scheme of reducing the data [11] constructs a sequence of terms such as S, which involve three counts, but alternates the sign and advances the order of the frequencies at which the counts are taken. To clarify *let* $n_1, n_2, n_3, \ldots$ be the photon counts taken with the frequency continuously alternating between the two sides of resonance, then the signal computer forms the following sum:

$$+(n_1 - 2n_2 + n_3),\ -(n_2 - 2n_3 + n_4),\ +(n_3 2n_4 + n_5) - \ldots\ldots \tag{10.17}$$

When a sufficient number of cycles have been completed to reach a desired ratio of fluctuation in the photon count to the total, that is, signal to noise, the center frequency is changed in small increments to sweep through the resonance curve. The photon count after conversion to analog form serves as an error signal for the feedback servo control of the low-noise microwave source that induces the transitions. For stability of the servo, the error signal must be passed through an integrator which produces a ramp voltage with positive or negative slope depending on whether the error signal is positive or negative. The output of the integrator is applied to a frequency control element in the microwave source.

The two most important figures of merit of any frequency standard are the fractional linewidth and the signal-to-noise ratio of its output. In the mercury ion standard operated with the ions in vacuum the radiation process is not limited in time by collisions and therefore there is no lifetime broadening of the spectrum. However there are two other fundamental factors that affect the linewidth: [1] the Doppler effect: there is zero *first*-order Doppler broadening since the ions are confined in a space comparable to the wavelength of the radiation (the so-called Dicke effect) and their mean velocity is zero with respect to the microwaves. On the other hand the second-order Doppler effect will cause a spectral shift amounting to the following:

$$\Delta\nu_D = \frac{1}{2}\frac{V^2}{c^2}\nu_0. \tag{10.18}$$

The ions in the older Paul trap designs were typically in a potential well of about 2.5 V depth. Since the ion population has energy distributed in a range up to that value it follows that the Doppler shift causes a spread of frequencies on the order of $1.4 \times 10^{-11}\nu_0$. As the kinetic energy is continuously distributed over the ions, the Doppler *line shape* will also depend on that distribution. The later development of the laser cooling of ions has radically reduced Doppler broadening and brought about the fulfillment of the original motivating ideal of observing an atomic ion, isolated and at rest. In fact the joining of the techniques of particle confinement and laser cooling has opened up far-reaching developments, not only in frequency standards but also in quantum computing. The spectroscopy of field-confined ions using laser optics has reduced spectral line widths to unprecedented levels. More will be said about these developments in Chaps. 11 and 18.

The other fundamental source of line broadening is the frequency shift due to the ambient magnetic field; extreme measures are taken to shield the ion frequency standard from the earth's magnetic field, for example, by using a superconducting shield. However a weak *uniform* magnetic field is required, and established through special coils parallel to the microwave magnetic field. Such a field causes a shift in the hyperfine frequency of mercury ions amounting to

$$\nu = \nu_0 + 9.7 \times 10^9 B^2, \tag{10.19}$$

where B is the magnetic field intensity in Teslas. Therefore any inhomogeneity in the field over the space of the trap will lead to some spectral broadening.

Finally we consider the signal-to-noise ratio attainable in the photon count of the fluorescence. In the absence of any systematic fluctuations, for example in the lamp output, the ratio of signal to pure shot noise on the photon count at resonance is given by $(N_r - N_0)/(N_r + N_0)^{1/2}$, where N_r and N_0 are the photon counts on resonance and off resonance. This leads to the root mean (Allan) variance $\sigma(\tau)$ varying as $(\tau)^{-1/2}$; thus,

$$\sigma_y(\tau) = \frac{1}{2\pi}\frac{1}{Q_A(S/N)^{1/2}}\tau^{-1/2}, \tag{10.20}$$

where Q_A is the ion resonance Q-factor, S/N is the signal-to-noise ratio, and τ is the sampling time. In [12] Cutler et al. give the value of $\sigma_y(\tau) = 1.2 \times 10^{-12}\,\tau^{-1/2}$ for their experimental standard. A preliminary value for the hyperfine frequency corrected for magnetic field and second-order Doppler shift is given as 40,507,347,996.9 ± 0.3 Hz.

10.5 Recent Developments: Trapping Individual Ions

As in other fields, the advent of the laser, which is discussed at length in the next chapter, transformed the whole aspect of ion confinement in ways which had a profound effect not only on atomic time standards but also on the quantum behavior of individual ions and its application to quantum computing. One of the main hurdles in realizing the advantages of a laser source in the spectroscopy of stored ions is the synthesis of laser frequencies in the ultraviolet region of the spectrum, where some of the interesting heavy ions have their resonance frequencies. Thus the resonance wavelength of the mercury ion is around 194 nm in the ultraviolet region of the spectrum. Since there is no known laser that produces this wavelength, it must be synthesized. For this purpose advantage is taken of the extremely high intensity of a laser beam to drive certain crystals into a nonlinear regime, resulting in output radiation at harmonic frequencies, or the sum (or difference) of frequencies of two input beams. Thus Berkeland et al. [13] were able to synthesize a continuous laser beam at 194 nm to observe the Hg^+ resonance. However, at present, a miniature RF mercury vapor discharge lamp is uniquely suitable for a portable ion frequency standard, but that no doubt will change as laser materials and techniques continue their rapid advance.

The physical dimensions and geometry of traps compatible with the use of lasers are on a microscopic scale, some on the order of a hundred microns. In such a trap, the ion oscillation amplitude is so small that the necessary saddle point in the potential of the electric field can be achieved over a sufficient volume using electrodes whose exact shape is unimportant. The simplest example might be a plane conducting sheet or an array of parallel sheets with circular holes: the field in the neighborhood of the centers of the holes will have the required quadrupole geometry [14]. One such arrangement is illustrated in Fig. 10.9, the so-called elliptical trap.

An important requirement is of course that a laser beam should have access to the ion(s), with minimum stray scattering by the electrodes. The extraordinary brightness and spectral purity of a laser beam compared with a conventional light source mean that scattering from *individual* trapped ions can and have been detected and their oscillatory motion in the trap slowed down. A laser-cooled ion trapped in a microscopic trap is inherently free from the three principal sources of frequency instability in a standard: the second-order Doppler effect, the second-order magnetic field shift, and collisions with buffer gas atoms. It also means that the energy *quantum levels* of the ions in the harmonic potential well of the trap can be resolved and transitions between them observed as in an atom! This has enabled

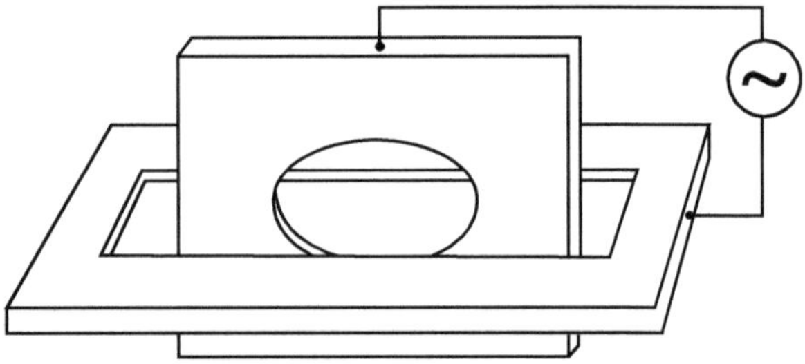

Fig. 10.9 Plane electrode geometry of the "elliptical" ion trap

the study of coherence between the internal (for example hyperfine) quantum states of the ion and its motion in the trap. It is this capability which fulfilled the promise of achieving a logic gate for a quantum computer.

The previously mentioned linear trap configuration has been adapted to be compatible with solid-state devices using micro-electromechanical systems techniques (MEMS). The stimulus for this comes from a number of important applications such as quantum computing which require arrays of micro-traps that can be integrated into solid-state devices and are scalable, allowing replication into large numbers of units in a finite volume. The most promising development has involved trapping individual ions *above* a flat surface on which linear conducting strips have been laid using semiconductor fabrication techniques. In a recent paper, J. Britton et al. of the NIST group [15] describe such a surface electrode micro-trap made from bulk silicon by photolithographic techniques. The plane electrode geometry consists of the two parallel strips carrying RF, while the other two strips, called the control electrodes, are segmented and carry DC-positive potentials at their ends to produce a confining axial field. The pseudo-potential describing the mean ion motion in the RF field confines the ions in the radial direction completing the three-dimensional confinement of the ion. The silicon wafer is given structural support by bonding to a substrate of borosilicate glass chosen to have a coefficient of thermal expansion matching the silicon.

The studies at NIST were conducted using Mg^+ ions formed by laser photoionization of Mg vapor. The shorting out of the electrodes by condensation of Mg vapor was avoided in the detailed construction design by the absence of a bridging surfaces between critical electrodes. The ions were formed by a two-photon process in which a quantum of laser radiation at a wavelength of 285 nm was first absorbed by an atom raising it to an excited state, followed by the absorption of a second photon resulting in its ionization. Parallel to the ionizing laser beam is another laser beam with the wavelength set at 400 MHz below the ion resonance wavelength to achieve Doppler cooling (see the next chapter). Individual ions are resolved by a CCD camera detecting the fluorescent radiation excited by the laser. The success of this experiment is clearly a significant advance in the pursuit of a quantum computer using trapped ions.

References

1. J. Stratton, *Electromagnetic Theory* (McGraw Hill, New York, 2008). reprinted Adams Press
2. J.R. Pierce, *Theory and Design of Electron Beams* (van Nostrand, Princeton, 1954), p. 41
3. W. Paul, Rev. Mod. Phys. **62**(3), 531 (1990)
4. J.P. Schermann, F.G. Major, Appl. Phys. **16**, 225 (1978)
5. F.G. Major, NASA Tech. Rep. X-521-69-167 (1969)
6. F.G. Major, G.N. Georghe, G. Werth, *Charged Particle Traps* (Springer, Heidelberg, 2005)
7. F.G. Major, H.G. Dehmelt, Phys. Rev. **170**, 91 (1968)
8. H.G. Dehmelt, K.B. Jefferts, Phys. Rev. **125**, 1318 (1962)
9. F.G. Major, G. Werth, Phys. Rev. Lett. **30**, 1155 (1973). Appl. Phys. **15** 201 (1978)
10. S. Mrozowski, Phys. Rev. **58**, 332 (1940)
11. D.W. Allan, Proc. IEEE **54**, 221 (1966)
12. L.S. Cutler, et al. Proc. Annual PTTI Meeting Washington, DC (1981)
13. D.J. Berkeland et al., Phys. Rev. Lett. **80**, 2089 (1998)
14. F.G. Major, J. de Phys. Lett. **38**, L221 (1977)
15. J. Britton, et al., arXiv:quant-ph/0605170v1 19 May (2006)

Optical Frequency Oscillators: Lasers

11

11.1 Introduction

The name laser is obviously derived from maser; instead of *microwave* we have *light* wave amplification by stimulated emission of radiation. Although the physical principles are identical since they both deal with electromagnetic waves, the practical problems of creating the conditions to observe them are very different. In common usage the terms laser and maser are used to denote an oscillator rather than an amplifier. It is a remarkable fact that the realization of a laser does not involve any physical theory or indeed any practical technique that had not been known long before laser action was actually observed. The concept of stimulated emission was well established, there was abundant spectroscopic data on the wavelengths and line intensities of spectral lines, and the practice of light wave interferometry was well advanced. Furthermore the study of electrical discharges through rarefied gases and the theory of crystal optics had been studied since the nineteenth century. It is little wonder that as soon as Maiman reported laser action in a ruby crystal in 1960, there was an explosion in the number of published reports of laser action in different materials. Lasers are playing such a transformational role in advances in the ultimate shape of atomic clocks and gyroscopes, and hence of navigation in the future, that this entire chapter is devoted to them.

11.2 Optical Cavities

Like the high-Q cavity a maser needs to sustain electromagnetic oscillation, an optical frequency "cavity" is required for laser action. However, there is an important physical difference between the cavities in the two cases, a difference arising from the radically different wavelengths. Since a maser operates usually at wavelengths in the 1–20 cm range, a cavity with comparable dimensions must be totally closed with strongly reflecting walls to achieve a high-Q value. Such a cavity

F.G. Major, *Quo Vadis: Evolution of Modern Navigation: The Rise of Quantum Techniques*, DOI 10.1007/978-1-4614-8672-5_11,

will have its lower resonant modes of oscillation at well-separated frequencies, one of which must be tuned to the atomic transition frequency. In contrast, when the wavelength is very short on a laboratory scale, even a widely open "cavity" consisting essentially of a pair of slightly concave mirrors exactly aligned some distance apart can have a significant Q value. The reason is that the light is reflected back and forth between the mirrors and the only loss, aside from absorption by the mirrors, is due to the diffraction of light at the mirrors, which will be small provided the mirrors are large compared with the wavelength of the light.

The arrangement of two optically flat mirrors mounted at some distance apart parallel to each other to form an optical resonator has historically a precursor called the Fabry–Pérot interferometer, whose introduction in 1889 marked the beginning of the era of modern wave optics. It relied on multiple reflections between the two highly reflecting mirrors to enhance the amplitudes of just those wavelengths that satisfied a certain resonance condition; by varying the distance between the mirrors it acted as a high-resolution spectrometer. The degree to which it is able to separate close spectral lines is specified by a quantity called its *finesse*. To understand the significance of this quantity, imagine a monochromatic light wave having an infinitely sharp spectral width entering the interferometer along a perpendicular to the planes of the two mirrors. Let the distance between the mirrors D correspond to the nth mode, that is, n is the number of wavelengths contained in the total distance from one mirror to the other and back to the first, that is, $n\lambda = 2D$ Since the time to traverse that distance is $\tau = n\lambda/c$, it follows that if the waves maintain their amplitude and coherence over an average of F traversals between the mirrors, that is, a coherence time of $T = Fn\lambda/c$, then the frequency uncertainty (or width) is $\Delta\nu$ where $\Delta\nu T = 1$. This leads to the following important result for the resolving power:

$$\frac{\nu}{\Delta\nu} = Fn. \tag{11.1}$$

The frequency interval $\Delta\nu$ is clearly the closest the frequency of two spectral lines can have and still be seen as distinct . The quantity F, which depends mainly on the reflectivity of the mirrors, is the figure of merit called the *finesse* of the cavity; clearly the quantity Fn corresponds to the Q of a microwave cavity. In terms of the (power) reflectivity of the mirrors R, the finesse is given by

$$F = \frac{\pi\sqrt{R}}{1 - R}. \tag{11.2}$$

The design and manufacture of laser multilayer mirrors have progressed to the point of achieving values of R reaching an impressive 99.998 %, corresponding to $F = 157{,}000$.

As a spectrometer in which the mirror spacing is scanned to display the spectrum, each wavelength produces a multiplicity of resonances corresponding to the different modes, and therefore it is necessary to know what range of wavelengths

Fig. 11.1 Diffraction in a two-mirror optical cavity

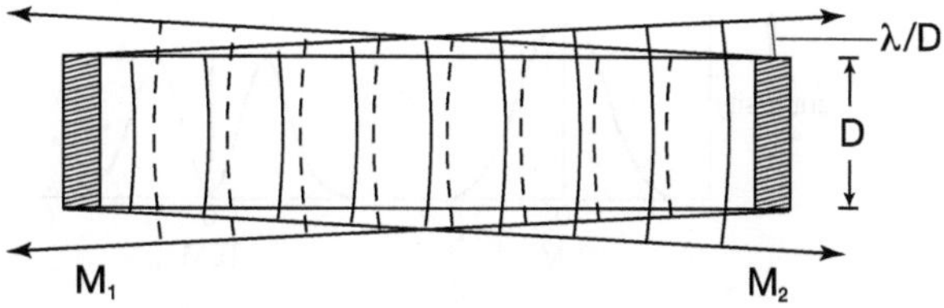

fall between successive modes in order to avoid ambiguity. For a spacing of L between the mirrors, the difference between two adjacent resonant frequencies is simply given by the following:

$$\nu_{n+1} - \nu_n = \frac{c}{2L}, \tag{11.3}$$

or in terms of wavelength:

$$\lambda_{n+1} - \lambda_n = -\frac{\lambda^2}{2L}. \tag{11.4}$$

When the Fabry–Pérot interferometer was first developed, spectral lamps had limited wave coherence lengths, and the distance between the mirrors was typically in the millimeter range; now gas laser cavities can extend to tens of centimeters.

Unlike a completely closed cavity with reflecting walls as we have in a hydrogen maser, an open "cavity" consisting of two mirrors, strictly speaking, has *no* discrete resonance frequencies. However, prior to the actual demonstration of laser action in a two-mirror cavity, theoretical calculations had indicated [1] that modes of the optical field do exist that are concentrated along the axis between the mirrors. We can arrive at a simplified derivation of the Q factor of such an optical cavity by assuming that the only loss of optical energy is due to diffraction of light waves at the mirrors. Assuming the mirrors are circular with diameter D large compared to the wavelength, and set at a distance L apart, a plane light wave of wavelength λ falling on one will be diffracted (spread out) at an angle of λ/D radians, as illustrated in Fig. 11.1.

If we make the crude assumption that the energy of the wave is spread evenly over the expanded wavefront, then on reaching the other mirror a fraction $4\lambda L/D^2$ of the wave will fall outside the rim of that mirror and be lost. That fractional loss occurs on reflection from each mirror and repeats at intervals of L/c, the transit time of the light between the two mirrors. Thus the average *rate* of *fractional* loss of optical energy R is given by $R = 4\lambda c/D^2$. Of practical interest is the Q factor for the different resonant modes of the cavity; this we can immediately obtain from the rate of energy loss using the definition:

$$Q_n = \frac{2\pi\nu_n}{R} = \frac{\pi D^2}{2\lambda_n^2}, \tag{11.5}$$

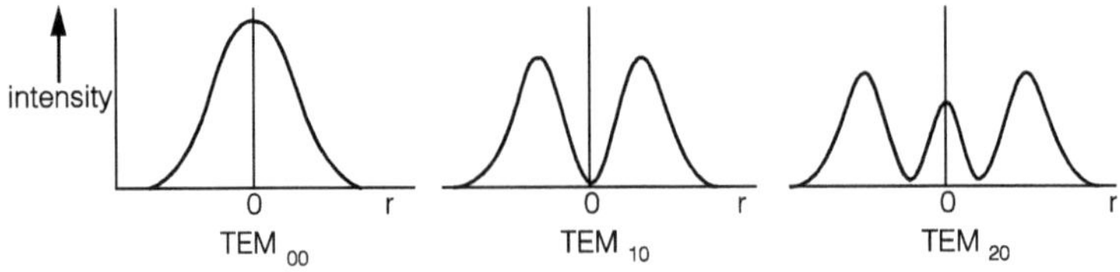

Fig. 11.2 The radial intensity distribution for some low order radial modes

where λ_n is the wavelength of the nth mode. As a numerical example suppose $D = 2.5$ cm and $\lambda = 632.8$ nm, the common red light from a He–Ne laser, then the diffraction determined Q is on the order of 5×10^9.

A more realistic description of the light wave distribution surrounding the two-mirror cavity is considerably more complex than we have assumed. The intensity profile of the wavefront reflected from a mirror is far from uniform; it has a radial distribution that can be analyzed in terms of certain modes, somewhat analogous to the vibration pattern on a drum head. In fact the classical ray optics used to design a lens system using ordinary (noncoherent) light does not apply in laser systems. What is required is a formulation of electromagnetic wave theory to suit the boundary conditions of a laser system. In fact such a solution of the wave equations has been constructed which fits the boundary conditions of the two mirror system, a solution which has symmetry about the axis of the mirrors, but with radial TEM (transverse electromagnetic) modes having characteristic radial intensity distributions. The beam profiles for some of the lower order radial modes are shown in Fig. 11.2.

The lowest order mode designated as TEM_{00} has one maximum on the axis with an intensity distribution that is Gaussian, that is, it varies as $\exp(-r^2/r_0^2)$; a light beam having this property is called a *Gaussian beam*. The theory of the design of optical systems involving such beams is called Gaussian optics, which differs fundamentally from the ray-tracing methods of classical optics; the classical formulae for mirrors and lenses are not valid where the wave nature of light is of the essence.

The existence of solutions of the wave equation for a two-mirror cavity having low order radial modes is what made it possible to observe laser action in gases. A three-dimensional cavity made up of reflecting surfaces on an ordinary scale would have an almost infinite number of modes of vibration of the light field, all competing for excitation. In fact it is shown in the theory of black body radiation that in a cavity of volume V the number of modes of the radiation field of frequency ν in the range $\Delta\nu$ is given by

$$\Delta N = \frac{8\pi\nu^2 V}{c^3} \frac{\Delta\nu}{\nu}. \tag{11.6}$$

This result applies to a cavity of any shape provided the linear dimensions of the cavity are much larger than the wavelength.

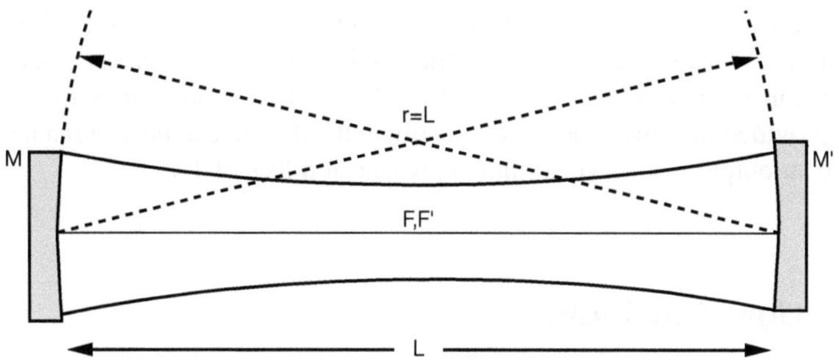

Fig. 11.3 A confocal optical cavity

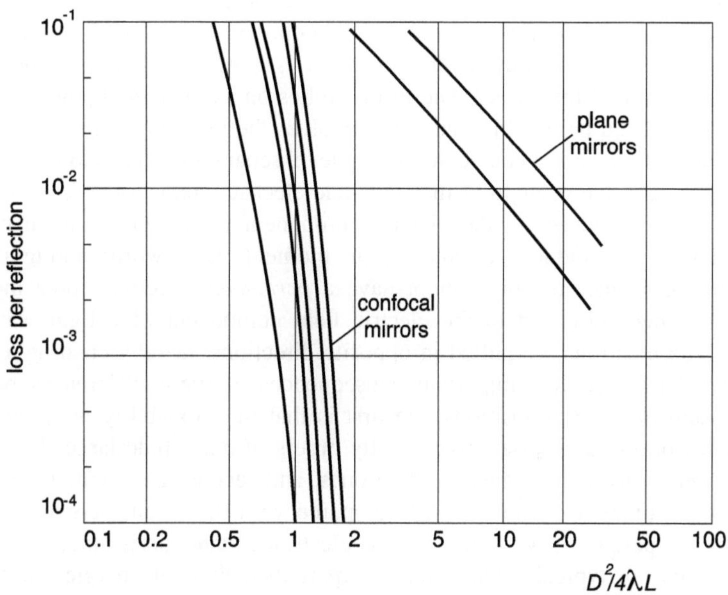

Fig. 11.4 Diffraction loss in two mirror cavities as a function of $D^2/4\lambda L$ [2]

The fundamental problem in the design of the cavity is to determine the factors that limit its Q and how to optimize those factors to achieve the highest Q. The presumption is that, as borne out in practice, the optical gain achievable in a gas will always be small; it is fortunate therefore that theory shows that diffraction loss can be radically reduced, and optimum Q achieved, if instead of plane mirrors, concave mirrors are used in a confocal configuration, as shown in Fig. 11.3.

We recall that a spherical concave mirror of radius R focuses a parallel beam of light to a point on the axis $R/2$ from the mirror. It follows that a confocal cavity has the distance between the mirrors $D = R$. The diffraction loss in a two-mirror cavity has been computed for plane and concave mirrors; plots of the loss versus $D^2/4\lambda L$ are reproduced in Fig. 11.4 for different cavities [2].

According to our approximate theory the graph for plane mirrors should be a straight line with a slope of -1. From the graphs we see that even for plane mirrors the loss can be as low as 3×10^{-4} if $D^2/4\lambda L = 100$, a not impossible figure to attain, provided the mirrors are perfectly aligned. Of course to be useful a laser must provide an output beam, which may represent of a loss of 1 %.

11.3 Light Amplification

We are already familiar from our study of the hydrogen maser that in order to have sustained oscillation in a resonant cavity it is necessary to have more atoms being stimulated to emit radiation than those absorbing radiation, in order to make up all the losses and produce a net gain in the field energy. However, the nature of the microwave transitions in the hydrogen maser is quite different from the optical transitions required for a laser. The former relies on a *magnetic* dipole interaction between atom and field, while the latter involves the far stronger *induced electric dipole* interaction; and of course the energy level separations in an atom also reflect the difference in the strength of magnetic and electric interactions. Electric dipole transitions are induced by the electric component of a light beam driving an oscillation of the induced atomic dipole moment. It is worth noting that on symmetry grounds, an atom cannot have a *permanent* electric dipole moment; however, when subjected to the electric field component of a light wave, the nucleus and electrons are pulled in opposite directions, thereby creating a dipole moment. There are two important consequences of these differences between electric and magnetic transitions: the first is that the probability of *spontaneous* emission from an upper state is generally orders of magnitude larger for electric dipole transitions than magnetic dipole ones, and second the magnetic hyperfine levels are split by an amount small compared with the mean thermal energy at ordinary temperatures, while the various electronic states determined by electrostatic forces are typically much farther apart than thermal energies. It follows according to Boltzmann's law that in thermal equilibrium at ordinary temperatures the atoms or molecules are almost totally in the ground state. Thus if N_1 is the state population of an upper quantum state and N_2 the population of the lower state, then Boltzmann's law states the following:

$$\frac{N_1}{N_2} = e^{-\frac{E_1 - E_2}{kT}}. \tag{11.7}$$

Since the absolute temperature T is positive, it follows that in thermal equilibrium $N_1 \ll N_2$ and so to achieve oscillation which requires $N_1 > N_2$ we need to create a nonequilibrium condition, namely an *inversion* of state populations.

A crucial parameter in the action of a laser is the transition probability for *induced* emission and its ratio to spontaneous emission. This we get from what are called the Einstein A and B coefficients:

$$\frac{A_{\mathrm{nm}}}{B_{\mathrm{nm}}} = \frac{8\pi h\nu^3}{c^3}. \tag{11.8}$$

These coefficients are defined as follows: the probability per atom of absorption or *stimulated* emission of radiation of spectral density ρ_ν is given by $B_{\mathrm{nm}}\rho_\nu$, while A_{nm} gives the probability of spontaneous emission. It is the stimulated emission which is driven in phase with the optical field already present, and represents an amplification of it, the basic mechanism for laser action. If numbers are substituted for the conditions that obtain in a typical discharge lamp it is readily verified that the rate of spontaneous emission, which is incoherent, vastly exceeds any stimulated emission. This accounts for the fact that laser action is not normally found in a classical discharge lamp, or indeed in any common light source such as the sun or tungsten lamp. What we call "spontaneous" emission does not arise from nothing as the name might suggest: field theory attributes it to "zero point" random fluctuations of the electromagnetic field and has random phase. We will take up the subject of coherence of light a little later.

Again using Einstein's coefficients we can obtain the optical gain of a gas, given a degree of population inversion produced by some unspecified mechanism. Thus let n_1 and n_2 represent the number densities of atoms in the upper and lower quantum states whose energies differ by $h\nu_0$, and let a light beam of frequency ν be directed through the gas along a direction chosen as the z-axis. Then it can be shown that the transmitted intensity I_ν is given by:

$$I_\nu(z) = I_\nu(0)\exp(\gamma z) \quad \gamma = \frac{1}{4}(n_1 - n_2)\lambda^2 \Delta\nu_n g(\nu - \nu_0), \tag{11.9}$$

where the function $g(\nu - \nu_0)$ is included to account for the spectral line shape of the atomic absorption. On the simplifying assumption that the beam is weak and the populations are nearly constant, the solution to the equation is a simple exponential function; under normal thermal equilibrium conditions when $n_1 < n_2$, the intensity of the beam *decreases* exponentially with the distance z, as required by the classical Lambert's Law of light absorption, but if by some pumping process a nonequilibrium state is created in which the populations are inverted, then the intensity *increases* and the gas is a light *amplifying* medium. The exponential gain constant may be usefully written in terms of Einstein's A-coefficient since that is directly related to an observable quantity, the natural spectral width $\Delta\nu_n$ of the transition between the states.

In analyzing the functioning of a laser it is important to draw a distinction between two essentially different circumstances that give rise to the broadening of spectral lines, represented by the line shape function $g(\nu - \nu_0)$: they are called *homogeneous* and *inhomogeneous* broadening. The terms have their origin in pre-laser magnetic resonance spectroscopy, and apply to a system consisting of many atoms or molecules. Homogeneous broadening is one that applies equally to all members of the system, typified by the broadening of a transition frequency due to the finite lifetime of the radiative process. Inhomogeneous broadening, on the other hand, is seen in an ensemble when each member has a little different frequency from the rest by virtue of a different Doppler shift or position in a nonuniform electric or magnetic field. The form of $g(\nu - \nu_0)$ that arises from a finite radiation time is the familiar Lorentzian we encountered in the previous chapter. As for inhomogeneous broadening, the most universal, particularly in gases, is due to the Doppler effect arising from the thermal motion of the particles under study. If the particles are in thermal equilibrium, then their velocities, although continually changing as a result of collisions, nevertheless have a statistical distribution known as the Maxwell–Boltzmann distribution. Since each individual particle will experience a different Doppler shift in the frequency of a light beam, the overall effect on the absorption spectrum of a large number of atoms is a distribution of frequencies that produces the line shape. Using the Maxwell–Boltzmann distribution of velocities, the Doppler line shape function $g(\nu - \nu_0)$ giving the probability that the frequency is between ν and $\nu + d\nu$ can be written in the following simplified form:

$$g(\nu - \nu_0) = \frac{1}{\nu_0\sqrt{\pi}}\left(\frac{c}{V}\right)\exp\left[-\frac{c^2}{V^2}\left(\frac{\nu - \nu_0}{\nu_0}\right)^2\right], \tag{11.10}$$

where for brevity we have written V in place of $(2kT/M)^{1/2}$. The form of $g(\nu - \nu_0)$ is that of a Gaussian function centered on ν_0.

We will illustrate the basic design principles of a laser using a simple model consisting of a Fabry–Pérot cavity filled with a gas as the amplifying medium. As with any oscillator, two conditions must be satisfied for oscillation: the amplification must be sufficient to make up for all optical losses and it must be regenerative, that is, reinforce the existing optical field. We begin with the condition on loop gain in the optical intensity as the light completes a round-trip between the two mirrors. If R_1 and R_2 are the reflectivities of the two mirrors, and γ, α the amplification and attenuation factors of the medium, then to achieve oscillation the amplification γ must satisfy the following threshold condition:

$$R_1R_2\exp[(\gamma - \alpha)2L] \geq 1. \tag{11.11}$$

The condition on the phase , namely that the light wave return with the same phase to a given point after having made a round-trip between the mirrors, is more difficult to assess because the phase velocity of the light wave depends on the frequency, particularly near a resonance. We can define a frequency-dependent refractive index as $n(\nu)$; then the condition on phase in a homogeneous gas can be stated as follows:

$$\frac{2n(\nu)L}{c}\nu = m, \tag{11.12}$$

where m is an integer. In the absence of the gas we would have $n(\nu) = 1$ and $\nu = m(c/2L)$, that is, L would correspond to the mth resonance frequency of the cavity ν_m.

To derive the function $n(\nu)$ would take us too far into dispersion theory; however, we will quote an approximate result that demonstrates an important property of atomic oscillators. The actual frequency of oscillation is given by the following [3]:

$$\nu = \nu_m + \frac{\Delta\nu_C}{\Delta\nu_A}(\nu_A - \nu_m), \tag{11.13}$$

where ν_m is the passive frequency of the *m*th cavity mode, and $\Delta\nu_C$ and $\Delta\nu_A$ are the frequency widths of the cavity resonance and the atomic transition providing the amplification, respectively. We note the same *frequency pulling* result that we encountered with the hydrogen maser: again the atoms and the cavity represent two coupled resonant systems that are closely tuned in frequency, leading to that characteristic phenomenon. However, in this case, unlike the hydrogen maser, it is the *cavity* that has the higher Q, or narrower resonance width.

11.4 Laser Output Power

Having met the threshold conditions for sustaining oscillation, the next question is: precisely how does the oscillation begin, and to what power level does the optical field buildup in the cavity. The threshold conditions address only the question of reinforcing an existing oscillation; the question of how the whole process begins must still be answered. In the case of a gas laser such as the He–Ne laser, the inversion in the populations of the chosen states is achieved by an electrical discharge, which causes the excitation of many atomic states, including ones involved in the laser transition. After the initial phase, the further development of the amplitude of oscillation and the ultimate power level achieved must take into account the effect of the amplified light *back* on the population difference itself that produced it. If the population inversion exceeds the threshold for oscillation the light intensity will grow rapidly, and the upper state atoms are induced to radiate and return to the lower state, thereby reducing the population inversion and therefore the gain back down to the threshold value.

In the case of a homogeneously broadened laser transition, *all* particles will contribute stimulated emission at a single frequency, and the gain at all frequencies within the spectral line profile saturates, falling back to the threshold value. On the other hand, for inhomogeneously broadened transitions, those particles whose frequency is resonant with any of several modes of the cavity will simultaneously contribute to the laser action and become saturated; also by sweeping the length of the cavity, laser action may be obtained on different modes. In fact different modes of the cavity can oscillate simultaneously without mode competition; the latter is a phenomenon in which the buildup of the optical field at one frequency affects the conditions for oscillation at another frequency. The phases of the different modes are in general random and the output intensity is nearly constant. However, it is possible to cause the phases of the different modes to be tightly correlated leading to an amplitude that sharply peaks at regular intervals, a condition called

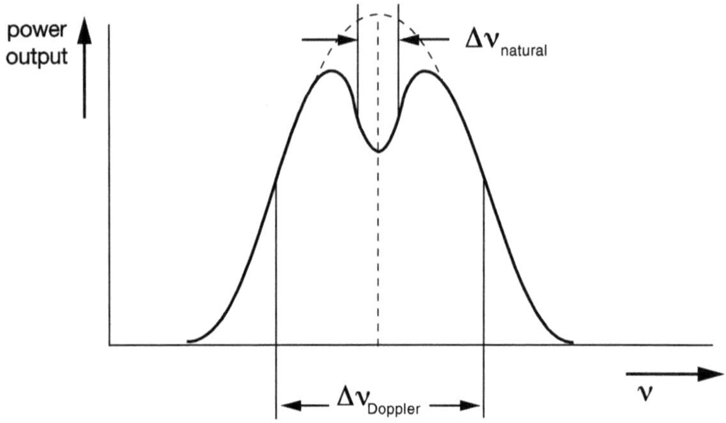

Fig. 11.5 The Lamb dip in the output of a gas laser

"phase lock," an extremely useful phenomenon for producing ultrasharp pulses using broadband laser amplifying crystals.

The full quantum theory of the interaction of the optical field with atoms that goes beyond the semiclassical approximation so far implied was given in 1964 by Lamb, who published a paper [4] on the quantum theory of the laser which takes the interaction of the radiation field with atoms to the third order of approximation. He derived the following "equation of state" for the electric field amplitude E_n of the nth mode:

$$\frac{dE_n}{dt} = \alpha_n E_n - \beta_n E_n^3. \tag{11.14}$$

This leads to a steady state of the field intensity proportional to $E_n{}^2$ given by

$$E_n^2 = \frac{\alpha_n}{\beta_n}. \tag{11.15}$$

Of particular interest is the result that the coefficient of the linear term, α_n, as a function of tuning has the characteristic Gaussian expected of a gas in thermal equilibrium, while the coefficient of the nonlinear "saturation factor" β_n has typically a far sharper frequency dependence, determined by the natural lifetime of the transition. The result is that the output of the laser as a function of tuning exhibits a relatively broad Gaussian profile with a sharp Lorentzian minimum in the center now called the *Lamb dip*. This feature has proved extremely useful in stabilizing the frequency of lasers and made possible extremely refined distance measurements between their mirrors. Figure 10.6 illustrates the Lamb dip in the output of a gas laser (Fig. 11.5).

11.5 Laser Output Spectrum

Laser radiation is distinguished by its extraordinary coherence length reflecting extremely low phase noise and hence great spectral purity. When its frequency is locked to a suitable atomic or molecular transition frequency, its long-term stability and reproducibility are such that it can serve as a frequency standard in the optical region of the spectrum; as such its role would be similar to that of the Gunn oscillator in the rubidium standard. In practice the phase noise in the output of a laser arises from two types of sources: the first is *artificial* fluctuations in the dimensions of the cavity and the optical properties of the amplifying medium in it, and the second is fundamental quantum fluctuations in the interaction of light with the atomic system.

As to the first of these, great progress has been achieved in isolating the cavity from mechanical vibration and ultimately using servo systems to lock the frequency on resonances of suitable molecular transitions. Thus, for example, certain spectral lines in methane happen to fall within range of the He–Ne laser and are used to stabilize it. The drive to achieve the ultimate phase stability, and therefore spectral purity, was spurred on by the work of Townes and Schawlow [5] who had predicted, before laser action was first observed, that the fundamental limit to the spectral purity of laser radiation was still far beyond what had been achieved with existing lasers. By analyzing the radiation processes involved in laser action, they showed that there is a fundamental minimum to the output spectral width. This limit is set by the inevitable presence of *spontaneously* emitted photons in addition to the stimulated emission on which amplification depends. We recall that the spontaneous emission probability is independent of the presence or absence of photons and that the emitted photons have random phase, uncorrelated with the phase of photons emitted by other atoms or even photons emitted by the same atom at different times. This is in contrast to an atom in the upper state being *induced* to emit a photon by an existing optical field, which it does in proportion to the field intensity, but more significantly, with a definite phase correlated to the inducing field. This means that if a number of atoms are subjected to the same optical field and are induced to emit radiation, they will all do so coherently, amplifying the same light wave.

As might be guessed, it is the presence of the incoherent spontaneously emitted radiation that limits the phase coherence of the laser output. It can be shown that the square deviation in the phase $\Delta\varphi$ of the laser radiation field caused by N_{sp} spontaneous transitions is given by the following:

$$(\Delta\varphi)^2 = N_{sp}\frac{E_{sp}^2}{2E_{tot}^2}, \tag{11.16}$$

where E_{sp} is the electric component of the cavity field due to spontaneous emission and E_{tot} is the total field. On an amplitude-phase diagram, the total optical field vector is the resultant of a relatively large rotating vector, the coherent part, and a smaller vector with random magnitude and relative phase, that is, angle, producing a resultant field whose amplitude and phase fluctuate randomly, as shown in Fig. 11.6.

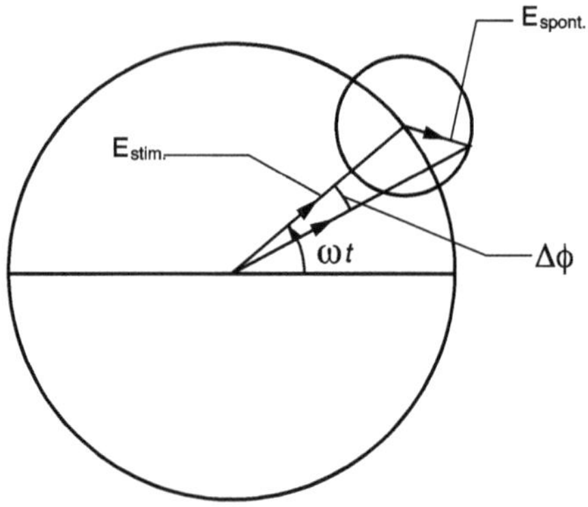

Fig. 11.6 The phasor diagram showing the coherent and spontaneous fields

If we assume that the laser oscillation level is stabilized through saturation of the gain, for example, leaving the phase unconstrained, so that the mean intensity of the optical field is constant, there will be nevertheless a small fluctuation in phase due to the spontaneous component. The maximum change in the phase of the total vector clearly occurs when the spontaneous contribution is in quadrature to the coherent vector, at which point the phase change is maximum $\Delta\varphi_{\mathrm{m}} = E_{\mathrm{spont}}/E_{\mathrm{tot}}$. On the other hand if the spontaneous vector is parallel to the coherent vector there is zero shift in phase, so that with the proper averaging we get the factor 2 in the denominator of Eq. (11.16). The change in phase, of course, is equally likely to advance the phase or retard it: that is, $\Delta\varphi$ is equally likely to be positive or negative, presenting us with the statistical problem we have already encountered: the *random walk*. On that basis, the result stated above immediately follows, with the number N_{sp} raised to the first power. It has been shown by Townes and Schawlow that the laser radiation linewidth is given by

$$\Delta\nu_{\mathrm{L}} = \frac{\pi h\nu_{\mathrm{L}}}{P}\frac{N_2}{N_2 - N_1}(\Delta\nu_{\mathrm{C}})^2, \tag{11.17}$$

where ν_{C} is the passive cavity resonance linewidth, P is the optical power output, and N_1 and N_2 are the populations of the ground and excited states, respectively. As a rough example, a He–Ne laser with a 1 m long cavity and 1 % output mirror, emitting 1 mW of radiation at $\lambda = 632$ nm wavelength, would have $\Delta\nu_{\mathrm{L}} \approx 1.6 \times 10^{-3}$ Hz, which corresponds to a fractional linewidth of 3.2×10^{-18}! This quantum limit is so small that when it was first derived it was assumed that it was of no practical consequence; however, recent advances in laser stabilization in the pursuit of an optical standard of frequency suggest that it may someday become relevant.

11.6 Gas Laser systems

11.6.1 The Helium–Neon Laser

Although solid-state lasers increasingly play a dominant role in modern technology being uniquely suited to integration in modern electronics, nevertheless the helium–neon laser, now ubiquitous with its thin red beam, occupies a special place in the history of the laser. We will therefore devote some space to it, along with some other gas lasers, particularly the argon ion laser, because of the important part they played in the early development and applications of lasers.

After the ruby laser, which we will discuss later in this chapter, the helium–neon laser was the first to successfully demonstrate in practice the emerging ideas on how to extend the principle of the maser to the optical region of the spectrum. It is a testament to the exceptional experimental skills of the investigators at the Bell Telephone Laboratories in 1961 [6] that they were able to align mirrors a meter apart with the needed precision to achieve oscillation when presumably the only way to judge whether the mirrors were approaching the correct alignment was by the onset of laser action itself! The first success was on the infrared line at $\lambda = 1.1523$ μm; the familiar red beam at $\lambda = 632.8$ nm was to come later. Immediately after the news broke of the successful demonstration of an optical oscillator, there was a flood of reports of successful laser action in other systems... so much so, that some wondered why it took so long to discover laser action in the first place!

The active element in the helium–neon laser is neon, one of the noble gases, but the "magic ingredient" is another noble gas: helium. To understand the role of helium we must recall the necessary condition for light amplification, which is at the heart of building a laser. It occurs when the neon quantum state populations deviate from the normal thermal equilibrium status, in which the populations of states *decrease* with their energy level; what is required is an *inverted* population distribution in which a higher level has greater population than a lower one. This clearly requires a mechanism for exciting atoms from a low-energy state into one with a higher energy, a process that is well known to occur in an electrical discharge in a gas, such as that which produces a glow in a neon tube, a clear indication that atoms are indeed being energized to higher levels, from which they return to a lower energy by radiating light. Some have wondered how many neon signs have been lasing in the past without anyone's knowledge! The historical field of "gaseous electronics" which was subsumed by plasma physics had been studied since the nineteenth century and a wealth of data existed giving electron collision cross sections with atoms and radiative lifetimes of different excited states in atoms. It may very well be that without any other special provisions, some population inversion has occurred in discharge tubes in the past. However, the genius of the method used by the Bell group is that instead of relying on electron collisions, as in a normal discharge, to excite the neon atoms, they used helium atoms to deliver the energy. The problem with relying on electron collisions is that they excite atoms indiscriminately to all states, perhaps not equally, but the differences in population, even if some are inverted, are expected to be too small for laser action.

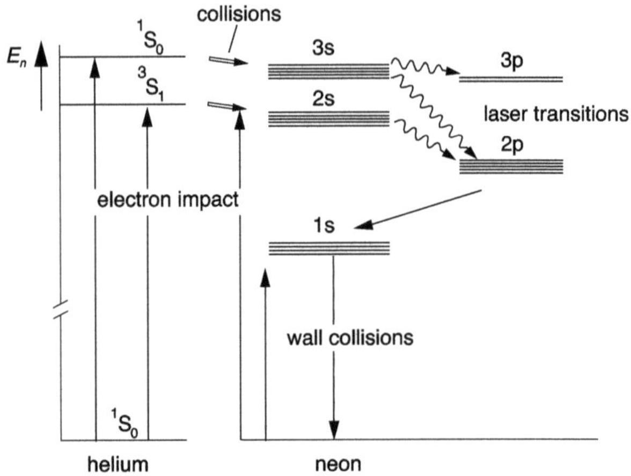

Fig. 11.7 Energy levels of helium and neon relevant to the helium–neon laser

We can understand the rationale for the inclusion of helium in the laser by looking at the quantum level structures of neon and helium; the relevant levels are shown in Fig. 11.7.

Neon is a noble gas of atomic number 10 with a closed shell structure whose ground state is designated as $^1S_0$ having zero spin and orbital angular momentum. It has a filled outer $n = 2$ shell with 2 s-electrons and 6 p-electrons. Since excitation of the atom necessarily requires a change in the principal quantum number to $n = 3$, a large amount of energy is required; this fact, as well as the zero angular momentum, accounts for its inertness, just as in the other noble gases. The excitation of one electron from the 6-p electron subshell leads to energy levels that are broadly similar in spacing to the hydrogen levels because the nucleus is almost completely shielded by the nearly complete $n = 2$ shell. Thus these excited configurations may be labeled as ($2p^5 3s$), ($2p^5 3p$),... etc. and ($2p^5 4s$), ($2p^5 4p$),... etc., but the incomplete $n = 2$ shell is left with a vacancy with angular momentum which must couple with the $n = 3$ electron to produce a rather complicated set of energy terms. For our purposes what is important is simply the existence of these states, their energies, and the fact that the lowest of these states, which we have labeled ($2p^5 3s$) is very long-lived because of selection rules governing (electric dipole) transitions.

The case of helium is in some ways unique. It has atomic number 2 with two electrons in the $n = 1$ shell, which, because of quantum symmetry requirements (Pauli principle), can exist in one of two possible spin states of opposite symmetry: one state is symmetric with respect to an exchange of the electrons and has total spin one, and the other antisymmetric (changes sign) with spin zero. Because selection rules governing optical transitions forbid transitions between these two spin states, helium quantum levels may be separated into two sets that are not connected by any optical transition; it is as if we had two distinct forms of helium: ortho-helium with spin $s = 1$, and para-helium with spin $s = 0$. The ground state is para-helium with zero spin and zero orbital angular momentum. The first two excited states both have zero orbital angular momentum (S-states) and therefore electric dipole transitions from them to the ground state (also an S-state) are forbidden, so they are called *metastable* states. It follows that in an electrical discharge through helium gas, atoms that are excited by electron collisions to many of the upper quantum levels will radiate energy in transitions leading to these two metastable states. Thus in an electrical discharge in helium gas we expect to have a high concentration of atoms in these so-called metastable states with energy above the ground state, energy which they will deliver, by a process called *collisions of the second kind*, to any another atom or molecule that has a nearly "resonant" state, that is, one at nearly the same energy above the ground state. It is called collision of the second kind to distinguish it from collisions of the first kind in which the *kinetic energy* of one particle in a collision raises the energy level of the other particle. If we compare the energy levels of helium and neon we see that indeed there is near resonance between the two metastable states in helium and states in neon. Presumably that is precisely why the combination of these particular gases was chosen. This type of collision, in which there is a transfer of excitation from one atom to another in a collision, had been a well-established phenomenon, studied for a long time both theoretically and experimentally, for example, in the quenching of the emission lines of mercury. Using the published data for the excitation energies of states in the two gases we find for example the following reaction:

$$\mathrm{He}^*\left(2^1\mathrm{S}_0\right) + \mathrm{Ne} \rightarrow \mathrm{He} + \mathrm{Ne}^*(3\mathrm{s}) - 0.05\,\mathrm{eV}.$$

The beauty of using excitation transfer between helium and neon is not only because it *selectively* excites the desired neon level, but also that the helium pressure and hence the collision rate with the neon atoms can be very much higher than for example with electrons.

The physical design of the laser has over the years been developed to the point of becoming a small mass-produced portable item commonly used as a pointer! As such it would have the cavity mirrors permanently sealed on the ends of the discharge tube. For larger lasers the mirrors may be external to a sealed off tube equipped with nonreflecting Brewster windows. We recall that these windows, which have found wide application in laser optics, have the property of not reflecting light polarized in the plane of incidence if the angle of incidence θ, with respect to the perpendicular to the plane of the window, satisfies the condition $\tan\theta = n$. The original experimental set up [6] is shown schematically in Fig. 11.8. It used a 1-m long quartz tube sealed at the ends to bellows carrying the high reflectance plane mirrors to form the Fabry–Pérot cavity. The electrical discharge was produced by a 28 MHz rf field generated by the output of an rf power oscillator applied between electrodes at the ends of the tube, placed symmetrically about a central electrode. The gas pressure ratio of helium to neon was about 10:1, that is, about 100 Pa of helium to 10 Pa of neon. The output power was on the order of 15 mW at the infrared wavelength $\lambda = 1.15\ \mu\mathrm{m}$.

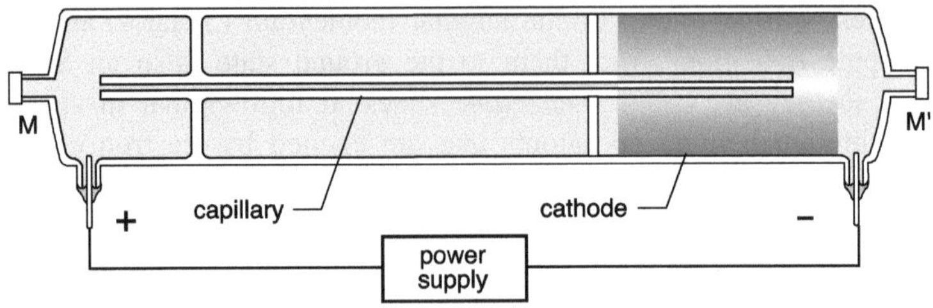

Fig. 11.8 A schematic diagram of a He–Ne laser

The helium–neon laser has found universal appeal as a relatively simple, low-cost laser with good spectral purity and beam quality. However, it has two drawbacks: first, it is virtually a fixed wavelength laser, tunable only over a very limited range of wavelengths, and second, it is fundamentally a low power laser: the practical limit is around 100 mW, but more typically the power is in the 1–10 mW range.

11.6.2 The Argon Ion Laser

The observation of laser action in a high power electrical discharge in argon gas was first reported in 1964 by Bridges [7]. Since then lasers using the ions of noble gases such as argon and krypton as the active medium have been built to generate output beams of high CW power at several different visible wavelengths, making them extremely useful for optically pumping other lasers, such as tunable liquid dye lasers.

Argon is a noble gas with a complement of six electrons in its 3p subshell belonging to the $n = 3$ shell. To ionize argon, that is, to remove an outermost electron and produce an argon ion Ar^+, requires 15.7 eV, a relatively large amount of energy. This implies that to sustain an electrical discharge, in which ions are produced by electron collisions, requires high electric fields and great power dissipation. The singly charged ion has one electron missing from the complete 3p subshell leaving a vacancy with an angular orbital momentum of a p-electron, which is one unit of angular momentum. There is therefore a spin–orbit magnetic interaction leading to a narrow fine structure splitting of the ground state of the ion. To reach the higher 4p quantum levels in the $n = 4$ shell, from which laser transitions can take place to the lower 4 s level, requires a further excitation energy of about 20 eV; this means an argon atom requires a total of nearly 35.7 eV of energy to reach the upper lasing states of the ion. Some of the relevant transitions are shown in Fig. 11.9.

The laser transitions occur between the fine structure sublevels in the $3p^4$ 4p ($^4D_0$) state and the $3p^44s$ (^2P) state. Clearly the complexity of the argon ion level structure provides ample material for spectroscopists to rationalize; however, we can safely limit ourselves to the empirical results. Although the radiative lifetime of

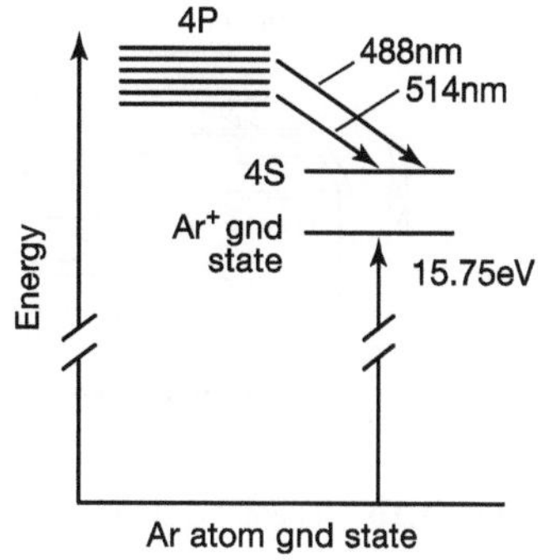

Fig. 11.9 Energy levels relevant to the argon ion laser

the upper states is quite short (about 9 ns), that of the lower state is even shorter as it makes transitions down to the ground state, with the emission of radiation at $\lambda = 72$ nm in the extreme ultraviolet. The argon ion laser can be made to deliver powerful beams at several wavelengths ranging from about $\lambda = 350$ nm to about $\lambda = 530$ nm: the most prominent lines are at wavelengths $\lambda = 488$ nm and $\lambda = 476.5$ nm in the blue part of the spectrum and $\lambda = 514.5$ nm in the green. The laser is capable of *broadband* operation with many of these wavelengths oscillating at once; however, it can also be operated on a single wavelength by placing a rotatable prism in front of one of the mirrors to deflect all but the desired wavelength. Alternatively an *étalon* consisting of a fused quartz plate with optically flat surfaces acting as a Fabry–Pérot cavity will have transmission peaks that select particular wavelengths.

The physical design and construction of the laser is dictated by the extraordinary amount of power that must be dissipated. This is necessitated by the shortness of the radiative lifetime of the lasing states and the consequent need for rapid excitation of those states. Of course this in turn means that the possible output power is considerably greater than, for example, the He–Ne laser. A typical argon ion laser may operate on a D.C. electrical discharge current of 40 A requiring 200 V potential difference, that is, 8 kW of power. All but a small fraction of that power appears as heat and must be safely dissipated, typically with water cooling. An integral part of the design is a current-carrying solenoid coaxial with the tube to produce a magnetic field parallel to the discharge column in order to constrict the plasma along the center of the tube, away from the walls. The discharge is further concentrated in a capillary tube made of a refractory material such as beryllium oxide (beryllia) because of the extremely high temperatures involved. Igniting and maintaining the discharge are made easier and more predictable by incorporating a hot electron-emitting cathode. To attain the highest possible mean electron energy, it is important to maintain a high degree of gas purity, particularly avoiding molecular gases such as nitrogen and oxygen which have the effect of degrading the electron energy. Finally, the high current DC discharge can actually cause a pressure difference along the tube making it necessary to provide a return path for the argon gas outside the discharge. A schematic diagram of an argon ion laser is shown in Fig. 11.10.

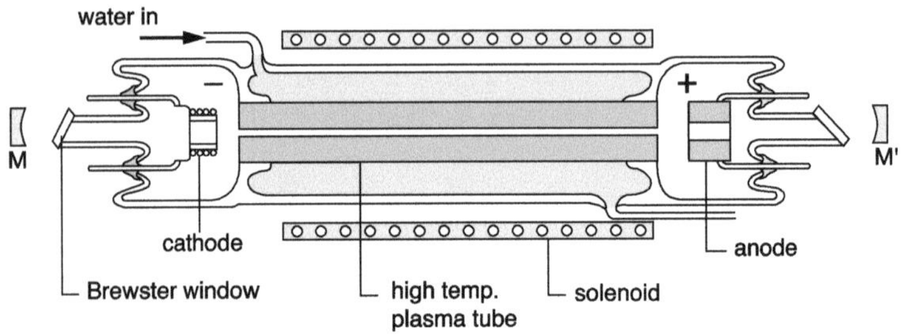

Fig. 11.10 Schematic drawing of an argon ion laser

11.7 Semiconductor Lasers

For a fundamental understanding of solid-state lasers we begin with a brief review of the concepts and properties of the energy states of electrons in solids, and their interactions with photons. We recall that a crystalline solid is composed of a three-dimensional periodic array of atoms or ions in three dimensions. The quantum description of the motion and energy of electrons throughout the solid is no longer determined just by the central Coulomb force of any one nucleus, but rather by a three-dimensional array of positively charged nuclei, each with its attractive Coulomb field. To anticipate what the quantum solution for the electron motion might be for such a periodic electric field there are two approaches: we might imagine the crystal lattice being assembled one atom at a time, or we may model the crystal as a simple, periodic electric field and analyze the motion of electrons in it.

In the first approach, we are guided by what we know when just two atoms are brought together; when the nuclei are far apart, their energy levels are identical, and when looked at as one diatomic system with one set of levels, there are two electrons in each level. As they are brought closer together to form a molecule, since there are in each state two electrons which must be described by properly symmetrized wavefunctions, they will have slightly different energy according to whether the electron spins are parallel or antiparallel. If a third atom is brought near the first two, again there will be three electrons which at first coincide in energy but as the atoms come together diverge because of the different symmetry. In the limit of very large numbers of atoms coming together to form a crystal (here we mean numbers that may reach 10^{23}), there will be a continuum of states, that is, an energy *band*. Thus we are led to the fundamental characteristic of electron states in a solid, namely they occur in bands. Just as individual atoms may have states that are not occupied by an electron, so the bands in a solid may not be filled. Figure 11.11 compares the allowed states for two potential wells and a periodic set of many wells.

The second approach to describing electron behavior in a solid begins by assuming a simple model of the Coulomb field around the nuclei as regularly spaced *square potential wells* a distance a apart. Using this model it becomes reasonable, considering that the energy of a (free) electron is related to its wave

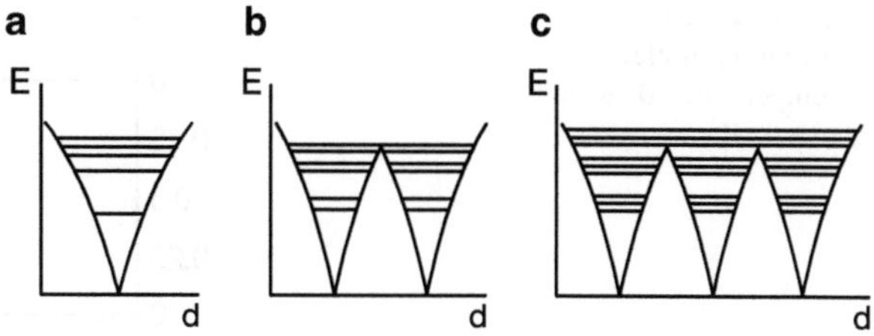

Fig. 11.11 Formation of energy bands: allowed electron states for (**a**) one nucleus, (**b**) two nuclei, and (**c**) three nuclei

vector $k = 2\pi/\lambda$ by $E = (hk)^2/2m$ that there should be gaps in the available energy spectrum arising from reflection of electron waves diffracted from successive atoms/ions, much as X-rays are diffracted in a crystal. Quantitatively, if the spatial period is a, and the wave vector of an electron is k we would expect reflection to occur if the Bragg condition is satisfied, that is, $ka = n\pi$. For values of k away from these special values, it can be shown that the energy depends on k as it would for a free electron.

On the basis of the energy band structure of a solid, we can broadly draw distinctions between a conductor, an insulator and a semiconductor. In the lowest state of a solid, the inner shells of the atoms are filled and only the outermost electrons are involved in the band structure. If the outermost band containing electrons is unfilled, then there is a continuum of energy states available above the outermost occupied state to which an electron can be elevated by an external electric field, and we have a conductor; hence the last unfilled band is called the *conduction band*. For example, an isolated Rb atom has one electron in its outermost shell, which can accommodate two electrons with opposite spin directions; therefore the corresponding energy band of N rubidium atoms can accommodate $2N$ electrons, but N atoms have only N outer electrons, which fill only one-half of the band; therefore Rb is a good conductor, as are the other alkali metals. On the other hand, if the outermost band is completely filled in the sense of the Pauli Principle, and the next empty band is so much higher in energy that at normal temperatures no electrons can reach it through thermal agitation, then it is an insulator. Finally we have the important class in which the outermost (valence) band is filled, but the next higher (conduction) band is within reach by normal thermal excitation, then we have a semiconductor. The vacancies left in the valence band by the departure of electrons going to the conduction band are called *holes*; they behave like positive electrons, so that when a hole moves under the action of an electric field, what is really happening is that *all* the real electrons move in the opposite direction. From the way they are formed, it is obvious that the number of holes left in the valence band must equal the number of electrons appearing in the conduction band. A less obvious question is: exactly how does the number of electrons in the conduction band, and therefore holes in the valence band, depend on the temperature of the semiconductor. The answer comes from the energy distribution law first derived by Fermi for particles like electrons that obey the Pauli exclusion principle. The distribution law is the analogue of the classical Maxwell–Boltzmann distribution law for the energy of particles in thermal equilibrium. We quote the law because of its central role in semiconductor theory:

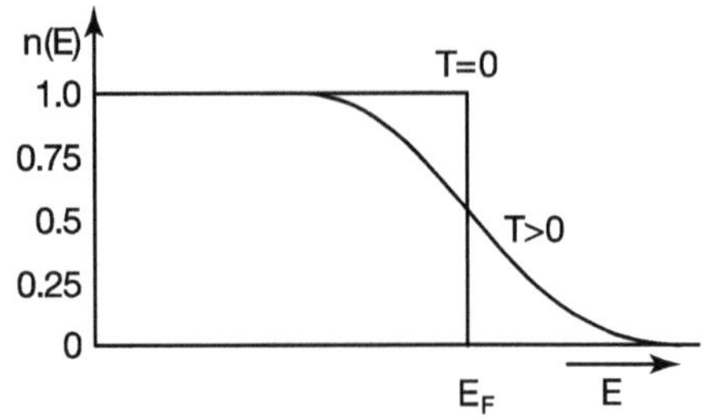

Fig. 11.12 The Fermi distribution of electrons at temperatures 0° K and above 0° K

$$F(E) = \frac{1}{\exp\left(\frac{E-E_F}{kT}\right) + 1}, \tag{11.18}$$

where $F(E)\, dE$ is the number of electrons in the range E to $E + dE$, T is the absolute temperature, and E_F is called the Fermi energy. To appreciate the significance of E_F we note that as $T \to 0$, $F(E) \to 1$ for $E < E_F$, but $F(E) \to 0$ for $E > E_F$. In discussions of the workings of semiconductor devices, the value of E_F in relation to the boundaries of the energy bands is very significant. Figure 11.12 shows the Fermi energy distribution law for electrons.

The discussion of semiconductors so far has assumed ideally pure materials with impurities well below parts in a million; in fact it is not an exaggeration to say that the whole semiconductor industry was not possible before the development of techniques of reaching such high levels of purity, and tight control of impurities. The process of adding controlled amounts of "impurities" to the melt during the growth of the semiconductor crystal is called *doping*. The product is called an *extrinsic* semiconductor in which, by doping with the appropriate material, the number density of electrons may be chosen to be greater than the density of holes (n-type), or by a different choice of dopant, the density of holes is greater (p-type). To understand the consequences of doping, we must look to the atomic structure of the elements involved. The two most common semiconductors in electronic devices are silicon and germanium, while gallium arsenide crystals find applications involving photons. Silicon and germanium have a valence of four and crystallize like carbon in a diamond crystal, with each valence electron shared in a single covalent bond with a valence electron from each of the four nearest atoms; thus all bonds are satisfied and the valence band of states formed by the many atoms of the crystal is completely filled at $T = 0$. Now suppose that through doping a lattice site is occupied not by a host atom of valence four, but by an atom of valence five, such as arsenic. We would reasonably expect that four of the arsenic valence electrons would tend to form covalent bonds with four of the host atoms leaving the fifth arsenic electron weakly bound to the remaining ion. Therefore we might expect that this extra electron would be bound to the ion by a Coulomb field reduced by the dielectric constant of the crystal, and as such would require, as would all the other similarly displaced electrons, a set of hydrogen-like quantum states to describe them. These states close to the continuum of free electrons are called *donor* states since at temperatures above absolute zero, they are raised to the conduction band,

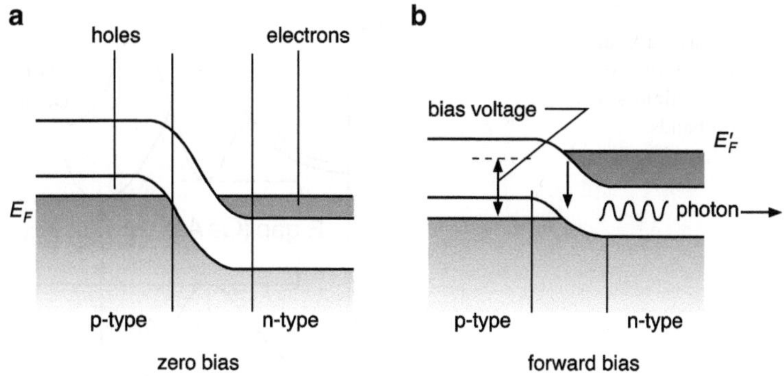

Fig. 11.13 The energy bands in a p–n junction: (**a**) isolated and (**b**) forward voltage applied

making the crystal what is called *n-type semiconductor* since it conducts electricity through negative charges. If, on the other hand, the crystal is doped with an element that has a valence of three, such as aluminum or gallium, then at a crystal site occupied by an atom of one of these elements there will be only three electrons available to satisfy the four covalent bonds of the host. If an electron is drawn from the valence band to complete the four covalent bonds, it would create a negative ion and leave a positive "hole" in the valence band, that is, one single electron state would be empty. We will accept without proof that this hole moving in the field of the negative ion would have hydrogenic energy levels, again reduced by the crystal dielectric constant, that extend in the positive direction above the valence band.

Laser diodes are based on the junction between p-type and n-type semiconductors, described briefly as p–n *junction* diodes. Before the development of light-emitting diodes and lasers, diodes were commonly used solely in electrical circuits as rectifiers that can operate at high current and low voltage. As simple compact sources of radiation, coherent or otherwise, in the red and infrared regions of the spectrum they have found many important applications including communications using fiber optics. We begin with the properties of the basic p–n junction, and the consequences of applying an external voltage across it. In Fig. 11.13 is shown the energy band diagram of an isolated p–n junction in thermal equilibrium, with the Fermi energy at equal level on the two sides of the junction. To reach this equality of E_F level, there must be a redistribution of electrons and holes across the junction to result in the energy bands being appropriately displaced.

This requires the formation of a dipole layer, called the transition region, a layer that has a preponderance of positive charge on one face and negative charge on the other, producing an electrical potential gradient within the layer. The effect on the potential is completely analogous to being on one level and going down a grade to a lower level. The formation of the dipole layer is a little complicated, but involves the movement of electrons and holes resulting in (positive) donor ions being formed on one side, and (negative) acceptor ions on the other. Suppose we now apply a voltage across the junction, and assume it is a *forward bias*, that is.

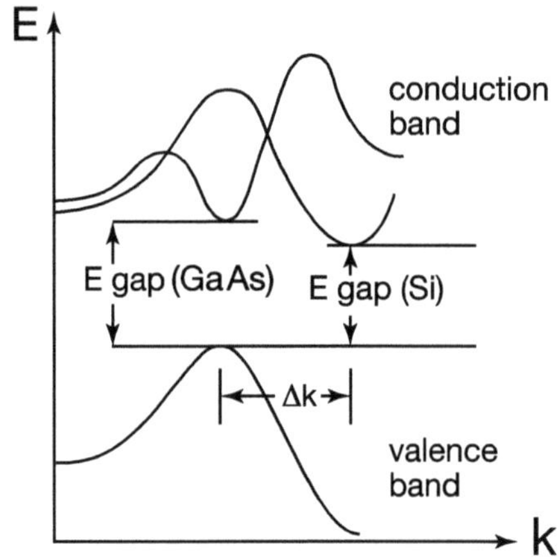

Fig. 11.14 Energy band boundaries for (**a**) GaAs and (**b**) Si showing the difference in k between the minima of the conduction bands

The p-side is positive and the n-side negative. The result is that the junction is no longer in equilibrium with the Fermi level aligned across it, but the potential step in the transition region is reduced and the diffusion and drift of charge carriers across it greatly increased. If, on the other hand, the diode were *reverse biased* the opposite effect occurs, the transition region would present a larger potential step, and the flow of current would be severely reduced. The details of the process are considerably more complicated, but that is essentially how the characteristic rectifying property of a diode is explained.

Having been inured to seeing, in the context of free atoms, transitions by electrons accompanied by the emission of radiation, the absence of radiation in a silicon diode calls for an explanation. After all, a forward-biased silicon diode will have electrons in donor states at the bottom of the conduction band and at the same time, at a lower energy, holes in the acceptor levels above the valence band. Energetically it would be possible to have photon emission, but unfortunately energy is not the only thing that has to be conserved: linear momentum must also be conserved. It happens that, except under very special conditions (the interested reader should look up the *Mössbauer Effect* in γ-ray spectroscopy), the crystal lattice does not recoil to conserve momentum, and therefore the emission process is allowed only if the transition does not entail a change in linear momentum. In order to find such transitions in different crystals the boundaries of the energy bands are plotted as functions of the electron *wavenumber*, k, which we recall is related to the linear momentum p by the equation $p = hk/2\pi$, where h is Plank's constant. Such plots for silicon and gallium arsenide are shown in Fig. 11.14.

Noting that in the conduction band the occupied states are near the minimum point on the boundary, we see that in GaAs the minimum occurs at the same value of $k = 0$ as the maximum of the valence band, a property that distinguishes it from crystals such as Si in which the stationary points occur at different k values; the former are referred to as *direct* semiconductors while the latter are *indirect*. A radiative transition between the two energy bands in Si is ruled out because it would require a change in k (that is, linear momentum), which we have already

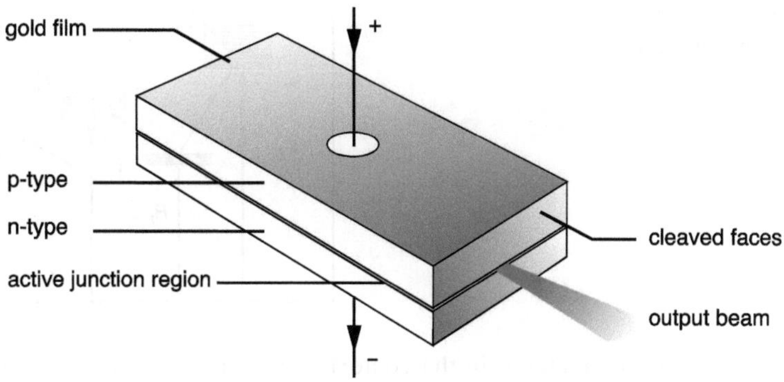

Fig. 11.15 A gallium arsenide diode laser

emphasized is not taken up by the lattice. On the other hand a transition in GaAs requires no change in k, and indeed that crystal is commonly used as a light-emitting diode, with light output in the near infrared at $\lambda = 870$ nm.

To convert the light-emitting diode into a laser requires only an optically resonant cavity to be formed around the diode. Since the density of electron–hole pairs is high, the gain is correspondingly high, relaxing the need for a high-Q optical cavity, so that the cleaved faces of the crystal are adequate. A basic diode laser is extremely small, typically no more than fractions of a millimeter, consisting of a gold-plated rectangular chip with polished ends, as shown in Fig. 11.15.

The refractive index is so high that the reflectivity of the surfaces is adequate without coating; however, for efficient operation the temperature must be kept low even at the extremely high current densities required for laser action.

Lastly we should mention the heterojunction lasers that are designed to reduce the threshold current for oscillation by adding one or more layers of different band gap, formed epitaxially, on either side of the GaAs diode.

11.8 Crystalline Solid Lasers

11.8.1 The Ruby Laser

The first observation of superradiance, an indicator of stimulated emission and the precursor of laser action, was made by Maiman on a ruby crystal in 1960; it was a groundbreaking discovery reported in the British journal Nature, under the title "Stimulated Optical Radiation in Ruby Masers" [8], after it had been initially turned down by the Physical Review Letters [9]!

The gem ruby is in fact "doped" crystalline alumina (Al_2O_3), which in pure form is colorless, but when small amounts of *chromium* are added as powdered chromium oxide to the powdered alumina fed into the crystal growing furnace, the result is an artificial ruby, whose depth of red color depends on the relative concentration of chromium. Maiman was well acquainted with the properties of ruby since he had

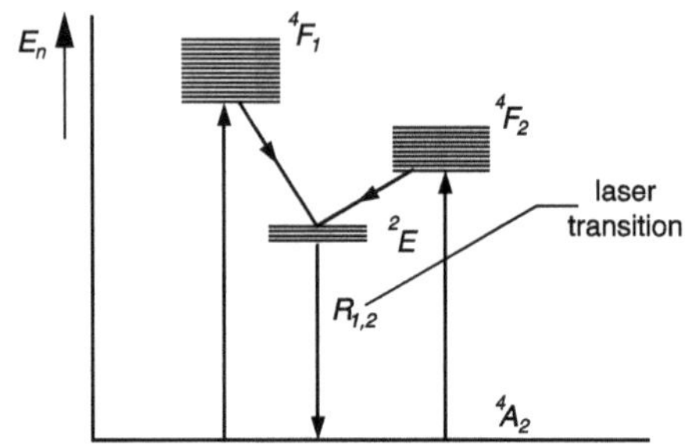

Fig. 11.16 Energy levels relevant to the ruby laser

pursued studies in that material in the context of maser design in the microwave region of the spectrum. We assume that in a moment of inspiration he realized that it is possible for an atomic ion such as that of chromium, although subjected to the strong binding forces in a crystal, can yet have an *optical* spectrum that is not totally broadened. The fact is that chromium is among the so-called *transition elements* because of their position in the periodic table, where they interrupt the natural progression of elements within the table; they begin occupying the 4s shell *before* filling up an *internal shell* the 3d shell. The trivalent chromium ion in the ruby crystal has three electrons removed from the 3d shell while retaining the outer 4s shell, which somewhat acts as a shield from external fields, whether in a crystal or in solution. The transition element ions in crystals are generally colored because of optical absorption in the visible part of the spectrum. Their spectra are somewhat broadened by thermal vibration in the crystalline field, which may dictate operating at low temperatures.

In ruby the trivalent chromium ions Cr^{3+} occupy lattice points in the host alumina crystal, where their quantum level structure is altered by the strong crystalline fields, but fortunately it is not necessary for our purposes to pursue the theoretical basis for the spectroscopic classification of the states; we will use the spectroscopic notation only as labels for the states. In Fig. 11.16 are shown the essential quantum levels involved in the optical pumping of the lasing states and the stimulated emission that results from it.

The laser transitions R_1 and R_2 occur between the two upper levels labeled as ^2E and the lower level labeled $^4A_2$ at $\lambda = 694$ nm. The pumping (absorption) transitions are from the lower level $^4A_2$ to two broad upper energy bands $^4F_1$ and $^4F_2$. These latter transitions occur in the green to violet region of the spectrum and account for the red color of the crystal, although the laser crystal has a lower density of chromium and is pink in color rather than the deep red that we expect of the gem. For the purposes of analyzing the operation of the laser we may regard the two F-bands as two equivalent components of the same pumping transition, so that the laser operates really with *three* functionally different levels. For the efficient production of population inversion between the upper and lower laser levels, we must satisfy two conditions: first, the absorption rate by the two F-bands must be very high, and second their radiative lifetime must be very short compared with the upper laser level. The availability of a relatively broad absorption spectrum

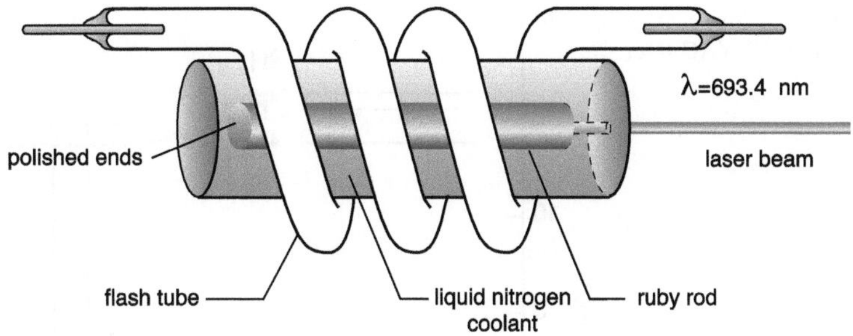

Fig. 11.17 Schematic drawing of Maiman's historic ruby laser

explains the use of a powerful flash lamp surrounding the ruby crystal in Maiman's classic laser design, as shown in Fig. 11.17. As the pumping proceeds, the population of the upper laser level builds up until the critical level is reached for laser oscillation, when the light intensity suddenly increases and with it the induced emission, leading to the upper level becoming depleted, and oscillation stops; the result is an output spike.

The ruby laser is remarkable, not just historically, but because of the way it fulfills the essential requirements of a laser: The ion number density is much higher than can be achieved in a plasma, it can be pumped optically with a broad spectrum, and by polishing the ends of the crystal, a stable optical cavity is formed.

11.8.2 The Nd^{3+}YAG Laser

The YAG laser is another optically pumped solid-state laser that has become one of the most widely used sources of coherent light; it finds applications in such diverse fields as medicine, defense, and industry, in addition to pure research. It is based on the ion Nd^{3+} of the *rare earth* element neodymium as a dopant in the crystal lattice of yttrium aluminum garnet or YAG ($Y_3Al_5O_{12}$), in which it substitutes for the yttrium ion Y^{3+}. Large garnet crystals can be grown artificially to produce pure, homogeneous, optical quality components, free of strain and transparent from a wavelength of about 4 μm in the infrared to 300 nm in the ultraviolet regions of the spectrum. Although commonly doped with neodymium, YAG can accept other trivalent ions of the rare earth atoms.

As a laser, it is classified as having four levels with the lower laser level $^4I_{11/2}$ so far above ground in energy that in equilibrium its population is near zero. The pertinent energy levels of the laser are shown in Fig. 11.18.

The laser is usually pumped to the $^4F_{5/2}$ level with red light at $\lambda = 808$ nm from which transitions lead to the upper laser level $^4F_{3/2}$. As the population of that level builds up, population inversion is quickly created with respect to the lower lasing level, since the latter is nearly empty; the consequence is the early onset of stimulated emission at $\lambda = 1{,}064$ nm and loss of the population inversion. If a more intense output is desired it is possible to delay oscillation by purposely

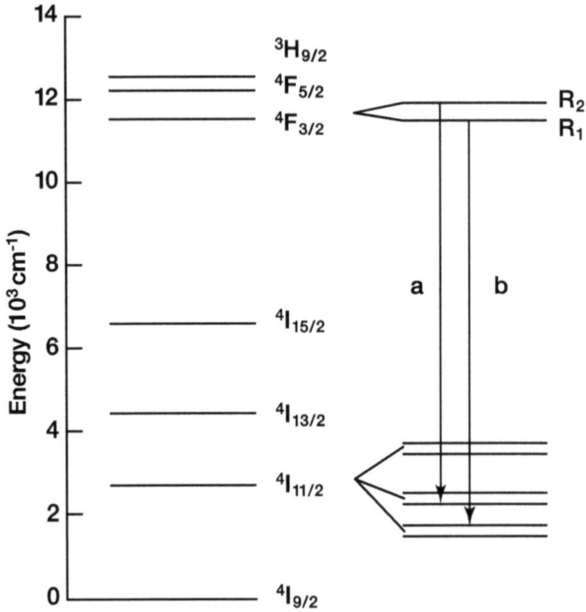

Fig. 11.18 The four energy levels in the Nd^{3+} YAG laser

preventing feedback during the buildup using a controlled misalignment of one of the reflecting surfaces, a technique called *Q-switching*. The reduction of the Q raises the threshold for oscillation allowing the population inversion to rise to a level beyond what is otherwise achievable. When the population inversion reaches its peak value and the gain per pass is at its highest value, the Q is suddenly switched resulting in an extremely rapid buildup of the output radiation. There are two possible sources for optically pumping the laser: flash lamp or a stack of diode lasers. By far the most efficient method is to use a stack of diode lasers aligned with the axis of the crystal and tuned to selectively populate the upper lasing level directly. In this manner this laser is capable of producing up to 250 W of coherent radiation.

11.8.3 The Ti^{3+}:Sapphire Laser

By far the most versatile laser, combining high output power and wide wavelength range, is the Ti^{3+}:Sapphire laser. It was introduced in 1982 by Peter Moulton [10] and has since enjoyed widespread commercial development and become the *tunable* laser of choice for many applications. It combines the excellent physical and optical properties of sapphire with a tuning range that extends from a wavelength of 660 to 1,050 nm, thereby supplanting messy dye lasers.

Like the chromium ion in the ruby laser, the active element in this laser is the trivalent ion of titanium, another transition element. It has a single 3d electron outside a closed shell structure. The host crystal is sapphire (corundum) Al_2O_3 with

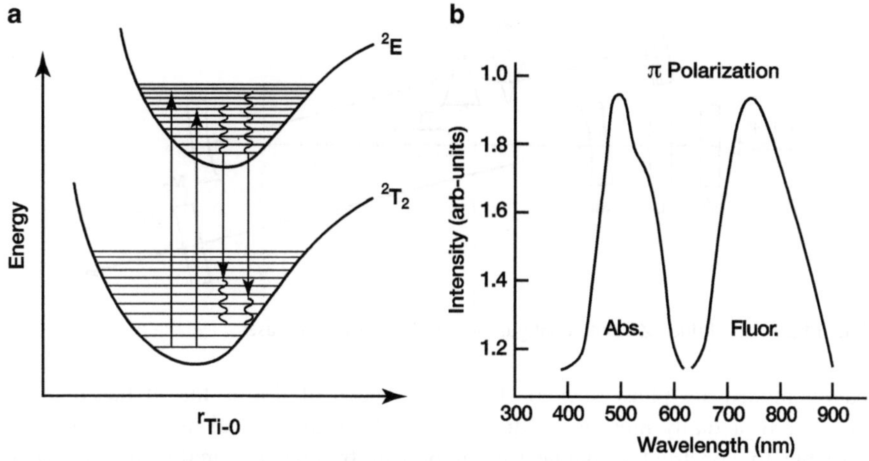

Fig. 11.19 (a) The $^2T_2$ and ^2E energy bands of Ti^{3+}:sapphire and (b) its absorption and fluorescence bands at room temperature

trigonal symmetry, in which the titanium ion occupies an aluminum site. The crystal field splits the otherwise five degenerate 3d states, resulting in among others two vibrational bands conventionally referred to as the $^2T_2$ ground electronic state and an excited state labeled ^2E; as a laser medium Ti:sapphire is classed as a four-level system. Figure 11.19 shows the $^2E \rightarrow {}^2T_2$ absorption and fluorescence bands at room temperature.

The crystal exhibits absorption bands in other regions of the spectrum, for example, in the ultraviolet and infrared regions of the spectrum, but the one of present interest occurs in the blue-green region of the spectrum, as shown in Fig. 11.19. The fluorescent emission at longer wavelength that follows the absorption of photons was observed by Moulton to decay exponentially with a time constant of about 4 μs at low temperature, falling slightly at room temperature, independently of the Ti concentration.

The basic optical layout of a Ti:sapphire laser is shown in Fig. 11.20. The crystal, measuring several millimeters across with ends cut at the Brewster angle with respect to its axis, is placed at the intense beam waist of the two concave mirrors forming part of folded Z four mirror configurations made necessary to correct astigmatism caused by the crystal and the concave mirrors, with the two arms of the cavity bounded by plane mirrors. The pumping beam is introduced through an antireflection-coated converging lens. The threshold pumping intensity is relatively high, in the range of 4–5 W at $\lambda = 532$; however, there are several options for a source, for example, a flashlamp-pumped dye laser or a frequency doubled Nd:YVO_4 laser, or more commonly a diode-pumped solid-state laser.

Among all the solid-state laser amplifying materials Ti:sapphire has an extraordinarily wide frequency bandwidth, which means that it is capable of delivering extremely short light pulses, or a comb of extremely sharp frequencies, when the

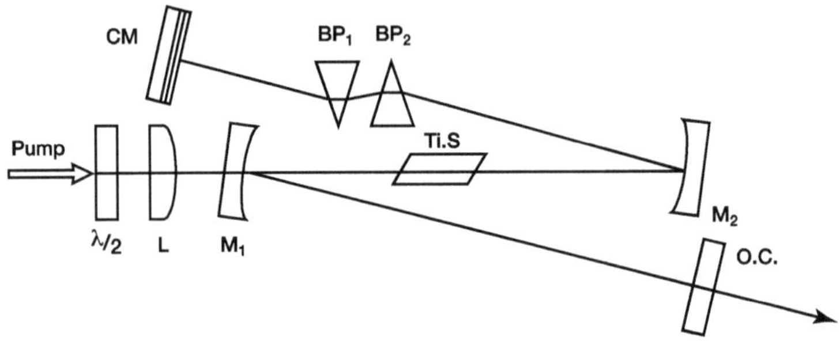

Fig. 11.20 The folded Z cavity configuration of the Ti:sapphire laser

cavity modes are phase locked. The significance of this is that it has made it possible to actually measure frequency directly in the optical range; that is, it is possible to construct a synthesizer at optical frequencies analogous to one at microwave frequencies [11].

11.9 Laser Cooling of Atoms

The central point in understanding how a laser beam might be used to cool atoms is that the absorption of photons by atoms does not only excite them to higher internal states, but since photons carry linear (as well as angular) momentum, the absorption of a photon must, according to the law of the conservation of momentum, result in a change in momentum of the atom. Long before the famous $E = mc^2$ of Einstein it was known through Maxwell's electromagnetic theory that light reflected from a surface exerted a pressure on it. Even before that there was the debate as to whether light is made up of particles or waves, and it was proposed that if light indeed consist of particles, then it should be possible to measure the pressure it exerts on a reflecting surface. To prove this point, attempts were made in fact to shine powerful light beams on suspended surfaces in the hope of seeing a deflection. A product of this line of thinking was the familiar Crookes radiometer, which consists of a glass bulb inside which light metallic vanes, blackened on one side and reflecting on the other, were mounted to spin freely, which in fact they do when light shines on them. However, it is now accepted that the dominant reason for the rotation is really the thermal imbalance in the air surrounding the vanes. Even if it could be proved that the light exerted a mechanical pressure on the vanes, it still would not have disproved the wave theory, as was pointed out long before Maxwell's theory of light by the great Swiss mathematician Leonhard Euler.

In terms of Maxwell's theory of light as an electromagnetic wave, it is not difficult to show that a light beam falling perpendicularly on a metal surface must exert a pressure on it proportional to the intensity of the beam. The conducting surface is subjected to mutually perpendicular electric and magnetic fields oscillating at optical frequency. The electric field causes the flow of electric current which is then acted on by the magnetic field to produce a force directed at right angles to both, that is, in line with the direction of the beam. In the original Maxwell theory such forces were interpreted through stress in a medium, which Einstein's theory dispensed with, leading to the concept that the field itself carries momentum. If a light beam of intensity $I(\mathrm{W/m^2})$ falls on a perfectly absorbing surface so that the directed momentum of the beam is converted to random thermal motion with zero total momentum, then there is a continual change of momentum which by Newton's law must be taken up by the absorber in the form of radiation pressure which can be shown to be given by $P = I/c$. This result can easily be derived using de Broglie's relationship for the momentum of the photons making up the light intensity. Since $I = jh\nu$, where j is the flux density of photons we have $P = jh/\lambda = jh\nu/\lambda\nu = I/c$. The numerical value of radiation pressure is extremely small for ordinary sources of light; for example, the intensity of direct sunlight on the earth's surface is about 1.6 $\mathrm{kw/m^2}$; this produces a pressure of only 5×10^{-6} $\mathrm{newtons/m^2}$.

The first direct observation of the mechanical effect of laser light is attributable to Ashkin who around 1970 reported on an experiment in which a focused argon ion laser beam was directed at a suspension of small latex spheres in water contained in a glass cell. He was able to manipulate the beam and observe with a microscope its interaction with a selected sphere. He explained the motion of the latex spheres using ray optics, although the diameters of the spheres were only about twice the wavelength of the laser light and therefore a full solution in terms of the wave theory is required. This work is noteworthy mainly because it was the first to address the problem of the dynamic interaction of laser radiation with matter. During the subsequent decade a far more fruitful line of research was pursued by many laboratories: namely to use laser radiation to *cool* atoms and ions to very low temperatures. The motivation originally was to reduce the Doppler broadening of spectral lines. The pioneering work in this endeavor is attributable to several groups including those led by Hänsch, Dehmelt, Wineland, Tannoudji, Phillips, Toschek, and Chu, among others. The first experimental success achieved by Wineland et al. came in 1978 with the laser cooling of *ions* confined in a magnetic Penning trap to a temperature of about 40° K. This was followed in 1982 with the successful cooling of neutral atoms by Phillips et al. and by the late 1980s the lowest temperature theoretically achievable by the Doppler technique, which we shall discuss below, had been reached. In 1988 Phillips et al. using a technique demonstrated by Chu et al. on Na atoms were surprised to find their temperature measurements indicate that the lowest temperature attainable was below the Doppler limit. Finally a theoretical analysis by Cohen-Tannnoudji, a professor at the Ecole Normale Superieure in Paris demonstrated that the supposed barrier to cooling the atoms further, can be breached, taking the temperature limit down to the micro-Kelvin level; he shared the Nobel Prize in physics with Phillips and Chu in 1997.

The treatment of the exchange of linear momentum between a laser beam and an atom or ion requires of course a quantum description. We are already familiar with the radiation process called resonance fluorescence in which a monochromatic incident light beam causes strong fluorescence of the atoms if the photon energy is equal to the difference in energy between two quantum states of the atoms.

Provided certain selection rules governing quantum numbers of the states are satisfied the photons are strongly absorbed then reemitted. The net transfer of linear momentum is obtained by applying the momentum conservation law between the initial incident photon-atom state and the final reemitted photon-atom state. The angular distribution of the reemitted photons, that is, the *radiation pattern*, is governed by the change in the angular momentum state of the atom and the type of transition involved: whether it is generated by an oscillating electric dipole or quadrupole, etc. However, the *total* linear momentum carried away by the reemitted photons also depends on the type of transition, but in the case of spontaneous emission the total momentum will be zero since, by symmetry, the intensity along any given direction must equal the intensity in the opposite direction; the atom has no memory of how it was excited. This does not mean the scattering is isotropic but rather that the photon has equal probability of emerging in any given direction as in the diametrically opposite direction. This means that in order to conserve linear momentum the atom must receive an increment of momentum or an impulse in the direction of the beam. As the atom continues to absorb and reemit photons it will experience a succession of impulses in the direction of the beam tending to either accelerate it or slow it down depending on the initial relative direction of the beam and the motion of the atom. To get an idea of the magnitude of this process and see if it has practical implications or is merely of academic interest, let us assume a beam of intensity I incident on an atom moving in the opposite direction to the beam, and presenting an absorption cross section σ, so that $(I/c)\sigma$ is the force exerted on the atom tending to slow it down. If the atom belongs to a gas at an absolute temperature T, then its mean kinetic energy along the direction of the light beam is $(kT/2)$. Therefore if the atom continues to absorb and reemit photons, it will be on the average stopped after moving a distance D given by the conservation of energy:

$$\frac{I\sigma D}{c} = \frac{1}{2}kT. \tag{11.19}$$

Inserting practical numerical values $I = 10^2$ W/m^2, $\sigma = 2.5 \times 10^{-13}$ m^2 (for resonance absorption), and $T = 300$ K, we find D to be on the order of 2.5 cm, a very practical value. In this rough estimate we have neglected the finite radiative lifetime of the upper atomic state which of course will limit how fast photons can be absorbed and reemitted. In fact if the light intensity reaches the point where the absorption rate is so high that the upper state lifetime allows the buildup of atoms in that state, another radiation process becomes significant, namely induced emission, which has a very different radiation pattern of emission, and the above argument does not apply.

So far we have a mechanism for imparting momentum to an atom using a laser beam; this in itself does not cool the atoms. What is required is an interaction with the laser beam that opposes the motion of the atoms in whatever direction they are moving, whether toward the laser source or away from it. This may be accomplished by using the Doppler effect. If the atom has a strong resonance line in its

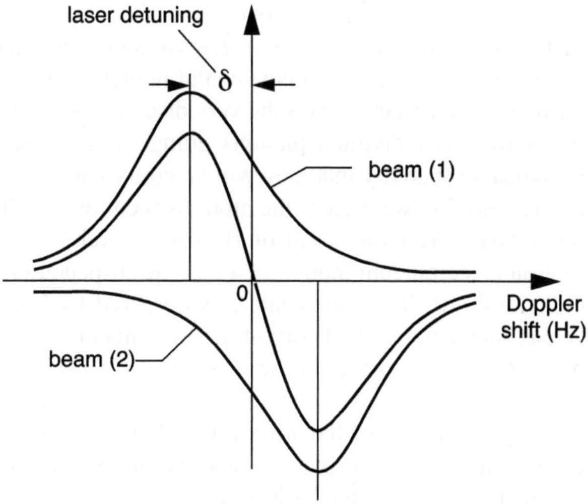

Fig. 11.21 The force on an atom in two opposing laser beams tuned below resonance

spectrum, as many atoms do, the trick is to use a laser beam whose frequency is set on the *low* side of the peak absorption frequency. This simple change has the effect of making atoms moving against the beam absorb more than those moving in the same direction because the *Doppler effect* shifts the laser frequency closer to the peak resonance for atoms moving against the beam, and farther from resonance for the atoms moving away. Therefore the momentum and kinetic energy are reduced for some atoms without increasing it for others, with a net reduction in kinetic energy, and therefore of temperature. The idea can be extended to *two* collinear laser beams traveling in opposite directions and tuned to the low side of the atomic resonance peak. Now we have the atoms being slowed down coming and going. Thus ideally an atom subjected to the combined effects of the opposing laser beams will move as if acted on by a frictional force. If we assume that the atom absorption line has the ideal Lorentzian shape characteristic of natural lifetime broadening of spectral lines, we can construct a plot of the force acting on the atom as a function of its velocity. Such a plot is shown in Fig. 11.21.

The optimum choice of detuning from the center of resonance is half the linewidth, $\Delta\nu/2$, and an atom with a Doppler shift between $+\Delta\nu/2$ and $-\Delta\nu/2$ in the field of two opposing lasers will experience the equivalent of a viscous force opposing its motion. If the instantaneous velocity of the atom is V and the scattering cross sections for the two beams are represented by σ_+ and σ_- it can be shown that for a Lorentzian resonance line shape we have the following result:

$$\langle F\rangle = \langle\sigma_+ - \sigma_-\rangle\frac{I}{c} = -\sigma_0\frac{IkV}{c\pi\Delta\nu}, \tag{11.20}$$

where k is the wave number of the laser beam and V the velocity of the atom.

The logical next question to address is: how far can this cooling process be continued? We have already hinted in the historical introduction that there are certain barriers that must be overcome. One obvious limitation with the Doppler method is the natural lifetime of states and the consequent spectral linewidth, even for atoms at rest. A more fundamental limitation to any process relying on photon absorption and reemission is the very discreteness of the photon; the emission must be viewed as individual photons going off in different directions with only a statistical probability that they will be emitted in a particular direction. As a result of the emission we picture the atom as recoiling in all directions in a statistically distributed way reminiscent of Brownian motion. We may recall that Brownian motion is the random motion of a particle suspended in a liquid caused by random collisions with liquid molecules, as observed by botanist Brown under a microscope. Incidentally the theory of the Brownian motion is notable as one of the few non-relativity papers published by Albert Einstein in 1905. It was in fact an important paper in which the kinetic theory of molecular heat was put to the test. In the present case of the atom it recoils $(I\sigma_0)/h\nu$ times per second, each time its momentum is changed by (h/λ), and therefore from random walk theory the mean momentum *gained* after 1 s is given by

$$\langle p^2 \rangle = 2\left(\frac{h}{\lambda}\right)^2 \frac{I\sigma_0}{h\nu}. \tag{11.21}$$

From this we are led to the limiting condition when the rate of laser cooling is equal to the rate of heating through random recoil. This occurs when the mean temperature T has been reduced to the value given by

$$k_B T_{\min} = \frac{1}{4\pi} h\Delta\nu_n, \tag{11.22}$$

where $\Delta\nu_n$ is the natural spectral line width. Any increase in the laser beam intensity beyond that point merely increases the intensity of the scattered photons without cooling the atom to a lower temperature. If typical experimental values are inserted we find the limiting temperature is around 120 μK. At this temperature the mean momentum of the atom is at the same order of magnitude as the momentum of a single photon! In the case of atomic ions confined in an electromagnetic trap such as the Penning trap, the absorption spectrum of the ion is a discrete line spectrum due to the periodic motion of the ions, and the laser must be tuned to a lower *side band*. But this may affect only one mode of oscillation in the trap, and therefore a special design is required to ensure that cooling takes place along all three orthogonal directions.

To achieve efficient cooling of a gas in three dimensions requires the use of three mutually perpendicular pairs of laser beams. Such a configuration was used successfully by a group at Bell Laboratories to cool a cloud of free neutral sodium atoms. The reduction in the velocities of the atoms results of course in a slowing of the diffusion rate, equivalent to a restraint on the motion of the atoms, a condition dubbed "optical molasses". At first these observations seemed to agree qualitatively with the theory as outlined, but careful measurement by W. Phillips and his coworkers at the NIST in the USA proved that the temperature reached was in fact *below* the Doppler limit. There was naturally skepticism at first since the processes assumed in its derivation were well known—usually a theoretical limit is an ideal limit seldom attained, let alone exceeded! Clearly some other mechanism was at work. It was Cohen-Tannoudji at the ENS in Paris [12] who discovered an optical pumping effect he dubbed the Sisyphus effect. The name suggests a process in which you are forever facing an uphill struggle. Sisyphus was the legendary king of Corinth who was condemned in Hades to push a heavy rock up a hill for all eternity only to have it roll back. Cohen-Tannoudji's early achievements were in the area of the quantum theory of optical pumping and so he was naturally inclined to do a formal quantum analysis of the photon interactions with all the magnetic sublevels of the sodium atom. Previously we had encountered optical pumping between hyperfine states in the context of the rubidium frequency standard. But equally important is Kastler optical pumping of Zeeman substates using polarized resonance light, which produces a global atomic polarization, that is, a magnetic moment. The mechanism that Cohen-Tannoudji and his coworkers examined was one that had been extensively studied in the past, namely the energy displacement of the atomic quantum levels by the interaction with the laser light. The "light shift" had long been recognized as one of the causes of instability in the optically pumped Rb frequency standard. They found that under suitable conditions a rapid spatial variation in the polarization of the laser light can lead to a lower temperature than the Doppler limit. It is assumed we have two *overlapping* laser beams linearly polarized at right angles to each other, traveling along the same line in opposite directions. Let us focus on one point along the beams: the optical electric field of the two waves may have any phase difference between them. At points where there is a $\pm 90^\circ$ phase difference between the waves, the resultant is circularly polarized. Now it was known that continued absorption–reemission cycles (pumping) by circularly polarized light of opposite sense produce energy level shifts in opposite directions and further that there is a net transfer of population among the magnetic substates. Without getting further into the details of the theory, the result is that the atoms are always being pumped into the substate whose energy is on the *rising* side of the energy curve, independent of the direction the atom is moving. The result is a loss of kinetic energy, that is, a lowering of the temperature. A rather unexpected property of this cooling mechanism is that it is favored by having the lasers tuned farther from the center of the atomic resonance line than the Doppler method. The ultimate limit to the achievable temperature is set by the recoil of the atom in the photon emission process; for Rb the limit is on the order of micro-Kelvins.

In Chap. 18 we will discuss the more recent development of cooling methods that can take us to even lower temperatures. They involve nonlinear Raman type of processes resulting in "recoilless" cooling.

References

1. A.G. Fox, T. Li, Bell Syst. Tech. J. **40**, 453 (1961)
2. G.D. Boyd, J.P. Gordon, Bell Syst. Tech. J. **40**, 49 (1961)
3. W. Demtrőder, *Laser Spectroscopy* (Springer, Heidelberg, 1981)
4. W.E. Lamb Jr., Phys. Rev. **134**, 1429 (1964)
5. A.L. Schawlow, C.H. Townes, Phys. Rev. **112**, 1940 (1958)
6. A. Javan, W.B. Bennett, D.R. Herriot, Phys. Rev. Lett. **6**, 106 (1961)
7. W.B. Bridges, Appl. Phys. Lett. **4**, 128 (1964)
8. Maiman Nature, **187** 493 (1960); also Phys. Rev. Lett.**4** 564 (1960)
9. Physics Today, Oct. 1988
10. P. Moulton, J. Opt. Soc. Am. B **3**, 125 (1986)
11. S.T. Cundiff et al., Rev. Sc. Instrum. **72**, 3749 (2001)
12. T. Cohen, J. Dalibard, J. Opt. Soc. Am. **B6**, 2023 (1989)
13. Physics Today, op. cit
14. G. Werth, V.N. Gheorghe, F.G. Major, *Charged Particle Traps* (Springer, Heidelberg, 2009)

The Mechanical Gyrocompass 12

12.1 Gyroscopic Motion

We are of course all familiar with examples of spinning bodies: from a child's spinning top to the spinning earth and heavenly bodies in general. Such a body spinning freely in space will continue doing so indefinitely about a fixed axis; indeed the absence of any obvious forces of interaction between the parts of such a body must have convinced the ancients that the circular motion is "natural" and involves no forces. In fact the dynamical behavior of rotating bodies provides proof of the correctness within human experience of Newton's laws of motion and the role of forces; Newton's theory fully describes the often unintuitive dynamics of spinning bodies.

The central postulate of his theory is the existence of a frame of reference in which a body will preserve its state of rest, or uniform motion in a *straight line,* unless it is acted on by a force to change its motion; that is his first law of motion. Such a frame of reference is called an *inertial frame*. But where is such a frame of reference to be physically found? There is no absolute frame of reference; certainly not if Einstein's theory of relativity is to be believed. A frame of reference fixed in the earth is clearly *not* inertial; thus if a long-range cannon is fired in a certain direction, the path of the cannon ball relative to a frame of reference fixed in the earth will certainly not be a straight line because of the latter's rotation. The longer the path of the cannon ball, the greater will it deviate from a straight line with respect to the earth's frame of reference. In free flight the cannon ball travels in a straight line with respect to an inertial system (such as the fixed stars) but with respect to a rotating earth it will deviate from a straight line. This deviation is often artificially attributed, in the spirit of Newton's laws, to a force called the *coriolis force*. We shall return to the mathematically difficult question of describing the dynamics of a rotating body referred to a rotating system of coordinates later in this chapter.

The gyrocompass is based on the gyroscope, which is essentially a rapidly spinning rigid body, called the rotor, mechanically isolated in such a way that its

F.G. Major, *Quo Vadis: Evolution of Modern Navigation: The Rise of Quantum Techniques*, DOI 10.1007/978-1-4614-8672-5_12,

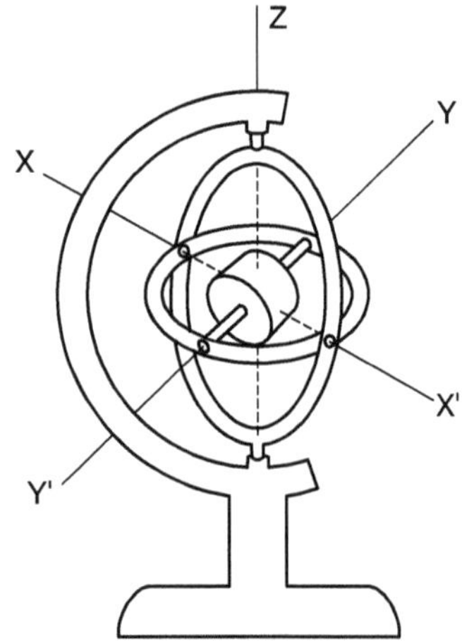

Fig. 12.1 Gyroscope mounted on gimbals

axis of rotation is free to point in any direction with a minimum of friction. The drive to perfect the gyrocompass as a navigational instrument has been directed at reducing that friction and maintaining a horizontal position along the meridian. Mechanical gyrocompasses achieved isolation through the use of *gimbals* to allow the axis of spin to be free to assume any orientation; these are nested rings with bearings at opposite ends of diameters as shown schematically in Fig. 12.1. Other more robust low-friction methods of suspending the rotor that have evolved in the course of development of the gyrocompass will be discussed later.

The property that makes a *freely* spinning body an indicator of spatial orientation is of course the conservation of its (vector) angular momentum about its center of mass in an inertial frame of reference, that is, one defined by Newton's laws. In general the angular momentum of a body about a given point is separable into the angular momentum due to the motion of the center of mass and the spin angular momentum about the center of mass. The spin angular momentum is a vector quantity with a magnitude and direction along the axis of spin; therefore constant angular momentum means in particular a constant direction of that axis. It is worth noting that this conservation law does not imply that there are no forces present; indeed there are strong forces holding the body together and providing the net centripetal force to keep the parts in circular motion. In fact the description "rigid body" assumes that these forces are so strong that we may neglect any elastic properties the body may have. The point is that these forces obey the law of action and reaction and cancel out in the summation over all parts of the body, yielding zero torque on the body. Such would not be the case, however, for example, in a system consisting of charged particles, where the angular momentum of the electromagnetic field must be taken into account.

In discussing the motions of a spinning body it simplifies matters to regard the body as made up of a discrete set of particles whose positions are given in terms of two coordinate systems: one an inertial system XYZ and the other a system xyz imbedded in the rigid body, and moving with it, whose origin is chosen at the center

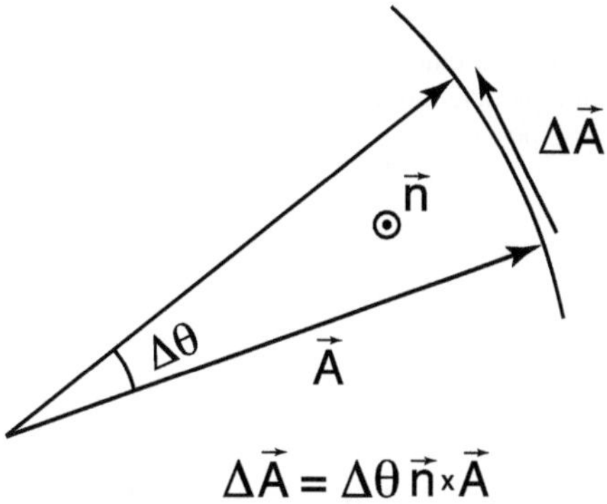

Fig. 12.2 The change in a vector due to rotation

of mass of the body. Since we are concerned with only rotational motion, we assume that the origin of the body system xyz is fixed at the origin of the inertial system XYZ. The orientation of the body axes with respect to the inertial system is completely specified by three angles, and the body is said to have three degrees of rotational freedom. In classical theory there are more than one possible set of angles; one widely used set consists of the *Euler angles*, named for the famous Swiss mathematician. The interested reader will find ample discussion of this topic in standard texts on classical mechanics [1]. We will assume without proof that the angular velocity of a body, that is, the rate at which it turns about a given axis in radians per second, usually represented by the Greek letter omega Ω, is a *vector* having magnitude and direction, as do linear velocity and force. We will further assume the fundamental relationship giving the rate of change of a vector with respect to the inertial system in terms of its value with respect to a body-system rotating with an angular velocity of Ω. Thus if $\vec{A}$ represents any vector in the body system which varies in time *solely* by virtue of the rotation of the body through an angle $\Delta\theta$ in a time interval Δt, the change in the vector $\vec{A}$ is the vector product $\Delta\vec{\theta} \times \vec{A}$, as illustrated in Fig. 12.2.

By dividing by Δt and proceeding to the limit $\Delta t \to 0$ the following fundamental result in gyroscopic theory is obtained:

$$\left(\frac{d\vec{A}}{dt}\right)_{\text{in}} = \vec{\Omega} \times \vec{A}, \tag{12.1}$$

in which $\vec{\Omega}$ is the angular velocity vector in radians per second. We have used the vector notation for a vector product, defined as the vector $\Omega A \sin\theta\, \vec{n}$, where θ is the angle between the vectors $\vec{A}$ and $\vec{\Omega}$ and $\vec{n}$ is a unit vector perpendicular to the plane containing those two vectors. The use of vector analysis in dealing with rigid body rotation is almost inevitable; but we will limit ourselves to the description of salient results without any pretense at logically developing the theory; the reader

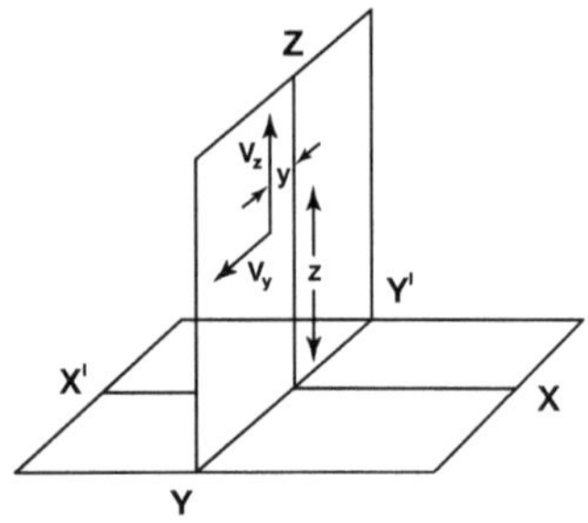

Fig. 12.3 The angular momentum along the X-axis of a particle moving in the Y–Z plane

may wish to skip the more mathematical content in this section. If the vector $\vec{A}$ is simply the position vector $\vec{r}$ of a particle, then the above equation takes the form:

$$\left(\vec{V}\right)_{\text{in}} = \vec{\Omega} \times \vec{r}, \tag{12.2}$$

where $\vec{V}$ is its velocity vector in the inertial frame.

To find an expression for the *angular momentum* of the body, consider the angular momentum of just one constituent particle of mass m, whose position is given by the radius vector $\vec{r}$, moving with a velocity vector $\vec{V}$, as shown in Fig. 12.3.

The angular momentum about an axis is the length of the radius vector $\vec{r}$ of the particle multiplied by the component of the linear momentum $m\vec{V}$ perpendicular to that radius vector. If the angle between $\vec{r}$ and $\vec{V}$ is θ then the required angular momentum is $mrV \sin \theta$ and has its direction perpendicular to the plane containing $\vec{r}$ and $\vec{V}$. But this, by definition, is the vector *cross* product of the vectors $m\vec{r}$ and $\vec{V}$.

For a rotating system of particles, of which the body is assumed to be composed, the total spin angular momentum $\vec{M}$ is given by the following:

$$\vec{M} = \Sigma m_i\left(\vec{r}_i \times \vec{V}_i\right) = \Sigma m_i\left\{\vec{r}_i \times \left(\vec{\Omega} \times \vec{r}_i\right)\right\}. \tag{12.3}$$

If the only forces present in the system are those of interaction between the particles, that is, for a body moving freely in space, the vector $\vec{M}$ is constant in magnitude and direction. This can be shown to follow from Newton's laws of motion, but more fundamentally from the invariance of the system under a change in orientation in space; that is, the isotropy of space.

Without getting any further into the mathematics we will assume as given that there is a choice of coordinate axes *in the body* called the *principal axes of inertia* (x_1, x_2, x_3) with respect to which the components of the angular momentum have the *diagonal* form:

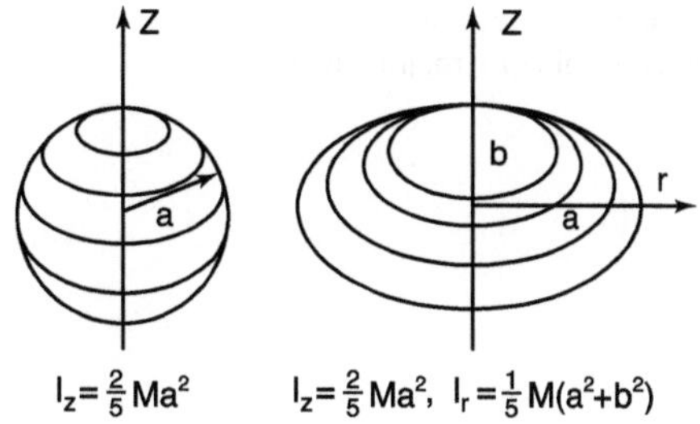

Fig. 12.4 Principal moments of inertia of a sphere and ellipsoid of revolution

$$M_1 = I_1 \Omega_1 \quad M_2 = I_2 \Omega_2 \quad M_3 = I_3 \Omega_3, \tag{12.4}$$

where I_1, I_2, and I_3 are called the *principal moments of inertia*, and Ω_1, Ω_2, and Ω_3 are the components of the angular velocity along what are called the *principal axes* of the body. The principal moments of inertia of a sphere and an ellipsoid of revolution are illustrated in Fig. 12.4.

As a numerical example, consider a solid steel sphere of radius 10 cm. It has a mass of about 32 kg and a principal moment of inertia about an axis through its center (2/5) MR^2 of about 0.128 kg m^2. If it is set spinning at the rate of 3,600 rpm its kinetic energy of rotation, (1/2) $I\omega^2$, would be about 9.1×10^3 J, equal to the kinetic energy of falling under gravity from a height of about 29 m. Its angular momentum $I\omega$ would be about 48 kg m^2/s.

In general the angular momentum vector $\vec{M}$ is *not,* counter to what one may be inclined to assumed, in the same direction as the angular velocity vector; but in the special case where the rotation is about one of the principal axes, they *are* in the same direction. A body in which all three principal moments of inertia are different is called an asymmetrical top, while if only two principal moments are equal we have a symmetrical top. The common example of the latter is a child's top. If all three moments are equal, we have a spherical top, for which the angular momentum is simply $\vec{M} = I\vec{\Omega}$ for any axis of rotation.

From the definition of $\vec{M}$ it can be shown using Newton's laws that the following relationship holds:

$$\frac{d\vec{M}}{dt} = \Sigma_i (\mathrm{r_i} \times \mathrm{F_i}) = \vec{\Gamma}, \tag{12.5}$$

where $\vec{\Gamma}$ is a vector called the torque. The application of this result to a spinning body in which a component of the torque is perpendicular to the axis of spin leads to a solution in which the axis of rotation swings around generating a cone. This motion is called *precession* and is illustrated in Fig. 12.5 which shows the spatial geometry of the various vector quantities associated with that motion.

The dynamic conditions that produce precession in systems with angular momentum are encountered in not only gyrocompasses but as diverse systems as

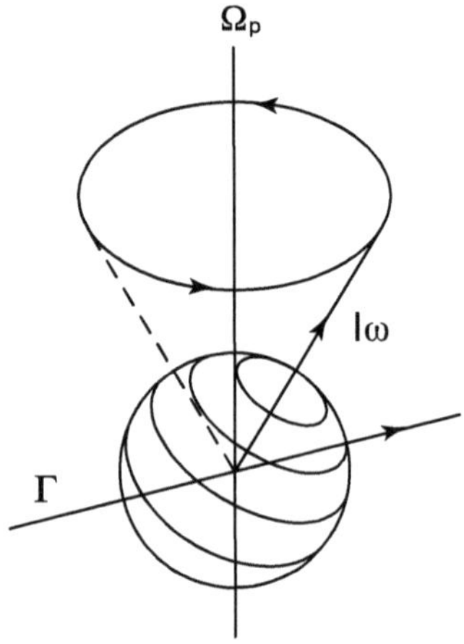

Fig. 12.5 Precession of the axis of spin of a rotating body

the spinning earth and atoms in a magnetic field. In the case of the earth, the precession of the equinoxes which was mentioned in an earlier chapter is due to torques acting on the spinning Earth due mostly to the gravitational pull of the sun and moon acting on departures of the earth from a perfect sphere. In atoms, the application of a magnetic field produces a torque on both the orbital and spin angular momenta of the electrons producing important changes in the atomic spectrum.

In order to describe the *free* motion of the body in *inertial space* we need to make the transformation from a system defined by the principal axes of the body to an outside inertial system with respect to which the body is rotating with the angular velocity $\vec{\boldsymbol{\Omega}}$. We have already introduced the required transformation equation which applies to any vector; in the case of the angular momentum vector it is as follows:

$$\left(\frac{d\vec{\boldsymbol{M}}}{dt}\right)_{\text{space}} = \left(\frac{d\vec{\boldsymbol{M}}}{dt}\right)_{\text{bodysys}} + \vec{\boldsymbol{\omega}} \times \vec{\boldsymbol{M}}. \tag{12.6}$$

But $\vec{\boldsymbol{M}}$ is constant in the inertial system, that is $\left(d\vec{\boldsymbol{M}}/dt\right)_{\text{space}} = 0$, and therefore the right hand side of the equation is zero, leading to three equations for the three components of $\vec{\boldsymbol{M}}$ called the Euler equations. In terms of the components with respect to the principal body axes they have the form:

$$I_1 \frac{d\Omega_1}{dt} = \Omega_2 \Omega_3 (I_2 - I_3), \quad \text{etc.} \tag{12.7}$$

We will now attempt to understand the basic motions of a rigid body as determined by these equations of motion. The solution is particularly simple in the case of the spherical top which has all three moments of inertia equal. Thus we have the time derivatives of all components of Ω equal to zero and therefore the angular momentum vector is independent of time. This means of course that it remains constant both in magnitude and direction.

The case of the symmetrical top is only slightly more complicated; here $I_1 = I_2 = I$ (say) and $I_3 = I_{sym}$ along the axis of symmetry. It can be shown that for a prolate spheroid, that is one elongated along the axis of symmetry, $I_{sym} < I$. The solution to the three Euler equations in this case can be shown to yield the following results:

$$\Omega_1 = \Omega_0 \cos(pt) \quad \Omega_2 = \Omega_0 \sin(pt) \quad \Omega_3 = \text{const}, \tag{12.8}$$

where $p = \Omega_3(I_{sym} - I)/I$. The last equation shows that the angular velocity about the symmetry axis remains constant, while the component of the vector $\vec{\Omega}$ *transverse* to the body symmetry axis rotates with angular velocity p about that axis. Thus we find that as long as the moment of inertia about the symmetry axis is different from that about a transverse direction, there is a possible solution in which the angular velocity vector describes a cone about the symmetry axis.

However of greater interest to us is the motion of the body with respect to the inertial system; in this case the body executes what is called *regular precession* [2]. Again considering a symmetrical top, we find that the law of the conservation of angular momentum and the consequent fact that the angular momentum is fixed in space, is sufficient to predict free precession. In general, as already noted, the angular velocity of a body may not be in the same direction as the angular momentum. By resolving Ω into components along M and the body axis, it can be shown that the symmetry axis will precess about the total angular momentum vector $\vec{M}$, describing a cone. The rate of precession Ω_{prec} is as follows [2]:

$$\Omega_{prec} = \frac{M}{I}, \tag{12.9}$$

where I is the principal moment of inertia *perpendicular* to the axis of symmetry. This should not be confused with the precession that results from the presence of an *external* torque acting on the spinning body as we have with a top.

We have been discussing the dynamics of rotating bodies and some of the counterintuitive behavior, particularly the fact that a force applied in one direction appearing to produce motion in another direction. We recall that it was Newton's inspiration that a particle moving in a circular orbit was in fact doing so under the action of a force toward the center of the circle. Change in the motion of a particle as referred to in Newton's second law meant not only a change in the magnitude of the velocity but also its direction. For rotational motion the classical $F = ma$ formula when applied to rotational motion takes the analogous form:

$$\vec{\Gamma} = \vec{\omega} \times \vec{M}. \tag{12.10}$$

The most familiar example of the consequence of this relationship is the behavior of a spinning top in which on account of the torque exerted by the gravitational pull tending to tip its axis, its axis rotates around the vertical direction describing a cone.

12.2 Motion on a Rotating Earth

Imagine we have a gyroscope consisting of a rapidly spinning rotor mounted on frictionless gimbals that allow its axis of spin to freely assume any direction. Since it is assumed that there are no torques acting on it, its angular momentum vector is constant in magnitude and direction with respect to an inertial frame of reference such as the "fixed" stars. Thus if the axis of the gyro is directed at a particular star, it should continue to point at that star as the ground on which it stands rotates with the daily rotation of the earth about its axis. But because of this rotation all the stars appear to rotate around the north celestial pole; therefore the axis of the gyro follows its star as the latter takes part in the apparent rotation of the celestial sphere. In other words the gyro axis swings around describing a circular cone whose axis is the celestial *polar* axis. If for example we take an extreme case in which the gyro axis is directed initially toward a star near the horizon in the due E direction at a given time, and observed as the day wore on, we would see the spin axis *rise* toward a vertical direction that reaches the zenith 6 h later (see Fig. 12.6), and continue to the due *W* direction 12 h later. During the following 12 h it would continue, possibly below the horizon, to complete the circle in 24 h. In fact a precision gyro could be used to monitor the orientation of the earth in space and hence used as a clock giving sidereal time!

In order to quantitatively describe such a motion of the gyro axis requires the specification of *two* angles: one about the vertical axis (*azimuth*) and the other in the meridian plane (*elevation*). For the more nautically inclined reader these would be called the *drift* and *tilt* of the compass. Ideally of course we want the gyro axis to be always horizontal and to settle in the meridian, that is, point due *N*. If the gyro axis is initially set to point at the north celestial pole, it will maintain that direction as the earth rotates underneath it; however of course, at points on the surface of the earth which are not on the equator, the gyro axis would not be horizontal, but rather inclined at an angle equal to the latitude, as is made clear in Fig. 12.7.

It would not be appropriate here to embark on the highly mathematical solution of the problem of the motion of the gyro in general; this would involve the transformation of the equations of motion from the body system to an inertial system. Instead we will be satisfied with a more qualitative, approximate analysis. If we call θ and φ the elevation (tilt) and azimuth (drift) angles of a free (uncontrolled) gyro precessing as a result of the earth's motion, then it can be shown that they vary approximately as follows:

$$\frac{d\varphi}{dt} = \Omega_{\mathrm{E}} \sin\lambda \quad \frac{d\theta}{dt} = \Omega_{\mathrm{E}} \cos\lambda \ \ \sin\varphi, \tag{12.11}$$

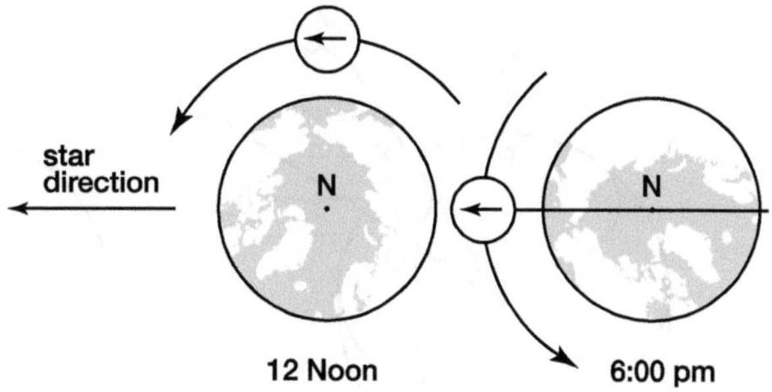

Fig. 12.6 Apparent rotation of gyro from due East at $t = 0$ to vertical direction 6 h later

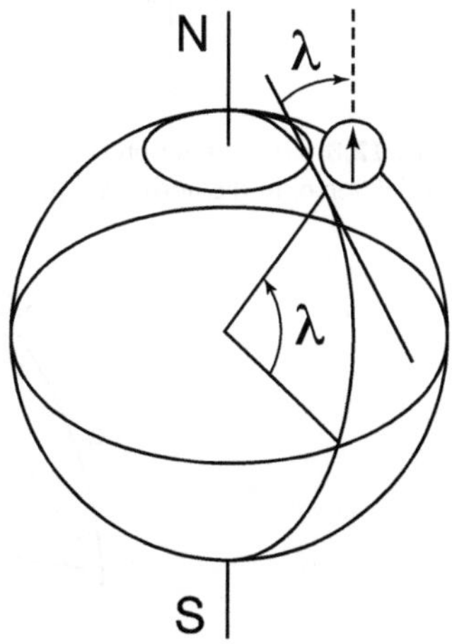

Fig. 12.7 Orientation of free gyroscope axis on the Earth at latitude λ

where Ω_E is the angular velocity of the earth. These approximate equations are of limited use however, particularly the azimuth equation which is only approximately valid for very small elevation θ. It is nevertheless useful to become familiar with the angular movement of the gyro axis under different conditions. The description of the orientation of the gyro axis with respect to the horizontal system of coordinates, based on the observer's zenith (vertical) axis and horizontal plane, requires a little spherical trigonometry, as does celestial navigation in general, of course. We will take only the simplest example of the precession of the gyro axis due to the earth's rotation [3]. For example, given an ideal gyro located at a given latitude, say 45°N, and set with its spin axis *horizontal* and pointing at a small angle east of the meridian at 020°N, let us compute how long it would take to precess to the meridian (ignoring its tilt). The problem is illustrated in Fig. 12.8 in which (a) represents the actual spatial geometry of the problem, while (b) is a projection on the horizontal plane, so that the zenith is in the center and the axial pole appears at a point north of the zenith by the colatitude angle. If we imagine the gyro direction is fixed on a particular star, then the geographical position of that star will describe a parallel of latitude as the earth rotates.

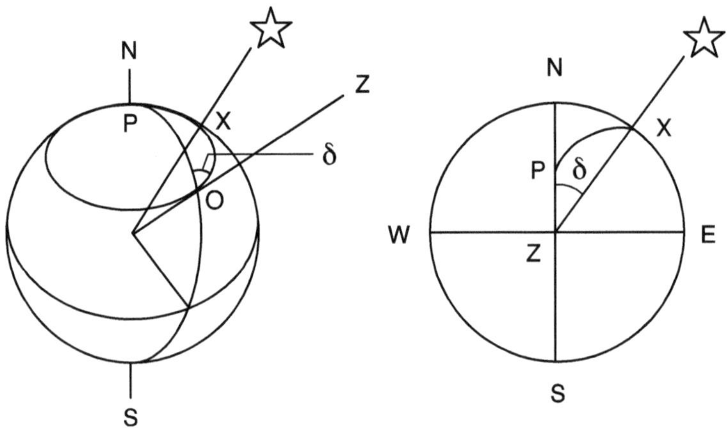

Fig. 12.8 Gyroscopic precession in the horizontal coordinate system. (**a**) spatial geometry (**b**) projection onto the horizontal plane

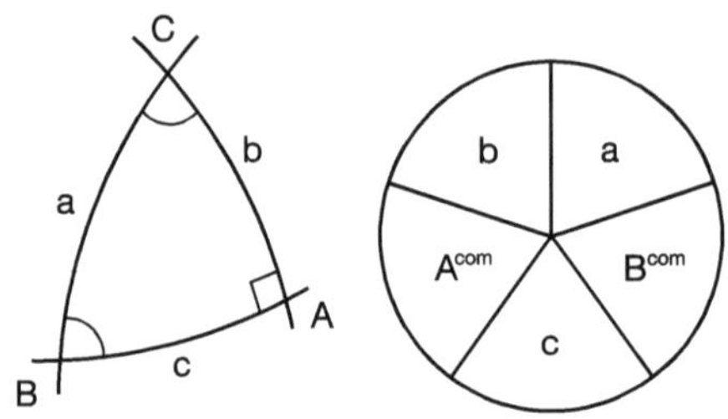

Fig. 12.9 Napier's rules for spherical right-angled triangle

To carry out the computation requires the solution of a spherical triangle in which one angle is a right angle; the solution of such spherical triangles is typical of astronomical navigation; the mathematics involved was developed by Napier and embodied in what are called *Napier's rules*. This is the same Napier who is given credit for inventing logarithms. We include this simple calculation to hint at the kind of problems the motion of a gyro entails; we will not pursue the formal theory of gyroscopic motion any further. The Napier rules regarding *right-angled* spherical triangles are shown in Fig. 12.9.

We shall need the formula cos $\mathbf{c} = \cot \mathbf{A}.\cot \mathbf{B}$; the spherical angles involved in our gyro problem are identified in Fig. 12.8a. An approximate projection on the horizontal plane [3] shows more simply the geometry of the problem. The celestial polar axis is marked as P, the zenith (vertical axis) at the latitude of the gyro is marked as Z, and the deviation (shift) of the gyro axis makes an angle δ with the meridian. Applying the above formula to the spherical triangle PZX, we have

$$\cos\lambda = \cot\delta \;\; \cot P. \tag{12.12}$$

From this we are able to compute the value of the angle P around the pole that the gyro must rotate to reach the meridian. Substituting the numerical values of the latitude and deviation we can compute the angle P; then knowing the angular velocity of the earth about the polar axis (15° per hour) we can compute the required time.

12.3 Control of a Gyrocompass

It might be thought that all we would have to do to make a gyroscope function as a compass would be to make a vertical projection of the axis onto a horizontal plane to obtain the line of the meridian. However that is ruled out for applications onboard moving platforms such as ships or airplanes whose motion inevitably causes their attitudes to fluctuate. What is required is a way to cause the gyroscope to seek a *horizontal* position in the meridian as a *stable* equilibrium condition. This can be accomplished by introducing an appropriate *control torque* on the spin axis and a *damping torque* to reduce oscillation. The first tends to tilt the axis in order to produce a transverse precession that counters the precession due to the earth's rotation, and the second is designed to shrink the range, or amplitude, of the excursions in azimuth and altitude. The detailed manner in which the controlled torque is realized varies among manufacturers. We begin with the design that is simplest to describe, namely one in which a weight is attached to the rotor casing; it can be placed either on top, vertically above the center of the rotor, giving a *top heavy* control, or on the bottom giving, what else, a *bottom heavy* control. We must distinguish the two possibilities since the sense (clockwise or counterclockwise) of the torque produced when the gyro axis tilts is opposite for the two choices. In the further description we will assume a top heavy control, as shown in Fig. 12.10.

If for example the gyro axis is pointing *east* of the meridian and tilted *above* the horizon, the weight of the added mass is no longer vertically above the center of the rotor and a torque is produced tending to turn the axis of rotation about an axis perpendicular to both the axes of torque and rotation. That we recall is the characteristic behavior of gyroscopic motion; thus the gyro axis would precess in a horizontal direction. The sense of the gyro spin, that is the direction of the angular momentum, must be such that the torque is in the direction to produce a horizontal precession that corrects the assumed easterly drift of the gyro axis. If the same gyro, with the same spin direction were bottom loaded, it would precess in the wrong direction, increasing the easterly drift; the correct placement of the added mass gives rise to a countervailing motion of the gyro axis toward the meridian.

Let us follow the motion of the gyro axis from a starting point where it is pointing at some angle east of the meridian. As the earth rotates the axis begins its circular path in the *upward* direction bringing into play the torque produced by the control mass, a torque that is designed to produce a precession in the *opposite* direction to the effect of the earth's rotation. The axis continues through the meridian where the tilt and therefore torque and consequent precession rate is maximum, to the farthest point in the west where the axis is again horizontal and

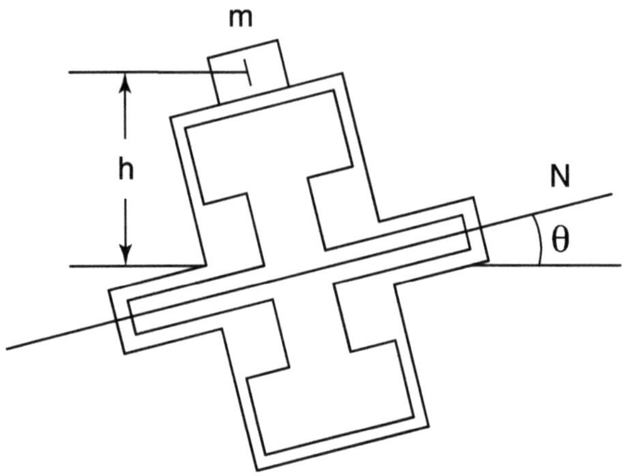

Fig. 12.10 A gyroscope with a mass attached to the top of the rotor casing

the torque passes through zero; the movement continues back again below the horizon to the starting point. Because of the variation in precessional angular velocity caused by the control torque of the added mass, the path traced out by the gyro axis in space is a cone with an *elliptical* cross section. We can readily compute the precessional rate, given the magnitude of the attached mass, the dimensions of the casing and the angular momentum of the gyro. Thus using the basic formula we have already introduced for the precession Ω_{pre}, namely

$$\Gamma = \Omega_{\text{pre}} \times L, \tag{12.13}$$

where Γ is the torque and L is the angular momentum of the rotor, and the fact that the three vectors are mutually perpendicular, we find:

$$\Omega_{\text{pre}} = \frac{mgh}{L} \sin\theta, \tag{12.14}$$

where m is the added mass, h its height above the center of gravity and θ is the angle of tilt. For small angles of tilt $\sin\theta \approx \theta$ and we see that the rate of precession is approximately proportional to that angle.

The use of an added solid mass to control a gyro, while conceptually simpler to explain, has in practice drawbacks that tend to favor alternative methods of stabilization. These drawbacks are particularly serious on moving platforms, on which gyros have found their principal application. A preferred system uses a *liquid ballistic* method in which a liquid, usually mercury because of its high density, flows between two reservoirs attached to the bottom of the gyro casing, one on the north side and the other the south side, along a N–S axis through the center of gravity of the rotor, thereby creating a torque when the gyro axis tilts from the horizontal direction. The reservoirs are connected by a tube between their bases to allow unrestricted flow of mercury between them; there is also a tube connecting

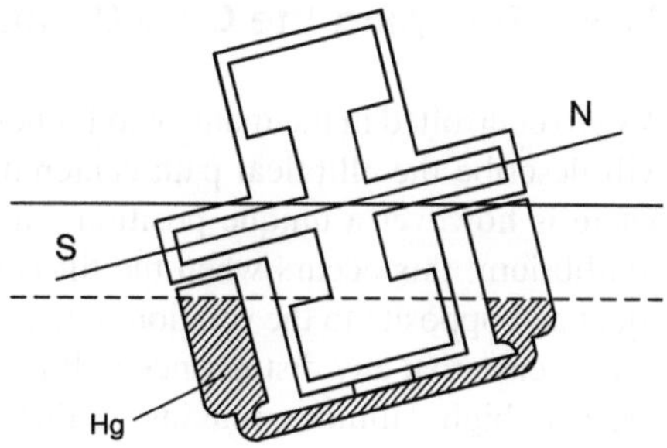

Fig. 12.11 The effect of tilt on the liquid ballistic control

the upper spaces above the mercury levels (not shown in the figure) to maintain equal air pressure in the two reservoirs, thereby ensuring that the mercury levels truly define the horizontal direction. Ideally the whole ballistic system must be symmetric about the horizontal E–W axis, and its center of gravity coincident with that of the rotor when the gyro axis is horizontal and the mercury levels at equal height. In this quiescent condition the mercury has no effect on the gyro since the weight of the liquid is acting through the center of mass of the rotor, and there is no torque acting on the latter. In Fig. 12.11 we show the effect of a tilt *upwards* of the north seeking end of the gyro.

The tilt causes a flow of the mercury toward the south reservoir resulting in an imbalance of the weight of mercury on the two sides of the rotor and a consequent torque about the E–W axis. The sense of rotation of the rotor must be such that this torque causes a westerly precession when the north end tilts up, a condition imposed by the sense of rotation of the earth and the definition of north. The computation of the rate of precession resulting from a given tilt follows the same argument as used for the solid mass, except here the torque is $2mgL$, where m is the mass of mercury transferred and L is the distance between the center of gravity of the rotor and that of the transferred mercury. Again the torque depends on $\sin\theta$, where θ is the tilt angle and thus for small tilt the torque is approximately proportional to the angle in radians.

The time it takes a controlled gyro to complete the elliptical cycle is expected to depend on the magnitude of the control precession rate, which in turn depends on the angular momentum of the rotor and the control torque, but it also depends on the drift and latitude. Commercial gyro compasses may have periods in the range of 80–120 min. We give without proof the expression for the period of oscillation of a gyro compass, that is, the time to complete one trip around the ellipse:

$$T = \sqrt{\frac{L}{\Gamma_{\mathrm{m}} \Omega_{\mathrm{E}} \cos\lambda}}, \tag{12.15}$$

where L is the angular momentum of the rotor, Γ_{m} is the maximum control torque which depends on the quantity of mercury and the dimensions of the control design, Ω_{E} the angular velocity of the earth, and λ is the latitude.

12.4 Damping the Gyro Oscillation

A gyro controlled in the manner so far described will not remain in the meridian but will describe the elliptical path centered on the meridian that we have described. There is however a unique position where the gyro is stationary, but in *unstable* equilibrium: this occurs when the tilt is such as to produce a precession precisely equal and opposite to the rotation of the earth. But this position is not only unstable in the sense that any disturbance will make it start to oscillate, but its tilt becomes large at high latitudes making it difficult to establish the horizontal bearing, particularly on a moving platform. What is required is, as previously stated, a *stable* equilibrium with the gyro axis horizontal, so that any disturbance will decay and the axis settle in the stable horizontal direction. The use of the term "damping" in the present context is misleading since the term is almost always used in mechanics to mean the dissipation of kinetic energy as heat in order to reduce the amplitude of an oscillation. In the present case the aim is not to dissipate the energy of movement of the gyro, but rather to force it to keep its alignment in a given meridian as the earth rotates; this requires its axis to remain horizontal and rotate at the same rate as the earth. Hence, far from dissipating some of its energy as heat, the gyro must *gain* the energy of rotation to keep up with the earth. Of course the angular velocity of the earth is so small that this observation is only of academic interest. Here "damping" is achieved by providing another torque in such a manner that it counteracts the elliptical oscillation of the controlled gyro and causes the amplitude of the oscillations to gradually diminish until the gyro has settled horizontally in the meridian.

To achieve this requires a torque around the *vertical axis* to appear when the gyro axis is tilted from the horizontal; this would lead to a correcting precession in the vertical plane. In the Sperry Mark 20 gyro compass [2] a weight is placed on top of the rotor casing at some distance perpendicular to the axis of the rotor as shown in Fig. 12.12.

The weight is set at some distance away from the point directly above the center of the rotor because when the gyro axis is pointing in some arbitrary direction, the force vector due to gravity acting on the weight will then not necessarily pass through the vertical line through the rotor center. This has the crucial result that the torque it exerts on the rotor will have components not only about a horizontal axis but, more importantly, also about the *vertical axis*. Such a torque is necessary to counteract the tilt of the gyro axis and the consequent horizontal drift induced by the control torque. It acts against the tilting when directed away from the horizontal and with it when directed toward the horizontal. It requires a highly developed sense of three-dimensional spatial relationships or a three-dimensional model to fully visualize the complicated interplay of torques and rotations in a gyro compass. We can attempt to understand the behavior of a gyro simultaneously controlled and damped by analyzing the torques on it as its axis progresses along its elliptical path. It will be shown that if properly designed, the gyro axis will follow a spiral path to a *stable* end point.

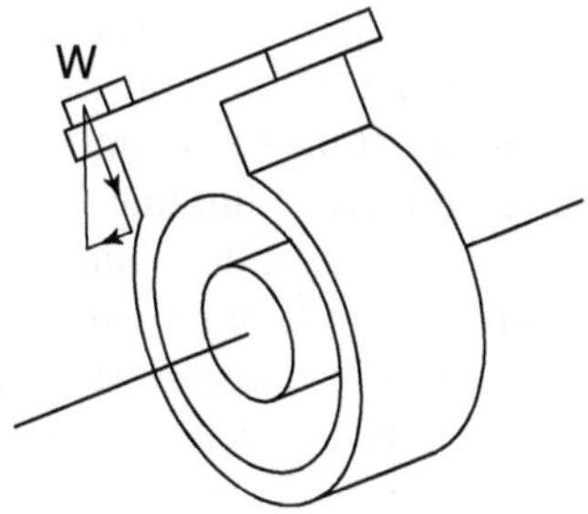

Fig. 12.12 The placement of the tilt-damping weight in the Sperry Mark 20

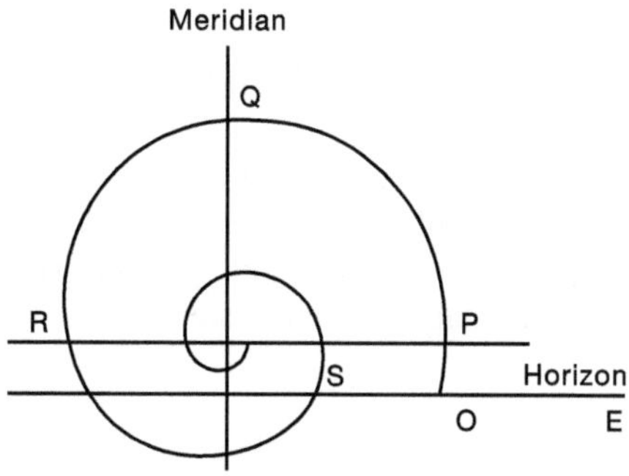

Fig. 12.13 Spiral convergence of the gyro axis to a stable point [3]

In Fig. 12.13 let us assume we begin with the gyro axis set in a horizontal direction at some point in the northern hemisphere, marked as point O. With the axis in a horizontal direction, the control and damping mechanisms produce no torques, but the earth's rotation will lift the north end of the axis and cause a drift toward the east. This will lead to the appearance of both a corrective control precession toward the west as well as a damping downward precession toward the horizon. This continues until point P is reached where the control westerly precession equals the opposing easterly drift due to the earth's rotation, resulting in a momentary vertical movement of the axis. The rate of tilting will be reduced at this point by the precession due to the damping torque. As the tilting continues, the control torque increases to the point where its effect exceeds the drifting and the axis moves west toward the meridian. With the increased tilt comes also an increased damping torque ultimately leading to a turning point of the tilting, after which it declines reaching a momentary stationary point when the axis reaches the meridian at the position marked Q. Beyond that point the change in tilt becomes negative under the influence of the damping torque, and the axis rapidly moves toward the horizontal direction. This action of the damping torque causes the amplitude of the swing in azimuth to be radically reduced compared with what the control torque acting alone would have produced. The point of maximum drift, marked as R, is where the control precession equals the drift due to the earth's rotation; thereafter the axis starts to move back towards the meridian. The control and damping precessions will continue to decrease, reaching zero as the axis assumes a horizontal direction. The axis then tilts below the horizon, inducing a control precession to the *east*, that is, in the same direction as that due to the earth's rotation. At the same time, the damping precession is directed upward toward the horizon thereby limiting the downward tilt of the axis and causing it to return to the horizontal at the point marked S, considerably closer to the meridian than the starting point O of the cycle. To summarize, the damping precession is opposed to the control precession when the latter is directed away from the horizon, but acts with it when directed toward the horizon; the consequence is that on successive cycles the swings in azimuth continue to decrease converging toward the meridian.

12.5 Principal Errors of the Gyro

In the following sections on gyro errors we will implicitly be dealing with gyros subjected only to relatively slow motions, such as may be encountered on ships. This obviously leaves out gyro compasses that are designed to use on aircraft or to serve as stabilizers on rockets.

12.5.1 Settling Error

The final *settling* position of a damped gyro, that is, the final stable condition to which it converges, is not precisely on the meridian, nor even in the horizontal direction; the equilibrium is reached where the control precession annuls the horizontal drifting and the damping precession does its part against the vertical tilting. However, a small residual tilt of the axis is inevitable; for navigational purposes the departure from the meridian is important, even though it is small. What makes it particularly troublesome is the fact that the error varies with latitude; in fact it is called, quite logically, the *latitude error*, or the settling error. We can estimate the magnitude of this error and its dependence on latitude by using the Eqs. (12.11) and (12.14), the latter being modified to relate on the one hand to the control torque and the other to the damping torque which involve different geometrical factors; thus writing $m_1gh_1 = J_1$ and $m_2gh_2 = J_2$ we have

$$\frac{d\varphi}{dt} = -\frac{J_1}{L}\sin\theta = \Omega_\mathrm{E}\sin\lambda \quad \frac{d\theta}{dt} = -\frac{J_2}{L}\sin\theta = \Omega_\mathrm{E}\cos\lambda\sin\varphi, \tag{12.16}$$

where L is, as before, the angular momentum of the rotor, θ is the angular displacement of the axis, λ is the latitude, and φ the azimuth. It follows that by dividing these equations one by the other we obtain

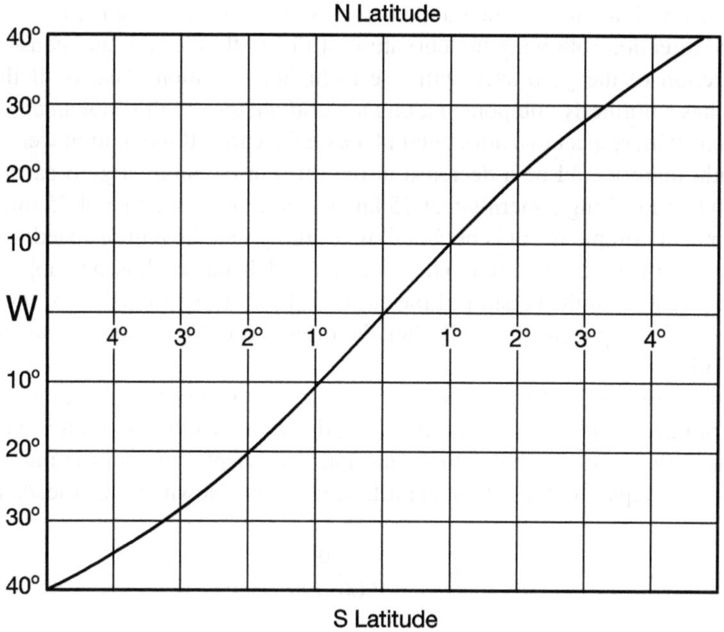

Fig. 12.14 The damping error of a gyro as a function of latitude for $J_2/J_1 = 0.1$

$$\sin\varphi = \left(\frac{J_2}{J_1}\right)\tan\lambda. \tag{12.17}$$

If the azimuth angle φ is small, sin φ may be approximated by φ (radians) or $(\varphi/\pi) \times 180°$; typical practical values of the azimuth error are in the range of 2–3°. Theoretically the error diverges to large values at high latitudes, approaching infinity at the pole. In Fig. 12.14 is shown the graph of the settling/damping error as a function of the latitude.

12.5.2 Gyro Body Motion Error

As a navigational instrument on a platform moving on or near the earth's surface and following its curvature, it will experience motion of its center of mass which must be included in the angular momentum equation on which gyroscopic theory is based. We can distinguish two types of motion: there is the fluctuation in the movement of the platform, the rolling, pitching and yawing, and its steady motion following the curvature of the earth, a rotational motion. We will consider the latter curved motion over the spherical earth first. The consequence of this motion is made clear if we imagine the gyro axis as fixed on a particular star in the meridian plane, an alignment that is guaranteed to remain constant by the law of the conservation of angular momentum. As the gyro is carried along the meridian in a northern direction following the curvature of the earth, the star, and consequently the direction of the gyro axis, will rise to higher elevation. That is, if the gyro motion has a northerly component along a meridian, there will result an *upward* tilt of the axis with respect to a horizontal plane on the earth. If the motion were toward the south, there would be a decrease in the tilt. For example, a gyro carried on a vessel traveling along a meridian at 25 knots would tilt at the rate of 25 min of arc per hour. This simple result is obtained by recalling that the unit of speed, the knot, is defined as the speed required to travel a standard distance, chosen to approximate 1 min of arc around the equator of the earth, in 1 h of time. In general if the course has an E–W component that will alter the effective E–W component of the gyro axis rotation.

The equilibrium settling reached by the stabilizing system imposes a displacement from the meridian such that the tilting due to the earth's rotation is equal and opposite to the spurious tilt due to the motion. This displacement in azimuth can be computed by equating these two quantities; it can be shown that this leads to

$$\sin\varphi = \frac{V\cos\Theta}{900\quad\cos\lambda}, \tag{12.18}$$

where V is the speed in *knots* and Θ the course. Note that the angular velocity of the earth is 900 arc minutes per hour. Since the error in azimuth varies inversely as the cosine of the latitude λ, it becomes increasingly large at higher latitudes. We note also that this result does not depend on any instrumental design parameters of the gyro but depends only on the latitude and course. If the velocity V is high and has a significant E–W component, then the effective angular velocity of the earth is modified thus $\Omega_E = 900\cos\lambda \pm V\sin\Theta$.

When the platform on which the gyro is mounted changes its course, the gyro will respond by pursuing its damped spiral convergence to the new settling position. However if rapid, repeated maneuvers are carried out faster than its settlement time, then obviously it becomes totally unreliable. Understandably Sperry and other manufacturers have devoted considerable effort to design instruments that are as immune as possible to fluctuations in their gyros resulting from maneuvers. Indeed there are ways of introducing a precession that compensates for the spurious tilting due to the body motion of the gyro. Of particular interest is the exploitation of what is otherwise a problem called *ballistic deflection*, turning it into a stabilizing effect that in effect puts the gyro in its settling position without going through its spiral. The term ballistic deflection refers to the flow of the stabilizing mercury as a result of *acceleration* of the gyro. As a liquid it normally establishes the horizontal direction, but when the gyro system is accelerating by changing its vector velocity, even in a horizontal plane, the mercury tends to build up on one side by inertia, and the level tilts indicating a false horizontal and a corresponding false vertical. If for example the mercury is driven toward the south end because of a northerly

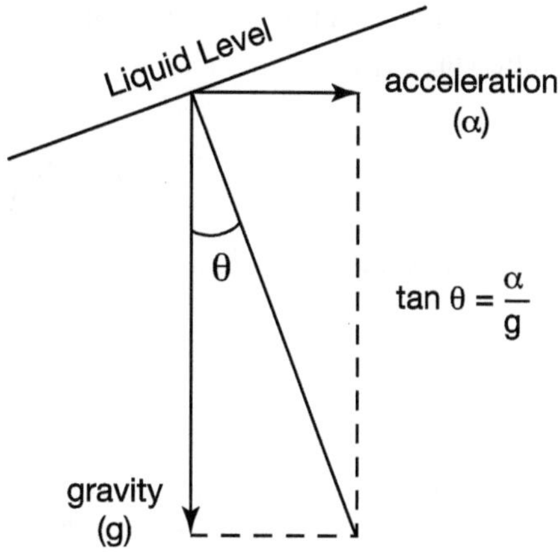

Fig. 12.15 The tilt in the level of Hg resulting from acceleration

acceleration, that tilt in the horizontal will cause a precession in *azimuth* in the direction east to west. The rate of precession at every instant can be shown to be proportional to the acceleration. It follows that the total angle of precession is proportional to the total change in the velocity vector. Figure 12.15 illustrates the tilt in the stabilizing liquid level as a result of acceleration of the system.

The mercury assumes a level inclined at an angle $\ddot{o}$ such that $\tan \delta = \alpha/g \approx \delta$, where α is the acceleration. This tilt of the horizon gives rise to a precession given by

$$\omega_{\mathrm{pr}} = -\frac{\alpha}{g}\left(\frac{J}{L}\right). \tag{12.19}$$

Fortunately the ballistic deflection resulting from the flow of the stabilizing liquid is in the right direction to make it possible for the gyro axis go directly to the settling position without going through the convergent spiral. This is achievable by the detailed design parameters of the control system; for example the dimensions, the quantity of mercury, the angular momentum of the rotor, etc. These are in fact the same factors that determine the oscillation period of the spin axis describing the end spiral. In order for the ballistic precession ω_{pr} to equal the rate of change in the speed and course we must have

$$\frac{\alpha T}{g}\left(\frac{J}{L}\right) = \frac{\alpha T}{\Omega_{\mathrm{E}} R_{\mathrm{E}} \cos \lambda}, \tag{12.20}$$

where α is the body acceleration of the gyro and R_{E} is the radius of the earth. It follows that

$$\frac{R_{\mathrm{E}}}{g} = \frac{L}{\Omega_{\mathrm{E}} J \cos \lambda}. \tag{12.21}$$

But the period of oscillation of the gyro axis can be shown to be given by the following [3] :

$$T = 2\pi\sqrt{\frac{L}{J\Omega_{\mathrm{E}}\cos\lambda}}. \tag{12.22}$$

It follows that

$$T = 2\pi\sqrt{\frac{R_{\mathrm{E}}}{g}}. \tag{12.23}$$

If the gyro is designed to have this period of oscillation, called the *Schuler Period*, the ballistic deflection will be just right for the gyro to avoid the spiral path to the final position.

12.6 The Sperry Mark 37 Gyrocompass

A late model gyrocompass in the Sperry range of marine systems is the Mark 37 [4] of advanced miniature design with emphasis on simplicity and reliability. It is notable in that it is based on a *gyrosphere* which floats with neutral buoyancy with its center of gravity coincident with its center of buoyancy. This is designed to mechanically isolate and thereby minimize accelerations acting on the gyrosphere due to motions of the platform; further, by appropriately restricting the ballistic flow of the liquid, it minimizes in the short term the effects of these motions. It is by these means that the Mark 37 gyro compass is made relatively immune to severe maneuvering conditions, and errors kept within tight tolerance.

12.7 The Ring Laser Gyro

Almost immediately after the first announcement of the achievement of optical laser action in the early 1960s, many researchers realized that one of the areas which would greatly benefit from the remarkable coherence of laser light was optical interferometry, that is, the use of the interference patterns of light waves to do fine spatial measurements. It was predictable that it would not be long before someone wanted to repeat the experiment of Georges Sagnac (pronounced "Sanyac") on the propagation velocity of light. Sagnac was a French physicist who worked during a time when there was a great deal of interest in the velocity of light and its central role in Einstein's revolutionary theory of relativity. In 1913 he reported on the now famous light interference experiment in which a light beam from a bright lamp (no lasers yet) is split by a half-silvered mirror into two beams which are sent in

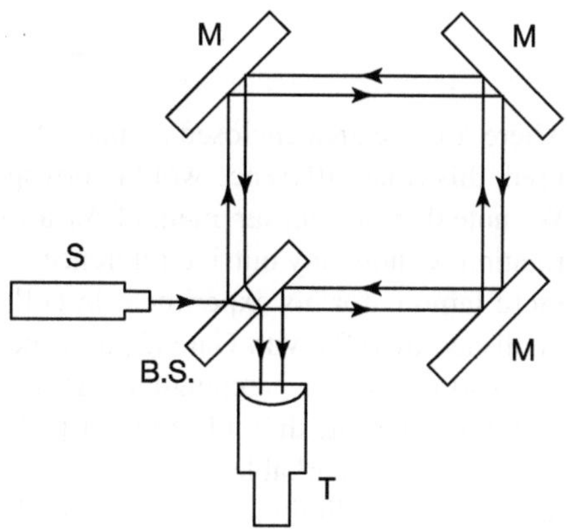

Fig. 12.16 Sagnac's classical ring interferometer

opposite directions around a square path with plane mirrors at the corners as shown in Fig. 12.16.

The counter propagating beams are recombined at the beam splitter to obtain interference fringes; these consist of alternately light and dark lines resulting from reinforcement or destruction of the wave amplitude according to the phase difference between the two overlapping light waves. He was able to show that if the mirror system is made to rotate at an angular velocity Ω then a path difference between the two beams is created leading to a shift in the interference pattern. This is commonly called the Sagnac effect. Sagnac used the classical theory of light propagation to calculate the effect of rotation on the time it took light to go around the mirrors. As a problem involving a non-inertial system with accelerated (circular) motion, we now know it requires General Relativity, but his work preceded the publication of Einstein's General Theory. However, the Special Theory of Relativity requires the velocity of light to be constant, therefore using Sagnac's classical argument we can look on the rotation of the mirrors as causing a difference in the *distance* the two beams travel to complete the path around the mirrors, resulting in a difference in *phase* when they are recombined. Thus for practical rotational speeds in which the linear velocity of the mirrors is much less than the velocity of light, and therefore the increments in distance small, we find the two classical propagation times for the two directions of wave travel to be

$$t_C^{\pm} = \frac{L}{c \pm R\Omega}, \tag{12.24}$$

for a square arrangement of mirrors, where R is the radius from the center to a mirror and L is the perimeter, Ω is the angular velocity of the rigid mirror assembly, and c as usual is the velocity of light. From the point of view of Einstein's General Relativity, it is the scale of time, that is, the clock rate itself that changes as we go around a closed loop; in fact according to that theory the difference in time Δt is as follows [5]:

$$\Delta t = \pm \frac{4\Omega}{c^2} A, \tag{12.25}$$

where A is the area enclosed by the path. If monochromatic light of wavelength λ is used, this time difference would correspond to a phase difference of $\Delta\varphi = c\Delta t/\lambda$. We note that the measurement of $\Delta\varphi$ allows you, in principle, to know your state of rotation without any outside reference! It is interesting historically that Michelson, more famous for his experiment in collaboration with Morley on the *Ether Drift*, published in 1925 with Gale [3] an experiment to measure the absolute rotation of the earth using a rectangular Sagnac arrangement of mirrors, a rectangle 2,000 × 1,100 ft; they observed a path difference of only a fraction of a wavelength. This remarkable experiment was achieved with a conventional discharge lamp with very limited coherence as a light source.

The situation was radically changed with the advent of laser technology. It became possible to effectively have the light beam circulate around a closed path an extremely large number of times, either using a fiber optic coil with an outside laser source, or providing an optical amplifying medium inside the cavity to produce counter propagating light waves. Both methods have been under intense development; but ultimately the final performance reflects the amount of engineering effort that is applied to perfecting them.

Let us first consider briefly the method of increasing the Sagnac effect by multiple turns of an optical fiber around a closed loop. Clearly the phase difference that develops on each turn is accumulated so that after N turns the phase difference is $N\Delta\varphi$, where $\Delta\varphi$ is the phase difference for one turn.

In the internal laser gyro the mirrors are used to make up the optical cavity of a laser oscillating simultaneously with running waves in the two opposite directions. That such a mode of operation exists had been previously demonstrated: it is possible to have *simultaneously* two oppositely directed waves running independently, that is, with different amplitudes and frequencies, but with certain serious limitations. We will later learn that there is a tendency for the two modes to lock to a common frequency, a condition called *lock-in*, as any two closely tuned oscillating systems will do when there is coupling between them. In the lock-in condition the laser gyro is of course insensitive to rotation. However as might be expected intensive development efforts have been dedicated to extend the lower limit of the angular rate scale at which the lock-in condition sets in. The motivation is that the laser gyro is so highly desirable in every other respect, particularly the absence of any moving parts that wear.

We recall that the oscillation condition for the nth mode of a laser of length L is simply:

$$\nu_n = n \frac{c}{L_C}, \tag{12.26}$$

where L_C is the optical length of the cavity. It follows that in a rotating cavity where the effective length depends on the direction of propagation of the light wave there are two modes with different frequencies. A small difference in the length ΔL_C corresponds to a difference in frequency $\Delta\nu$ given by

$$\frac{\Delta\nu}{\nu} = \frac{\Delta L_C}{L_C}. \tag{12.27}$$

Let us apply this result to a rotating square cavity using Eqs. (12.23) and (12.25) for the difference in propagation time; we find

$$\Delta\nu = \frac{4\Omega A_C}{\lambda L_C}. \tag{12.28}$$

The constant of proportionality between the frequency difference output and the angular velocity in Eq. (12.28), that is $4A_C/\lambda L_C$, is called the *scale factor*, which must obviously be kept constant. This requires more than dimensional stability since the laser has an amplifying medium with a refractive index, and any gas flow would affect the optical length of the cavity by what is called the *Fizeau Effect*.

To form a concrete idea of the numbers involved, assume a square cavity having the sides 15 cm long, rotating at the earth's angular velocity of 15° per hour. With the laser oscillating at a wavelength of 633 nm, the red line of the He–Ne laser, the gyro would have a difference in frequency between the counter-rotating modes of about 17 Hz. The possibility of actually measuring this difference in frequency between two light beams would have been unthinkable before the advent of the laser and the extraordinary coherence of its light. Today it is well within modern laser techniques to measure the beat frequency output of a detector receiving the two beams. Since the two laser modes share the same cavity, the beat frequency should be highly stable and less susceptible to environmental conditions.

Almost from the beginning the ring laser was conceived as a navigational device and indeed has since developed into a superior inertial guidance instrument not only for the guidance of ships, aircraft, rockets, etc., but also such surveying tasks as guiding drilling direction for oil wells and tunnels. These applications can rely on a level of gyro performance in stability and reproducibility in the range of 0.01° per hour and noise fluctuation around 0.03° per hour for an averaging time of 100 s. In Fig. 12.17 is shown schematically the basic design of a ring laser gyro in which the active laser medium is internal to the Sagnac cavity.

Because of the stringent requirements on the mechanical and thermal stability of the laser, its body is constructed from materials such as Cervit that had previously been developed to have an extremely low coefficient of thermal expansion, in the range of 10^{-8} per °C. Chemically also, these materials are extremely resistant and have good vacuum properties such as low desorption. The size of the cavity perimeter is controlled by using piezoelectric elements attached to the backs of two of the mirrors. These control elements are commonly made of PZT (Pb Zirconate–Titanate), a mineral having a strong piezoelectric effect and are part of a servo feedback loop which uses as error signal the departure from maximum laser oscillation intensity; that is, the system locks on to the peak of the laser field intensity.

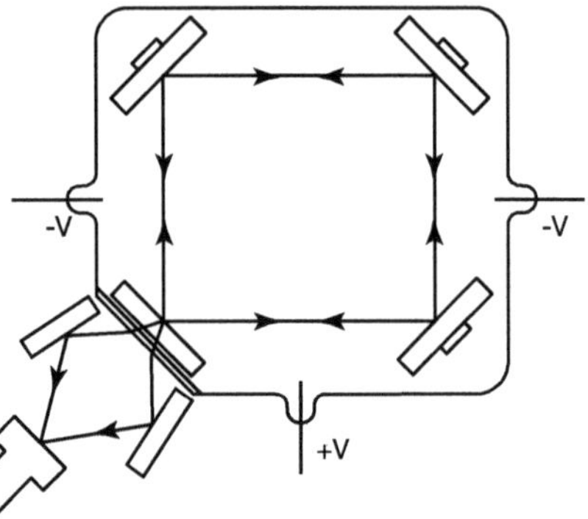

Fig. 12.17 Schematic drawing of a basic ring laser gyro

The reason for this choice is that the optical gain of the medium has a Doppler profile centered on the natural frequency of the atoms. Thus the maximum intensity occurs when the cavity is tuned to the atomic transition which is constant. With proper design, the perimeter of the cavity can be controlled to within a small fraction of a wavelength of light, with corresponding stability in the scale factor. In order to suppress higher spatial modes of the laser oscillation, an aperture of such diameter and position as to block the higher radial modes of the laser is placed at a point that preserves the symmetry of the laser. Symmetry is important in order that the counter-propagating beams be subjected to identical propagation conditions in order that their effect may cancel out in the beat frequency. Thus the electrical discharge is applied symmetrically between a central anode and two cathodes, as shown in the figure.

The components that are most crucial to the performance of the laser gyro are its mirrors. They determine the *finesse* of the optical cavity formed by the four mirrors, that is, the highest spectral purity (smallest line width) in order to be able to detect the smallest difference in frequency between the opposing beams. To achieve a high finesse it is necessary to have the highest possible reflectivity (99.99 % reflectivity is not uncommon) and highest degree of polish as specified by the "scratch and dig" specifications. With current state-of-the-art fabrication techniques a smoothness better than $\lambda/1{,}000$ is possible, where λ is the wavelength of visible light. This requirement on polish is common to most laser applications, but it is particularly important in the case of the laser gyro since any backscattering from the surfaces of the mirrors would couple into the counter propagating mode and cause the dreaded lock-in. A degree of smoothness has been reported giving a total scatter of less than 1 ppm [6].

The gases that constitute the amplifying medium are a mixture based on the He–Ne laser design which we described at some length in a previous chapter. However in this case it has been found necessary to have *two* isotopes of neon in addition to the He gas, ^{20}Ne and ^{22}Ne at a pressure around 600 Pa with the ratio of He/Ne around 10:1. Naturally the precise pressures depend on the other actual design parameters chosen to achieve stable laser action. It is found experimentally that with just one isotope of Ne, stable independent laser radiation in both directions simultaneously is impossible. This is due [6] to severe laser mode competition in which different modes within the Doppler broadened gain curve build up at the expense of gain for other modes. The incidence of effects of this type, called *hole burning*, was noted very early in the history of lasers by William. Bennett. It happens that in a mixture of the two neon isotopes the holes burned in the gain curve for the 632.8 nm transition do not overlap.

As we saw in the chapter on lasers, there is a limit to the spectral purity of laser radiation set by the presence of spontaneously emitted photons. Whereas photons resulting from stimulated emission are coherent with the photons doing the stimulating, those resulting from spontaneous emission have random phases, unrelated to any existing photons. Their existence is inevitable according to Einstein's theory of emission and absorption of photons, and the A and B coefficients. We noted in the last chapter that it was shown by Townes and Schwalow in 1958 that in the quantum limit laser radiation should have a fundamental Lorentzian frequency distribution with a spectral line width $\Delta\nu$ given by the following:

$$\Delta\nu = \frac{\pi h\nu(\Delta\nu_C)^2}{P}, \tag{12.29}$$

where $\Delta\nu_C$ is the *passive* (absence of lasing medium) cavity resonance frequency width and P is the laser output power. It was on the basis of this expression, first derived prior to the actual achievement of a working laser, that the extraordinary spectral purity of laser radiation was predicted. If we substitute actual numbers for a He–Ne laser operating at $\lambda = 633$ nm with output power of 1 mW and a cavity resonance width of 0.5 MHz, we find the spectral width of the laser radiation to be about 2.5×10^{-4} Hz! This is of course a theoretical lower bound, and in practice fluctuations in the amplifying medium cause the spectral widths of laser radiation to be seldom less than 10^3 Hz unless extraordinary steps are taken to improve it. This would correspond to a minimum detectable rotation of 1.6×10^{-3} rad/s. for a square laser whose sides are each 0.1 m long. This resolution falls short of measuring the earth's angular velocity of 7.2×10^{-5} rad/s, indicating the extent to which spectral resolution must be improved.

12.8 The Fiber Optic Gyro

The alternative to a ring laser, as already pointed out, is to use an external laser source to provide the counter-propagating beams in a fiber optic coil of many turns. A commercial example of such a gyro is the Sperry Marine NAVIGAT 2100. It is designed as a compact instrument for fast maneuver vehicles. It uses a combination of three fiber-optic coils and a dual-axis level sensor to establish the axis of rotation of the earth from which the direction of true north is established. It is a totally solid-state device, readily integrated with solid-state electronics, and, as with the laser gyro, it is directly mounted on the vehicle without the need of any delicate mechanical system. It can also be used in vehicle stabilizer systems, and it is not susceptible to shock and vibration.

12.9 MEMS Vibratory Gyroscopes

The acronym MEMS stands for micro-electromechanical systems, or simply microsystems; the reference here is to solid state microsystems that derive their gyroscopic action from the *Coriolis force* which, as we found earlier in the chapter, is invoked to describe the motion of particles referred to the rotating earth coordinates. It has a similar origin to *centrifugal force* which also arises when motion is referred to the non-inertial rotating reference frame of the earth. Recall that it is only in an inertial frame does a free particle move in a straight line; when the motion is referred to a rotating system of coordinate then it behaves as if acted on by a centrifugal force and a velocity-dependent Coriolis force. The magnitude of this force on a particle of mass m moving with respect to the earth coordinates is obtained from

$$\vec{F}_{\mathrm{C}} = 2m\vec{V} \times \vec{\Omega}_{\mathrm{E}}, \tag{12.30}$$

where $\vec{V}$ is its velocity vector, and $\vec{\Omega}$ is the angular velocity vector of the earth.

It follows that an otherwise free particle constrained to oscillate about a point on the surface of the earth will oscillate along a line that appears to rotate with respect to the earth. A spectacular demonstration of this fact for many years was on display in the form of a Foucault pendulum at a Smithsonian museum in Washington, DC. It is of course the earth that is rotating, not the line of oscillation of the pendulum. This same phenomenon applies to the high frequency vibration of a body such as a driven piezoelectric plate, for example, one similar to a quartz crystal in an oscillator.

Because of their natural compatibility with integrated electronic devices there has been in recent years intense interest in developing sensitive gyroscopes based on the vibration of solids. One of the earliest is the Draper Laboratory tuning fork gyroscope illustrated schematically in Fig. 12.18 which was first produced in 1993 for the automobile industry. It was designed as a yaw (rotation about a vertical axis) sensor as part of a vehicle antiskid mechanism.

Many other forms of vibrating solids have been published, notably vibrating rings and rectangular plates. There are also different proposed methods of exciting and stabilizing the level of vibration as well as the manner of sensing and the type of output, whether rate of rotation or the integral of the rate, that is the angle of rotation. An interesting proposed design for a rate integrating MEMS gyroscope [7] separates the functions of driving the oscillation and sensing it and thereby notably achieves dynamic amplification.

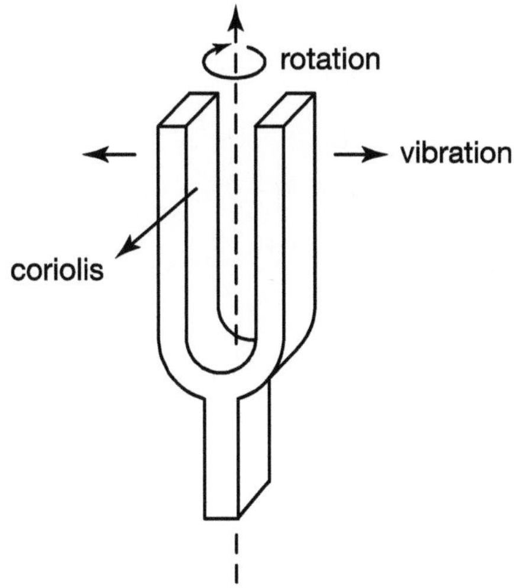

Fig. 12.18 Draper Laboratory tuning fork gyroscope

References

1. H. Goldstein, *Classical Mechanics*, 3rd edn. (Addison Wesley, Reading, MA, 2001)
2. L.D. Landau, E.M. Lifshitz, *Mechanics*, 3rd edn. (Butterworth and Heinemann, Oxford, 1976)
3. A. Frost, *Marine Gyrocompasses for Ships' Officers* (Brown, Son & Furguson, Ltd, Glasgow, 1982)
4. A.L. Rawlings, *The Theory of the Gyroscopic Compass*, 2nd edn. (McMillan, New York, NY, 1944)
5. L.D. Landau, E.M. Lifschitz, *Classical Theory of Fields* (Addison-Wesley Press, Cambridge, 1951)
6. W.W. Chow et al., Rev. Mod. Phys. **57**, 61 (1985)
7. C.C. Painter, A.M. Shkel, Nanotech, vol. 1 (2003), http://www.nsti.org. ISBN 0-9728422-0-9

Radio Navigation 13

13.1 Introduction

The use of radio as a navigational aid was recognized early in its history. Its role now extends far beyond the initial basic function of direction finding, although that of course is important; in fact, it would be accurate to say that the electromagnetic spectrum is involved in all modern means of navigation; these include nondirectional radio beacons (NDB), long-range navigation (Loran-C), VHF omnidirectional radio range (VOR), and of course the Global Positioning System (GPS). Thus in this chapter and succeeding chapters we will learn of the different ways in which radio is involved. We begin with radio direction finding and the long-range radio navigation system called Loran-C which had been in use since 1958; prior to that, less accurate forms of long-range radio navigation had been in use since the second World War. On February 8, 2010, the US Coast Guard announced the termination of Loran-C services in the USA; it marked the end of a half a century of service that Loran-C had provided to mariners. The termination date for the Canadian portion of Loran-C was August 3, 2010. It was expected that other nations may retain local versions of radio beacons and Loran-C, since they are popular with fishermen and recreational mariners. The reason for the demise of Loran-C in North America is of course the revolutionary change in the technology of position finding brought on by the Global Positioning System, which will be described at length in succeeding chapters.

13.2 Radio Direction Finding

The first significant application of radio beyond communication was in marine navigation where radio reception enabled a navigator to obtain bearings in all weather conditions with respect to a known transmitter broadcasting within his range in the frequency band of his receiver. It is a relatively unsophisticated means of navigation that is popular among fishermen and recreational boat operators;

F.G. Major, *Quo Vadis: Evolution of Modern Navigation: The Rise of Quantum Techniques*, DOI 10.1007/978-1-4614-8672-5_13,

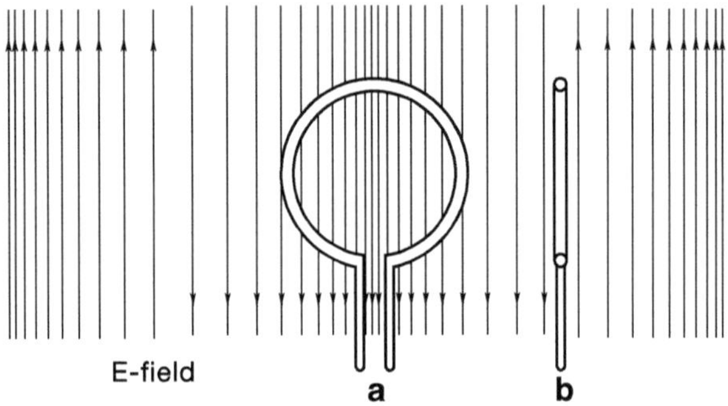

Fig. 13.1 A loop antenna oriented with its plane (**a**) in the direction of a plane radio wave and (**b**) perpendicular to it

today it has been mostly superseded by the Global Positioning System (GPS). We distinguish between two possible modes of operation: in the first and currently the one generally deployed, the broadcast is from an *omnidirectional* antenna, that is, one broadcasting equal power in all directions, in which case the user must establish the bearing of the transmitter by using a highly directional antenna; or alternatively, as was attempted in the 1930s [1], the transmitter broadcasts a rotating *directional* beacon sweeping all bearings in a precisely timed manner allowing the user to use an omnidirectional antenna relying simply on the time of the received signal to deduce his bearing relative to the transmitter. This latter method is only of historical interest, presumably because of the intrinsic weakness that the time of reception also depends on the navigator's distance from the beacon.

The operation of the radio direction finder is based on the directional property of a loop antenna, shown as a circular conductor in Fig. 13.1. Assume a plane electromagnetic wave, that is, one with plane parallel wave fronts, passes through an antenna whose plane is parallel to the electric component of the wave, and compare the radiofrequency currents that would be induced around the antenna as it is turned around the electric field as axis into the two possible orientations (a) and (b) with respect to the wave direction, as shown in the figure.

Let us first consider the orientation (a) in which the plane of the antenna is *parallel* to the direction of the wave. The different parts of the antenna in this case are at different phases of the oscillating electric field, resulting in a nonzero electromotive force around the loop. In other words, if we imagine an electric charge moving around the loop, there will be a net amount of work done on or by the charge that varies as the wave passes by; thus a radiofrequency signal is picked up by the antenna. On the other hand if the plane of the antenna is parallel to the wave front, the electric field is at all times exactly the same at all points around the loop and there will be zero net electromotive force *around* the loop. Therefore by monitoring the strength of the received radiofrequency signal as a function of antenna angle, the ship's bearing relative to the broadcast station can be determined. This information of course is useful only if the transmitter is identifiable from its broadcast signal, and its geographical location is marked on the navigation chart. For more remote stations corrections must be made for significant refraction by atmospheric and surface conditions along the signal path. Since the signal from a simple loop antenna clearly cannot distinguish between radio waves arriving in opposite directions, there is a critical ambiguity that must be resolved by other means. In the absence of other independent navigational information, the opposite bearing may be mistaken for the true one, with possibly disastrous consequences. A historic example of the seriousness of such an error is recorded in US naval history: in September 1923 seven destroyers ran aground off Honda Point in California under conditions of poor visibility and reliance on the radio navigational aid, which was new at the time. It was not long before a simple solution was found to resolve this ambiguity between "reciprocal angles": for example, by simply using an auxiliary receiver with an *isotropic* antenna to decide which side of the loop has the stronger field intensity.

A more subtle source of error is the curvature of the earth's surface. Radio signals travel in the ground wave mode along great circles (*geodesics*) which do not plot as straight lines on a Mercator chart. To make such a plot requires the conversion of the observed bearing which is an angle between great circles into an angle on the plane Mercator projection. To calculate the correction is a mathematical problem in spherical trigonometry which fortunately the average navigator is spared the trouble; he may either resort to tables of corrections, as published by NGA (National Geospatial Intelligence Agency), or use a mechanical cam attachment on the radio direction finder (RDF) to give the correct angle on the marine chart. Other sources of error include refraction of radio waves traveling obliquely across a coast line, and polarization effects observed in the early morning and evening hours.

To fix a *line of position* (LOP) using a radio direction finding station a navigator must tune in to broadcast stations marked on a navigation chart, stations that are identified by coded signals. As the name suggests, a line of position is simply a straight line drawn on the chart passing through the station in the direction the signal is received from the vessel and communicated to it. To obtain a fix using RDF requires of course the reception of directions from two or more such radio stations organized as a group, whose lines of position will intersect at the navigator's position, as shown in Fig. 13.2.

In certain coastal areas, particularly where there is dense shipping traffic, as in ports, there are systems of radio direction finding stations called *marine radio beacons*, which can communicate with a vessel to mutually identify each other, and the shore-side station will transmit to the navigator its radio bearing. To obtain this service a vessel must make the request using QTE (code for "what is the true bearing from you"), to the RDF station on the designated frequency and indicate the frequency it will use to enable its bearing to be taken. A sequence of communications ensues following a certain protocol which we need not elaborate here. The control station may respond with the coordinates of the vessel and a determination of the accuracy or *class* of the fix. The class ranges in alphabetical order from A to D, A being an accuracy within 5 nautical miles and D within 50 nautical miles. The control station may also transmit the bearings observed by each of the stations in the group. While two radio bearings are geometrically

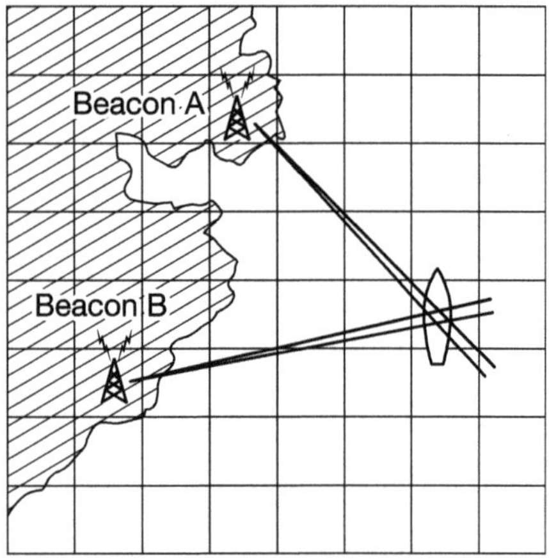

Fig. 13.2 Intersecting lines of position define a fix

enough to define the position of the vessel, in practice, in the absence of independent navigational data, it is deemed necessary to have *three* radio bearings to define a fix.

A common mode of navigation favored by fishing boats and some recreational boats is to use an RDF to go out on a constant heading established with respect to a given station and then *home in "on the beam."* The broadcast frequencies are typically in the medium frequency band (285–395 kHz) as well as 500 kHz, the international distress frequency for radiotelegraphy and 2.182 MHz for single side-band telephony. The single side-band description refers to the spectrum of a *modulated* radio wave which normally would consist of the base carrier frequency plus upper and lower side-band frequencies. In order to narrow the spectrum occupied by the broadcast, only one side-band frequency is broadcast, knowing that the full spectrum can be synthesized at the receiving end. The once universal availability of radio direction stations presumably will eventually die out as GPS becomes more dominant. However RDF is such a basic technology compared with GPS that remnants will no doubt survive at some locations.

A very important function performed by RDFs even today when other sophisticated navigational aids are available is to allow aid ships to home in on ships in distress, when a distress signal is received. There are by international agreement also shore-based radio stations around the world, designated as providing a maritime navigational service, stations that are assigned specific frequencies for radio direction finding. Nautical charts showing the location of these stations are made available in the USA by the National Imagery and Mapping Agency (NIMA) in the publication *Radio Navigational Aids* and by NOAA/National Oceanic Service. The reception frequencies of these stations are in the range of 1.605–2.850 MHz in addition to the international call and distress frequency 2.182 MHz for single side-band transmissions.

13.3 VOR Aircraft Navigation

VOR is acronym for very high frequency omnidirectional radio range, a highly accurate short range aircraft navigation system which operates in the very high frequency radio band 108–118 MHz. Unlike the radio beacons described earlier, the VOR signal conveys signal phase information that the navigator can use to determine his compass bearing. The underlying principle is to have the VOR station broadcast *two* signals: first the reference signal (REF), a continuous wave that radiates out from the antenna uniformly in all directions at an assigned carrier frequency, and a lower frequency subcarrier modulated at 30 Hz. The second signal (VAR) is variable with a strongly directional figure eight radiation pattern, which is *electronically* rotated about a vertical axis at the rate of 30 rev/s. The idea of creating a rotating pattern without an actual rotation of the antenna is easily understood when we recall that the combination of two harmonic oscillations at right angles to each other and 90° apart in phase results in a uniformly rotating resultant field. The combined radiation pattern which resembles a spherical balloon with a deep dimple along one radius, called a *limaçon* (French for snail), rotates at 30 rev/s. This rotation produces at the VOR receiver a 30 Hz amplitude modulation whose *phase* relative to the reference is calibrated to be the *angle* the radiation pattern has turned from the direction of north. The REF and VAR signals are processed through separate channels and the phase difference between them, determined in a phase comparator, gives directly the bearing of the line from the VOR station to the aircraft, called the radial.

13.4 Radar

Since its historic implementation in Britain at the beginning of the second World War, when it was used to provide critical early warning of the approach of German war planes, radar (the first letters of *ra*dio *d*etection *a*nd *r*anging) has spawned the whole field of microwave electronics and the development of many specialized radars to meet various military and civilian applications. Of particular importance is the development of Doppler radar, in which the relative velocity of the target is determined in addition to its range. The basic principle of radar is akin to the way bats perceive their surroundings in the dark, that is, it is an echo-location method. Instead of a sonic pulse, a radar transmitter emits a powerful electromagnetic burst of very short duration (sub-microsecond) which, in order to have directional resolution, must have a much shorter wavelength than ordinary radio waves, in the centimeter range rather than the hundreds of meters of radio waves, hence the name microwaves. Originally the short 10 cm (X-band) wavelength was common since it permits the antenna, which may be a round parabolic dish or a narrow parabolic arc, to have a convenient size and yet have the desired directionality in its radiation pattern. We recall that the angular diffraction of a wave front limited by an aperture depends on the ratio of the wavelength to the diameter of the aperture, or in this case, antenna. The latter is usually rotated about a vertical axis to direct its pulses repetitively in all directions. The radiation backscattered by the target, that is, the echo, is usually picked up by the same antenna, downconverted to intermediate frequencies (IF) using nonlinear mixers with local oscillators, amplified, and detected, that, is converted to a low frequency pulse signal suitable for display on a PPI (plan position indicator) screen. In such a display a polar diagram is produced in which the radial distance of a bright dot produced by the echo is a measure of the echo delay and therefore the distance of the target. The direction of the echo signal around the center follows the rotation of the antenna. The velocity of the microwave pulse is the velocity of light and is a constant, equal to 3×10^8 m/s (in vacuo) so that each microsecond increment in the time that it takes the echo to return corresponds to a distance increment of 150 m to the target.

In a so-called *monostatic* radar system, the same antenna is used for both transmission and reception and therefore faces an engineering problem that was a formidable challenge in the early days of radar. In the transmission phase, the antenna must be illuminated by a powerful radiation pulse, while in a matter of microseconds it must convert to the reception mode in which it must detect a weak echo signal with the least possible interference from the transmitter. This is possible only if the receiver is sufficiently isolated from the transmitter during the receiving phase. Clearly what is required is a very fast switch to effectively switch the connection of the antenna from the transmission mode to the reception mode and do it in a fraction of a microsecond. The required device is called a *duplexer* which "duplexes" two signals into one channel, providing alternating connection to that channel from two independent sources. In radar systems that have waveguide antenna feeds, waveguide junction *ferrite circulators* can be used as duplexers. Ferrites are complex nonmetallic polycrystalline solids consisting of lattices of ferric oxide usually with divalent oxides of other elements making them ferrimagnetic (as distinguished from common ferromagnetic substances such as iron). This type of magnetism is characterized by two sets of molecular magnetic moments lining up in opposite directions to each other. The property that makes them useful as microwave components is that the transmission of microwaves through a ferrite in the presence of an external magnetic field displays *nonreciprocal* properties, that is, the transmission in one direction is different from the opposite direction. The duplexer must have a minimum of three ports, as shown schematically in Fig. 13.3.

The output of the microwave power generator enters at port A and exits at port B which leads to the antenna to be transmitted. An echo signal entering port B from the antenna will exit at port C to a receiver protector stage before reaching the receiver. In practice duplexers are based on four-port circulators, the fourth port serving to avoid the transmitter power from being reflected back into the generator.

Doppler radar requires that the *phase* of the transmitted microwave be preserved as a reference. An important application of such phase coherent radars is as a *moving target indicator* (MTI) which can discriminate between a moving target, which reflects a Doppler shifted frequency, and incoherent background clutter which appears at the same range. The velocity resolutions of such radars are clearly limited by the signal-to-noise ratio of the echo and the phase stability of their local

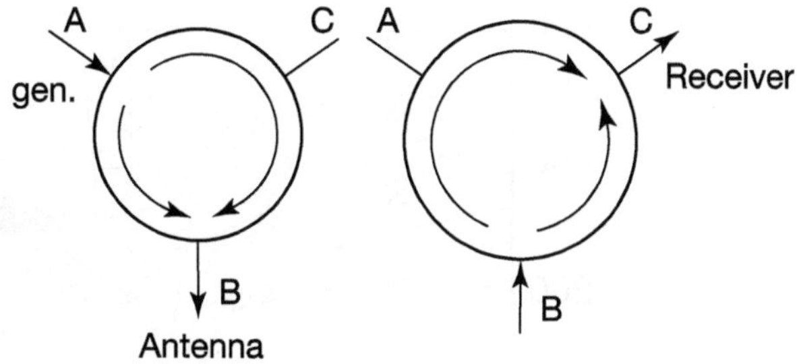

Fig. 13.3 Schematic drawing of three-port clockwise ferrite circulator

frequency reference. However it is only the short-term stability of the local oscillator that is important since the time delay of the echo is, for example, only 1 ms for a distance of 150 km. Thus a high-quality quartz oscillator is sufficient for most terrestrial distances; satellite tracking is another matter.

13.5 LORAN-C

Loran, an acronym for *lo*ng-*ra*nge *n*avigation, was designed to be a highly accurate *regional* navigation system servicing areas where navigation may be potentially hazardous because of physical factors or high traffic. Loran was a precise time-based system that, unlike radar in which the user must actively broadcast and receive a signal, consists of networks, called *chains* of powerful radio transmitters of precisely known geographical location that maintain precise synchronism among their members. These transmitters broadcast coded time signals on a regular schedule, and by observing the *difference* in time between the reception of signals from at least three different stations, a fix is obtained by the navigator. The original Loran-A operated on a carrier radio frequency in the range 1.75–1.95 MHz; this was superseded ultimately by Loran-C which operated at the lower frequency of 100 kHz. This gave it a greater useful range since the lower frequency travels farther in the *ground wave* propagation mode used by the system for the constancy of its propagation velocity. This is important of course since the distance to the transmitter is calculated from the transit *time* of the radio signal from the station to the receiver. In the ground wave mode the radio waves travel, as the name suggests, along the surface of the earth; this however defines only the geometric path, the precise propagation time is affected by surface conditions on the earth and to a much lesser extent atmospheric conditions; this involves a knowledge of the temperature and humidity of the atmosphere as well as the conductivity and dielectric constant of the surface, whether sea water or solid terrain. By applying the necessary corrections satisfactory accuracy can be achieved using the ground wave. The sky wave modes, in which the wave bounces off ionized layers in the upper atmosphere, are deemed unsuitable in spite of their longer range. The reason is that the heights of these layers can vary between day time and night time and fluctuate unpredictably when there is solar activity. The sky wave signal arrives

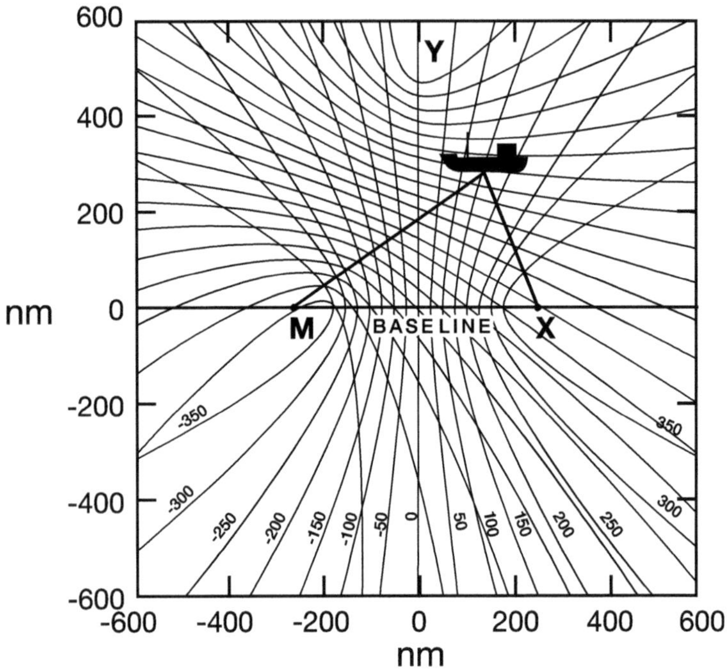

Fig. 13.4 Hyperbolas of constant difference in distance from two points A and B [2]

later than the ground wave and must be carefully discriminated against in the interpretation of the received signals.

For the system to have universal accessibility the user vessel or aircraft cannot be assumed to have a clock stable enough to use one way ranging, that is, simply carrying a clock that has been synchronized with the Loran stations and simply note the delay time for the signal to arrive at the receiver. Such a one-way ranging system would have a very limited number of users since they would have to carry atomic clocks as sophisticated as the stations themselves! Nevertheless for decades the idea of one-way ranging had been seriously explored for aircraft at busy airports, since it requires only one-way time communication between aircraft to avoid collisions. The argument for it is based on the fact that the *number* of frequency channels required increases only as the first power of the number of aircraft instead of the square of that number. The genius of the Loran system is that it requires the user to have only a clock good enough to measure the *difference* in the time of arrival of two signals from two widely separated but closely *synchronized* stations. This is very different from requiring every user to maintain absolute synchronism with the stations for long periods of time. For surface navigation we can show that signals from *three* widely separated stations are required to fix a position. The theory of reducing the time delay data to obtain a fix is simple if we neglect the curvature of the earth's surface; otherwise, to plot the position on a marine chart, usually a Mercator projection, would require the conversion of spherical coordinates to the plane chart. In Fig. 13.4 we have drawn the locus of all points *in a plane* that satisfy the condition that the distance to point A *minus* the distance to point B is a given constant number. With a little algebra we can show that points satisfying that condition lie on hyperbolas that depend on that constant difference in distance.

Hyperbolas are in the broader family of conic sections, one member of which is the ellipse, which we met in connection with Kepler's law of planetary motion. There we noted that the ellipse has an analogous property to the hyperbola: in that case the *sum* of the distances from its foci to any point on the ellipse is constant, rather than the difference. Given a measured delay between time signals received from stations M and X in Fig. 13.4, the navigator knows that he is somewhere on the hyperbola corresponding to the constant difference in distance from those two stations. The perpendicular bisector of the *base line* segment MX is called the *center line*, it corresponds to equal distances from the stations, and therefore the signals arrive at the same time, that is, zero delay. The lines obtained by extending the segment MX in either direction to infinity are called *base line extensions*; they give the loci of points where the time delay is constant equal to the propagation time from M to X. In order to get a fix it is necessary to receive a time signal from a third station Y preferably far removed from the first two. If the time delay between signals from stations X and Y is measured, then the navigator knows that he is somewhere on another hyperbola belonging to the set for stations X and Y. The point at which the hyperbola for stations M and X intersects the one for X and Y fixes his position. Since hyperbolas are quadratic curves, that is, the coordinates of points on them satisfy a quadratic equation, they will theoretically intersect at *two* points, and therefore there is an ambiguity as to the actual position of the ship; this can be generally resolved by an approximate knowledge of the ship's position obtained, for example, by dead reckoning. There are regions however where there would be great uncertainty: for example, in the baseline extension regions where double intersections are more likely and the errors large.

The implementation of a time-based navigational system obviously requires widely separated broadcast stations at precisely known geographical positions to be able to maintain their clocks constantly in precise synchronism. This requires the stations to maintain state-of-the-art atomic clocks and to closely coordinate their timekeeping with each other and with the US Naval Observatory. But beyond that, the transmitters themselves must produce and broadcast signals accurately pulse-modulated within the prescribed tolerances. The satisfactory operation of the stations requires constant monitoring and adjustment of the electronic equipment, including the atomic frequency standards and the phase of the broadcast time signals. The essential functions of monitor and control are achieved by organizing the system stations in chains, each covering a certain geographical region, with one station designated as the master station (M) with two or more other secondary slave stations (W, X, Y, Z...), any one of which may be paired with the master to form a master–slave pair. They broadcast a carrier frequency of 100 kHz which is modulated in groups of pulses that are repeated at a rate that identifies the chain The optimal frequency of the transmission and type of modulation to use for optimal signal-to-noise ratio are crucial parameters: the object of course is to

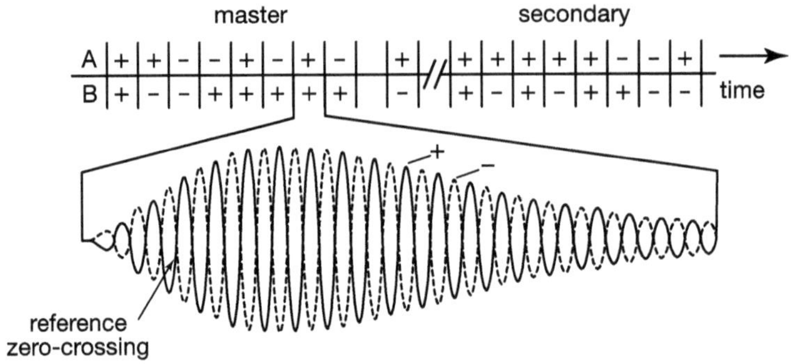

Fig. 13.5 Waveform of a pulse in Loran-C [2]

transmit time signals as accurately as possible and as far as possible. Furthermore the multiplicity of stations making up the chain all broadcasting the same frequency dictates that an orderly system be implemented to avoid confusion. The type of modulation used is pulsed amplitude modulation with a strictly defined envelope referred to as a *t*-squared pulse, since it involves a quadratic function of the time *t* given by $t^2\exp(-2\ t/65)$ where *t* is in microseconds. The pulse shape is designed in order that 99 % of the radiated power lies within the frequency band allotted to Loran-C of 90–110 kHz. The third zero crossing of the carrier is used to mark the time; this occurs 30 μs into the pulse, a point chosen because it is late enough to ensure an adequate signal strength and yet too early for skywave contamination.

The signal format for the master station consists of eight phase-coded pulses 1 ms apart followed by an additional pulse 2 ms after the first eight to identify it as master, while the secondary stations broadcast only the first eight pulses. The time separating the pulse groups of the master station is called the group repetition interval (GRI). The group repetition interval in units of 10 μs is called the *group designator*; thus, for example, the north east United States (NEUS) chain is designated 9960 since the repetition time is 99,600 μs. This interval is chosen in order that the signals have time to propagate throughout the chain well before the beginning of the next group of pulses. To further ensure discrimination against sky wave contamination and to further enable received signals during one GRI to be distinguished from another GRI, *phase coding* is used in which the phase of the *rf carrier wave* is switched between 0° and 180° between successive pulses according to a binary code, the two phases corresponding to 0 and 1. In fact there are two binary codes designated as A and B and the broadcasts alternate between them as shown in Fig. 13.5. The code repeats after every two GRI intervals, a period called *phase code interval.*

The cycle of broadcasts from the members of a chain begins with the master station broadcasting first, followed after appropriate delays by the other stations in order according to their letter designations: X (X-ray), Y(Yankee), AND Z (Zulu). To avoid ambiguity, the first slave station begins its broadcast after a time interval equal to the sum of the *baseline delay*, the time required for the master signal to reach it, plus a further *secondary coding delay;* then the next slave station broadcasts after similar delays and so on along the chain. Since the interval between pulse groups is on the order of 100 ms and the propagation times are typically less than 10 ms, there is no ambiguity in identifying pulses from the different stations. The extra ninth pulse unambiguously identifies the master, and the slave signals are certain to arrive later except on the baseline extension on the side of the slave where the signals would coincide were it not for the coding delay. This delay not only helps to separate broadcasts but also can be changed at will for security purposes.

The instrumentation in the Loran-C stations is capable of fractions of a microsecond in time resolution and by radio communication the clock at the master station communicates corrections to the other station clocks to maintain microsecond synchronism. To maintain such a degree of synchronism independently for say 1 week without correction, would imply medium term stability of 1 part in 10^{12}; this is well within present day capability of atomic or ionic clocks. Such close synchronism is in principle required only of stations within a chain; however in order to provide a precise time distribution service in addition to navigation, the master oscillator was tracked relative to the official time standard at the US Naval Observatory and periodic adjustments made to maintain global synchronism within a set tolerance.

Considerations of accuracy in fixing position at sea lead us to estimate the effect of an error in the measurement of the time difference or more precisely the difference in the distances to the two stations. By plotting constant delay hyperbolas on a navigation chart, with the delay incremented by equal amount from one hyperbola to the next, we would see that the spacing between neighboring hyperbolas varies throughout the geographic area defined by the range of the radio signals. The plotted pattern has the appearance of a bundle of lines squeezed at the base line and fanning out in all directions. Since the time increments are the same between hyperbolas, the distance between consecutive hyperbolas is called the *gradient* since it is the rate of change of position with respect to delay. The gradient directly indicates the expected error in position that is attributable to a given error in the time measurement. It is clear therefore that the most precise positioning occurs in the vicinity of the baseline where the hyperbolas are closest on the chart. If we overlap the two sets of hyperbolas for base lines MX and MY, then where a pair of consecutive hyperbolas belonging to MX intersect a pair belonging to MY, a figure resembling a curved parallelogram is defined, whose area is a measure of the error in the location of the receiver (see Fig. 13.5). An actual plot would show that the greatest errors would occur in the neighborhood of the baseline *extensions*; these are also the regions where the two points of intersection of a pair of hyperbolas may not be so far apart as to permit easy resolution. In general we should expect that the error in position is small where the two sets of hyperbolas cross at right angles and increases as the angle between them decreases.

13.6 Loran-C Charts

Loran-C was so widely used that special nautical charts were available with lines of constant time difference (TD) overprinted on them for different Loran-C chains [3]. Such a chart, showing, for example, the hyperbolas for a master and two slave stations, would be overprinted with two sets of intersecting hyperbolas (variously referred to as LOPs, TDs, or rates) to form a grid. On some US charts different colors are used for the secondary hyperbolas, for example, the color blue may be used to print the lines for the W (Whiskey) secondary, magenta for X (X-ray), and so on; however it is the GRI that unambiguously distinguishes them. If the chart covers only a relatively small geographical area, say 200 nautical miles across, the segments of hyperbolas would appear as two sets of slightly curved parallel lines generally inclined to each other. These lines are plots of geographical position derived from the computed TDs, taking into account all appropriate corrections to the propagation times. The point where these TDs intersect gives the navigator a fix of his position, except in the rare case where a second intersection occurs near the baseline extension. The particular manner in which these TDs are labeled on the charts may differ among manufacturers; thus the secondary stations may be identified numerically, for example, 1 = Whiskey, 2 = X-ray, and so on. As noted earlier these hyperbolic TDs being second-degree curves, one belonging to one pair of stations may intersect at *two* points with a curve belonging to another pair. This gives rise to an ambiguity that must obviously be resolved through dead reckoning or by receiving signals from another master-secondary pair of stations. Some sophisticated Loran-C receivers will sound an *ambiguity alarm* and automatically seek TDs from another secondary station.

We have already noted the pattern of hyperbolic TDs in the coverage area of a given chain, and in particular the importance of the spacing of TDs drawn at equal increments in time delay. To describe the pattern more quantitatively we recall the definition of the gradient as the rate of change of position with respect to time delay. Thus if the chart shows lines 10 μs apart and the distance between them is, for example, 5 nautical miles, then the gradient would be 5/10 = 0.5 nm/μs. In terms of accuracy this gradient would imply of course that an error in the time delay of 1 μs would correspond to an error in position of 0.5 nm. Clearly, if it can be assumed that the time error is nearly constant throughout the coverage area (a somewhat dubious assumption), the greatest accuracy is found where the gradient is least. It is not difficult to show that along the baseline the hyperbolas for equal increments of time are equally spaced, and therefore the gradient there is constant. If a time error of 1 μs is assumed, then the 0.25 nautical mile accuracy of the Loran-C system would clearly require a gradient less than about 0.25 nm/μs. It is evident from the general pattern of hyperbolas that the gradient increases rapidly as we go away from the stations and becomes unacceptably large in the neighborhood of the extensions of the baseline.

These observations concerning the significance of the gradient in determining the position accuracy on the chart, and the angle of intersection of the hyperbolas, raises the question of how best to choose the stations in order to maximize accuracy

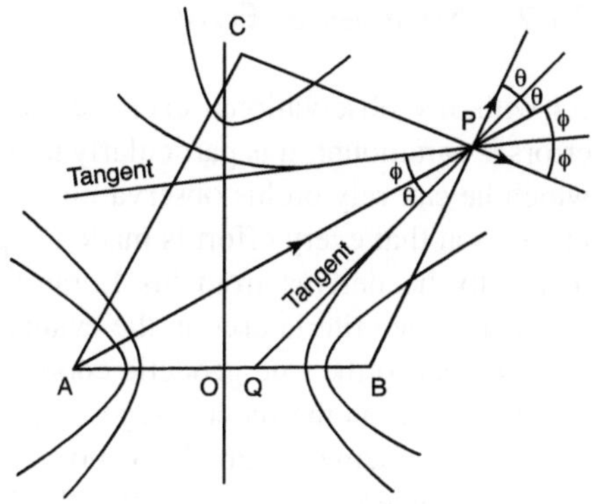

Fig. 13.6 Geometric determination of the angle between pairs of TD's

within a defined coverage area. The intuitive answer turns out to be the correct one: simply make the baselines as long as possible and place the baselines to the secondary stations as nearly perpendicular to each other as possible. These guiding principles are valuable of course, but the ultimate realization of the physical chain depends on many other practical factors, not least of which is the radio transmission power that would be necessary for an extended range. Given the uncertainty in the true value of the TD caused by various sources of error we shall be discussing in detail, the graphical representation on a marine chart of the TD curves should be bands whose width represents some mean error. From this it is evident that there are two criteria for the choice of TD to optimize the accuracy: first, choose TDs that are closely spaced on the chart since this indicates a small difference in position for a given time error, and second, the TDs should be as nearly perpendicular to each other as possible to minimize the area of the intersection region, which gives the extent of the error in the fix.

Thanks to the geometrical properties of the hyperbola; the theoretical determination of the angle of crossing of two DTs requires only a knowledge of the bearings of the stations relative to the navigator. Again the hyperbola shares a geometric property analogous to the well-known reflection property of the ellipse: we may recall that an ellipse has the property that light rays emanating from a point source at one focus will converge on the other focus when reflected by a surface along the *inside* of the ellipse. In the case of the hyperbola, light rays radiating from one focus of a hyperbola will, upon reflection by the *outside* of the second branch of the hyperbola, radiate out as if from a point source at the other focus of the hyperbola. By applying the laws of optical reflection we immediately deduce the angle relationships that we need in order to find the angle of intersection between two sets of TDs. Figure 13.6 illustrates the geometry of the problem: the tangents to the two intersecting TDs make an angle of $(\theta + \varphi)$, where $2\varphi =$ angle MPX and $2\theta =$ angle MPY.

In practice the crossing angles of the TDs are obtained from the overprinted charts, enabling the choice of those secondary stations with crossing angles as near as possible to 90°.

13.7 Sources of Error

As with any observational exercise the question of the nature and magnitude of errors is paramount. It is particularly important for a navigator to know the extent to which he can rely on his observations; his life may literally depend on it. It is for this reason that every effort is made to apprise the navigator of the potential errors in the fix he derives from his Loran-C receiver. To that end we distinguished between three criteria used in the evaluation of errors: these are *absolute* accuracy with respect to the true geodetic coordinates of the fix, reproducibility, or *repeatability*, which as the name suggests, is the accuracy with which a navigator can return to the same geographical position he had previously determined with the same equipment, and finally the third measure of accuracy is called *relative* accuracy; this relates to the accuracy with which one navigator can locate the position of another navigator given the latter's location using an identical Loran-C system. This latter is a reflection of the closeness of tolerance in the construction and operation of Loran-C sets and their relative insensitivity to the conditions of operation. Relative accuracy would be important, for example, in the case of search and rescue missions where the rescue ship may rely on a Loran-C receiver similar to the one the ship in distress used to give its position. Loran-C has proven to be consistently accurate with respect to reproducibility, which suggests that any systematic errors that may exist are at least constant, hence its continued existence, at least in some quarters.

Of the three categories, the principal concerns in practice are absolute and/or repeatable accuracy. As we saw in the discussion of the stability of atomic clocks, for example, the limits of absolute accuracy are set by two kinds of errors: statistically random and systematic errors. The most universal form of random error in electronic instruments is electrical noise; some of the basic types of noise were discussed in an earlier chapter on the stability of atomic clocks. The most basic types of noise are thermal (or Johnson noise), shot noise, and flicker noise, each having its own characteristic frequency dependence; to these we must add the universal presence of noise due to mechanical vibration, all contributing to random errors.

The discussion so far has assumed errors in the measurement of the propagation *times* of the signals, when what is really the function of the system is to determine spatial *position* on a spherical earth and its representation on a marine chart. To achieve precision in the conversion to distance requires knowledge of the propagation velocity of the radio waves in the ground wave mode. At the level of precision demanded of the system we must take into account the dispersion properties of the medium through which the radio waves must travel. The actual propagation time is arrived at by successive approximations applied to the velocity of light in vacuum (2.9978×10^8 m/s, or 161,829 nm/s): the starting value is the time for the waves to travel in a clear atmosphere at normal temperatures called the primary phase factor (PF); this in fact is a close approximation, though not good enough for Loran-C when the path is over land and sea.

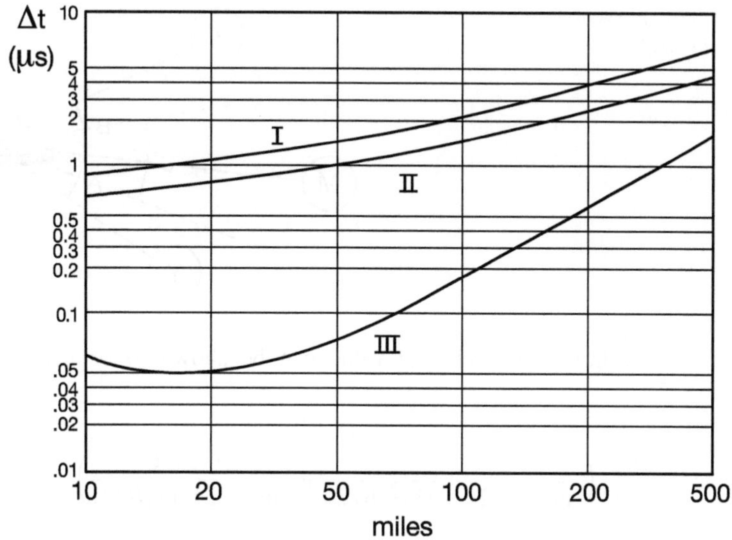

Fig. 13.7 Radio wave propagation delays over different surfaces. I and II over different terrain, and III over seawater [2]

The first correction, which reflects the overseas range of the system, is to account for propagation over seawater called the secondary phase factor (SF). Then we have what is called the additional secondary factor (ASF) which corrects for a mixed land–sea propagation path. In summary, first the time is computed for propagation in air (PF) to which are added the corrections (SF) and (ASF). The physical properties of the medium that are relevant to the propagation of electromagnetic waves are electrical conductivity and dielectric constant. In the case of mixed land–sea propagation, an exact treatment is not only theoretically difficult but the modeling of the physical properties of the different parts of the path would also be dubious. An approximate method called the *Millington method* is used in which the path is broken up into homogeneous sectors for which the required phase delays can be estimated. Since the greatest difference in propagation velocity occurs between land and seawater, the sectors naturally are defined by coastlines. The US National Institutes of Science and Technology publish circulars giving the "phase of secondary factor" for radio propagation over seawater and different types of terrain. In Fig. 13.7 are reproduced curves taken from NBS circular 573. Using this information and a suitable map showing the coastlines along the Loran-C radio propagation path, the delay corrections can be estimated.

The numerical specification of the accuracy of a Loran-C fix is quoted technically in what is referred to as *2d-rms* which is shorthand for twice the root mean square error in distance. It is defined as the radius of the *smallest* circle about the receiver's stated position, such that, in at least 95 % of fixes, the receiver's position

Fig. 13.8 Definitions of the angles *A*, *B*, and *C*

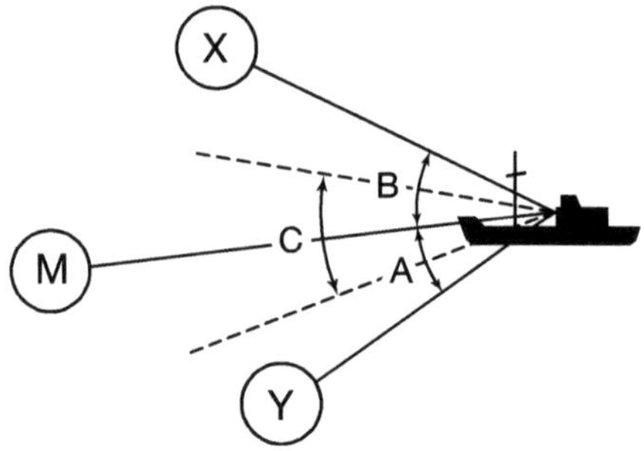

is somewhere within that circle. Its value can be shown to be given by the expression [2]:

$$2\text{drms} = \frac{2K\sigma}{\sin C}\left(\frac{1}{\sin^2(A/2)} + \frac{1}{\sin^2(B/2)} + \frac{2\rho\ \cos C}{\sin(A/2)\sin(B/2)}\right)^{1/2}, \qquad (13.1)$$

where the angles A, B, and C are the angles marked in Fig. 13.8 and $C = (A + B)/2$, K is the baseline gradient, σ is the standard deviation about the mean of the measured values of the TDs, generally assumed to be 0.1 μs, and ρ is the correlation coefficient between the measured TDs, usually assumed to have a value of 0.5 for the purpose of this calculation.

Equation (13.1) clearly shows that, for a given statistical error σ in the measurement of the TDs, 2d-rms is minimized by having the angles A, B, C as close to 90° as possible and K as small as possible. This latter condition is satisfied by points in the vicinity of the baseline. In practice navigators are spared the onerous task of actually calculating *2d-rms* for each choice of secondary stations; even before integrated circuits came into common use, tables and charts had been published by the US government maritime agencies: the US Navy and the US Coast Guard. Navigators have at their disposal handbooks giving the precise coordinates of Loran-C stations in terms of the latest World Geodetic System (WGS), last revised in 1984. We will encounter again the World Geodetic System, which defines coordinates on the earth's surface with reference to a standard ellipsoid, when we deal with the Global Positioning System (GPS). One such publication is the document COMDTINST M16562.4 which contains the specifications for the Loran-C chains including transmitted signal specifications and theoretical plots of contour lines of constant *2d-rms* on maps covering different chains around the world. In Fig. 13.9 is reproduced a map showing the northeast US chain (NEUS) which has a GRI of 99,600 μs and therefore the designator 9960. The accuracy contours are shown for the M–X and M–Y pairs at *2d-rms* = 0.25 nautical mile or about 1,520 ft, the error that actually is the Loran-C specification. The contours are based only on geometrical factors and do not take into account actual range limits.

The absolute accuracy of Loran-C fixes, after being appropriately corrected for possible overland propagation, is on the order of 0.1–0.25 nautical miles. The absolute error was officially specified as not to be greater than ±0.25 nautical

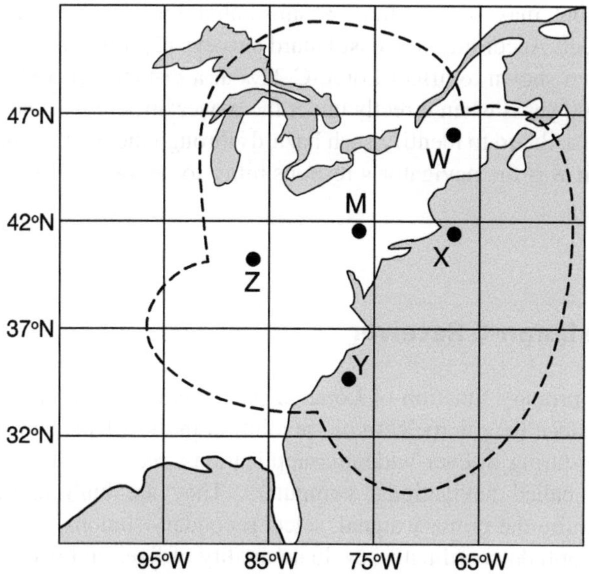

Fig. 13.9 Error boundary of Northeast US Loran-C chain. M: Seneca, NY, X: Nantucket, MA, Y: Carolina Beach, NC, Z: Dana, IN, W: Caribou, ME

miles in the designated range of the chain. The repeatable accuracy which is really more a measure of stability rather than accuracy is, as expected, greater than the absolute accuracy and is measured in feet: it can range from 60 to 300 ft, depending on the navigator's location in the coverage area. With this level of accuracy Loran-C was qualified for navigation support in critical areas such as harbor approaches.

The most serious uncontrollable source of instability in the propagation time of the radio waves on which the system is built, is the atmosphere. In addition to long-term instability caused by varying conditions of pressure, temperature, and humidity, other more violent events, such as electrical storms with heavy rain, and lightning strikes and solar flares are always possible somewhere over an extended geographical region. Atmospheric fluctuations increase in importance with range, adding significantly to the electrical noise in the receiver, and with the inverse square law fall-off of the radiation field will ultimately set the limit to the range of useful reception. In addition to variable atmospheric conditions, radio propagation over land masses is also subject to a so-called *seasonal effect*, a seasonal variation in propagation due to the changing moisture content on the ground, and hence its electrical conductivity. Seasonal variation also occurs in atmospheric *static* noise: its level is generally higher in the summer than the winter months at a given geographical location.

Another class of objects affecting radio propagation not mentioned so far are objects that may distort radio wave fronts and therefore the arrival time at a receiver that includes large steel structures, such as bridges, pylons, power lines, etc. These structures may not be included in the additional secondary factor (ASF) corrections to the TDs, but may in fact have a substantial local and possibly dangerous distorting effect. According to Coast Guard surveys [2], for example, some power lines have been shown to affect Loran-C TDs at a distance of 500 yards, causing errors up to 200 yards when directly under the power lines. It was the responsibility of the US Coast Guard to identify such hazards through the publication of bulletins; nevertheless it was the navigator's responsibility to be aware of the possibility of such hazards.

13.8 The Loran-C Receiver

Although the primary function of Loran-C receivers is obviously to receive radio signals, ever since the microelectronic revolution in the 1970s, and the ubiquitous microchip appearing in ever widening applications, receivers have become more appropriately called navigational computers. They are built to perform many functions: acquire the primary signal, select secondary stations, convert from TDs to latitude–longitude coordinates, evaluate quality of fixes and warn of any loss of integrity in the system. In fact, receivers and navigational equipment in general, have become so sophisticated that only a specialist can hope to fully appreciate their inner workings and architecture. Like all modern electronic gadgets built around integrated circuits, to use them requires little knowledge beyond recognizing an overheated component: in fact there is little incentive to try to understand how they work beyond native curiosity to gain such knowledge. Simply replacing ICs hoping that will fix the problem is curiously unsatisfying.

A Loran-C receiver is a specialized radio receiver with signal handling capability to ensure that accurate, undistorted waveforms are received from which the precise timing of pulsed radiofrequency signals can be extracted. Although the actual design of receivers on the market is of course proprietary, we may safely assume that the front end design conforms to the same well-established radiofrequency techniques. In this case we are not interested in recovering an audio modulation on a radiofrequency carrier as in ordinary radios, but rather the strictly defined series of rf pulses we described earlier, pulses whose third zero crossing marks the time of arrival of the signal. The detailed design of how the timing information is retrieved may differ among manufacturers, but the basic requirement remains to amplify that signal without distortion and with as little noise as possible. One approach is to work with the pulse envelopes relying on the slope of their leading edges; this requires a high gain, low-noise, *band-pass* radiofrequency amplifier that is linear over a 20 kHz bandwidth extending from 90 to 110 kHz. In that low frequency (LF) radio band there is liable to be a good deal of amplitude noise due to "static" interference, the main reason for the development of FM radio. Any attempt to reduce noise by narrowing the bandwidth would cause distortion of the signal, hence several notch filters are essential to suppress the many sources of radio static; these filters are sharply tunable inductance–capacitance resonant circuits that short out the offending spurious signals.

The main figures of merit of any amplifier are the noise figure, which determines its ultimate sensitivity, and dynamic range, that is the maximum input signal strength that can be amplified without distortion. One definition of the noise figure, F, on the logarithmic *decibel scale* is the following:

$$F = 10 \log_{10}\left[\frac{(S/N)\mathrm{in}}{(S/N)\mathrm{out}}\right], \tag{13.2}$$

where $(S/N)_{out}$ and $(S/N)_{in}$ are the signal-to-noise ratios (on the usual linear scale) at the output and input of the amplifier, respectively; F is a measure of the extent to which the amplifier has degraded the signal-to-noise ratio. An ideal amplifier would have $F = 0$, that is, it does not add any noise. An ideal amplifier with a bandwidth $\Delta\omega$ in which the only source of noise is the fundamental thermal (Johnson) noise at room temperature of (say) 290 K will have a base noise level of -174 dBm/Hz and therefore the effective input noise level in decibels is given by

$$N_0 = -174 + F + 10 \log(\Delta\omega). \tag{13.3}$$

Thus an ideal receiver with a 20 kHz band width would have $N_0 = -131$ dBm.

The functions of the receiver may be divided into four areas: receive and filter the signal; search, decode, and phase-lock on to the master station signal; identify the secondary station signals and extract the time delays between master and secondary; and finally compute the longitude/latitude coordinates of the position of the receiver. We will consider these in order. Rather than describe the details of any specific receiver design, since they are in any case proprietary, we will highlight some essential types of circuits that are commonly found in Loran-C receivers.

The radiation mode employed by Loran-C is, as already noted, the ground wave mode, in which the electric field is generally polarized in the vertical direction, with possible local distortion due to the presence of large structures. Therefore the first obviously important constituent of the receiver is its antenna. This should be mounted vertically, as far away from perturbing conductors as possible. Since the wavelength of the radio wave is 3 km, having a resonant $\lambda/4$ antenna is of course out of the question, but it does mean that the greater its length is, the better.

The signal from the antenna first passes through a preamplifier, which preferably is located at the site of the antenna and its output connected to the main body of the receiver by a 50 Ω coaxial cable. The next important circuit elements are a band-pass filter centered on 100 kHz and notch filters to "clean" the signal. The band-pass filter may take any form depending on the manufacturer's taste; for example it may be a Butterworth filter which can be designed with the requisite band pass of 30 kHz. A defining property of a Butterworth filter is that the gain function $G(\omega)$ have the following form:

$$|G(\omega)|^2 = \frac{1}{1 + (\omega/\omega_c)^{2n}},$$

where n is the order which determines how sharply the gain curve varies at the cut off frequency ω_c.

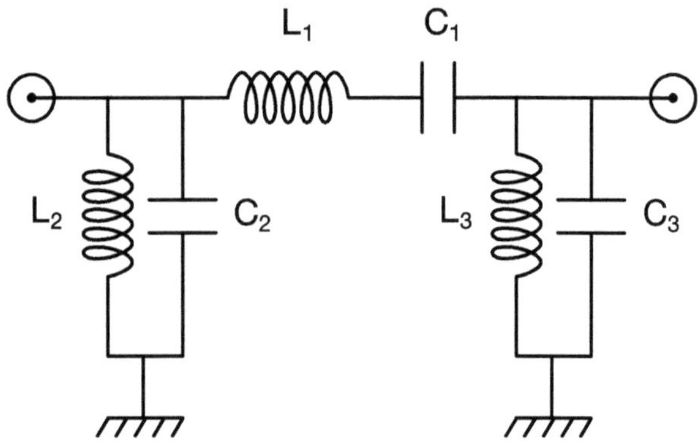

Fig. 13.10 A three-pole Butterworth band-pass filter

A circuit diagram of three-pole passive low-pass analog filter is shown as an example in Fig. 13.10; the values of L and C must be appropriately chosen to have the properties of a Butterworth filter.

In practice such a schematic is only of academic interest; the reality is more a question of what *integrated circuit* (IC) is available to perform the required function. Indeed subsequent amplification stages can be assumed to be in the form of integrated circuits giving typically130 dB of adjustable gain. Recalling the phase inversion code imposed on the Loran-C signal we should expect an analog (electronic) switch called the *signal gate* to switch polarity of the signal according to the Loran-C phase-reversal code that identifies the chain and whether master or slave; this function, along with identifying the GRI, is the principal means of searching and acquiring the Loran-C stations. Since receivers may use the inversion point on the leading edge of the *envelope* of the pulse rather than the third zero-crossing of the rf carrier to mark the time, another integrated circuit acts as an in-phase/quadrature (I/Q) demodulator, a sophisticated way of detecting, that is, retrieving the pulse envelope. It involves dividing the signal into two signals, one in phase with the carrier and one in quadrature (shifted 90°). The in-phase and quadrature signals are connected to separate amplifiers and then are sampled by analog-to-digital converters, in order to digitally carry out further processing.

Analog-to-digital converters (ADC) applied to time varying signals digitize by sampling the signal at a regular sequence of discrete intervals of time. A principal specification is the *sampling rate*: it must sample the Loran-C signal at a frequency of at least 1 MHz to ensure a faithful representation of the signal waveform and avoid the phenomenon of *aliasing*. As the name might suggest, aliasing is producing a false digital representation of an electrical signal. Its presence is easily confirmed by converting back with a digital-to-analog converter; an aliased conversion would lead to a different waveform from the original. The phenomenon is analogous to the stroboscopic effect: it was common to see in old movies a wagon moving forward while its wheels were turning backwards! According to the *Nyquist rule* the sampling rate must be at least twice the highest frequency component in the signal; the conversion of a 100 kHz signal requires a minimum sampling rate of 200 kHz and in practice a sampling rate of 1 MHz is favored in order to faithfully represent the signal with greater resolution.

A critical function of the receiver is to establish the phase of the signal from the master station and find the relative delay to that of identifiable slave stations. This requires a stable voltage-controlled oscillator (VCO) whose phase can be servo-locked to the phase of the master signal. To achieve the lock requires a phase comparator circuit in a *servo* feedback loop. This is a much studied servo feedback circuit widely used in frequency synthesizers. To ensure its stability, certain frequency response characteristics must be met in the loop which are beyond our scope here; the interested reader can find ample literature on the subject. The phase comparator generates a DC signal that is proportional to the difference (including sign) in phase between the incoming 100 kHz signal and the output of the local (nominal) 100 kHz voltage controlled oscillator. This difference, the error voltage, is connected through a filter network to the control input of the VCO to correct its frequency and cause its phase to lock on to the incoming radio signal. The filter removes any radiofrequency signal present and ensures the proper frequency response characteristics for a stable lock.

13.9 Commercial Receivers

Since the recent termination of the Loran-C service in North America, the market for receivers has dried up, except possibly in other parts of the world. Previously there were many manufacturers with numerous models of Loran-C receivers on the market designed for marine and aircraft use. Since the advent of micro-miniaturization of electronics, receivers had become ever more sophisticated with proliferating features and functions, in physically more and more compact units. Integrated circuits have not only made receivers more portable but also more affordable leading to wider use on recreational aircraft and vessels.

Naturally the receivers that were available on the market varied greatly in sophistication and price. The essential capability is to acquire and lock on to Loran-C stations and at least display the identity of the stations, the time delay, and the conversion to latitude/longitude coordinates. Many units had a navigational mode in which the user can monitor progress along a selected course and make corrections if necessary. Some more expensive units had the capability of *automatic transmitter selection*, in which the operator simply entered his estimated coordinates and the unit automatically selected the "best" GRI and station pairs. Once the selection had been made, the unit went through the sequence of search, acquire, and lock on to the signals from that chain. The length of time this process took depended on the signal-to-noise ratio, but was on the order of minutes.

On February 7, 2008 the US Department of Homeland Security announced that it intends to pursue an enhanced version of Loran-C (which has been dubbed eLoran) as part of an international positioning and timing system complementary to the different Global Navigation Satellite Systems (GNSS) such as the Global Positioning System (GPS). It would provide a backup in the event of a failure of GNSS in order to minimize the damage to safety, national security, and the economy. The Loran system will be upgraded and modernized with enhanced power and accuracy in order to be able also to provide support in different environments where GPS is unable to penetrate. An important enhancement of performance is realized through an additional data channel that conveys vital information on the integrity of the eLoran signal and any corrections or critical warnings. It is expected that the system will facilitate navigation under critical conditions such as when ships are entering a harbor or an aircraft landing.

13.10 The Omega System

An important limitation of Loran-C is the limited range (1,500 km) of the ground wave and its limited penetrability below the surface of water. This makes it necessary to have a complicated network of chains for global coverage, and consequent complexity in the design of the receivers. Omega has a *global* reach with a network of only eight stations that broadcast in the very low frequency (VLF) band between 10 and 14 kHz. The stations are spread out across the globe from Hawaii to Norway and they transmit strictly according to a precise schedule. For example, station A (Norway) begins its transmission format on 10.2 kHz for 0.9 s, then remains silent for 0.2 s, comes back on at 13.6 kHz for 1 s, is silent again for 0.2 s, then transmits at 11.33 kHz for 1.1 s, and finally falls silent for the balance of 10 s before repeating the same sequence of broadcasts. Omega is a hyperbolic system based on the difference in the time of arrival of radio signals, but unlike Loran-C which uses the time separation between short pulses, the Omega system uses the *phase difference* between long stretches of radio waves received from different stations. The main complication in using phase is that it repeats periodically and therefore an independent determination of the whole number of cycles must be made. This means that the position of the user must be known within ±30 km; otherwise if the user starts from a known position but can *initialize* the phase reading, the Omega receiver will track the phase throughout the subsequent movement of the user.

The use of signals in the VLF region of the radio spectrum assures long-range propagation in a waveguide mode between the ground and the lower D layers of the ionosphere. Unfortunately the height of the ionospheric layer is subject to daily variation from 70 km during daylight hours to 90 km during the night, with a consequent effect on the propagation time of the radio waves. Sunspot activity, which has a periodic component with an 11.4-year period, also affects the propagation time; there is also the possibility of different modes of propagation with different velocities.

With the power and sophistication of the Global Positioning System a reality, the Omega system became redundant. It ceased operation on September 30, 1997.

References

1. H.A. Thomas, Jnl. Inst. Elec. Engineers, 77, 285 (1935)
2. United States Coast Guard Loran C User Handbook, COMDTPUB P16562.6 (1992)
3. Richard R. Hobbs, *Marine Navigation*, 3rd edn. (U.S. Naval Institute Press, Annapolis, MD, 1990), p. 539

Navigation by Satellite: The Space Segment 14

14.1 History

The basing of transmitters on satellites for communication, time dissemination, and positioning has the obvious advantage of line-of-sight radio communication over wide geographical areas. Surface communication, as we saw in the last chapter, suffers from uncertainties in the propagation velocity of radio waves over different surface conditions on land and sea, and the possibility of sky wave contamination. Indeed as far back as 1945, the futurist Arthur C. Clarke had published an article in the British magazine *Wireless World* [1] explicitly describing global radio communication using three artificial satellites in geostationary orbits. He made no claim of having "invented" stationary satellite orbits, that is, ones in which a satellite remains apparently fixed overhead above some point on the equator as the earth rotates about its axis; such a circumstance might easily have been noted by Newton. Rather he published his vision of their future application as overhead platforms for communication transmitters with global reach. What makes this article transcend mere science fiction, for which he is noted, is that he used his knowledge of orbital theory to compute the actual parameters of a geostationary orbit and what it might take to place an object in that orbit. That such a feat might not be so far-fetched he may have concluded from witnessing the powerful German V2 rockets come over London during the latter part of Second World War. Of course long before the war, going back as far as the Middle Ages rockets and rocket making began to appear in the history of warfare. In the USA in the early 1920s, 20 years before the V2, Robert Goddard carried out truly pioneering work on rocket propulsion using *liquid* propellants [2], work which presaged the large boosters later developed by NASA. So Clarke's speculating about a geostationary communications satellite was not far-fetched.

In the years after the war, there was intensified interest in the development of powerful rockets in the Soviet Union and the USA, as well as in the other major powers. Initially the motivation was military superiority, but later also for civilian applications. There were determined efforts by the victorious Western powers,

F.G. Major, *Quo Vadis: Evolution of Modern Navigation: The Rise of Quantum Techniques*, DOI 10.1007/978-1-4614-8672-5_14,

particularly the USA, to make sure that the services of former German rocket scientists were secured. One man in particular, Wernher von Braun, the young man mostly associated with the German V2 rocket development, was given every encouragement to advance the U.S. space research effort. It was not long, therefore, after the world was stunned by the first successful launch into space of a man-made satellite, *Sputnik I*, in 1957 that the U.S. military establishment embarked on an accelerated program of development in space technology. The U.S. Naval Research Laboratory (NRL) was charged with the project *Vanguard* to put a small observation satellite into orbit in time for the International Geophysical Year (IGY). Unfortunately the first attempt failed, a bitter blow coming as it did soon after the Russian triumph. A second launch, carried out by the U.S. Army, successfully put the first U.S. satellite, Vanguard I, in a highly eccentric orbit in 1958. In that same year the National Aeronautics and Space Administration (NASA) was established by an act of the U.S. congress, expanding a previous agency charged with aeronautics (NACA). The civil service personnel assigned to the Vanguard project were transferred to NASA and were relocated to the new Goddard Space Flight Center (GSFC) in Greenbelt, Maryland. Manned space flight began in 1961 with the flight of Yuri Gagarin on board Vostok. Thus began the "space age" culminating in the Apollo 11 lunar landing with humans walking on the moon in 1969 (the subject of Chap. 17) and the launch of the Hubble space telescope in 1990.

Just one year after Sputnik, a prototype of the first U.S *navigational* satellite called *Transit* was developed under contract to the U.S. Navy by the Applied Physics Laboratory of Johns Hopkins University. It was launched in 1959, but unfortunately did not reach orbit. A second satellite *Transit 1B* was successfully launched in 1960 and was delivered to the U.S. Navy in 1964. The transit system consisted ultimately of six satellites in *polar orbits* at an altitude of about 1,100 km, orbits which were completed in a little over 100 min as the earth rotated under them, giving them global coverage. The principle on which the system's navigational function was based is the constancy of its satellites' orbits and therefore the exact predictability in time of their position with respect to geographical coordinates on the earth. Assuming that a satellite transmits a radio wave with a constant frequency, a ground receiver could integrate the Doppler-shifted frequency to determine the distance traveled. It is instructive to compare the functions of Loran-C and Transit: instead of measuring the time delay in radio propagation from two different stations as is done in Loran-C, the Transit system in effect measures the time delay between successive known positions of the *same* station; that is after all what causes the Doppler shift. Of particular value is the observation of the point in the satellite's orbit when the Doppler shift goes through zero and changes sign, for at that point the satellite is closest to the receiver. Noting the exact times when this occurs and knowing the satellite's ephemerides provide the data necessary to fix the position of the receiver. In this system a frequency/phase standard (clock) with long-term stability is not required of the receiver since the Doppler frequency is measured only for a relatively short time. Although the greatest demand is on the short-term stability of the satellite frequency standard, the longer term stability is also somewhat important in order to lessen the need for frequent corrections to be applied.

The satellites were equipped with high-quality quartz standards with a drift rate less than a few parts in 10^{11} per day; atomic clocks for onboard spacecraft use were still in the future. The long-term ("aging") drift of these quartz oscillators is not truly random and can be modeled mathematically to allow corrections to be applied as required to keep the system within tolerance. The broadcast frequencies used were in the UHF part of the spectrum at 150 and 400 MHz allowing smaller antennas to be used. The satellites were tracked by widely separated stations of known location to update their orbits. Since the Doppler shift gives only the *range rate,* that is, the velocity, integrating that velocity with respect to time yields the *change* in the distance of the satellite from the station that occurred during the integration interval; in order to obtain the actual satellite positions requires independent knowledge of at least one satellite position in order to determine the integration constant, as a mathematician would put it.

Transit was designed as an all-weather surface navigation system for navy surface vessels as well as submarines, although eventually civilian use was authorized and the U.S. Coast and Geodetic Survey used it to create accurate survey maps. It was designed primarily for surface navigation and as such it had global coverage, but it was not continuous, as a satellite may pass overhead every 100 min or so and users had to interpolate their position by dead reckoning between passes. It was therefore unsuitable for high-speed aircraft or missiles, where continuous three-dimensional coverage is necessary.

The earliest explicit proposal to establish a navigational system based on time comparisons between synchronized satellite-borne clocks and the navigator's clock was submitted by Roger Easton of the U.S. Naval Research Laboratory around 1955. It was called *Timation* and, as the name suggests, it was a navigation system based on the measurement of propagation time between navigator and satellite using the newly deployable atomic clocks, whose high stability allows signal propagation times to be accurately determined and therefore one way or *passive* ranging to be practical. With transmitters on a large constellation of geosynchronous satellites the system would be capable of *continuous global* coverage, but unlike Loran-C for example, the number of users would be severely limited because it assumes users are equipped with atomic clocks to maintain synchronism with the clocks on board the satellites. The original proposal of Timation anticipated by several years the idea of using the new portable Rb clocks and later Cs clocks for one-way ranging. That idea as applied to air traffic control, in fact, became one of the principal justifications for the continued support by NASA of atomic clock development; the other justification was tracking deep space probes. At the time Timation was first proposed, emphasis was placed on the fact that a navigator could, by making observations on only *one* satellite of known ephemeris, get a fix: namely, by measuring the propagation delay (its range) and Doppler shift (its range rate) along its orbit. Easton et al. had shown that the methods of star navigation (see Chap. 4) could be directly applied to get a fix by constructing the range-azimuth lines for more than one position of the satellite. The procedure begins by computing the satellite altitude angle from the observed range, as shown in Fig. 14.1; a sextant is ruled out, of course, because of the rapid movement of the satellite.

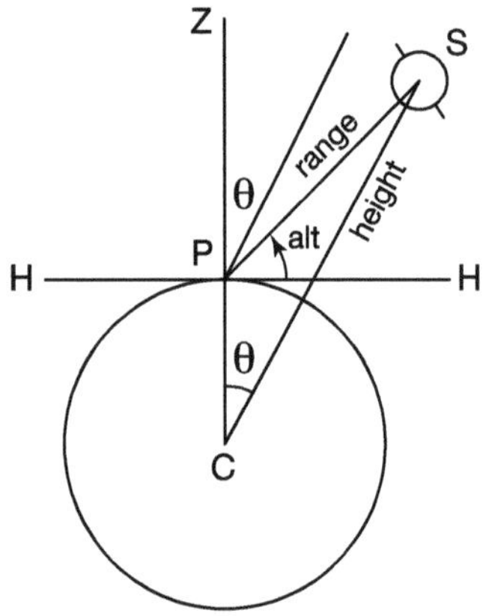

Fig. 14.1 Determination of satellite altitude from range measurements

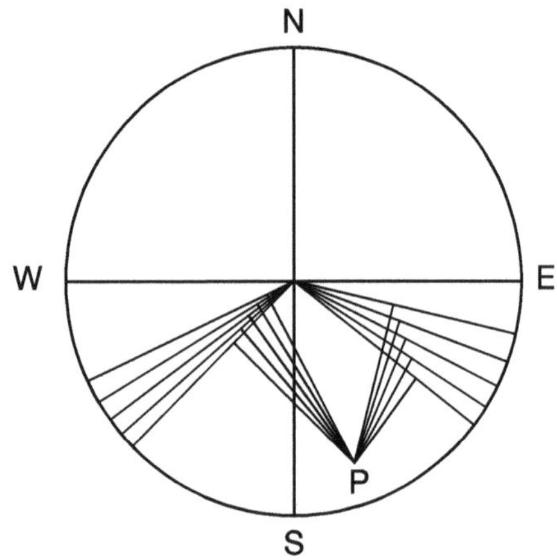

Fig. 14.2 Convergence of lines of position

To proceed, an approximate position must be assumed by the navigator, and given the geographic position of the subsatellite point (the foot of the perpendicular from the satellite), and the hour angles of the navigator and satellite at the time of observation, the spherical triangle can be solved to obtain the direction (azimuth) of the range vector. The observed propagation delay is marked at a point along this range vector (which rotates as the satellite continues in its orbit). The line of position is drawn at *right angles* to the range vector at the time delay marker. As more observations are made of the orbiting satellite these perpendicular lines should cross at a common point, namely, the position of the observer, as shown in Fig. 14.2.

About this time the Aerospace Corporation had been established to support the U.S. Air Force in the military exploitation of space. Aerospace Corporation initiated what was referred to as Project 57, later given the equally cryptic designation as *Project 621 B* (621*A* is rarely mentioned). There is little doubt that this project had many facets that are found in the ultimate GPS design, as does the NRL

Timation project for the distribution of precise time using atomic clocks on board a satellite. To extrapolate on the basis that "success has many fathers" one could plausibly say that the timing section under Andrew Chi at the Goddard Space Flight Center was also engaged in research on space-borne atomic clocks relevant to GPS; in fact the mercury ion frequency standard research the author conducted was justified specifically as a spacecraft clock.

The U.S. support for a space-based navigational system initially came as we have seen through military funding, first through the U. S. Navy at the Johns Hopkins Applied Physics Laboratory and by the U. S. Air Force at the Aerospace Corporation. While this channeling of research funds clearly indicates the importance the military attached to space technology, it characterizes the broader national inclination to use military funding for research in general, including basic research. In fact government funding has historically been critical for technical developments that have no immediate commercial return. A good example is the development of the hydrogen maser; private industry would not have developed it because the market for it is too small to defray the developmental cost. After its initial invention at Harvard University it was developed by NASA at Goddard Space Flight Center, and under NASA funding first by Varian Associates, and later by the Hewlett Packard Company.

The period 1964–1966 was one of great activity and optimism in the application of satellites to many different scientific and civilian areas in addition to the military, from geodesy, to meteorology, to communication, etc. The three research/development efforts on satellite navigation at APL, NRL, and Aerospace Corp. were joined by another program pursued by the U.S. Army called *Secor*. This divergence of approaches no doubt gave rise to a healthy degree of rivalry and subsequent claims as to who came closest to the ultimate, successful Global Positioning System. It is clear that in 1972 B. Parkinson, newly assigned to manage Project 621 B, determined that these efforts should be coordinated as part of a concerted effort to develop a system primarily based on the Aerospace Corp model. Consequently he sought the approval of such a combined effort by the U.S. Department of Defense, and so in 1973 these projects were combined under the title Defense Navigation Satellite System which later was given the acronym NAVSTAR. In September 1983 after incorporating the ability to control the resolution of the system, NAVSTAR was made available to the general public and became known as the Global Positioning System (GPS), the "must have" application on every iPhone!

During the latter part of the 1960s and early 1970s, Russia, which was then part of the USSR, also began in earnest the development of navigation by satellite. It began a military program with the launch of a constellation of six navigational satellites orbiting the earth at an altitude of 1,000 km; its positioning accuracy of ships and submarines was rated at 80–100 m. The favorable reaction of the Russian navy commanders to the performance of this system stimulated interest in the further development of an accurate navigation system available to all users. Flight tests for the new medium earth orbit navigation system that goes by the acronym GLONASS began in 1982 with the launch of the satellite Kosmos 1413. To achieve simultaneous visibility of at least four satellites from any point on the earth requires at least 18 satellites; the number was increased to 24 to give some redundancy. The satellites are in circular orbits at altitudes of 19,100 km in three equally spaced orbital planes with an inclination of 64.8° (see below). In 2009 the Russian Federation government enacted laws regulating the use of GLONASS in transportation and official land surveys, etc. In 2011 a series of satellites, designated GLONASS-M space vehicles, began to be flight tested; this series provides for all the functions that have come to be expected of GPS, namely, navigation in all environments, surveying, time synchronization, and personal use, etc. .

Responding to a desire to have an independent system, although somewhat late in the game, ESA, the European Space Agency, was charged in 2003 with developing a superior satellite-based navigational system called GALILEO for general civilian use. It is designed to have *real-time* positioning accuracy down to the 1 m range anywhere on earth and to have the capacity to warn users within seconds of any failure in the system, qualifying it to be applied to guide cars, trains, and aircraft in critical times, for example, when they are landing. ESA's first experimental satellite dubbed GIOVE-A was put into orbit by a Soyuz launcher in 2011 to be part of in-orbit validation, and a second satellite GIOVE-B was scheduled for launch in 2012. Unfortunately this ambitious effort is behind schedule and over budget; the original financing scheme relied on a government–industry partnership, but evidently industry became skeptical of the program. Nevertheless the program continued and in March 2013 GALILEO signals from four satellites were tested for accuracy in fixing a point in three dimensions, latitude, longitude, and altitude, and found to be well within design expectations. This expands the coverage of the worldwide global navigation satellite system (GNSS) with an increasing number of receivers equipped to take advantage of that expansion.

Currently at the back of the pack, but probably not for long, is the regional Chinese satellite positioning system called BeiDou. This is a two-way system that enables mobile users to receive position information. In our further discussion of satellite navigation we will limit ourselves to the U.S. GPS system unless these other navigational systems differ in some significant technological way that highlights the advantages or disadvantages of the GPS system.

14.2 GPS: The System Design

Originally intended for military use only, GPS was then made accessible to the general public with diminished resolution, and now it is generally available with unrestricted accuracy. It is a time-based system, the space analog of the Loran-C system, made possible by the availability of ultra-stable atomic clocks small enough to be carried in a spacecraft. Starting in the year 1978, after an initial failure, ten GPS satellites were launched over a period of a decade, making up a constellation called Block I. Before this phase was completed the U.S. Air Force awarded a large contract to Rockwell International to build altogether 28 improved satellites designated Block II and Block IIA. The first of the Block II satellites was put into orbit in 1989; a further 27 were eventually launched, the last 19 Block IIA satellites being upgraded versions of Block II. By December 1990 GPS operations had been fully tested and by 1995 the U.S. Air Force declared the Block II constellation to be fully operational. This led the U.S. Federal Aviation Administration to announce that GPS would be integrated into the U.S. air traffic control system.

The GPS system (excuse the redundant use of "system") has the advantage over earlier navigation systems of having continuous *three-dimensional* and truly *global* coverage using transmitters outside the atmosphere. The crucial technical advance that made GPS practical, aside of course from putting a sizable object in a geosynchronous orbit, is the development of extremely stable *portable* atomic clocks. As with Loran-C, where the precise locations of the transmitters must be known, the position of the GPS satellites in their orbits must be known precisely as a function of time; this is called in an astronomical context the *ephemerides* of the satellites. In order to ensure that the system permits users to fix their position continuously in all weather conditions anywhere on or near the earth, a constellation of many satellites is used, each carrying Rb or Cs atomic clocks. The ranging method is again based on the measurement of the *difference* in the propagation delay of radio signals received from different satellites. The beauty of this system is that by having signals received from precisely synchronized sources it is unnecessary for the receiver to also be synchronized with the satellites and is therefore not required to carry around an atomic clock! We recall that in the case of Loran-C, where only surface two-dimensional positioning is enabled, a minimum of three transmissions must be received; for three-dimensional positioning we require at least *four* transmissions (and an electronic computer!). It follows that for global coverage, at least four satellites must be observable at all time from any point on the earth; this raises an interesting design problem, one that researchers at the Naval Research Laboratory had thought about early in the development of the subject. Since the user on the ground cannot keep strict synchronism with the satellites, the ranges to the different satellites computed on the basis of the observed delays are called *pseudoranges*; the true range must be computed after the clock error has been deduced. In order to fix the receiver in three dimensions, latitude, longitude, and altitude, requires of course three *true* ranges and hence the need for a fourth measured delay to calculate the receiver clock error. We can imagine the radio waves radiating out as spheres from the satellites with radii equal to the pseudoranges; two spheres will intersect in a circle and the third will intersect that circle at two points. However the sphere from a *fourth* satellite with radius equal to *its* pseudorange will in general *not* intersect exactly at either point. But if the satellite clocks are exactly synchronized and their assumed positions correct, then a *single* correction to the *user's* clock will make all spheres intersect at one point, namely, the user's position.

We will follow the usual practice of separating the description of the GPS into three segments or sectors: the satellite system, the ground monitor system, and the ground reception by users and applications; we will give a brief account of each of these in turn.

14.3 GPS Satellite Orbits

A fundamental premise in the use of satellites as platforms for radio transmitters in a navigational network is that their ephemerides, that is, their orbital positions as a function of time, can be accurately predicted at all times and that this can be communicated to the spacecraft control. This premise is of course based on Newton's theory of gravitation and the results that flow from it, in particular, the Keplerian elliptical orbits with the earth at one of the foci. Simple motion in an ellipse is the beginning point in the process of successive approximations to refine the accuracy of the orbital data.

Starting with the satellite regarded as a point mass, in three-dimensional space we require six quantities to completely specify its motion when acted on by known forces; these may be chosen to be three coordinates in space and three components of velocity at some given time. The general solution to the equations of motion requires six parameters, or to put it mathematically, six integration constants to define it. A satellite moving under the central gravitational pull of the earth will describe an elliptical orbit in a plane passing through the earth's center provided the initial conditions of position and velocity for such a closed orbit are met. The center of the earth is at one focus of the ellipse; the point in the orbit closest to the earth is called the *perigee*, while the farthest point is the *apogee*. The period of time required for the satellite to complete one revolution, T, is given by Kepler's law as follows:

$$\frac{T}{2\pi} = \sqrt{\frac{a^3}{Gm_E}}, \tag{14.1}$$

where a is the semimajor axis of the ellipse, G the gravitational constant (6.67384×10^{-11} m^3 (kg)$^{-1}$ s^{-2}), and m_E the mass of the earth (5.9736×10^{24} kg). The GPS system is designed to have the satellites complete an orbit in 12 *sidereal* hours (see Chap. 4), which requires the semimajor axis to have the value $a = 26{,}560$ km, that is, about 20,182 km above the earth's surface; this period corresponds to retracing the same ground track twice per sidereal day. The position of the satellite in its orbit has been specified by different angles, one is called historically the *true anomaly* (sic); it is shown in Fig. 14.3 as the angle θ. As already pointed out the complete description of the orbit requires six numbers, or *parameters* to specify it: five to specify the ellipse and one to specify the *epoch*, that is, when the satellite is at a specified point in the orbit, for example, the perigee. These six parameters are illustrated in Fig. 14.3.

The parameters are as follows: (1) the semimajor axis of the orbit, (2) the eccentricity of the orbit, (3) the orbital inclination, which is the angle between the plane of the orbit and the earth's equatorial plane, (4) the argument of the perigee, which is the angle between the perigee and the (ascending) node, the point along the line of intersection of the plane of the orbit and the equatorial plane, (5) the right

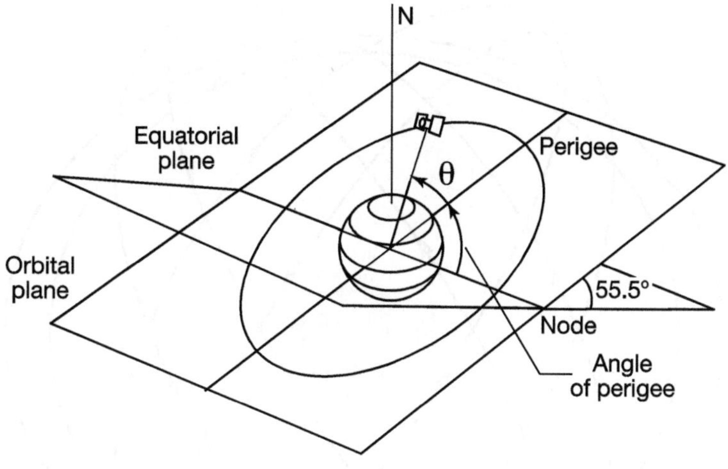

Fig. 14.3 The satellite orbital parameters

ascension (astronomical term for the azimuthal angle relative to *first point of Aries*) of the (ascending) node, described above, and finally, (6) epoch (time) of perigee passage. The parameter referred to by the intimidating term "the right ascension of the ascending node" is hard to visualize without a model; the reference to the first point of Aries is merely in order to state a longitudinal angle with respect to an astronomical reference outside the earth.

The number of satellites and the configuration of their orbits are dictated by the need to allow a user to receive signals simultaneously from at least four satellites at any point on or near the earth's surface, 24 h a day. The satellites are arranged in six equally spaced orbital planes around the earth's polar axis, each inclined to the equatorial plane by an angle of 55.5° to minimize the precession of the orbit (see below), as shown schematically in Fig. 14.4.

In each plane are four "slots," positions that may be occupied by a satellite, adding up to a total of 24 satellites in the constellation. Over the years the U.S. Air Force has continued to manage up to 31 operational GPS satellites, with 3–4 decommissioned satellites, to ensure that a constellation of at least 24 satellites are functioning properly 95 % of the time. The Air Force successfully completed a GPS constellation expansion in 2011, increasing the number of "baseline" satellites to 27.

The satellites that make up the constellation are a mix of new and old; their lineage can be traced to the *block* to which they belonged. Thus we have GPS Block II satellites that were launched in 1989–1990 and were the first to become operational as part of GPS; they are in fact second-generation satellites that followed the initial Block I satellites, which were not actually operational as part of GPS. Block II was followed by Block IIA, a more advanced version developed by Rockwell International whose aerospace division was sold to Boeing in 1996. The Block IIA series comprised a total of 19 satellites, and in the satellite numbering system, they

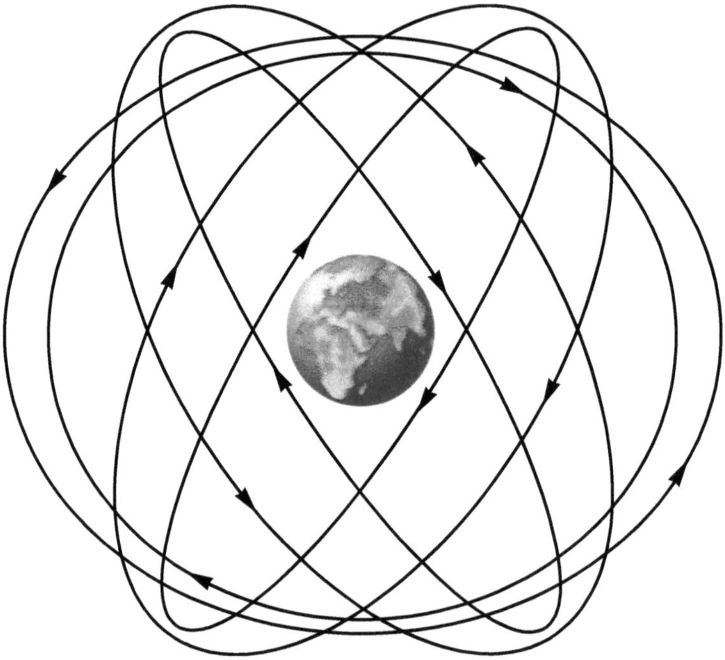

Fig. 14.4 The six inclined satellite orbits of GPS

are assigned Space Vehicle Numbers (SVN) ranging from 22 through 40. The first in this series was launched in 1990 and the last in 1997. By January 2012 there remained only ten Block IIA satellites remaining operational as part of the GPS constellation. The most significant advance in the onboard instrumentation of this series is the accommodation of civilian as well as military users by providing two classes of coding: one called a course-acquisition (C/A) code for civil users and another a precision P(Y) code for military users (see below).

The next series of satellites is referred to as Block IIR, the designation IIR being for replenishment of the aging IIA satellites, which, as the name suggests, are destined as replacements in the event of a failure among the IIA satellites or retirement due to old age. The contract for this block of 13 satellites was awarded to Lockheed Martin; the first successful launch was in 1997 and the last in 2004. Finally we mention Block IIRM (M for modern) made up of eight upgraded versions of IIR, which constitute the main thrust to modernize the navigational system and improve performance with an upgraded second civilian GPS signal devoted to improving commercial as well as military service; the first launch in this latest series occurred in 2005 and the last in 2009.

14.4 Orbital Perturbations

We have assumed reentrant (closed) elliptical Keplerian orbits for the satellites, but that presumes that they are in the gravitational field of a rigid spherically symmetric earth which acts on the satellites as would a point mass at its center. An actual earth satellite is subject to forces that depart from that simple model, departures that are called *perturbations*. Fortunately these perturbations are small compared with the gravitational force that gives rise to elliptical orbits, a fact that is exploited theoretically in refining the theory by the method of successive approximation. This is implemented by regular updating of the corrections required to maintain the specified accuracy of the orbital parameters through communications with the (ground) control segment. The application of these corrections may involve a complex array of sensors and actuators including the possible use of thrusters to physically correct the satellite motion or merely update the numerical tracking data and adjust the orbital parameters.

We can distinguish between two types of perturbations: those that are gravitational in origin such as the oblateness of the earth and contributions to the gravitational field from the sun and moon, and other perturbations that are not gravitational, such as solar radiation, solar wind, and air resistance. A reasonable place to begin in bringing some order into the bewildering array of perturbations that may be present is to assess the relative degree of impact each perturbation has on the accuracy of the orbital parameters and the limit it sets to the system's positioning accuracy in relation to the design objective. But first let us stipulate what error is tolerable in the navigational data that the system is to provide. If we say for example that the orbit must not deviate more than about 1 m during an orbit, then we would find that a constant perturbing force must not produce an acceleration greater than roughly10^{-9} ms^{-2}. To appreciate the scale of this result we calculate the acceleration of the satellite in its orbit $a = (2\pi/T)^2 r$; we find $a = 0.55$ ms^{-2} which is more than 5×10^8 larger than the maximum tolerable perturbation. In other words a fractional perturbation in the gravitational field of the earth amounting to only one part in 5×10^8 is sufficient to produce an error of 1 m in the satellite orbit.

Not surprisingly the largest perturbation comes from the departure of the earth's shape from the first close approximation, namely, a rigid sphere. In the next approximation it is an oblate spheroid (see Chap. 3), that is, a somewhat flattened sphere, or more accurately the equilibrium shape of a spinning plastic sphere, stretched out at the equator by the centrifugal force. It has an elliptical cross section through the polar axis. The actual oblateness of the earth is relatively small: the diameter from pole to pole is only about 43 km shorter than the diameter through the equator, that is, a difference of one part in 298. In the next higher approximation the shape of the earth is slightly pear shaped with a small north–south asymmetry. The detailed surface structure of the earth is of course quite irregular on a scale finer say than 9,000 m; the degree of detail that is significant depends of course on how far the satellite is above the surface. At an altitude of 20,200 km the GPS satellites are sufficiently far from the earth's surface that the oblate spheroid is considered a good working model; it introduces a correction of about one part in 10^4. A theoretical analysis of the effect of the earth's oblateness on the satellite orbit shows that it leads to a slow precession, that is, a slow rotation of the perigee at a rate proportional to $(5\cos^2\theta - 1)$ where θ is the angle the *plane* of the satellite orbit makes with the earth's equatorial plane. The truly significant result we can draw from the model is that if we choose θ to satisfy the equation $(5\cos^2\theta - 1) = 0$ we will have an orbital plane in which there is no precession; this occurs at $\theta \approx 63°$. Unfortunately that angle could not be achieved; in reality the satellite orbits are inclined at 55.5°.

The other gravitational perturbations are due to the sun and moon, and are called *tidal effects* since these are the same perturbations that cause tidal action on the earth. We can apply Newton's law of gravity to estimate the size of the gravitational disturbance caused by these two bodies; thus writing Newton's law for the peak-to-peak change in the gravitational acceleration due to the sun as the satellite goes around the earth we have

$$\Delta\alpha = \frac{GM_S}{(R_E - r_s)^2} - \frac{GM_S}{(R_E + r_s)^2}, \tag{14.2}$$

where M_S is the mass of the sun, R_E is the radius of the earth's *orbit* around the sun, and r_s is the radius of the satellite orbit. Since $r_s \ll R_E$, a first approximation to $\Delta\alpha$ is given by

$$\Delta\alpha = 4\frac{GM_S}{R_E^3} r_s. \tag{14.3}$$

If we substitute numerical values for the quantities on the right we find $\Delta\alpha \approx 4.2 \times 10^{-6}$ ms^{-2}; the corresponding number for the moon is about $\Delta\alpha \approx 7.7 \times 10^{-6}$ ms^{-2}. To put this in perspective we note that the satellite acceleration as it swings round the earth is on the order of 0.56 ms^{-2}. Lower down in the scale of significance are variations in the tidal actions of the oceans and the plastic deformations of the "solid" earth; these would range on the order of 10^{-9} ms^{-2}.

These gravitational perturbations, although quite significant, are at least theoretically predictable since they are governed by well-established laws. Unfortunately the many-body problem, which this is, involving as it does the satellite, the earth, the moon, and the sun, is notoriously difficult to solve mathematically; even the three-body problem has challenged mathematicians at least since the time of Newton. What has made the prediction of the satellite position even possible and GPS a viable means of positioning is the availability of fast modern computers.

To compound the problem of maintaining an accurate and stable "station" for a satellite are unpredictable influences that may fluctuate. It is true that the GPS satellites are far above the atmosphere and so surface weather conditions have no effect on them, but there is solar radiation which does exert radiation pressure on them. This is the life-giving sunlight that is paradoxically called blackbody radiation with a continuous spectrum peaking in the middle of the visible range. This radiation exerts pressure on any absorbing or reflecting surface, since radiation, whether radio waves, γ-rays, or visible light, carries momentum and therefore a satellite moving from light to shade will experience a change in the balance of forces. The scale of the solar perturbation can be calculated given the intensity of the radiation, that is, the *solar constant* S, the cross section presented by the satellite to the sun's rays, and the reflectivity of the surface, etc. Since the light scattering is very much dependent on the shape and optical properties of the surface of the satellite, the scattering will not be the same in all directions and the net force will not necessarily be in the direction of the incident light rays. The magnitude of S is on the order of 1.4 kW/m^2. To estimate the magnitude of the effect on a satellite we use the general result that the momentum carried by radiation is $p = E/c$ and therefore the acceleration of the satellite $\alpha = F/M = (\mathrm{d}p/\mathrm{d}t)/M = (\mathrm{d}E/\mathrm{d}t)/cM = SA/cM$ where A/M is the satellite cross-sectional area divided by its mass, and c as usual is the velocity of light. If we substitute numerical values, we find that the acceleration is on the order of 10^{-7} ms^{-2}, showing that radiation pressure is a significant factor that must be taken into account in the design of the satellite.

In addition to blackbody radiation the sun also emits less predictable streams of high-speed charged subatomic particles, mostly consisting of electrons and protons that engulf the earth in a manner determined by the earth's magnetic field and hence the preponderance of the aurora lights in the northern and southern polar regions. This so-called *solar wind* is subject occasionally to large abnormal fluctuations called *solar flares*. Near the earth's orbit the number density of the protons is on the order of 3×10^6 m^{-3} and their average speed is about 400 (km)/s; this corresponds to 1.2×10^{12} protons crossing 1 m^2 per second, each having a linear momentum of 6.7×10^{-22} kg/ms. If the proton momentum is assumed to be transferred to a satellite with $A/M = 0.03$ it would accelerate at the rate of 2.4×10^{-11} ms^{-2}, several orders of magnitude smaller than the other effects.

14.5 Systems Onboard Satellites

The equipment on board a GPS satellite is a complex, interactive set of sophisticated circuits and devices of which each subsystem plays a critical role. Perhaps the most critical for maintaining sub-microsecond synchronism throughout the system are the Rb and Cs atomic clocks, whose principles are the subject of Chap. 8. All Block IIR satellites carry 3 Rb frequency standards with improved "physics packages" and frequency stability of about one part in 10^{14}, that is, roughly 1 s in three million years! The other basic subsystems we will focus on are the satellite attitude control and the satellite orbit control, subsystems that ensure the proper orientation of the satellite and maintain a stable orbit conforming to the ephemerides provided by the control segment. The operation of the satellite is an exquisite example of advanced telemetry... everything is by remote control. The subsystems aboard the satellites necessarily involve a multitude of technologies, comprising on the one hand sensors, such as sun and earth sensors, and inertial measurement units, and on the other hand, actuators such as thrusters, reaction wheels, and magnetic torquers. Their overall function is to tightly control attitude, steer the solar panels, and implement orbital corrections based on received control data from the ground control segment (CS). In addition we have the *navigation payload* whose function is to generate and transmit the different navigational *ranging codes* and other data back to the ground user segment. The satellite receives updated navigational and control data by way of the tracking, telemetry, and control (TTC) communications link.

The frequency/phase of the ranging codes generated in the navigation payload is controlled by the Rb and/or Cs atomic frequency standards of which there are multiple backup units to ensure system reliability; at least two functioning standards are required to immediately detect a malfunction and allow a speedy recovery. The short beam space-hardened Cs standard was developed in the late 1960s under the leadership of Len Cutler at the Hewlett Packard Company, mostly under NASA contract for the Apollo program. The frequency/phase stability of atomic standards has reached new heights in recent years with the development of laser-cooled Cs atomic fountain standards and compact mercury ion standards. Since these atomic standards stabilize at their atomic hyperfine frequencies, low-noise synthesizers are needed to generate the desired time codes.

Rather than getting lost in the alphabetical jungle of acronyms used to describe the complex array of subsystems that make up a GPS satellite, and wade into the specific design of each subsystem (which in any case is not readily available to the general public), we will highlight the general principles on which the main devices depend, and their function in achieving the desired overall mission of the positioning system.

We begin with the determination of attitude: clearly the GPS satellites must preserve their attitude both with respect to the earth, since that involves the orientation of the antenna system for ground communication, and with respect to the sun, since that determines the orientation of the solar panels on which the satellite depends to supply electrical power. It often occurs that spacecraft missions require attitude control with respect to a true inertial system; in such a case the attitude with respect to the stars is required. The instruments by which the attitudes of the GPS satellites are determined with respect to the earth, sun, and stars are called earth sensor, sun sensor, and star tracker. Of the three by far the most accurate is the star tracker which establishes the attitude of a spacecraft with respect to the stars, a true inertial frame; however, as we have already stated, the main concern of the GPS satellites must be to maintain a controlled attitude with respect to earth and the sun. Fortunately, modern sensors whether earth, sun, or stars all have the same principles of design and can even use similar optical systems and detectors. The differences are in the form of the data, the information drawn from it, and the specific manner in which it is applied. Let us consider a typical star tracker design, since it is most familiar to amateur astronomers and has all the features essential for precision attitude determination. It is essentially a telescope having ideally good resolution and low stray optical noise, with some type of pixilated digital detector, and a memory containing star patterns. By using advanced nonspherical optics a relatively compact telescope can be designed to have good resolution without increasing the diameter and the focal length of the objective mirror. Thus telescopes using aspherical

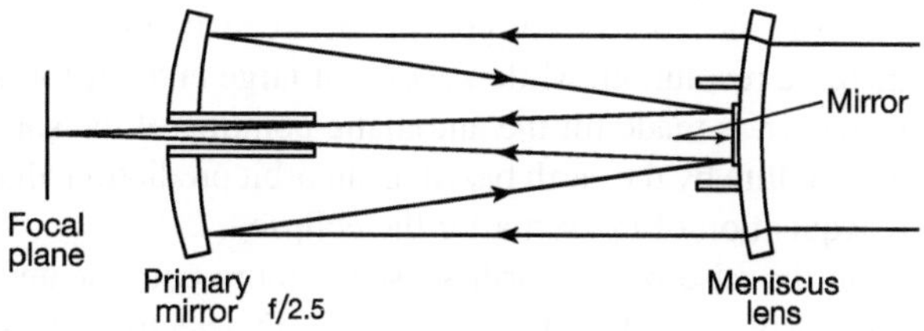

Fig. 14.5 Schematic drawing of Meade 7 in. Maksutov–Cassegrain telescope

refracting surfaces, such as in the Maksutov–Cassegrain design, have good resolution and contrast; Fig. 14.5 shows an outline of such a telescope made by the Meade Company. It is distinctive in having a meniscus lens in front of the telescope on which is mounted a small convex spherical mirror and an aspheric concave primary mirror in the back of the telescope with a baffle tube along the axis.

In order to align the axis of the telescope with a chosen star or field of stars, a modern star tracker would have a planar array of CCDs (charge-coupled devices) which are discrete photo-detectors arranged in a grid with the desired resolution in the focal plane of the telescope. Other commonly used types of detectors include APS (active pixel sensors) based on complementary metal oxide semiconductor (CMOS) technology, commonly used in digital reflex cameras. A more recent development is the enhanced signal-to-noise ratio attainable with *electron-multiplying* CCD (EMCCD) detectors. In order to lock the position of the star field with respect to the telescope reference system one requires a servo control loop in which the telescope is able to "search" for the stars and converge on a stable settling point coinciding with the best interpolation of the star images. The discreteness of the pixels presents a special problem when dealing with an image field consisting of discrete points of light: the output of the photo-sensor would vary discontinuously as the image field is modulated and the star images land on different pixels. This difficulty is overcome by a technique given the hybridized term *centroiding*: this consists of defocussing the telescope in order to convert the point images of the stars into discs, resulting in a large number of pixels being exposed. From the light intensity registered by each pixel in the star image, the centroid of the light distribution is computed and taken to be the position of the star image. In order to lock on to the star field one requires software that can compare the detected pattern of stars to a star catalog in the memory and determine the necessary displacement to make them coincide.

The earth and sun sensors carried on a GPS satellite consist of basically the same type of optical instrument, namely, a specialized camera. In the case of the sun sensor, since it is not used for precise navigation, and the sun is such a bright source, a pinhole perhaps 250 μm in diameter can be used in place of a lens to produce an image on a Position Sensing Diode (PSD). This type of photo-diode, as the name suggests, is designed to yield outputs, four in number, from a square detector surface divided into four equal quadrants from which the position of the sun's image can be determined using the appropriate software. Clearly the diameter of the pinhole must be a compromise between making its diameter small in order to increase resolution, while keeping it large enough for a good signal-to-noise ratio. Provision is made for the automatic pointing of the solar array during an eclipse of the satellite by the earth based on an orbit prediction algorithm to enable the smooth reacquisition of the sun after the eclipse.

In the case of the earth sensor, a more complicated *focusing* optical system is obviously needed, and a pixelated CCD detector to detect the image. The software in this case must use algorithms on the output data to detect the line of the horizon where a sharp transition occurs in the image between the earth and sky.

The attitude control devices include precision jet thrusters, reaction wheels, and magnetic torquers. The thrusters commonly use jets of hydrazine (N_2H_4), a compound that decomposes into ammonia and nitrogen and ultimately, on a catalytic surface, the ammonia breaks down into nitrogen and hydrogen gas according to the following reactions:

$$
\begin{aligned}
3N_2H_4 &\rightarrow 4NH_3 + N_2 + Q \\
&\rightarrow 3N_2 + 6H_2 - Q
\end{aligned}
$$

The liquid hydrazine is introduced into the thruster from the supply tank through a flow control valve and passes through a feed tube onto the catalyst "bed" and then through a constriction to the jet nozzle. The principle underlying the thrust delivered by a jet is of course Newton's third law of action and reaction, or the law of the conservation of momentum. For attitude control it must be possible to precisely calibrate the magnitude of impulses delivered by the thrusters over a wide range that extends to low values for fine adjustments; thus the spacecraft is fitted with numerous thrusters of different impulse strengths deployed in different parts of the spacecraft to apply attitude corrections about all axes.

The reaction momentum wheels are essentially relatively massive flywheels whose angular momenta are used to effect changes in orientation of the spacecraft, this time relying on the conservation of angular momentum law, as outlined in Chap. 11, where we were also introduced to the complexity of gyroscopic motion. By adjustment of the angular momentum of three wheels, each spinning about one of the three principal axes of the spacecraft, it is possible to correct for any rotation of the satellite.

Finally we mention the magnetic torque rods or "torquers"; these linear magnets or magnetic coils are subject to a torque due to the earth's magnetic field, a torque that can be communicated to the spacecraft frame. A bar magnet having a dipole magnetic moment of $\boldsymbol{M}$ in a magnetic field of intensity $\boldsymbol{B}$ is acted on by a torque $\boldsymbol{\Gamma}$ given by

$$\vec{\boldsymbol{\Gamma}} = \boldsymbol{MB} \sin\theta \, \vec{\boldsymbol{n}} \, , \tag{14.4}$$

where θ is the angle between the magnetic field and the magnetic moment, and $\vec{n}$ is a unit vector perpendicular to the plane containing the field and magnetic moment directions. Actually the direction of $\boldsymbol{B}$ with respect to the satellite varies as the satellite pursues its orbit in a plane inclined to the N–S axis; however the earth's magnetic field has been thoroughly mapped and the magnetic torques that can be generated for attitude corrections are predictable. Since the torque is always perpendicular to the magnetic field, it follows that corrective torques can be applied along only two axes perpendicular to the magnetic field direction.

14.6 GPS Satellite Signals [3]

The ranging signals broadcast by a GPS satellite are generated in the *navigation payload* where an atomic frequency standard provides the reference input for a phase-coherent synthesizer with output at the basic time code reference frequency of 10.23 MHz. This frequency whose mean instability is, as we have already stated, on the order of five parts in 10^{15} per day on later satellites, serves the essential role of determining the phase of the time code broadcast to the users on the microwave carriers whose frequencies are the 154th and 120th harmonic of the reference frequency. These two carrier frequencies, which fall in the so-called microwave L-band, are at $L_1 = 1.57142$ GHz and $L_2 = 1.22760$ GHz, with wavelengths about 19 and 24 cm, respectively. In addition to the ranging time codes these microwaves are also modulated with data including updated orbital data (ephemeris), satellite systems status, clock reading, etc.

The type of code used to modulate the transmitted microwave signal is called in the communications context a *pseudo-random noise* (PRN) code, based on the concept of *pseudo-random numbers*. Pseudo-random noise is not true noise in the statistical sense; it is a binary code which can be generated by a shift register in which at every clock pulse the bits advance to the right one place, the extreme right-hand bit becoming part of the output. The criterion for choosing the bit that replaces the far left now empty register is the key to the code: for example the criterion might be that the value of the bit should be the binary sum (modulo2) of the contents of the nth and mth registers. To illustrate the algorithm we show in Fig. 14.6 a simple example, in which the *chip* numbers 3 and 4 are taken as the defining pair.

Subsequent clock pulses in Fig. 14.6 will produce the digital sequence 1100110… Within a certain large set the binary numbers so generated appear quite random, but will actually repeat after a certain number of iterations have been reached. The degree of randomness of a PRN coded signal is readily verified by computing the autocorrelation function, recalling that a phase reversal of 180° is equivalent to a change of sign. Thus we assign the values +1 and −1 to the 1 and 0 bits and form the sum of the products of corresponding bits of the given signal with the same signal shifted any number of clock intervals. We find that the sum of the products is a maximum when there is *no shift* and rapidly falls even for a shift of one clock period. Moreover the correlation between the given coded signal and *any* other signal with a different PRN code is extremely small. Thus not only does the coding serve to make the signal secure by requiring the user to have the proper code, but also provides a sharp time-matching function between the (time-shifted) code received on the ground and the user's duplicate

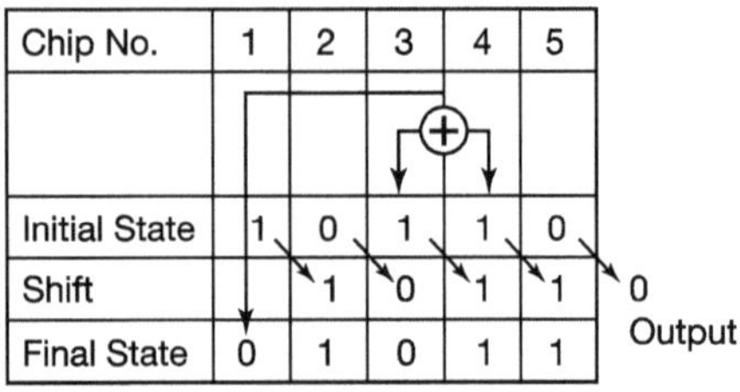

Fig. 14.6 Generation of PRN code by *tapped feedback* shift register

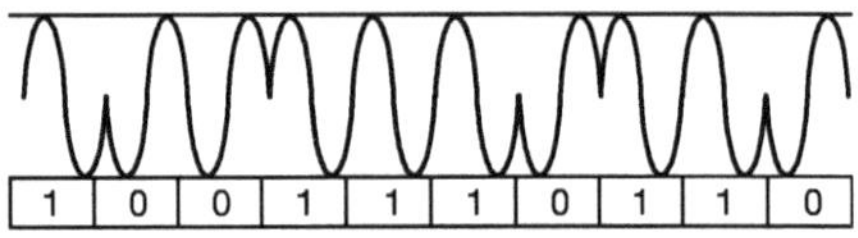

Fig. 14.7 Binary PRN phase modulation of the GPS signal

code. By assigning different PRN codes to the different satellites in the GPS constellation any ambiguity as to the identity of the source satellite is avoided. The binary PRN code is impressed on the basic RF broadcast frequency as a biphase modulation, that is, the phase for binary 0 is shifted by 180° with respect to a binary 1, as shown in Fig. 14.7.

Since the switching of phases to represent the binary bits occurs at a much higher frequency than the signal itself, it means that the system is operating in what is called a *spread spectrum* manner. This form of communication is now universally used, from secure military communications to "wireless" routers for computer systems, to digital telephone communications, etc. The first patent for a method of secrete radio communication that we would now describe as spread spectrum was awarded back during the Second World War in 1942 to an unlikely couple: an engineer and a woman who later became the Hollywood star, Hedy Lamarr! The invention was a radio transmitter in which the frequency was switched between different values in a coded sequence, making the message unintelligible unless the receiver tuning is switched in a corresponding manner. In terms of spectrum utilization the switching would of course broaden the occupied bandwidth, but *other* broadcasts using the *same* frequencies but a different code could use the same bandwidth without interference. The power spectrum of a spread spectrum broadcast looks like random noise and can occupy a bandwidth hundreds of times bigger than that of the information content. The method used in the cited patent is called frequency hopping, but there is a far more widely used method called *direct sequence,* which is used in the GPS system. In this case a high-speed code sequence is combined with the information being transmitted and both are used to directly modulate the carrier; it is the high-speed code sequence that sets the occupied frequency bandwidth.

As originally conceived the GPS was to serve the military and therefore there was great concern that it be secure, and the general public had access only to a low-

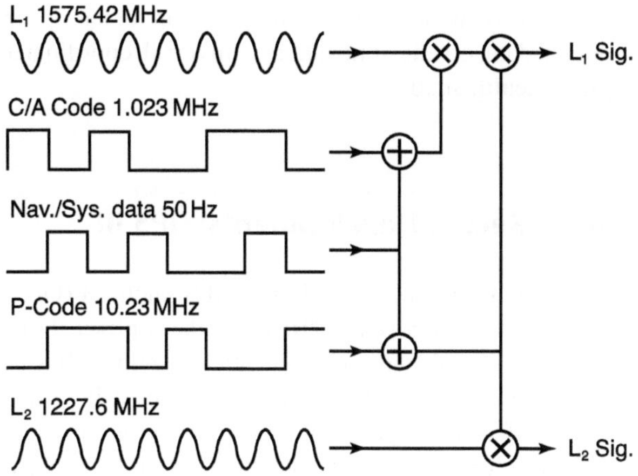

Fig. 14.8 Simplified diagram of the formation of the navigation signals L_1 and L_2

accuracy coarse acquisition (C/A) code. Now the general public has access to the precision (protected) P-code since the military has enhanced its security by incorporating an *anti-spoofing* capability; this required an additional W-code resulting in the Y-code. For those unfamiliar with dirty tricks, spoofing is sending out false navigational messages that mimic the GPS signals. In Fig. 14.8 is shown a simplified diagram to illustrate the complex jungle of codes.

The L_1 carrier signal is modulated by a code that is generated by adding modulo-2, that is, without carrying, the coarse acquisition (C/A) code and the satellite system/navigation data code. The clock rate for generating the C/A code is the tenth sub-harmonic of the standard, namely, 1.0234 MHz and coherent with it. The time interval between bits is about 1 μs and the whole code is repeated about every millisecond. In addition, the L_1 carrier is also modulated with the precise P-code, which is another PRN code generated at the standard frequency of 10.234 MHz. The time interval between consecutive bits is only 100 ns, corresponding to a difference in range between the satellite and user of 30 m. The generation of the P-code is somewhat complicated and we will follow the ways of forming it only so far as to demonstrate that it really is complicated. It is a combination of two PRN codes each generated by two shift registers, one containing over 15 million bits, while the other contains an *additional* 37 bits. The combined sequence has a repetition period of 266.4 days corresponding to about 225×10^{12} different bits; the total code length is divided into one-week segments, which are assigned to the different satellites as their identification PRN code. The satellite system/navigation data code consists of 1,500 bits subdivided into five subframes: the first subframe contains, among other things, the satellite identification number, coefficients to model the satellite clock correction, predictions of range accuracy, and the age of the data. The second and third subframes contain the satellite ephemerides. The contents of the fourth and fifth subframes, which are divided into pages, are broadcast by all satellites; some pages are reserved for military use while the rest contain data on the ionosphere, universal coordinated time (UTC), and satellite system health status.

14.7 Recent Developments: GPS III

The design of a new generation of GPS satellites (GPS III) by the Lockheed Martin Company was approved in 2010, and the contract for the construction of the vehicle and its navigation and communications payloads has been awarded by the U.S. Air Force. This step is in recognition of the increasing role of GPS as an integral part of modern life, quite apart from the military aspect. The design calls for significant modernization in the basic functions of positioning, navigation, and precise time dissemination with the introduction of a new generation of atomic and possibly ion frequency standards, and laser gyros. The first satellite is scheduled to be ready to launch in 2014. The design provides for multiple microwave transmitters including an additional signal using the L_1 carrier called L_1C to serve civilian users with high precision positioning. The projected signals will be three times more accurate than the Block IIR satellites currently provide and the design is expandable to take advantage of future technical developments, particularly in the area of atomic time standards. Figure 14.9 is a photograph of a GPS Block IIA satellite built by Boeing.

Fig. 14.9 GPS Satellite IIA (Boeing photograph)

References

1. A.C. Clarke, *Wireless World Magazine*, Oct 1945, p. 305
2. R.H. Goddard, *Smithsonian miscellaneous collections*, **95**(3), Mar 1936. Also *Scientific American*, Aug 1936
3. B. Hofmann-Wellenhof, H. Lichtenegger, J. Collins, *G.P.S. Theory and Practice*, 3rd edn. (Springer, Vienna, 1992)

Navigation by Satellite: Control Segment 15

15.1 Introduction

The functions of the control segment on the ground (CS) are crucial to the coordinated operation of the GPS: these functions include tracking the many satellites that make up the GPS constellation, monitoring their L-band navigation transmissions to determine the status of their onboard subsystems, particularly their atomic clocks, and analyzing their ephemerides. The operational control segment then transmits back commands and data to take corrective measures as needed with respect to orbital parameters and onboard atomic clocks, resolve any anomalous behavior, and generally require that all subsystems function according to specifications. These functions are carried out by a global network of ground-based installations controlled by a master control station and one alternate master station.

The master control station (MCS), or what might be more familiarly called the mission control center, is located at Shriever Air Force Base in Colorado; it is manned by the US Air Force which has the sole responsibility for the central control of operations of the entire system, 24 h a day. The ground control segment is supported by 17 globally distributed monitoring stations at geodetically precise locations chosen to give optimum coverage, although some predate GPS. They are made up of 6 stations run by the US Air Force and 11 are the responsibility of the National Geospatial (Intelligence) Agency (NGA). The geographical locations of these stations are shown in Fig. 15.1.

These monitoring stations are equipped with Cs atomic clocks of enhanced performance and geodetic quality GPS receivers. The field-hardened high beam intensity Cs atomic clocks were developed initially by the Hewlett Packard Company under NASA contract for the Apollo program; they are currently manufactured by the Agilent Company. The time reference of the monitor station at the US Naval Observatory (USNO) is coordinated with the USNO hydrogen maser ensemble and the US Department of Defense master clock. At all monitor stations around the world, atomic clocks provide standard 5 MHz signals as

F.G. Major, *Quo Vadis: Evolution of Modern Navigation: The Rise of Quantum Techniques*, DOI 10.1007/978-1-4614-8672-5_15,

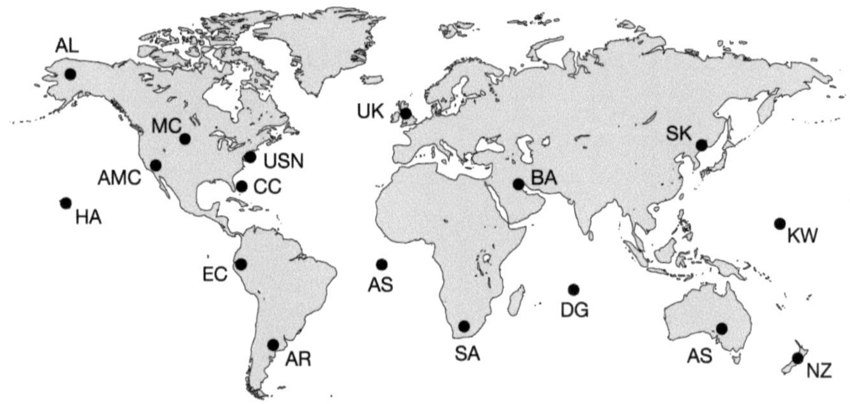

Fig. 15.1 GPS monitoring stations. *AL* Alaska, *AMC* auxiliary master control, *AR* Argentina, *AS* Ascension, *AU* Australia, *CC* Cape Canaveral, *BA* Bahrain, *DG* Diego Garcia, *EC* Ecuador, *HA* Hawaii, *KW* Kwajalein, *MC* master control, *NZ* New Zealand, *SA* South Africa, *SK* South Korea, *UK* United Kingdom, *USN* US Naval Observatory

references for their GPS receivers; the maintenance and upgrades of their equipment are carried out on an annual basis.

The master control station regularly receives satellite data from the 17 monitoring stations relating to the satellites' ephemerides and atomic clock readings, as well as data relating to the onboard systems, information that is analyzed to determine firstly the health and integrity of the satellites' onboard systems and secondly to determine their precise orbital positions and whether corrective action may be necessary to ensure that they are all on station. In the event of an emergency such as a satellite failure, prompt action can be taken to reposition the satellite in order to maintain the integrity of the system.

The master control station uploads this information to the satellites using the network of 12 command and control installations, 8 of which are part of the US Air Force Satellite Control Network (AFSWCN) and 4 are dedicated GPS ground antennas. The latter are located at the sites of the monitoring stations at Kwajalein Atoll, Ascension Island, Diego Garcia, and Cape Canaveral. These ground antennas support S-band microwave (2–4 GHz) communications and are equipped to be operated remotely from the master control station.

Perhaps the most science-fictional aspect of the GPS satellite navigation system, one that is not fully appreciated by this technologically jaded age, is to think that so many man-made satellites, speeding around the earth in their assigned orbits, can follow those orbits so precisely that they can be used to navigate on the ground! To help appreciate the workings of the system we must address several fundamental aspects: first the determination of the precise geodetic locations of the monitor stations; this is clearly necessary since the ultimate aim is to enable positioning with respect to earth-based geodetic coordinates. Second, the determination of the ephemerides of the satellites using the observations on the ground: the precise position of a satellite in its orbit must be known at every precise moment in time.

Lastly, since the system is based on the propagation delays of signals from the satellites, the satellite clocks must be kept in close synchronism with the control station clock. To convert the signal delay to a measure of distance, it is necessary to be able to model the dispersion of the microwave signal by the atmosphere and update that model as necessary. All this requires the ability to communicate selectively with each satellite and, through telemetry, to continually maintain in a timely manner the proper working of the satellite subsystems and maintain the ephemerides of all the satellites in the presence of inevitably random fluctuations.

15.2 Geodetic Position of Monitor Stations

We begin with the geodetic location of the monitor stations. Modern observation techniques have advanced to the point that subtle changes in the earth's topography or even in man-made structures, such as bridges and dams, are detectable by remote sensing. Aside from GPS itself and other Global Navigation Satellite Systems (GNSS), there are two principal techniques of precise geodetic measurement: Very Long Baseline Interferometry (VLBI) and Satellite Laser Ranging (SLR). These systems are able to locate precisely, often down to millimeters, the positions of the observers, forming precise reference points of a geodetic survey. Moreover these techniques provide independent but complementary additions to the overall geodetic picture. Thus VLBI is uniquely able to provide the orientation of the earth's axis with respect to the celestial reference frame, SLR most accurately tracks satellites and precisely defines the origin of the earth coordinate system, that is, its center of mass, and GNSS provides numerous readily accessible reference points.

Through these various techniques, geodetic reference systems of great accuracy have been established and international coordination has ensured global uniformity of definitions. A standard geodetic model of the earth used in cartography, geodesy, and navigation is referred to as WGS84, the World Geodetic System adopted in 1984. The definitions making up the system were last revisited in 2004. In the zeroth order of approximation, the earth's surface is basically an oblate spheroid pursuant to Newton's original theory of a rotating deformable solid. To model the actual gravity field on or near the earth's surface in order to compute the satellite ephemerides, a series expansion in spherical harmonics is needed reaching the order $n = m = 180$ to define what is quite logically called the earth Gravity (Field) Model (EGM); this is needed as reference to calculate altitudes with respect to the *geoid*, the name given to the mean global sea level used to measure elevations. However, for high GPS satellite orbit predictions, less detailed variation is required to be modeled and hence only 41 terms give sufficient accuracy. The coordinate axes are defined as follows: The origin is the center of mass of the earth; the z-axis is in the direction of the Conventional Terrestrial Pole (CTP) as defined by the Bureau International de l'Heure (BIH). The x-axis is the intersection of the (CTP-defined) equatorial plane and a plane through the z-axis; the definition of the zero meridian is determined by the International earth Rotation service (IERS) and set by the BIH;

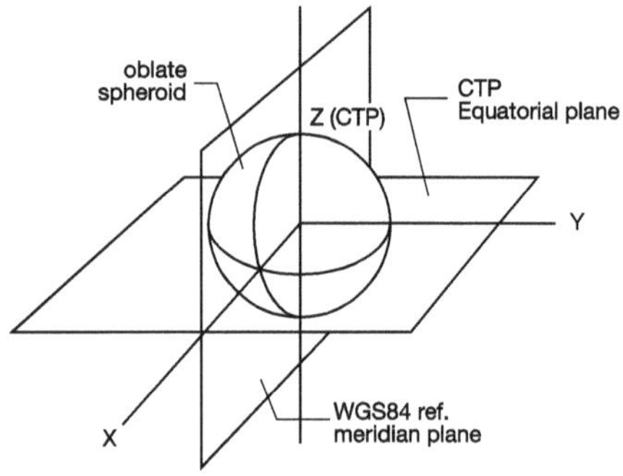

Fig. 15.2 The WGS 84 coordinate system

Table 15.1 WGS 84 primary parameters

Semimajor axis	$a = 6{,}378{,}137$ m
$(a-b)/a$ "flattening"	$f = 1/298.25722$
Angular velocity	$\Omega = 7.292115 \times 10^{-5}$ rad/s
Geocentric grav. const	$GM = 398600.5\ \text{km}^3/\text{s}^2$
Norm. 2nd deg. zonal coeff.	$C_{2,0} = -484.16685 \times 10^{-6}$

recently it was 5.31 arc s East of the Royal Observatory at Greenwich. The y-axis completes the right-handed Cartesian coordinate system. Figure 15.2 illustrates the WGS 84 reference system, also described as an earth Centered earth Fixed (ECEF) system.

The primary parameters of the rotating ellipsoid model are shown in Table 15.1.

The precision of positioning in the system is quoted as having a standard deviation of ±1 m in the horizontal direction and ±1–2 m in the vertical. These errors reflect the uncertainty in the location of the center of the earth, as well as instrumental errors. At the time WGS 84 was first proposed, the principal global surveying method was by satellite Doppler. Today the coordinates of GPS monitoring stations provide greater accuracy than local geodetic networks. The precise satellite laser ranging method, which we will describe later, made possible a worldwide terrestrial coordinate system called International Terrestrial Reference System (ITRS).

The principle underlying Very Long Base Interferometry (VLBI) is identical to the Michelson (optical) stellar interferometer (see Chap. 4): they both seek to increase the (angular) resolving power of a telescope by increasing its effective aperture, not by using a bigger and bigger diameter dish (for microwaves) or mirror (for light) but by sampling the incident wave front at just *two* well-separated points with two separate antennas or mirrors. We recall that an incoming light wave from a distant object falling on the two antennas (or mirrors) at even the slightest angle to the axis of symmetry bisecting the baseline, results in a phase difference between the signals arriving at the two antennas, a phase difference that is magnified by increasing the distance between the two antennas. If the radio source being observed is an unresolved point source, then as the direction of the interferometer

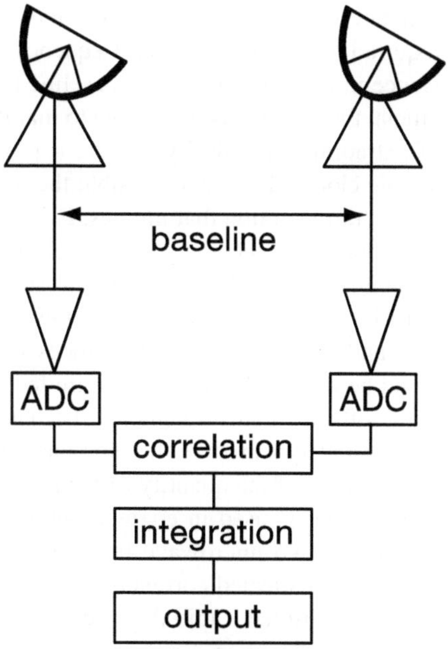

Fig. 15.3 A basic radiofrequency interferometer

rotates, for example with the earth, the signals from the antennas arrive at the receivers alternately in phase and opposite phase. The combined signal will exhibit the radio analog of an optical fringe system whose position varies with the direction of the source relative to the axis of the interferometer. It follows that if the source extends over a finite region of the sky then the signal from the interferometer is in effect a superposition of fringe systems with intensities corresponding to the intensity distribution of the source; but that is the Fourier spectrum of the intensity distribution of the radio source. In 1974 the Nobel committee recognized the contributions of two Britons, M. Ryle and A. Hewish for their development of the Fourier synthesis technique in the field of radio astronomy. Working at Cambridge they showed that using separate antennas moving with the earth's rotation, sufficient data are obtainable to mathematically construct high resolution images equivalent to having a very large aperture telescope; the process is called *aperture synthesis* [1, 2].

A conventional radio interferometer consists of two radio telescopes separated by a distance called the baseline, whose base-band output signals are separately recovered by mixing with a local oscillator (LO), amplified and sampled. The digitized signals are connected to a correlator, which produces what is called a *visibility function*. Such an interferometer gives far greater angular resolution of distant radio objects than a single antenna; the reason is, as already indicated, the increase of the effective aperture by the baseline separation of the antennas. A basic radio interferometer is illustrated in Fig. 15.3.

In the case of VLBI interferometers, the baseline may extend to thousands of *kilometers* and therefore an essential difference between a VLBI system and a conventional interconnected interferometer is that in VLBI the local oscillator required to downconvert the frequency cannot be shared between distant stations. Hence in order to extend the baseline to the level of thousands of kilometers, it is crucial that the receivers be able to maintain strict synchronism with time standards of extraordinary stability; in fact it is safe to say that it was the development of atomic clocks that made possible the use of baselines spanning continents and the degree of resolution that implies. Even with the use of atomic standards, long-term drift makes it necessary to model the variation and constantly check against ensembles of standards at national standards laboratories. The base-band signals must be time tagged and recorded using some memory medium such as magnetic tape and shipped to a central processing center, where at some later time, which could be weeks, the digital data are fed into a *correlator*, a sophisticated computer with software that implements the handling of the data. It must reconstruct the signals and synchronize the time bases of the different stations, then carry out the computation of the quantity called the *cross correlation* between them, a statistical concept that is used in engineering and optics: here it is really a measure of the product of two microwave fields as a function of the time delay between them. Highly sophisticated correlator software has been developed under the auspices of the (US) National Radio Astronomy Observatory (NRAO) to reduce data from the Very Large Baseline Array (VLBA). This is a global network of ten sophisticated radio telescopes, each having a 25 m dish antenna, controlled remotely from an operations center run by the US Army in the state of New Mexico. The network went into full operation in 1993. In addition to radio astronomical findings, the system has yielded highly precise data relating to geodesy and geodynamics, of particular relevance to GPS. The angular resolution depends on the wavelength of the radiation: for microwaves in the centimeter range a resolution of 0.001 s of arc are possible.

In 1999 a memorandum of understanding was signed between the USA and a European consortium to design and build an ultrahigh resolution millimeter radio observatory on a high plateau (elevation 500 m) in the Andes in northern Chile. It goes by the acronym ALMA, the Atacama Large Millimeter Array. It represents the largest international astronomical project involving initially 66 high precision antennas forming parts of a very large synthesized antenna with resolution exceeding all other existing telescopes, including the Hubble space telescope. Observations using the telescope began in 2011, but a formal inauguration ceremony was held in 2013.

With the degree of resolution made possible by ALMA, there is a wide range of studies that are practicable, including the study of distant extragalactic sources as well as more mundane problems such as subtle terrestrial surface movements associated with fault lines and earthquakes, volcano expansion prior to eruption, and continental drift, and of course the main reason the subject is brought up here: precise geodetic location of the GPS monitoring stations. Furthermore, precise radio astronomy provides essential data to the GPS relating to the orientation of the earth's axis in inertial space relative to the stars and any irregularities in its rate of rotation. For this purpose, observations on quasars, those extragalactic quasi-

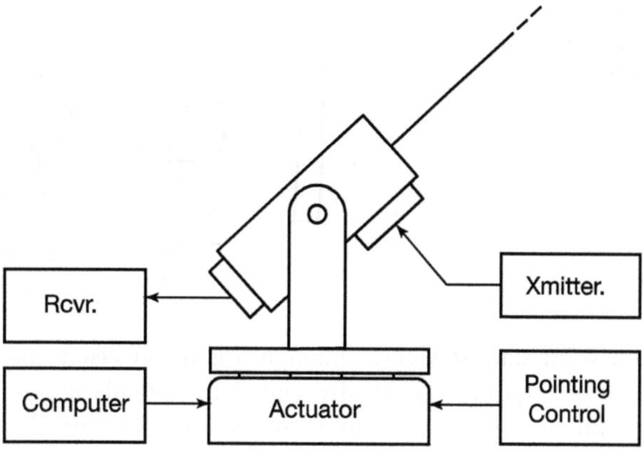

Fig. 15.4 The essential subsystems of an SLR station

stars that have high luminosity, even at radio frequencies, are ideal; this is important in order to tie the terrestrial system of coordinates used to locate points on the earth with the motion of the GPS satellites referred to the inertial celestial system.

In1964 NASA initiated a space geodesy program based on laser ranging of near earth satellites; numerous satellites were incorporated into a Satellite Laser Ranging (SLR) network [3]. The principle of laser ranging is the same as RADAR (see Chap. 3) except that optical (or infra-red) pulses are used with appropriate optics instead of microwaves [4]. We recall that the technique requires that a matrix of corner cube reflectors be placed on the satellite in order that the tracking laser beam will be reflected back precisely along its line of incidence, no matter in what direction it falls on the reflectors. The diffraction of the light waves causes broadening of the beam and consequent loss of intensity in the returned signal. The angle of divergence of the beam due to diffraction is on the order of λ/d, where λ is the wavelength and d the diameter of the output aperture of the transmitter. The phenomenon of diffraction is of course a fundamental property of wave optics and not a matter of poor design. Since the laser beam is typically generated in a relatively thin crystal rod, it follows that in order to minimize the broadening of the beam through diffraction, the transmitter optics must expand the beam diameter to fill a large exit aperture. If we assume a $Nd^{3+}YAG$ laser source with the wavelength $\lambda = 1.064\ \mu m$ and $d = 0.1$ m, so that $\lambda/d = 10^{-5}$ then the beam diameter at say a distance of $D = 2 \times 10^7$ m would be about 400 m and the intensity reduced approximately by a factor of 1.6×10^7. If we further assume ideally that the corner reflectors are perfect and intercept an area of 0.5 m in diameter, then the intensity received back on the ground would be reduced by a total factor of about 4×10^{11}. We note that for the assumed satellite distance, the time it takes for a light pulse to reach the satellite and return to the same SLR station is 133 ms, about 133×10^6 times a typical laser pulse length of 1 ns. Figure 15.4 shows a schematic diagram of a satellite laser ranging station.

The central component in a satellite laser ranging installation is of course the laser. As we learned in Chap. 11, laser technology has matured with many advances in the metrology of space and time. Currently the majority of laser devices are based

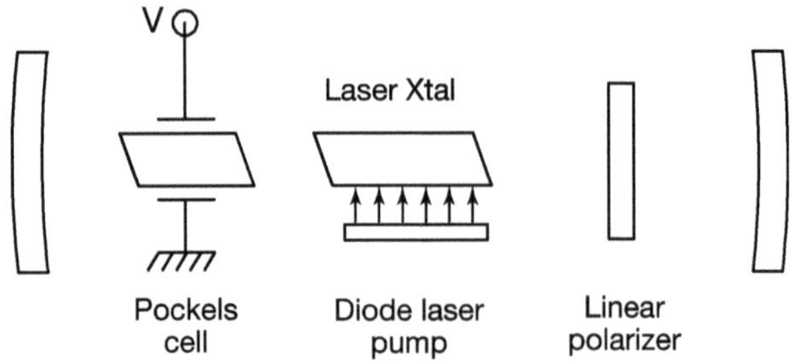

Fig. 15.5 Q-switching a Nd^{3+}YAG laser using an intracavity Pockels cell

on solid state devices, although argon ion lasers may still be used as pumping sources. Lasers are ideally suited to optical ranging of satellites; with special techniques pulse durations in the 100 ps (10^{-10} s) range are within reach corresponding to space resolution less than 3 cm! Furthermore with Q-switching (see Chap. 11) and mode-locking techniques, high power pulses with high repetition frequency are possible using a wide variety of designs. The least complicated design is the Q-switching, for example, of a Nd^{3+}YAG laser using a Pockels electrooptic cell. In the simplest design the Pockels cell is placed with a linear polarizer in the laser cavity with the laser crystal as shown in Fig. 15.5.

The Pockels cell is designed to produce a phase shift of $\pi/2$ between the orthogonal components of the linearly polarized laser field in the cavity during the optical pumping of the laser material. This prevents laser oscillation from building up since in traversing the Pockels cell back and forth a total phase difference of π is created between the orthogonal components of the field, that is, a reversal of direction of one of the field components, producing a resultant field polarized perpendicular to the original direction. This results in the polarizer blocking the light and preventing the buildup of laser action. After a period sufficient to allow the buildup of inversion in the populations of the lasing states in the YAG, the voltage on the Pockels cell is suddenly switched to zero allowing an intense laser pulse to be emitted. The crystal used in the Pockels cell must clearly have stable properties and be able to withstand intense laser fields. A commonly used crystal has been rubidium titanyl phosphate (RTP); however this is being superseded by another crystal: beta barium borate (BBO), which has a higher transmittance and lower insertion loss at visible and infrared wavelengths, and can be grown to larger sizes than RTP [4]. The defining specifications for the speed of any electro-optic crystal are the shunt capacitance and the voltage for quarter wave retardation. BBO crystals as large as 12 mm in diameter are reported to have negligible piezoelectric ringing effect in response to an applied 9.5 kV electric field pulse with a rise time of 11 ns.

More recent developments in the production of ultrasharp laser pulses in the picosecond (10^{-12} s) range utilize semiconductor saturable absorber mirrors (SESAM) to achieve mode locking (see Chap. 11), coupled with a regenerative amplifier controlled by a Pockels cell to switch the pulses on and off. The amplified pulses are fed to a second harmonic crystal to convert the primary infrared radiation into visible 532 nm radiation [5].

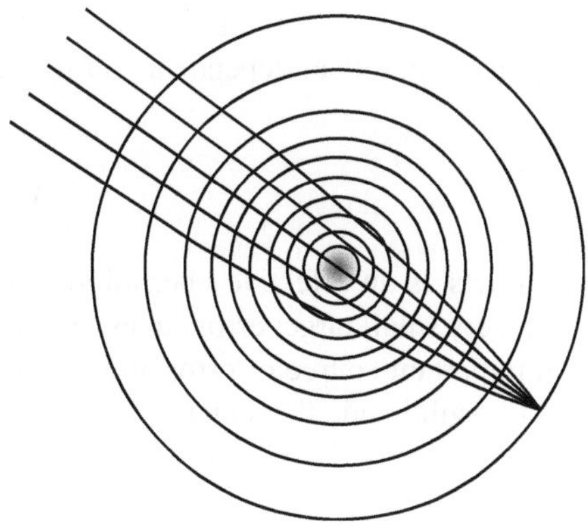

Fig. 15.6 The spherical Lüneburg retroreflector

In view of the precision that laser tracking is capable of, it may seem strange that, of all the satellites in the GPS constellations, only *two* were equipped with retroreflectors for laser tracking, and this in spite of the fact that laser tracking of satellites was being studied long before GPS became a reality [6]. The two GPS satellites belonging to Block IIA, space vehicle number (SVN) 35 and 36, were equipped to test whether system accuracy would be enhanced by the addition of laser tracking. The answer presumably was negative since no other satellite was so equipped. One has to suspect that the real reason was either cost or that there were more microwave engineers than laser engineers working in the space program. The corner reflectors used on the two quoted satellites were built by a Russian supplier and are similar to reflectors used on their GLONASS satellites.

The potential use of satellites for precise geodetic measurements was recognized early in the space age; thus the Laser Geodynamics Satellites, LAGEOS I (1976) and LAGEOS II (1992), were designed specifically for geodesy. They are spherical in shape with 426 corner reflectors covering their surface and were placed in circular orbits with perigee of 5,860 km and weighed 405 kg. Through a network of laser tracking stations, they provide the most precise geodetic data available on the earth's surface and are uniquely able to accurately define the center of mass of the earth. The system is so accurate that the data must be corrected for the penetration of the laser beam into the corner reflectors. The LSR stations must correct for return signal distortions by calibrating the system using the same intensity of the returned signals from a ground-based target at a measured distance.

A radically new design of a laser retroreflector was placed on the METEOR-3M weather satellite that was put into orbit in 2001. It is a layered glass sphere based on the optical Lüneburg lens. Rudolph Lüneburg was a German-born specialist in optical theory who described the remarkable property of a variable density, spherically symmetric lens that focuses incident parallel light rays on the opposite face of the sphere, provided the refractive index conforms to a particular function of radius. Since the lens is spherically symmetric, the same is true for parallel rays from any direction, and therefore we have an ideal retroreflector. Figure 15.6 shows a schematic drawing of the spherical retroreflector in which the Lüneburg lens is approximated by having spherical shells of different refractive indices. The required theoretical dependence of the refractive index n is given by [7]:

$$n = \sqrt{2 - \left(\frac{r}{R}\right)^2} \tag{15.1}$$

It acts as an effective aberration-free retroreflector over a wide incident angle with very small target error, an essential requirement for a precise ranging function. An important source of error however is the temperature which affects the optical path length inside the sphere.

15.3 Determination of Satellite Ephemerides

As we saw in the last chapter two microwave carrier signals L_1 and L_2 are emitted by the satellites, each carrying biphase PRN codes for satellite identification, clock readings, and navigational data. The fundamental modulation frequency observed on the ground is $f_0 = 10.23$ MHz, theoretically higher than what would be observed on board the satellite because of the relativistic effect of satellite motion and the reduced gravitational red shift at the orbit compared with the earth's surface. The satellite orbital motion according to Einstein's Special (i.e., Restricted) Theory Relativity gives rise to a dilation of time so that a clock (or oscillator) aboard the satellite runs *slower* than a stationary one by a factor given by

$$\beta = \sqrt{1 - \frac{V^2}{c^2}} \tag{15.2}$$

which numerically comes to about a fractional change of 8×10^{-11} for a GPS satellite. But there is also the gravitational red shift predicted by Einstein's General Theory of Relativity, a shift in which the clock runs *faster* in the satellite than on earth by an amount $\Delta\nu$, where

$$\Delta\nu = \frac{GM_E}{c^2 r_S}\nu, \tag{15.3}$$

where r_S is the mean radius of the satellite orbit; for a GPS satellite this is about 5.3×10^{-10}, that is about seven times larger than the velocity correction. The same relativistic corrections apply equally of course to the observed satellite and ground oscillator frequencies. Satellite clock readings are encoded on two carriers: the Coarse Acquisition (C/A) at $f_0/10$ and the Precision P- code at frequency f_0. It is the object of the receivers at the monitor stations of course to recover from the received microwave signals the encoded time and navigation messages and to process that information to determine the satellites' ephemerides.

A GPS receiver may be analyzed in terms of the following functional units: antenna, RF section, control unit, PRN code generator/microprocessor, memory unit, and power supply. Receiver technology has increasingly supplanted analog components and design with digital software; no longer are physical capacitors and inductors used to generate or filter sine-wave oscillations, rather such tasks are performed by digital gates and memories, as embodied in *software-defined receivers*. In a system that relies on the propagation time of signals between a satellite and the receiver on the ground, it is obvious that the radiation patterns at the antennas both in the satellite and the receiver play a significant role. Therefore the antenna design is critically important to ensure that the effective phase center of the antenna coincides with its geometric center, free of any fluctuation. GPS receivers commonly use microstrip antennas which are solid state with a simple planar shape; a measure of the quality of such an antenna is the degree to which its phase center is immune from changes in its orientation.

The microwave signal arriving at the receiver must first be detected to obtain the modulation signal and amplified prior to being decoded. As we saw in the last chapter, because of the spread spectrum of the encoded signal it looks like noise; it is only after mixing the signal with a PRN code appropriate to one of the satellites that the signal from *that* satellite is recovered. Therefore in a multichannel receiver, the signals coming from several satellites are separated into channels by using a different PRN code for each channel to retrieve the signal; these codes must of course be generated by the receiver. The detailed way in which they are constructed is necessarily complicated to accommodate the many satellites. We note for example that the C/A code is synthesized from *two* PRN codes added *modulo*-2. The signal phase can then be compared with that of the receiver oscillator, which is, within errors and drifts, in synchronism with the satellite clock; this yields the propagation delay and hence what is called the "pseudo-range" of the satellite at the time of the transmission. Since time plays such a central role we must be careful to distinguish its various measures; thus we have the readings of the satellite clock and those of the receiver clock, each having errors and drifts with respect to a common GPS system standard time. If t_S and t_R denote the clock readings of the satellite and receiver, respectively, then the time delay Δt derived from the decoding of the PRN modulation determines the pseudo-range of the satellite, $R = c\Delta t$. If e_S and e_R are the errors (biases) in the two clock readings with respect to the GPS system standard time, then we can write $R = c(t_R - t_S) + c(e_R - e_S)$. We note that the C/A code repeats every millisecond, during which time the signal travels 300 km, and therefore since the range of the satellites is on the order of 20,000 km, the integral number of milliseconds is indeterminate. However an approximate knowledge of the geodetic location of the receiver is sufficient to resolve this ambiguity. In summary, the pseudo-range is determined by the time displacement of the PRN code generated in the receiver that is required to retrieve the PRN-coded ranging signal from the satellite. Since the velocity of a satellite in its orbit is on the order of 1 kms^{-1} and the signal propagation time is about 0.09 s, it follows that, in the time it takes the signal to reach the receiver, the satellite has moved about 90 m.

The resolution of the range measurement using the PRN coding is on the order of 30 cm for the 10.23 MHz P-code, representing about 1/10 of a *chip* length in the code.

Far higher range resolution is theoretically possible using the phase of the microwave carriers L_1 and L_2 themselves. Since the wavelength is a small fraction of the range, there will be an unknown integral number of microwave cycles during the transit time of the signal; however it is still possible to incrementally track the satellite. Since the wavelength of L_1 is 19 cm, a measurable phase difference of (say) 1 % of a cycle would correspond to 1.9 mm!

The primary function of the control segment is to collate all the tracking data from the many monitoring stations and, among other things, compute updates to the GPS satellite orbits. This amounts to computing for each satellite the best Keplerian fit to the observed positions and velocities as a function of time, that is, the best ephemerides based on Kepler's theory, and then continue to update the orbital predictions as actual navigational data is received from the satellites. A classical approach to do this optimally, that is, with minimum error, would be to assume a mathematical function, for example, an elliptical orbit with the six parameters that define it treated as variable, then find the values of these parameters that best fit the observed data in the sense of minimizing the sum of the squares of the differences between the data and the assumed function. But this approach has long been superseded by what is called a *Kalman filter*, which is a recursive digital algorithm that computes the optimal updates of the values of the system variables in real time, based on their error variances and that of the new measurements. This digital computer-based method is ideally suited to the problem of satellite orbit determination and its continued update. It is an evolutionary process that does not require the storage of data beyond the current update. The predictions are statistically optimal in terms of minimizing the mean square error variance. The algorithm was published by R.E. Kalman in 1960 just as digital computer technology was being increasingly used in engineering and science. It has since been placed on a sound mathematical footing and is widely used in industry, navigation, satellites, and geodesy.

Since the Kalman filter deals with digital data it is appropriately formulated in the language of matrices and linear vector spaces. Readers who are not mathematically inclined may skip this section without losing a physical understanding of satellite navigation. To apply the Kalman filter to a system with many physical variables each with a measured value with possible error, the state of that system is represented by a column matrix, and operations relating one state to another are represented by matrices. Since the filter acts in an iterative manner with the state of the system progressing discretely from one state to the next, the state vector is labeled with an index, k. At each iteration the values of some of the vector components may be updated with the introduction of the results of new measurements. The components of the state vector are the physical variables that define the condition of the system. The equations of the Kalman filter are an expression of the state of the system after the kth iteration, x_k, in terms of the state just prior to that step x_k^- and a weighted difference between an actual measurement result z_k and a measurement prediction, Hx_k^-, thus

$$x_k = x_k^- + K\left(z_k - Hx_k^-\right). \tag{15.4}$$

The matrix K is chosen to minimize the covariance of the errors in x_k; it would take us too far into difficult terrain to do more than quote the result; the interested reader will find many published texts on the subject [8]. One solution is as follows:

$$K_k = \frac{P_k^- H^T}{HP_k^- H^T + R}, \tag{15.5}$$

where P_k^- is the covariance of the errors in the *estimate*, while R is the covariance of the errors in *measurement*. In broad terms this result predicts that if the errors in measurement are very small making R small, then K is large, and greater weight is given to the measurement over the estimate, and not surprisingly if the measurement errors are large, then less weight is given to them than the predicted values. In computing the predicted values in the case of a satellite orbit, a detailed theory is available to extrapolate the system variables, however in general, where the system may be nonlinear, a linearization approximation is necessary.

15.4 GPS Time Coordination

As we have mentioned at numerous points, the whole concept of the GPS is based on the ability of all the satellites to maintain close clock synchronism. This has been made possible to within a few microseconds per year thanks to the development of atomic clocks, which take the form of portable Rb or Cs clocks on board the satellites and larger hydrogen masers and Cs standards at the master control station. Even atomic clocks exhibit a small long-term variation in frequency which must be mathematically modeled to predict clock errors. These frequency standards define what is not surprisingly called atomic time (AT), whose unit, the second, now defines the international unit of time. However the civil time used by navigators and other users is closely tied to what is called universal time (UT) which is defined in terms of the *mean* solar day (the actual length of the day varies throughout the year as the speed of the earth varies along its elliptical orbit), that is, the average taken over the earth's orbit. However the time scale used by GPS is called Universal Coordinated Time (UTC); the unit in this system is the international unit of time, the atomic second. However because of the long-term drift between UT and AT, they are kept in step to within 1 s by inserting one *intercalary* or leap second, as required. This results in a piecewise uniform time scale that tracks universal time; fractional differences between the time scales are published periodically by national observatories. It was stipulated by the International Telecommunications Union (ITU) that if a positive leap second is to be added, that it should begin at 23 h 59 m 60 s and end at 0 h 0 m 0 s the following day. Dates of the application of a leap second are announced by IERS, the International Earth Rotation Service. It can be argued on physical grounds that the rotation rate of the earth is subject to slow decrease due to tidal friction and fluctuations due to atmospheric conditions, and therefore the need to add leap seconds is far more likely than to subtract them.

At the level of precision in time coordination that characterizes the GPS system, there are physical effects that are not normally of any practical consequence; for example, the gravitational red shift in frequency which has already been cited. Another phenomenon is the *Sagnac Effect* which we encountered in Chap. 12. We recall that there it was applied to design a laser gyroscope; but it is equally relevant to the propagation of radio time signals around the rotating earth. A coordinate system fixed in the earth is non-inertial since its rotation with respect to the "fixed stars" constitutes an acceleration, not of changing speed, but direction. Consequently the problem of the propagation of radiation with respect to the earth must be treated using Einstein's *General Theory of Relativity*. According to that theory, if two identical clocks that remain precisely in synchronism when fixed at the same point on the earth's equator, are separated by transporting one clock all the way around the earth along the equator, will display a *different* time when they are again at the *same* point. This is another example of the Sagnac Effect even though the original experiments of Sagnac used light beams and mirrors. If Ω represents the angular velocity of the earth and $S = \pi R^2$ is the cross-sectional area of the earth at the equator, where R is the earth's radius, the difference in the time reading between the two clocks according to Einstein's theory is given by

$$\Delta t = \pm \frac{2\Omega}{c^2} S. \tag{15.6}$$

Note that unlike the time dilation due to motion, the state of motion of the clock itself does not enter into this result; it is the scale of "time itself" that changes. If we substitute the numerical values $\Omega = 7.3 \times 10^{-5}$ rad/s and $S = 1.3 \times 10^{14}$ m^2, we find $\Delta t = \pm 1/5$ μs. This is a small time difference but is quite significant in the present context.

15.5 Signal Propagation Velocity

The underlying presumption in the idea of a time-based positioning system is of course that by measuring the propagation delay we can know the distance from the source to the receiver. This holds true only if electromagnetic waves can convey information at a constant velocity under all conditions. This would be strictly true only if the propagation were through vacuum; but of course the earth is bathed (fortunately) with an atmosphere and beyond that an ionosphere that extends to altitudes of several hundred kilometers. Thus to reach the user on or near the earth's surface, the microwave signals must penetrate the ionospheric layers as well as the troposphere. These media are dispersive, that is, signals having different frequencies travel at slightly different velocities, with the consequence that the modulations of the carrier frequency travel at different velocities. Although the velocities are little different from free space yet because of the long distances involved, the propagation times can be significantly different. It is mainly on account of the need to determine the amount of dispersion present that the two different frequencies L_1 and L_2 are used. On the basis of an approximate model in which the dispersion is assumed to be caused by the forced oscillation of free electrons in the ionosphere, the phase velocity V_p of an electromagnetic wave can be shown to be as follows:

$$V_p = \frac{c}{\sqrt{1 - \frac{80.6N_0}{\nu^2}}}, \tag{15.7}$$

where N_0 is the electron concentration in electrons per cubic meter and ν is the microwave frequency. The electron concentration is stratified *increasing* with altitude from the E-layer at 200 km with about 10^{11} electrons per m^3 to around 10^{12} electrons per m^3 in the F_2 layer at an altitude of 300 km. At the GPS carrier frequency of $L_2 = 1.22760$ GHz, for example, the phase velocity *increases* in the F_2 layer by about 2.7 parts in 10^5. We should recall that in describing wave propagation in a dispersive medium, care must be taken to specify exactly what the velocity refers to, since in general the *shape* of the wave changes as it propagates, so there is no identifiable point along the wave whose progress can be followed. However if the dispersion is weak and the frequency spectrum of the signal is contained in a narrow band centered on one frequency, as it is with the GPS signals, then the form of the modulation imprinted on the carrier is preserved and propagates with the *group velocity* (V_g). The group velocity is related to the phase velocity according to the following:

$$V_g = V_p + \nu \frac{dV_p}{d\nu}. \tag{15.8}$$

In the case of the ionosphere $dV_p/d\nu < 0$ and therefore the group velocity is smaller than the phase velocity, while the phase velocity is larger than the velocity in free space. These results indicate that the pseudo-ranges deduced from time code matching on the ground are different from those derived from the relative phase between the carrier wave and the local reference oscillator.

The concentration of free electrons in the ionosphere and its vertical distribution can vary significantly, particularly when there is sun-spot activity, and moreover the path of the microwave signals through the ionosphere varies as the satellite moves along its orbit. It is essential therefore that accurate corrections be applied to the timing signals and that they be continuously updated; that is the prime reason for using two different microwave frequencies L_1 and L_2.

The neutral troposphere at lower altitudes can be taken to be totally nondispersive, that is, there is negligible dependence of the phase velocity on frequency, and the modulation travels at the same velocity as the phase. However, that common velocity can vary according to the density and composition of the atmosphere and therefore variations in humidity, temperature, and density of the atmosphere can have significant effect on the velocity of the signal.

While on the subject of group and phase velocity we may note in passing that there are media in which the computed group velocity at which a signal travels can be greater than the velocity of light in free space; a phenomenon called anomalous dispersion. In the early days of Einstein's Theory of Relativity this was a matter of some concern until Sommerfeld and Brouillon showed that in fact the beginning of a radio signal always arrives at exactly the velocity of light and that the *signal* velocity is not the same thing as the group velocity in an anomalous medium.

References

1. J.G. Robertson, W.T. Tango (eds.), *Very High Resolution Imaging* (Springer, Berlin, 1993)
2. A.R. Thompson et al., *Interferometry and Synthesis in Radio Astronomy*, 2nd edn. (Wiley, New York, NY, 2001)
3. E.C. Pavlis et al., in *Proceedings of the 16th International Workshop on Laser Ranging*, Poznan, 2008
4. J. Kölbl et al., in *Proceedings of the 16th International Workshop on Laser Ranging*, vol 2, Poznan (2008), p. 429
5. U. Keller, Nature **424**, 831 (2003)
6. Y. Bar-Sever et al., in *ILRS Workshop on SLR Tracking, etc.*, Metsovo, Greece, 2009
7. J.P. Oakley, Appl. Opt. **46**, 1026 (2007)
8. C.K. Chui, G. Chen (eds.), *Kalman Filtering with Real Time Applications*, 4th edn. (Springer, Berlin, 2009)

Satellite Navigation: The User Segment 16

16.1 Introduction

Originally conceived as a global all-weather positioning system justified by its military applications, GPS was later opened up to the general public with the result that it has become an indispensable appurtenance of modern life. In addition to its familiar use to navigate on the highway, it has impacted fields as diverse as navigation, by land, sea, and air, surveying and geodesy, and large-scale construction projects such as mines, bridges, and tunnels. To utilize, the system requires of course a special multichannel radio receiver and microprocessor whose sophistication and cost are commensurate with the required capabilities and accuracy. There are professional receivers currently available on the market that go far beyond the units incorporated in handheld devices for the use of the general public. Among the more sophisticated units is the Proflex 800 [1] which uses advanced GNSS technology to enable it to obtain a precise fix using *any* available satellite signals, even under such adverse conditions that would preclude the use of other receivers. Many other prominent companies market "professional" instruments for satellite navigation of aircraft and ships; among them are names such as Garmin, Magellan, TomTom, etc., which also supply models for the general market, including motorists, hikers, and sailing enthusiasts, etc. GPS receivers are generally integrated into some form of computer or handheld wireless device.

A GPS receiver operates in the L-band microwave region of the spectrum, with a dedicated microprocessor and memory to enable it to display its position with reference to an appropriate geodetic grid. Although the principle of operation of the system is one-way ranging, in which the propagation times of the signals to arrive from the satellites to the receiver are used to measure the distance between them, the genius of the system is that, like Loan C, it does *not* require the receiver to carry an atomic clock! It achieves this not by the hyperbolic system used by Loran, in which the receiver measures the *intervals* between the arrival times of signals from *different* stations, but rather by the measurement of what are called *pseudo-ranges* to the satellites based on the actual propagation times measured using the

F.G. Major, *Quo Vadis: Evolution of Modern Navigation: The Rise of Quantum Techniques*, DOI 10.1007/978-1-4614-8672-5_16,

receiver quartz clock. The pseudo-ranges are so-named because it is expected that the receiver clock does not remain synchronized with the satellite clocks and therefore a clock error develops that must be eliminated. The satellite clocks on the other hand are chosen to be atomic standards precisely so that they will remain in strict synchronism with each other for long periods of time. Even they however will drift over time and it is the function of the control segment on the ground to apply corrections as necessary. The true ranges must be found by correcting for the receiver clock error. This is where the multiplicity of satellites is critical—not only does it make possible global coverage but also introduces a redundancy of erroneous pseudo-range vectors which in general will *not* converge on the position of the receiver, however, if the satellite clocks are in perfect synchronism, it should require *only one* correction, namely the correction of the receiver clock, to make them converge on one point. To fix the position of a receiver in *three* dimensions requires three numbers: longitude, latitude, and altitude and thus three *true* ranges are necessary. This may be visualized by constructing spherical shells about the three satellites as centers with the *true* ranges as radii: the first two intersect in a circle and the third must cut the circle of intersection in at most two points. This ambiguity is resolved in practice since the ranges from *four* satellites are used. In that case if the radii are *pseudo-ranges* because of an error in the receiver clock, then the spherical shell drawn about a *fourth* satellite as center will *not* intersect at the same point as the first three spheres. However, and this is the critical point, if it is known that the satellite clocks are perfectly synchronized and the errors in the pseudo-ranges are due *only* to the receiver clock error, then a single correction to the reading of *that* clock alone should make all the spheres pass through the same point, namely the actual position of the receiver. Of course in practice there are inevitable errors in all measurements, and the best one can hope for is that by including the pseudo-ranges to many satellites, a correction to the receiver clock can be found to minimize the spread in its computed positions. It is one of the principal functions of the receiver microprocessor to compute that clock correction. It follows that just as the accuracy of a measurement can be improved by repeating it, so the accuracy of a fix in position in three dimensions is improved by the acquisition of signals from many satellites, but the system is designed so that at least four satellites are in full view at all times from any point on the earth.

16.2 GPS Receivers

In the last chapter we outlined the essential functional elements of a GPS receiver, starting with an active microstrip antenna, a temperature compensated crystal oscillator (TCXO), frequency converters (mixers), a multichannel PRN code generator/correlator, and finally a microprocessor. This is obviously not your old-time radio receiver with a tuning dial and whip antenna. The integrated circuit revolution has been advancing apace in the design and manufacture of an expanding range of microdigital components including integrated mixers, low noise amplifiers, surface acoustic wave (SAW) filters, analog-to-digital converters (ADC) as well as

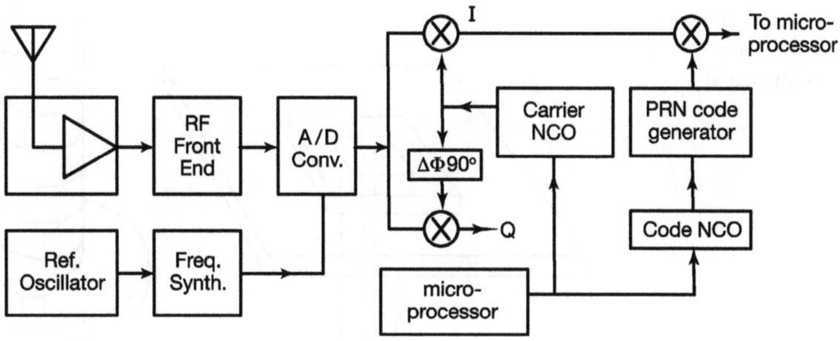

Fig. 16.1 General architecture of GPS receiver. The front end is analog, the rest is software defined. *NCO* numerically controlled oscillator, *A/D* analog to digital converter

programmable filters. In fact the ubiquitous GPS receiver that has become so commonplace is often in the form of a *chip set*, a solid state integrated circuit made to be compatible with other similar integrated circuits. Incredibly all the functions of a GPS receiver (apart from the antenna) are now available on the market as a module measuring $15 \times 15 \times 2\ \text{mm}^3$!—such is the degree of compact integration in present day electronics. This Micro Modular Technologies GPS module [2] is a complete 20 channel receiver with "cold start" acquisition sensitivity of −145 dBm; the tracking sensitivity is considerably higher at −159 dBm. We recall that the dBm scale is a logarithmic one defined as follows:

$$S(\text{dBm}) = 10\ \log_{10}\left(\frac{I}{I_0}\right), \tag{16.1}$$

where I_0 is 1 mW. Thus a sensitivity $S = -145$ dBm means that an input signal of about 3.2×10^{-15} mW is sufficient for the receiver to get a valid fix. The power of GPS signals received on the ground varies in the range −125 to −150 dBm, well within the capability of our micro-modular receiver. To gain some idea of the physical significance of these figures, we note that the basic *thermal noise* power in a frequency band of 1 Hz at room temperature is approximately −174 dBm. These theoretical numbers are not particularly useful in practice however, nor is it rewarding to attempt a rigorous analysis of the instrumental errors—the manufacturers themselves have devised definitions of capability that presumably bring out their products' strong points. An alternative equivalent definition of receiver sensitivity is "carrier-to-N-zero" (C/N_0) which, as the name suggests, is the ratio of the carrier power (dB) to the noise power in 1 Hz band width (dB). If the ratio is taken relative to the actual bandwidth instead of *I* Hz, we get the usual signal-to-noise ratio S/N.

The general architecture of a generic GPS receiver is shown schematically in Fig. 16.1. The object here is to break down the sophisticated engineering design of the receiver into functional units without attempting to describe specific circuits, which are mostly proprietary.

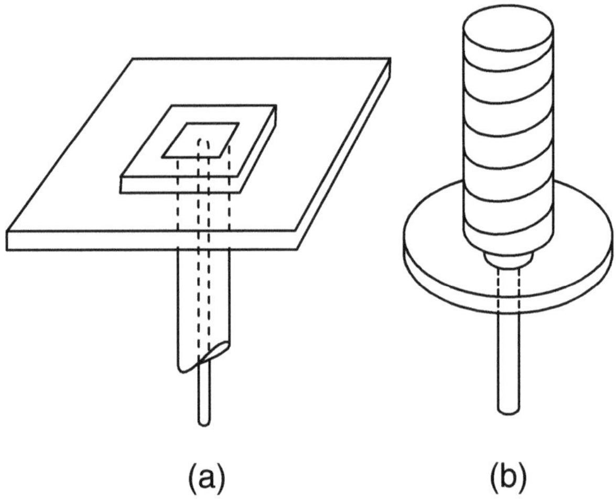

Fig. 16.2 (**a**) Microstrip microwave antenna. (**b**) Helix microwave antenna

As already emphasized, the *front end* of the receiver plays a crucial role in determining the phase stability and signal-to-noise ratio of the received signal; the rest is essentially digital data reduction. The critical components of the front end are the antenna and low noise amplifier (LNA). The types of antennas most compatible with a GPS chip set are the microstrip (or patch) antenna and the helix antenna. The microstrip consists essentially of a rectangular or circular metallic patch on the top surface of a flat dielectric slab resting on a wider conductive surface as shown in Fig. 16.2a. The structure acts as a relatively narrow bandwidth resonant cavity with the buildup of a strong microwave induced currents in the bottom plate. It is readily fabricated in an array and integrated with other solid state devices. The helix antenna consists of a wire coil wound on a low-loss dielectric cylinder, mounted on a metal base with the coaxial feed through the plate at the center of the coil, as shown in Fig. 16.2b. The helix antenna is found to be more immune to the perturbing effects of its surroundings and therefore more widely used in handheld devices. The most essential property of the antenna is that the *phase* of its output signal should track precisely the phase of the incoming microwaves; this is particularly crucial when the receiver is in motion or changing orientation, as in an aircraft. Moreover to receive signals from several satellites, whether simultaneously or sequentially requires an antenna whose response is not strongly dependent on the direction of the incoming wave, that is, it must be *omnidirectional*. On the other hand it should discriminate against low elevation signals which may be contaminated by spurious reflected waves. The user should recognize the importance of avoiding locations where there may be spurious reflections from buildings or other structures; the best antenna design will be of no avail in such a location.

A stable reference frequency $f_0 = 10.23$ MHz is provided by a temperature-compensated crystal oscillator (TCXO), which is a quartz oscillator in which the crystal is mounted in a constant temperature oven with proportional control (see Chap. 7). This provides the reference frequency to generate the L_1 and L_2 carrier frequencies at 154 f_0 and 120 f_0, and the C/A and P-codes at $f_0/10$ and f_0, respectively. The navigation message is transmitted/received at 50 Hz.

The one feature that sets a GPS receiver apart from other radio receivers is the presence of a PRN code generator and correlator. We met pseudorandom noise codes in Chap. 14; here we will consider in a simplified manner the function of a correlator. Simply stated, the correlator executes a mathematical operation drawn from statistical theory called *autocorrelation* which computes the average over time of an electrical signal (function of time) multiplied by the same signal displaced in time or more precisely:

$$C(\tau) = \int_{-\infty}^{+\infty} s(t)s(t-\tau)dt. \tag{16.2}$$

Now the *phase* of the received carrier wave is modulated between the two values 0 and π according to a PRN code identifying a particular satellite; this is accomplished using what is called binary phase shift keying (BPSK). The receiver also has a similar unit that can generate replicas of these PRN codes at variable epochs which are mixed with the signal to *demodulate* it and retrieve the carrier. If the received C/A modulated L_1 signal and a replica code generated in the receiver are connected to the correlator, then its output will rise to a coherent maximum *only* when the two signals match exactly in time: if they are different codes the output signal is essentially zero. However if the codes are the same, the output is a sharp maximum if they are matched in time but otherwise falls toward zero when there is even a two bit (or *chip*) displacement in time. If two identical PRN coded inputs to the correlator have a relative offset in time, a remarkable thing happens to the *spectrum* of the output: when the two codes are far from matching in time, the power spectrum is weak and spread out, reflecting the rapid phase modulation; however when there is a perfect match, the spectrum collapses to the carrier signal. Thus having singled out one satellite by its PRN using the correlator, the output is filtered and its phase compared with the receiver reference phase. The displacement in time/phase is the propagation delay of the signal, that is, the true delay plus or minus the error in the receiver clock. In the acquisition and delay lock loop (DLL), the receiver reference phase must be displaced until the output of a phase comparator passes through a maximum indicating a match; then the output can be used in a feedback servo loop to obtain a phase lock. The required circuitry is the same as the phase-lock loop (PLL) common in frequency synthesizers.

16.3 Differential GPS

The early success of GPS in the 1980s stimulated the search for ways in which the accuracy of the system could be enhanced and its applications broadened. This was given further impetus in 2000 when the S/A code that limited the public's access to the full accuracy of GPS was abandoned, opening up new possible applications. Clearly in order to improve accuracy the principal sources of error had to be mitigated: the most obvious of these is the dispersion in the troposphere and ionosphere and possible spurious multipath reflections on the surface. Since the

GPS satellites are on the order of 20,000 km above the earth, the signals received at points on the earth's surface (say) 100 km apart follow paths that are nearly parallel, and presumably through very nearly the same atmospheric conditions. Therefore if we have *two* receivers less than 100 km apart using the same four GPS satellites, and if the actual position of one, which we will call the reference, is by independent means very accurately known, then on the basis of its observed signal it can correctly determine the atmospheric correction and communicate it in real time to the second receiver. This should yield a far more accurate location of the second receiver than the usual single point positioning. Another mode of operation may in some applications be more appropriate, namely the second "roving" receiver transmits to the reference monitoring receiver the raw navigational data which the latter collects and processes; this obviates the need for the roving receiver to be equipped to process the data. In order to implement corrections in *real time* there are two procedures that may be followed (1) assuming the same four satellites are used, the reference receiver calculates its position on the basis of the satellite signals, and the difference between its known and calculated positions is conveyed to the roving receiver as corrections to improve *its* position, (2) the reference receiver compares the observed pseudo-ranges to the satellites with the computed values based on its known position and transmits the errors in pseudo-ranges to the roving receiver.

Aside from the use of GPS in the form of handheld devices by the general public, the most important applications of GPS are in the form of DGPS. These applications include close-in navigation around airports and harbors and inland waterways, critical positioning as in mine shafts, oil wells, and tunnels, and of course land surveys. In recognition of the importance of these applications to the national economy and security, extensive networks of DGPS stations have been established both within the USA and internationally. For example, in the USA we have the Nationwide DGPS (NDGPS) chartered by the US Department of Transportation and implemented by the US Coast Guard; it is a differential GPS positioning system scheduled to have 126 broadcast sites, providing double frequency surface coverage across the continental USA. The network incorporates sites of an existing US Air Force ground wave emergency network (GWEN) and a Coast Guard radio beacon system. Currently there are about 80 DGPS reference stations transmitting L_1 C/A code corrections to areas such as harbors and their approaches. The network is monitored and controlled by two Nationwide Control Stations. In a collaboration between the Coast Guard and the US Federal Highway Administration a development program is underway to establish a high accuracy nationwide DGPS (HA-NDGPS) with the objective of making available, as part of an advanced traffic safety system, a vehicular positioning capability with a dynamic distance resolution ultimately in the decimeter range, and a static resolution down to the centimeter range. The development program is currently in progress; preliminary tests have been conducted on radio links to multiple locations and achieved a horizontal accuracy better than 10 cm [3].

Another DGPS network run by the Jet Propulsion Laboratory for NASA consists of over 60 sites equipped with geodetic quality, dual-frequency receivers.

It provides real-time GPS tracking in its role as the GPS component of the International GPS Service consisting of about 100 stations, the largest worldwide *real-time* satellite tracking network. It is designed to provide real-time accurate positioning both vertically and horizontally with accuracy in the decimeter range using dual frequency GPS receivers. This has in the past been available over limited areas using a so-called Real-Time Kinematic technique (RTK) in which not only is the code signal used in pseudo-range determination but also the phase of the carrier. Since the wavelength of the carrier is 100 times smaller than that of the code signal, RTK resolution is 100 times greater than standard DGPS. However the shorter wavelength increases the possibility of error due to cycle ambiguity in the phase, that is, the inability to determine the whole number of wavelengths between satellite and receiver. It has been compared to having a ruler with a very fine scale, but no numbers to determine length. In an early time-transfer low frequency radio technique to synchronize the NASA satellite tracking station network (STADAN), the ambiguity was resolved by using *two* neighboring frequencies. The RTK relies on special computational techniques applied to GPS signals to resolve the carrier-cycle ambiguity. The network is a highly redundant system, by which is meant that a given GPS satellite may be observed by as many as 25 stations at any given time in order to guarantee reliability in support of some of the critical applications that depend on it; the multiple observations further check on the integrity of the GPS system as a whole. All the observation sites stream GPS data to the network's operation centers using several different modes of communication, including the Internet. Strict time synchronism is maintained throughout the network with the support of four national standards laboratories and the US Naval Observatory which maintain standard time; all other members are also equipped with atomic clocks.

The increasingly wide application of satellite positioning and navigation in fields that have an international impact has resulted in the need for cooperation on a global scale and the emergence of international cooperative organizations to advance and exploit the new technology. A prime example is the international Global Navigation Satellite System (GNSS) service, or briefly IGS. It is an international organization with around 200 different member agencies that pool their resources and share GPS and GLONASS station data and other products. The organization is committed to the highest quality data and the support of earth science research and education. It carries out rapid daily computations of GPS orbits and clock corrections at several IGS Analysis Centers (AC), producing a globally integrated frame consistent with the International Terrestrial Reference Frame (ITRF). In order to satisfy the demands of meteorological services, the Analysis Centers have instituted an enhanced ultrarapid response service in support of meteorological offices and Low Earth Orbiters. The ACs have undertaken to submit rapid tracking data thus reducing the *latency* (delay) of the data to a matter of hours. Over a period of about 8 years the precision of orbit determinations has improved from the order of 30 cm to 3–5 cm. The satellite clock estimates obtained by different analysis centers agree within 0.1–0.2 ns!

16.4 Combined GPS-INS

The many advantages of GPS as a navigational tool have overshadowed the role of Inertial Navigational Systems (INS), but the modern inertial sensors, usually in the form of micro-machined electromechanical systems (MEMS), which we encountered in Chap. 12, remain generally in use in aircraft as reliable backup systems. The classic gyroscopes with rugged high speed rotors can still be found as backups on ships. However, it has been pointed out [4] that inertial sensors have properties that complement those of GPS: in particular they have an effectively wider bandwidth, which means a quicker response, and lower short-term instability and noise. Hence if properly combined with GPS, an inertial sensor should give superior performance in applications where a rapid dynamic response is called for, such as avoidance of collisions between moving platforms, such as aircraft or automobiles. The limitations of a simple GPS receiver, in addition to low bandwidth due to the relatively low frequency of navigational signal updates, include the unpredictable quality of reception due to variable factors such as spurious reflections from the immediate environment, ionospheric disturbances, obstructed view of the satellites, and even possible outages. A MEMS inertial sensor unit on the other hand may have less high frequency noise, but suffers from long-term drift requiring periodic calibration against GPS.

Combining the two types of systems to produce optimum performance presents an interesting design problem: if the two systems operate independently, and each generates its own navigational solution of position, velocity, and attitude, how should the results obtained from them be combined to arrive at the most accurate final result? Clearly the standard deviation from the mean of the observation data and its dependence on integration time in each case is relevant; it also matters at what stage in the reduction of the observed measurements the inputs from the two systems are combined. Since GPS has the advantage of greater long-term accuracy, the way the data from the two systems are combined should reflect this; in the simplest "loose integration," the procedure is in effect using the INS for rapid changes in the dynamic variables, while the GPS maintains the longer term calibration of the INS. Having been calibrated to correct any biases it may have using the GPS as standard, the inertial sensor can be used alone in the event of an emergency or a GPS outage.

The enhanced speed achieved by combining INS with GPS in determining position, speed, and attitude of a platform makes it possible to initiate timely corrective actions on speed and attitude to avoid the imminent danger of collision or other hazardous conditions. It advances the idea of a possible collision avoidance system even for motor vehicles in highway traffic, a challenging prospect indeed. The simplest current collision avoidance system for motor vehicles depends on lidar (light detection and ranging, the optical version of radar), but the equipping of modern automobiles with GPS receivers has become commonplace, and it is only natural to assume that the next step beyond having a seductive voice giving driving directions is a collision avoidance system. But predicting the direction of successful technological development is a tricky business since advances are seldom in a straight line. A collision avoidance system to be effective must be able to respond to quickly changing emergency conditions, and GPS position updates are received at far too slow a repetition rate to avoid a collision. Furthermore the system must be able to handle the simultaneous presence of a large number of platforms; and of course it is just under such heavy traffic conditions that collisions are most urgently to be avoided. Lidar, an echo-location method in bat-like fashion, can provide the most basic collision-avoidance in automobiles, but lacks the broader ability that GPS affords of alerting the operator to other safety information. Aircraft have the option of using radar transponders in which a radar signal is transmitted by an aircraft and the signal received by another aircraft in the same air space, triggers a response shifted in frequency. Such a system would be increasingly difficult to implement in the high density of traffic characteristic of hub airports.

16.5 Applications of GPS

The explosive expansion of applications of GPS in recent years has impacted practically every aspect of modern life. The most obvious and direct application is of course in geodesy and topographical map-making, but that is just the beginning... we have applications in land, sea, and air transportation, precise time dissemination, geodynamics, agriculture, public safety, disaster relief, and, as propagandists like to say, the list goes on.

16.5.1 Surveying and Mapping

GPS had predictably its greatest initial impact on surveying and map making. It enabled the acquisition of data to be vastly more efficient and accurate than traditional methods of theodolites and triangulation. Where once it required a team of surveyors trudging through possibly rough terrain to find a suitably elevated position from which to make measurements, one surveyor with a GPS receiver could now accomplish a more accurate survey in a fraction of time. Unlike conventional surveying, GPS is not constrained by line-of-sight visibility between markers; much greater distances can be used, even over difficult terrain—all that is required is a clear view of the sky. The surface coordinates and elevations of topographical features can be accurately measured and electronically stored, analyzed, and displayed with respect to a geodetic system of coordinates to constitute a datum. The term datum is not used here with its common meaning of something given, but rather in the technical sense of a representation of the geometrical size and shape of the earth's surface using some system of coordinates, often over a defined area; in the case of GPS that area can encompass the entire globe.

In areas beyond the reach of land-based reference points such as coastlines and waterways, GPS plays a unique role on board survey vessels in which depth-sounding sonar is combined with GPS positioning to carry out hydrographic surveys from which nautical charts are prepared that give the topography of the sea bed and the precise location of any hazards (or sunken treasure) that may lie beneath the surface. The oil industry is another important beneficiary of the GPS revolution in making possible wide-ranging exploration offshore for oil, and the positioning of promising locations.

16.5.2 Time Dissemination

You may recall that we have already devoted an entire chapter to this topic—it was there called "the longitude problem," for indeed if Greenwich time is accurately disseminated around the globe by carrying a stable chronometer then you could compute your local longitude relative to Greenwich by simply observing the Greenwich time of the local noon. Hence the problem of determining longitude at sea with respect to Greenwich became the problem of constructing a clock that accurately distributed Greenwich Time around the world. Aside from the longitude problem, a somewhat later requirement for precision time and frequency was in the area of radio telecommunication which relied on stable carrier frequencies and sharply resonant circuits for frequency tunable reception. Furthermore in order to accommodate a large number of radio communication channels, each must be assigned a frequency bandwidth and must adhere to strict frequency stability requirements to avoid interfering with other channels. Of course as a time-based system, GPS has now taken timekeeping and navigation to a new level of sophistication; and while navigation is of course important, the dissemination of precise time is essential in numerous *other* facets of modern life, from digital communication of data and TV/radio, to power grids and financial networks, to railway systems, etc. In fact it would be hard to think of any regulated activity that does *not* involve keeping precise time or frequency.

GPS time is monitored by the US Naval Observatory in order to assure a coordinated time reference for the system. It monitors GPS time at two levels of accuracy the so-called Standard Positioning Service (SPS), which is incorporated as part of the Coarse Acquisition Code (C/A), and the Precise Positioning Service (PPS) incorporated in the Precision Code (P-code), described in Chap. 14. The multiplicity of highly sophisticated clocks of almost comparable accuracy in the GPS system raises the interesting question of whose clock is to define "GPS Time." The answer is that no individual clock, but a synthesis of all clocks, is created by processing pseudorange/time measurements. We recall that by receiving simultaneous signals from more than four satellites, not only can the clock error of the receiver be determined, but using knowledge of the satellite ephemerides, errors in the satellite clocks can also be made observable. Since there is no reason to single out one satellite clock as having a uniquely greater accuracy than the rest, but in reality are statistically of comparable accuracy, naturally leads to the application of a Kalman filter to optimize the ensemble clock reading at any given time. This is called the GPS Composite Clock. It is the function of the US Naval Observatory to steer GPS Time to remain within less than 1 μs of UTC/TAI (the letters TAI are the initial letters in Temps Atomique International, based on the atomic second). GPS Time is continuous and does not follow UTC in the insertion of leap seconds to track UT1, the former Greenwich Mean Time, which is based on the motions of the earth.

GPS has been used to link up national timekeeping organizations, such as the US Naval Observatory (USNO) and the German Physikalisch–Technische Bundesanstalt (PTB) to synchronize clocks on a global scale. In June 2010 PTB shipped to the USNO a GPS calibration system designed for time link calibrations, consisting basically of an integrated GPS timing receiver with a time interval counter and monitor. Prior to being shipped from PTB to another laboratory to compare time calibrations, the traveling GPS receiver was operated at PTB in close coordination with a fixed group of other GPS receivers as part of a common time standard. It is then shipped with its antenna and antenna cable to the other laboratory, where again it is operated in the new environment for several days. Then in order to ascertain that the internal delays have not changed, the traveling receiver is returned to its original location and compared with the time standard there. The results using a large number of such transfers of calibration showed a mean deviation on the level of an incredible ± 2 ns.

An alternative method of accurate time transfer uses GPS in the so-called *GPS common view* technique which, as the name suggests, is simply using GPS signals from satellites that are observable simultaneously at high altitude from the two locations at which the time standards are to be compared.

GPS has made precise time readily available to all users, whether in industry, commerce, scientific research, or the general public. There are listed on the US National Institute of Science and Technology (NIST) Web site about 40 companies that supply GPS timing receivers, with precision that was unheard of until recently: in the tens of nanoseconds range! The Trimble "Resolution T" GPS timing receiver circuit board has according to their Web site [5] a one pulse per second output synchronized to GPS or UTC within 15 ns (one standard deviation).

16.5.3 Aviation and Marine Navigation

The safety and efficiency of civil aviation has greatly benefitted from the introduction of GPS, which provides space-based three-dimensional positioning that supports all phases of flight: from take-off, to the execution of a flight plan, to the approach and landing, and finally surface taxiing at the airport. Dual frequency GPS is so accurate that it can form the basis of a collision avoidance system—not just aircraft–aircraft collisions, but also aircraft–terrain collisions. In fact it plays a central role in systems such as the Enhanced Ground Proximity Warning System (EGPWS). This type of safety system reduces the risk of plane crashes due to among other scenarios pilot disorientation, a tragic occurrence given the name Controlled Flight into Terrain (CFIT), where an airworthy aircraft under pilot control is unintentionally flown into a mountain, for example. GPS-based systems are critical in air traffic control and the avoidance of aircraft collisions, making it safe to permit the distance between planes to be reduced. For such applications the accuracy provided by the dual frequency receiver is required. We recall that the enhancement of accuracy using the second frequency comes from its use to correct for the ionospheric dispersion of the microwave signals. The US Federal Aviation Agency program office, which is charged with global GPS navigation services, provides satellite-based data for aircraft positioning, air navigation, and timing in support of all phases of aircraft operations.

The Wide Area Augmentation System (WAAS), as the name suggests, builds on the GPS system to provide more accurate navigational aid to aircraft in a very wide geographical area. It may be thought of as an extension of the differential GPS concept. It is based on a network of about 25 precisely positioned ground reference stations, forming the WAAS network that receives high resolution positioning signals from GPS satellites. This information is relayed to a master station where, on the basis of the known precise positions of the WAAS stations, GPS satellite orbit/clock errors are deduced and a correction message is prepared for uplink to three dedicated geostationary satellites. We recall that a geostationary satellite has a circular orbit in the earth's equatorial plane and moves in that orbit with the same angular velocity as that of the earth. The error message is then re-broadcast from the three satellites on the GPS L_1 frequency to user receivers on board aircraft and ships. The WAAS system is designed to improve basic GPS positioning accuracy to limit errors to within 7 m horizontally and vertically.

Currently the US Department of Transportation sponsors the development of a Traffic Alert and Collision Avoidance system (TCAS) for aircraft based on an onboard beacon radar to closely track all air traffic in a specified air space. On the basis of return signals, computer software applies complex logic to determine whether to alert the pilot of potential danger, and in the event of such danger advise the pilot on maneuvers to avoid it. The system has in the past been shown to significantly reduce the risk of midair collisions. Whether this type of system will remain effective with the increasing density of air traffic in the future remains to be seen. That question was addressed as early as the late 1960s when a precision time-based one-way ranging system was proposed. If each major airport is equipped as a reference station for DGPS and it broadcasts time corrections to aircraft coming into its airspace, their clocks could readily be synchronized to within 1 ns, corresponding to a radio propagation distance of 30 cm. Aircraft approaching an airport would correct their time and broadcast a time code that would be received by other aircraft to detect the length of the propagation delay and hence the distance separating them.

Another system that enhances GPS is the European Geostationary Overlay Service (EGNOS) which also increases the accuracy and broadens the availability of accurate navigational signals. Its most important aspect is that it makes GPS qualified for safety-critical applications such as aircraft take-off and landing and navigating ships through narrow channels. Like WAAS it has three geostationary satellites and a network of ground stations. The positioning accuracy of EGNOS is quoted as being within 1.5 m in Europe. EGNOS represents Europe's first program in satellite navigation and is a precursor to the full global satellite navigation system GALILEO. The EGNOS positioning data are freely available to anyone with an EGNOS-enabled GPS receiver.

16.5.4 Railroads

It should come as no surprise that GPS can help to "keep the trains running on time," the accepted hallmark of efficient government—after all, time is of the essence in GPS. It is used throughout the world to track in real time the movements of trains to improve safety, efficiency, and punctuality. In some countries a system called Positive Train Control (PTC) has been implemented to prevent accidents due to switching errors causing collisions and derailments, etc. It is a system that combines real-time tracking of trains with dynamic control of their movement. It can reroute train traffic by remotely operating switches and specify the location of faults in the rails to direct maintenance crews; taken together, these new capabilities have markedly improved safety and efficiency. The US Department of Transportation projects the implementation of PTC in the USA by the year 2015, using an enhanced form of nationwide DGPS that is sufficiently accurate to remotely detect rail switching errors. It will support track surveying and mapping and systematic rail inspections to allow the efficient positioning of safety equipment. In former times the standard method of rail inspection was by workers called "tie spotters" who walked the tracks looking for ties that need to be replaced, a laborious procedure in which they recorded their location by hand and noted the number of such faults in a given stretch of rail. However this was only the beginning, the second phase required the delivery of new ties as part of a maintenance cycle; this was carried out by sending out a second man to mark the locations where bundles of new ties are to be unloaded from the supply train. Clearly the lack of the ability to define position in a coordinate system made this exercise wasteful of time and manpower, not to mention money. This is just the kind of activity for which GPS supplies a ready answer: equip the spotter with a handheld GPS receiver! In fact Union Pacific Railroad has equipped spotters with handheld GPS receivers with an auxiliary logger which enables the operator to simply click when a bad tie is found and the event is recorded along with the exact coordinates of the location. If a simple computer is added to the receiver then any desired statistical analysis of the number of faults in different geographical regions can be quickly made.

16.5.5 Agriculture

The coming of GPS has transformed life even "down on the farm." Of course agriculture, the oldest technology known to man, was mechanized some time ago, but the new computer/satellite era has made yet another profound change in the management of food production. GPS-based technology has shaped new ways of managing farms and increasing productivity, while economizing on the use of resources. The use of GPS enhances productivity by quantitatively optimizing the use of the land and the efficient use of fertilizer and disease control. This has spawned a new kind of farming called *precision agriculture*, which involves the use of GPS (or DGPS) in field mapping, tractor guidance, soil chemistry surveys, disease surveys, etc. A basic problem in farm management is to correlate crop yield with the various factors that affect it, such as soil chemistry, pest infestation, water distribution, etc.; having these critical parameters mapped out using exact coordinates enables a systematic analysis of the farm productivity, and the ways to improve it.

GPS establishes such a coordinate system and therefore can play a significant role in a number of farm activities, such as fertilizing, planting, scouting (monitoring the health of plants), and spraying. Prior to fertilizing, a survey of the soil chemistry of a field would be made with respect to planar coordinates using a GPS receiver, or for greater accuracy, a DGPS receiver. On the basis of that map and the known needs of the crop, the required fertilizer is distributed by machinery adapted to be run by GPS-based software. This not only distributes fertilizer where it is needed but also avoids wasting it where it is not or indeed where additional fertilizer may be harmful to the plants. Thus the fertilizer may be dispensed at a variable rate at different locations according to the specific needs of the plants as plotted on the nutrients' distribution map: this is the essence of precision agriculture—it is the methodical use of technology, in this case to apply fertilizer where it is needed, and not to cover indiscriminately the entire field. Similar programmed control of the number of seeds sown per linear foot can be achieved by GPS, ensuring optimum distribution of the seeds, including shutting off the planter as necessary. Another critical function that can be performed under GPS guidance is insecticide spraying. A GPS receiver on the sprayer can be used to relay coordinates and velocity data to a control station computer which in turn communicates to the sprayer control commands according to a previously prepared spray map; this ensures that the insecticide is applied only where and to the extent it is needed. Accurate positioning and control of the sprayer is required to avoid the possibility of overlapping applications of insecticide. Similarly aircraft "crop dusters" can be guided using GPS receivers to efficiently cover a desired field. Finally GPS can be used to map and record for future reference the distribution of crop yield over an area or a region in order to correlate this with other factors such as soil chemistry and fertilizer use. GPS-based guidance systems are increasingly used on large farms to steer machinery, and several GPS manufacturers such as Trimble and farm equipment manufacturers such as John Deere have responded by making available on the market GPS-guided control systems for farm use.

16.5.6 Surface Transportation

The ever-increasing traffic congestion on highways around the world particularly in the crowded cities of the USA, Europe, and Asia clearly indicates the urgent need for different solutions for the transportation of goods and people. In the meantime the problem may be alleviated using modern technology to develop "smart" traffic management systems, the object of the so-called Intelligent Traffic Systems (ITS), the brain child of the US Department of Transportation's Research and Innovative Technology Administration (RITA). The program has a number of well-defined areas of interest directed at making highway travel safe and efficient: they include among others communications throughout the system including between vehicles and infrastructure, real-time data acquisition and management, coordination of modes of transportation, management of weather-related problems, integration of new technology, traffic management and roadside infrastructure, and finally advanced research on rail systems and transportation of the future.

Since GPS technology deals with such fundamental quantities as terrestrial position and time, it is hardly surprising that it should play some role as supporting technology in all of these areas. Even the most basic function of freeway traffic surveillance assumes the ability to record the flow of traffic at different times and different points in the highway system: When coupled with radio communication GPS allows the dissemination of information about the precise location of delays due to collisions, construction sites, hazardous curves, and railway crossings, thereby keeping motorists constantly informed of road and/or traffic conditions at specified locations in real time. The integration of the road maintenance function with the general highway management ensures the safety of workers on construction sites and the diversion of traffic around them. From a traffic management point of view, the availability of GPS means that traffic conditions can be plotted in real time on a map, giving a total picture of traffic flow patterns. On the basis of such surveillance, statistical studies can be made of the use of High Occupancy Vehicle (HOV) lanes, traffic congestion at ramps and repair sites, etc.

The area where modern electronics has inspired the greatest fantasy is that of automated vehicles that steer themselves or at least automatically avoid collisions and stay on the road. This indeed is the next step in the established technological progression: Not only do we already have the computational means of routing and rerouting travel along the highways, but we also have some of the computer-controlled mechanical functions of a vehicle, including braking, cruise control, and attitude stability control. Of course we are a long way from replacing a human driver, particularly in city traffic as things are at present, but it would not be far-fetched to imagine a futuristic city designed to be compatible with a computer in the driver's seat.

GPS also has a role to play in urban public transportation management. As reported recently by Umar Farooq et al. [6] in a paper on the subject, the authors propose a traffic management system based on GPS and radio communication for an urban transit system using buses in the city of Lahore, India. According to the authors Lahore has a public transportation system that is so poor that those who can afford it use private cars, leading to an aggravated air pollution problem. The proposed system is intended to make the system more efficient and "passenger friendly." Its electronic system consists of four functional parts: the central bus station module, modules on board individual buses, a base module, and modules at individual bus stops. The central bus station is equipped with computers and GSM (global system for mobile communication) modems that convert voice into the appropriate cell phone format. The bus station module initiates the process by sending via SMS (short message service) to the bus and the base station a number assigned to the bus plus its license plate number. The module aboard the bus using its GPS receiver and GSM modem now sends its identifying number and position along with other data such as seat occupancy to the base station. The latter is equipped with a computer and GSM modem to keep track of every bus and updates locations of buses at bus stops. Every bus stop has communication equipment together with a memory module and a display screen that can show the location of buses coming towards that stop.

16.5.7 Safety and Disaster Relief

Again GPS is a great enabling technology in another area where positioning and time are critical, namely, responding promptly to emergencies, whether on the scale of earthquakes and tsunamis or just local highway incidents. The global coverage of GPS means it can be used to map the boundaries of large disaster areas caused by hurricanes, earthquakes, or forest fires. As already noted, it can also provide critical data on the earth's crustal deformations to use in predicting earthquakes.

On an urban scale GPS is now almost universally relied on to navigate the streets to reach any desired address, a capability of immense help not only to emergency vehicles but also for commercial deliveries, not to mention visiting friends. GPS receivers embedded in communications equipment can be used to direct emergency assistance to specified locations in critical need of possibly life-saving action. This involves a response that requires efficient fleet management of ambulances and fire engines to make sure the required service is dispatched from the most advantageous point.

16.5.8 Recreation

When, as is now common, a GPS receiver is integrated into a microcomputer with a map display, it becomes an infallible guide to take along on a hiking trip into unfamiliar territory. Of course it takes away the challenge of path finding with a map and compass, thereby depreciating the value of one aspect of scouting. Lord Baden-Powell presumably would be able to think of other benefits of scouting. In fact the prudent hiker would carry a compass and map anyway, since batteries are known to rundown at the least opportune time, and even the receiver may at times fail to lock on to sufficient GPS satellites and the service itself may be temporarily interrupted. With a handheld set a hiker can know his/her location at all times and with the right receiver model can enter any point as a waypoint on a map to which he/she can return. GPS receivers on the market are available at all levels of sophistication and price. Garmin, one of the leading suppliers of GPS equipment, advertises about 40 different models for use on the trail. The level of sophistication depends of course on the desired functions: the cheapest models simply read out the coordinates of its position without reference to any map. The most useful unit would have a clearly visible color chart display on which position is plotted and retained in memory. Some models come with popular trails already in memory which the hiker may choose to follow. Currently GPS has found a new application in a form of hiking activity akin to treasure hunting called *geo-caching*, which apparently has been promoted as a family-friendly activity. Future development of GPS calls for the ability to receive signals under a forest canopy or in deeply forested areas, making it far more useful to hikers.

GPS is especially useful to fishermen and pleasure boatmen, particularly when venturing beyond sight of land to favorite fishing grounds. GPS receiver models marketed to fishermen have installed navigational charts that display coastlines and navigational hazards and have course plotting ability on which favored fishing grounds can be marked as waypoints. Of course recreational sailors enjoying coastal navigation can carry the same type of equipment as fishermen. One remarkable aspect of GPS that is sometimes overlooked when contemplating the flotillas of pleasure boats, each one no doubt equipped with a GPS receiver: the system is passive—there is no need for the user to broadcast any signal requiring an assigned a portion of the radio spectrum—hence there is no limit to the number of users.

16.5.9 The Environment

GPS provides the means of defining large geographical areas which, with the techniques of remote sensing, can monitor aspects of the environment on a global scale. Of particular interest is the impact on the environment of human activities such as forest clearing, strip-mining, oil spills at sea, as well as natural phenomena such as changing climate patterns, forest fires, sea level changes, etc. By melding the GPS-generated positional data into the Geographical Information System (GIS) software, the geographical distribution of environmental factors can be studied, and governmental decisions taken based on scientific data. GPS topographical coordinate data when correlated with aerial photography can yield detailed information on the distribution of vegetation and animal life, which can be correlated to human conservation or other activities.

Of considerable interest to the debate over the greenhouse effect and climate change is the extent of the rise in sea level. GPS observations of enhanced accuracy have been devoted to the measurement of tidal action and sea level changes at an increasing number of tracking sites. Other applications include the use of GPS receivers on helicopters to survey the spread of disasters such as forest fires and oil spills.

One application we have not thought about since Chap. 1 is the long-range tracking of migratory birds and animals, particularly those species that are endangered. These species are of particular interest since they are presumably more difficult to find, but once found and fitted with a GPS receiver, their fate may be traceable to a possible change in their environment. Thanks to micro-electronics, as exemplified by the small GPS module already cited, made by Micro Modular Technologies, it is possible to attach a receiver to medium size migratory birds and other animals to track their routes and details of their progress in time.

16.5.10 Space Applications

Last but certainly not least we mention the support GPS renders to activities related to satellites and space missions. The GPS constellation with its satellites in thoroughly studied orbits in effect form a reference system with respect to which the ephemeris of other orbiting satellites may be determined. At any given time there are countless satellites aloft on various missions: in addition to the common communications satellites there are specialized satellites, such as geodetic satellites and the cooperative US-ESA program called Jason I and II which study the ocean surface topology. In most of these satellite missions GPS plays an important role in providing information on satellite attitude and time, saving the client satellite the expense and weight of atomic clocks and the means of accurately ascertaining the satellite attitude.

References

1. Ashtech OEM, Sunnyvale, CA 94085
2. Micro Modular Technologies Pte Ltd, Huntington Beach, CA 92647
3. Fed. Highway Admin, Publ. No. FHWA-RD-03-039 (2003)
4. Santiago Alban, PhD Thesis, Stanford University, 2004
5. http://www.trimble.com/timing/resolution-t.aspx
6. Umar Farooq, et al. 2nd International Conference on Computer Engineering and Application, Bali, 2010

Navigation in Space

17

17.1 Introduction

Navigation in space can be said to have begun on October 4, 1957. Great transformations in human history often have debatable origins, but that date marks the defining moment when man took a first step to escape the "surly bonds of earth"; it was the beginning of the space age. On that date an aluminum ball about 58 cm in diameter called *Sputnik* (Russian for *satellite*) was successfully rocketed beyond the earth's atmosphere and into a low orbit circling the earth every 96 min. With a mass of only about 83 kg it was placed in a 700 km elliptical orbit with an eccentricity of $\varepsilon = 0.05$ and an inclination of 65°. As significant as this was as a demonstration of human accomplishment, the reality is that it was not initially prompted by some noble desire to advance human knowledge, but really was a by-product of military rocketry and ICBM development; the same can be said of the early US efforts in space. In fact although Goddard and other "rocket boys" did pioneering work in rocketry, it was the German V2 rocket development during World War II by young geniuses such as Wernher von Braun that made possible the sending of a Sputnik, and later the American Explorer I satellites into orbit. Be that as it may, the timing of the event was prompted by the International Council of Scientific Unions' resolution to call for the launching of artificial satellites to mark the International Geophysical Year (IGY). The Russian success sparked immediate reaction in the USA, resulting in intense rivalry dubbed the space race between the two cold war antagonists: the USA and the Soviet Union. One month after Sputnik 1 the Soviet Union successfully launched a much larger spacecraft designated Sputnik 2 which carried a dog on board; this was the second in a series that would be continued up to Sputnik 22. The first US satellite to be successfully placed in earth orbit was named Explorer 1 and was launched in January, 1958. In April 1961 the U.S.S.R successfully placed in orbit the satellite Vostok (Russian for East) with a human on board, Yuri Gagarin; 1 month later the American astronaut Alan Sheppard was sent aloft into a suborbital path aboard a craft named Freedom 7. It was not until a year later that John Glenn orbited the earth aboard the US spacecraft

F.G. Major, *Quo Vadis: Evolution of Modern Navigation: The Rise of Quantum Techniques*, DOI 10.1007/978-1-4614-8672-5_17,

named Mercury Friendship. The culmination of the US manned flight program was the landing on the moon July 20, 1969, still in the decade of the 1960s, the target that US president John F. Kennedy had exhorted the nation to achieve. On that date the world watched on television Neil Armstrong and Edwin ("Buzz") Aldrin step out of their Apollo 11 lunar module onto the moon's surface, while Michael Collins circled the moon in his spacecraft the Columbia.

It would not be appropriate here to attempt a history of space travel—there is more than ample material on that subject, for example, on the NASA and Jet Propulsion Laboratory (JPL) Web sites; however it *is* interesting to recall these events that defined the beginning of a new era within living memory of many people, including the author. In this chapter we will try to throw light on how, by marshaling and coordinating a national support system, such a feat of space navigation as the moon mission was accomplished; a feat that required providing navigational guidance and a benign environment for three men over a journey of millions of miles to a specific destination in space.

Over the half century of space exploration there have been inaugurated more numerous space missions than would be appropriate to list here. In addition to missions to all the planets and some of their moons, there are missions to asteroids, dwarf planets, comets, the Kuiper belt, the Hubble space telescope, and a shuttle service to the International Space Station. Our interest in these innumerable missions extends only to the navigational means that guided them to their targets. Clearly interplanetary travel toward a far-off destination requires prior knowledge of its location with respect to a frame of reference recognized by the navigational system of the spacecraft. Fortunately the distant stars are sufficiently far away that even for travel as far as the outer reaches of the solar system the degree of parallax is still small; therefore the stars still provide a means of navigation, and we shall encounter again the principles of astronomical navigation, applied here in space.

In the beginning, when the object was simply to get a body into orbit, there was little navigating involved beyond the critical control of attitude during the launch phase—a successful launch required at a minimum that the thrust be generally upward! Once the payload reached the desired altitude, thrusters were activated to give the satellite sufficient transverse momentum to insert it into an orbit of desired angular momentum and eccentricity. However the execution of a demanding space mission to a specified destination is another matter entirely; we will endeavor to draw the outlines of some important missions that illustrate different navigational challenges in their performance. In doing this we are in danger of getting lost in a forest of NASA's shorthand vocabulary. Government bureaucracies are often criticized for the overuse of "alphabet soup," but in the case of the Apollo missions, for example, it's hard to imagine how the astronauts aboard the spacecraft could communicate with mission control without some short hand way of referring to the multitudinous tasks that had to be performed. We will encounter abbreviations such as GET (ground elapsed time), CSM (command service module), IMU (inertial measurement unit), EPO (earth parking orbit), TLI (translunar insertion), S-IVB (Saturn IVB launch vehicle), TLC (translunar coast), DOI (descent orbit insertion), TPI (terminal phase initiation), and the list goes on.

17.2 The Apollo Program

The US program to "put a man on the moon," given the name Apollo, ranks in terms of the scale of national resources devoted to it, with the Manhattan Project to develop an atomic bomb. Its inspiration is commonly attributed to President Kennedy in an oft-quoted speech in which he challenged the nation to send a man to the moon and bring him back safely within the decade. Actually his motivation was less scientific or technological than international power politics; his science advisor is said to have been initially opposed to manned flight as too costly with too little scientific return; there are many who still believe that to this day. The idea of such a feat is of course at least as old as Jules Verne, the difference now is that the military ICBM projects had produced powerful rockets making the idea less science fiction; in fact the prospect was brought home in concrete terms by Yuri Gagarin, who had been successfully put in earth orbit by the U.S.S.R. This was the precipitating event for the USA: its competitive spirit had been aroused.

The US Apollo program had a tragic beginning on January 27, 1967. The command module of what is now designated as Apollo 1 was destroyed by fire, taking the lives of the astronauts Grissom, White, and Chaffee. It was during a preflight test in which the cabin was filled with pure oxygen gas, and it is assumed an electrical fault must have immediately engulfed the cabin with flames. Rescue efforts, which would probably have been to no avail, were hampered by the fact that access to the cabin was impeded by the excess pressure within it. The Apollo program was put on hold, and no spacecraft has been designated Apollo 2 or 3. It was an expensive lesson but led to important safety measures in a new module whose design has served well through many successful missions to the moon. The perfect record was marred by only one other major mishap: the service module of Apollo 13 suffered a catastrophic explosion of one of the tanks holding liquid oxygen, resulting in the rupture of an oxygen line and damage to a valve on another tank causing the rapid loss of oxygen, a literally life-threatening development. It is a testament to the mettle of the astronauts and the close coordination between astronauts and ground control that not only was there no tragic fatality but ways were also found to reprogram the mission and actually make a fly-by of the moon and return safely to earth!

17.3 Apollo Mission Design

In the beginning of the Apollo program it was presumed that the navigation would be controlled by the astronauts on board the spacecraft; but it was soon realized that, since close communication between the astronauts and the ground could provide the full support of the Manned Space Flight Network (MSFN) in a timely manner, the primary responsibility for the conduct of the mission was given to the Mission Control Center (MCC) in Houston, Texas. The detailed design of the lunar mission presents an interesting problem in classical mechanics, a problem on whose correct solution at least three human lives depended! It is a three-body problem in which fortunately the motions of the earth and moon are well known and unaffected by the presence of the spacecraft, which however is acted on by their gravitational field. To define the problem more quantitatively let us review the properties of the earth–moon system: the moon has a diameter of about 3,500 km, its distance from the earth is about 384,000 km, it has a nearly circular orbit with an eccentricity of 0.055, and a (sidereal) period of revolution of 27.3 days. The moon's rotation is "locked" in phase with its revolution around the earth so that we always see the same side of the moon. Its orbital plane is inclined to the ecliptic (plane of the earth's orbit) by a small angle of about 5° and its spin axis is inclined to its orbital plane by about 6.7°. The conduct of the mission must conform to at least two primary constraints: first, the spacecraft rocket engine is allowed a fixed quantity of fuel, and second, the positions of the initial and final points of the journey are predetermined within a certain range. Since the earth and moon are in constant motion, the times at which the journey is chosen to begin and end are important variables; this is the reason that there is so much discussion of *launch windows* and the need to analyze when they occur. Because of the daily period of the earth's rotation and the monthly period of the moon's motion it is not surprising that the launch windows repeat with the same periods. However these windows are of finite duration and together with the azimuth angle of the initial rocket thrust constitute important parameters that must be optimally chosen. The term azimuthal angle is used here to mean the angle measured in a horizontal plane from the direction of true north towards the east. By choosing the azimuth angle in the neighborhood of 90° guarantees that the rotation of the Earth will *add* to the initial velocity of the spacecraft with respect to inertial space. This factor has a significant effect on the size of the payload capability of the launch vehicle, the Saturn V; the reserve power requirement for the Apollo missions must be satisfied for azimuth angles in the range 72°–108°. The choice of angle determines the duration of the launch window which the Apollo missions commonly chose to make 4.5 h long. In the event that system failures necessitated a "scrub" or a hold that went past the daily window, the launch could be rescheduled for another month. There are good safety reasons for wanting the launch to occur in the daytime: in case of an emergency on the launch pad the rescue of the astronauts would be more difficult in the dark, also photographic record of the events during the blast off phase would help in analysis, and finally if an emergency arises which necessitates correcting the attitude of the spacecraft then a clear daylight view of the horizon is needed.

The flight to the moon and landing on its surface were programmed to be carried out in stages with constant communication by voice and other channels with mission control. We will describe the journey specifically with reference to that made by the historic Apollo 11 following closely the report on the mission submitted by the astronauts themselves [1]. The account given here will attempt to faithfully follow the sequence of events with occasional explanatory comments.

At 9:32 A.M. on July 16, 1969 at the Kennedy Space Center, the Apollo 11 mission was successfully launched using a Saturn V rocket engine, reaching an altitude of about 185 km in about 11 min. The Saturn V (the Roman numeral V refers to the combination of five F-1 rocket jets) has three stages: the most powerful is the first stage (S-1C) used for lift-off which has five F-1 jets and burns for 2.5 min, the second stage S-II has five J-2 jets and burns for about 6 min, and finally the third, designated S-IVB, has one J-2 jet which burns for a little less than 3 min, raising the spacecraft to the orbital velocity of 27,000 km/h. S-IVB was shut down with liquid propellant remaining unburned and remained attached to the spacecraft during the period in the earth parking orbit (EPO), to be reignited later for the translunar passage. This completes the first phase of the mission, namely to insert into circular parking orbit the spacecraft assembly which included the command/service module (CSM), the lunar module (LM), the S-IVB portion of the Saturn rocket, an instrument unit (IU), and finally the spacecraft lunar module adapter (SLA). The important functions of this phase in earth orbit prior to embarking on an interplanetary mission was firstly to provide flexibility in the launch timing, but more importantly to allow the astronauts a final check on the many crucial systems on which their lives depended. These included the communications systems, the various environmental monitoring and control systems, the command/service module propulsion system, the reaction and stabilization control systems, the electrical power system, and of course the command module computer. Of particular concern are the Saturn-IVB instrument unit, the inertial measurement unit (IMU) and the guidance, navigation, and control (GNC) system. There were limits imposed on the duration of the earth orbit phase by such factors as the S-IVB launch vehicle's capability to keep providing attitude control since its instrument unit (IU) platform may drift over time, and there was also concern over the S-IVB propellant boil-off—both good reasons not to delay what NASA calls *translunar injection*—a maximum of three earth orbits was therefore set as a limit. On the other hand it had been decided that there should be sufficient time for the spacecraft to be tracked by at least two manned space flight network (MSFN) stations and one command station prior to the translunar injection, a condition that in fact is satisfied by just one pass over the continental USA. Thus it was that the trans-lunar insertion (TLI) which marks the beginning of the trajectory to the moon, occurred at 2:44:26 GET (h:min:s ground elapsed time) over the Pacific Ocean after completing about one and a half revolutions around the Earth, taking the Apollo spacecraft with Armstrong, Aldrin, and Collins on board off to the moon.

The precise moment to initiate the translunar injection (TLI) and the later in-flight corrective burns of the S-IVB are critically important to achieving an optimal trajectory to the moon with the most efficient use of S-IVB propellant. Of major concern throughout the planning of the mission was the provision at every stage of an emergency abort procedure in the event of a failure in the system—it certainly must have helped relieve the astronauts of worry about their safe return! Fortunately there are such things as *free return orbits* that may be implemented during the phase of orbit transfer from circling the earth to circling the moon. The concept draws on the work of Walter Hohmann who in 1925 described a procedure for transferring a spacecraft between two coplanar circular orbits some distance apart with minimum expenditure of energy. In Fig. 17.1 assume we have a spacecraft in a circular Earth orbit and its rocket engine gives it a powerful kick of just the right mechanical

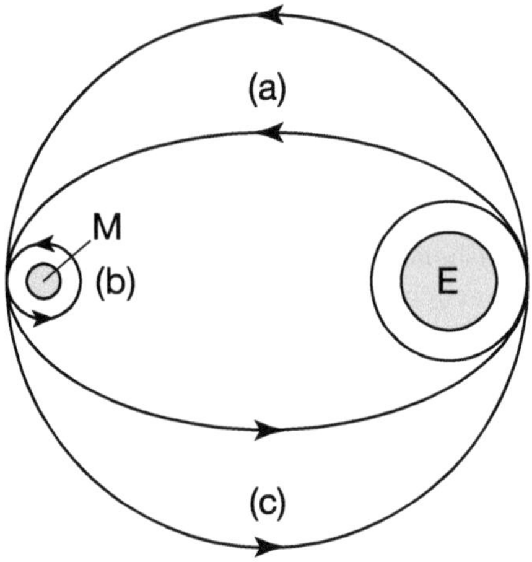

Fig. 17.1 Hohmann method of orbit transfer

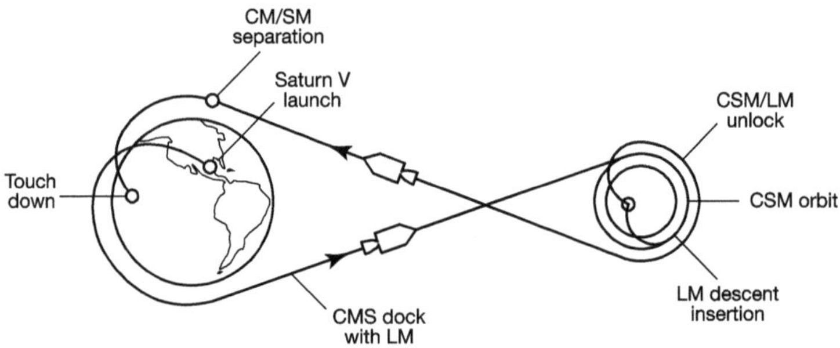

Fig. 17.2 Apollo 11 flight sequence

impulse ($m\Delta V = \int F dt$) to send it on an elongated ellipse that circumnavigates the moon, with the earth at one focus. If the spacecraft were allowed to freely complete its elliptical orbit then given a negative impulse equal to the initial positive impulse, it would return to its original orbit around the earth. On the other hand, if at the instant of perilune when it is passing closest to the Moon the spacecraft were given just the right *forward* impulse, it could be made to continue in a large circular orbit, as shown in the figure. But if it were given just the right *negative* impulse at the perilune point, it could be put in a circular orbit around the moon. However according to the published reports [2] this is not the orbit transfer procedure that was actually followed by Apollo 11; rather it was as depicted schematically in Fig. 17.2, where the revolution around the moon is in the *opposite* direction to the earth orbit. The figure shows the principal events during the 8-day mission, starting with the launch, the insertion into earth orbit, the restart of the S-IVB rocket engine after one and a half revolutions in the earth parking orbit, and reconfiguration for the long journey to the moon.

The control of the path of the spacecraft, whether during launch or insertion into a specified orbit around the earth or moon, or landing on their surfaces, clearly requires a critical sequence of orbital maneuvers, as will become evident later.

The timing of the go decision for translunar injection (TLI) is determined with reference to the position of the moon's antipode, which is the point where the line joining the centers of the earth and moon intersects the earth's surface. The geographic location of that point moves rapidly with the daily rotation of the earth and more slowly with the monthly revolution of the noon. The earth's rotation causes the antipode to trace a daily circle of constant latitude, while, because the orbital plane of the moon does not exactly coincide with that of the earth, the moon's revolution causes a monthly oscillation between higher and lower latitudes. In the meantime the spacecraft is pursuing its orbit in a plane determined by the azimuth of the launch; if the azimuth is not exactly 90° the projection of the spacecraft position on the ground will resemble a sine wave which will intersect the trace of the antipode at certain times and locations that depend of course on the azimuth angle of the launch. For a given azimuth therefore the launch must be timed so that the spacecraft will pass over the lunar antipode point at which the translunar trajectory injection should take place. Clearly this presents a three-body dynamical problem to determine the precise time of the TLI and the duration of the engine burn to give the spacecraft the necessary initial impulse, a problem that requires an electronic computer. In the 1960s, computers still relied on IBM cards and therefore data storage and computational power were primitive by the standards of the twenty-first century. Integrated circuits were still at a rudimentary stage, perhaps combining a few gates, and were very expensive. However, a concerted national effort was mounted to fully use all computational means available to design and carry out the lunar mission. This involved not only the use of available large mainframe computers and smaller minicomputers to solve the equations of motion and determine in detail the most suitable sequence of events but also the creation of a portable onboard spacecraft computer/controller to actually guide and control the space craft. The state-of-the-art portable computers at the time, it has been said, was not much more advanced than what is required to run an automatic toaster today! But with the impetus (and money) provided by the Apollo program that was about to change. In 1961 NASA signed a contract with what is now the Massachusetts Institute of Technology Draper Laboratory to develop the guidance and control electronics for a spacecraft environment. This applied physics laboratory was the foremost facility in the USA. in the field, having designed the Polaris and Poseidon missile guidance systems, although those projects used analog computers. Their design for the space borne units relied heavily on the development and use of the new field of integrated circuits. What began as a research project, ultimately led to an actual design for the Apollo program. The contract to build the Apollo guidance computer (AGC) was given to Raytheon using integrated circuits (IC) supplied by Fairchild Semiconductor and Philco-Ford, who had improved the IC manufacturing techniques and brought down the price to reasonable levels. ICs had not however reached the point of integrating the large number of gates that would be necessary to construct an adequate memory. At the time, bulk computer memory was generally recorded on magnetic tape; however, the designers of the AGC opted for a form of magnetic memory in which small rings of magnetic material with a special magnetization hysteresis curve makes them act almost in a bistable manner. The MIT group devised a new way of using these magnetic cores as *transformers* to form a read-only memory (ROM) called for some reason *core rope memory*. These small rings arranged in a matrix are intertwined with current carrying wires that either pass through the ring (digital 1) or bypass it (digital 0). Interaction between the astronauts and the machine was through a display keyboard (DISKY) with a very basic command structure, essentially a NOUN + VERB syntax with the commands assigned numerical values to be entered, for example, DISPLAY + ACCELERATION. Of course, as mentioned earlier it was up to the MSFN and mission control to monitor the actual trajectory and attitude of the spacecraft and to send up instructions to the astronauts to make corrections through the onboard computer in real time; to that end a telemetry channel allowed the onboard AGC to receive commands from computers on the ground.

A short time after the translunar injection the spacecraft was separated from the S-IVB, and after a short delay to allow the command/service modules to drift some distance away from the S-IVB, a transposition of the different parts of the space craft was conducted using a digital autopilot. This involved a rotating maneuver of the command/service modules to dock with the lunar module (LM) which was freed from its adapter in the former configuration. The umbilical connections between the CSM and the LM were made and the lunar module pressurized. After its separation, the S-IVB was vented of the remaining liquid propellant, putting it in what has come to be called a *slingshot* or gravity-aided trajectory that diverts it from a lunar trajectory and puts it in a solar orbit. The term slingshot or *gravity assist* simply refers to the acceleration experienced by a small craft as it swings past a larger body such as a planet. Presumably the slingshot referred to is the kind favored as a weapon by David rather than a handheld catapult!

During the lunar coast portion of the mission, there were two planned midcourse corrections involving the apparent horizon altitude and the optical system calibration: the first occurred at 6 h GET and the second at 24 h GET. Some difficulty was experienced during the first correction in locating the substellar points of the navigational stars, which as the name suggests are points where the stars are at the zenith, observations that are central to celestial navigation. A limitation of the onboard computer was that it did not account for the possibility that the lunar module structure may obstruct the view of a desired star. Added to this, the ground-supplied gimbal angles were not accurate enough to locate the precise substellar point, therefore additional maneuvers were conducted to align the craft using a sextant. The second correction at much greater distance from the Earth proved to be much simpler to execute. During the subsequent long translunar coast to the moon, which lasted a little over 3 days, worries about the possible overheating of the interior of the spacecraft by solar radiation prompted the use of *passive temperature control* in which solar overheating at isolated spots is avoided by a constant slow roll of the spacecraft about its longitudinal axis, set to point towards the ecliptic north, that is perpendicular to the orbital plane of the earth, the ecliptic. It proved successful in maintaining the spacecraft attitude and avoiding intolerable temperature conditions. Even under routine conditions, establishing an environment where humans can live and work requires an elaborate system of environmental control. The principal requirements to sustain a manned mission are potable water, edible provisions, temperature control, oxygen control, and waste disposal. To provide for these, several subsystems are required. The oxygen subsystem ensures that a controlled flow of *pure* oxygen (except on the launch pad where it is a fire hazard, a lesson learned from the Apollo 13 experience) is maintained at a regulated pressure of about 0.33 atm in the cabin. In the event of a catastrophic loss of oxygen pressure, an emergency cabin pressure regulator responds with a rapid supply of oxygen to maintain a somewhat lower pressure for at least 15 min to give the astronauts time to put on their pressure suits. The oxygen supply circuit of these pressure suits must be purged of carbon dioxide and moisture and the cabin atmosphere must similarly be reconditioned. Control of the temperature is also critical and requires both heating and cooling: the temperature control system relies on the use of a *water–glycol* solution for cooling and heat transfer. The chemical ethylene glycol is familiar to the general public as the former common antifreeze additive for car radiators. In more recent times the automobile industry has developed new and no doubt better chemical concoctions to serve as antifreeze. The addition of 50 % (by weight) of ethylene glycol to water to form an aqueous solution depresses the freezing point to about −45 °C and raises the boiling point to about 110 °C. The solution is commonly used in heat transfer systems where the temperature is below the freezing point of pure water; its specific heat however is slightly less than pure water and therefore a greater flow of liquid is required for the same transfer of heat. The cooling mechanism used by the temperature control system is the physical process of evaporation, in which molecules of the water–glycol solution with kinetic energy greater than the mean escape from the surface thereby lowering the mean energy of molecules in the liquid, as shown by a drop in its temperature. It is assumed of course that the evaporated molecules are removed and do not reenter the liquid with equal probability. The water–glycol subsystem consisting of two evaporators provided all the cooling needs, including the pressure suit circuit, the potable water cooler, as well as all the electronic equipment.

The insertion into lunar orbit occurred at about 76 h (GET) at an altitude of 148 km above the moon's surface with a first retrograde burn at the perilune (point of closest distance to the moon) by the service propulsion system (SPS) amounting to $\Delta V = 891$ m/s that put the spacecraft in a 273 × 98 km elliptical orbit according to the onboard computer. A second circularization burn also at the perilune, which occurred two lunar revolutions later, was intended to put the spacecraft in a 186 × 87 km ellipse which it was expected to evolve into a circular 100 km orbit due to irregularities in the lunar gravitational field. However, the onboard computation during a later phase indicated that the orbit was not exactly circular, showing that the ellipticity decay rate was less than expected.

At about 57 GET, the day before descending to the lunar surface, the lunar module was entered for the first time to inspect and photographically record the state of the lunar module interior and ascertain that all systems were properly set and the cameras functioning properly. The following day the crew was awakened on schedule and after putting on specially designed biomedical harnesses and cooling outfits they sat down to what the astronauts described as a "hearty" breakfast. One can only imagine the state of mind of these intrepid men, but one can assume that palpitations would be normal under the circumstances! Aldrin, the lunar module pilot, made his way to the lunar module to initiate its systems before returning to the command module to suit up. Then he returned to the lunar module, installed the *drogue* and *probe*, and closed the hatch. The drogue and probe are parts of the mechanism by which the lunar module aligns with the command module to dock. The probe was mounted in the command module docking tunnel, while the cone of the drogue was mounted in the docking tunnel of the lunar module. As the two modules approached each other to dock, the command module would extend the probe to be guided by the drogue into alignment in the docking tunnel and captured by latches designed for that purpose. The activities that followed the activation of the systems in the lunar module had been rehearsed many times and were accomplished without any surprises. The lunar module computer and inertial measurement unit (IMU) were powered up prior to loss of communication with the ground on the far side of the moon. This enabled the ground to calculate the gyro settings for aligning the lunar module platform to that of the command module. The landing gear deployment, descent propulsion, and rendezvous radar systems were checked out. We should note that at every critical phase of the mission there was an emergency abort system in place, including an abort guidance system, a testament to the extraordinary planning of the Apollo missions.

One of the important navigational activities conducted by the Apollo missions was to optically mark the positions of known features on the lunar surface such as craters and derive their selenographic (moon) coordinates. Features that had previously been accurately surveyed and have known coordinates could be used to verify the position of the spacecraft, that is, function as navigational aids. There were two periods of observation: the first consisted of five observations or *marks* on the crater Kamp in the Foaming Sea (sic). A second set of five observations were conducted on a small crater in the wall of a larger crater designated as No. 130. The landmarks were tracked optically using the sextant or the scanning telescope in the command module. The first five Apollo missions tracked a total of 19 landmarks including 6 on the far side of the moon. The landmarks were relatively small craters ranging in diameter between 100 and 1,500 m in diameter and their locations computed on the basis of optical angle measurements conducted from the command module. The observation of each landmark required a sequence of five precisely timed sightings equally spaced as the spacecraft passed over it at the nadir (vertically down). The first sighting is made when the spacecraft is at 35° altitude relative to the landmark and the other sightings are made at equal angles symmetric about the nadir. The data consisted of three angles that define the direction of the optical line of sight with respect to the three axes of the inertial measurement unit and two angles that define the direction of the optical line of sight with respect to the line from the spacecraft to the landmark. Computation of the coordinates of the feature was carried out in two parts: first, assuming precise knowledge of the command module orbital position at a specified time, that is epoch, the orbital position at the time of the mark is obtained by integrating the radar Doppler tracking data, obtained from the MSFN, from the orbit epoch to the time of the mark. The MSFN radar tracking stations accumulate Doppler frequency-shift data and are geographically distributed such that the spacecraft is in full view of at least two tracking stations when not occluded by the moon. The integrated data are used to compute the satellite orbit in selenocentric Cartesian coordinates; this presumes a precise knowledge of the moon's ephemerides, which are available from the Jet Propulsion Laboratory. The second part of the computation of the coordinates of the landmark used the observed angles of the landmark sightings, assuming the spacecraft position to be that derived from the radar data. The principal source of error in determining the selenographic coordinates is the imperfect mathematical modeling of the irregular gravitational field of the moon; the lunar gravitational model designated as L_1 is used for all MSFN orbit determinations and trajectory integrations because it has proven to yield more consistent results for landmark locations from one revolution to another. Nevertheless it is a source of error which affects the MSFN prediction of the orbit to a far greater extent than the optical sighting data used. The relative accuracy of the latter is attributable to the fact that the inertial platform of the spacecraft is constantly checked and any drift corrected, while clock error is kept in check by constant re-synchronization with the ground clock if the drift exceeds a certain limit.

After completion of the lunar landmark tracking by the command module, the lunar module was maneuvered to enable a steerable antenna to receive state vector updates for the primary guidance computer. The list of actions to prepare the lunar module for the descent is long: the landing gear must be deployed, the reaction (thrust) control must be checked, the descent propulsion system must be checked, and the rendezvous radar system must be activated and checked. With the completion of all preparations the crucial moment had arrived: it was time to begin the undocking of the lunar module from the command module and prepare to take the plunge! Great care was exercised in the undocking operation to make sure that the lunar module guidance computer (LGC) maintained accurate display of position and velocity. The undocking began by imparting a 0.12 m/s velocity to the lunar module as measured by the latter's primary guidance system; the resulting reaction velocity of the command module was maintained until the separation reached 12.2 m when the velocity was nulled with respect to the lunar module. At this point the lunar module executed a 360° yaw to enable the command module pilot (Collins) to visually check that the lunar module landing gear is properly deployed. Alignment procedures were then followed in preparation for the descent orbit insertion; this included the alignment of the inertial platform and that of the radar boresight with the optical sight. The rendezvous radar lock-on condition between the two spacecraft, a feature of servo feedback systems, was achieved under manual control. The maneuver to descend was made with the engine in the manual throttle mode set at the minimum thrust, which resulted in a gradual acceleration. After 15 s the throttle was carefully opened to increase the thrust to the planned level of 40 % of maximum. At the programmed cutoff, the velocity components along the three coordinate axes were equal to the planned values within tolerance. Using the reaction control system thrusters the module was put in an orbit computed to be an ellipse with axes measuring14.6 × 91.9 km, slightly different from the predicted value. The radar dedicated to aid in the rendezvous of the two spacecraft confirmed that the relative velocity at this point was well within the predicted range.

The next phase is what all the prior Apollo flights were really leading up to, namely the powered descent to the moon's surface. Before attempting that "leap into the unknown" as the boosters of space travel would say, a final check of possible inertial platform drift was made. To do this the observed angle between a selected axis of the inertial platform and line of sight of the sun was compared with the onboard computer prediction. We can assume there was general relief to find that all angles were within acceptable agreement. Estimates of the altitude of the module at the ignition point for descent based on line-of-sight rates of landmarks put it at 15.8 km above the lunar surface. At the appointed time ignition began initially at the lowest thrust and gradually increased to the programmed maximum after 26 s. At an altitude of 14 km, about 4 min after ignition a yaw maneuver was executed to bring the module to the landing orientation, at which point the landing radar began reassuringly to receive echoes immediately; at that altitude a small discrepancy was found in altitude between the computer and radar of about 850 m. What must have been an alarming development occurred at 5 min 16 s after ignition when "the first of a series of computer alarms indicated a computer overload condition." This resulted in the computer display being "occasionally precluded." The thrust reduction occurred as programmed on time, but it was assumed the landing point would be off due to the computer malfunction.

The next phase is that of final approach in which control was transferred from *automatic* to *attitude hold* mode in order to correct attitude errors then automatic control mode was resumed. At this point it was evident that on the present course the landing would be among large boulders, therefore manual control was again assumed and the descent slowed to extend the trajectory until a more suitable site is found. Finally the module was maneuvered down range to a tolerably flat area surrounded by boulders on one side and craters on the other. As the craft descended down to about 30 m, so much dust was blown up that it obscured the surface, making it difficult to judge height and speed relative to the surface. In spite of all that the Eagle landed! Figure 17.3 shows the Grumman-built lunar module at that historic moment on the surface of the moon, July 1969.

Meanwhile Collins, in the command module orbiting the moon, had a relatively relaxed time; with systems operating normally; and with implicit faith in Mission Control, he was able to *sleep* at the allotted time. It is remarkable the way these manned missions integrated highly intelligent human beings as extensions of an artificial computer-based system to achieve extraordinary feats of navigation. It is also a testament to the faith these astronauts had in the competence of each other and the people at Mission Control. All attempts by Collins to make out distinctly the

Fig. 17.3 The Lunar module on the surface of the moon, 1969 (NASA photograph)

lunar module from 110 km above the landing site on the moon's surface below evidently failed in spite of the fact that he had several minutes in which to make the observation. This is attributable to the uncertainty in the location of the lunar module being considerably larger than the area Collins could view in any one pass.

The activities of the astronauts on the moon's surface were of course the main objectives of the mission and certainly the most spectacular for the general public; however our interest is mainly in the navigational accomplishments of the mission. From that point of view there are really two crucial phases we should address: the rendezvous of the lunar module back with the command module and the insertion of the combined spacecraft into the homeward trajectory. Preparations for the ascent to the command module begin with a check on the orientation of the lunar module. For this purpose the crew is provided with a guidance and navigation (G&N) manual which lists 41 objects that are suitable for establishing the orientation. The lunar module guidance computer (LGC) stored a list containing the coordinates of 37 stars plus the earth, sun, and moon, plus another suitable object to be defined. A special purpose telescope, called the Alignment Optical Telescope (AOT), was used to establish the angular position of the chosen celestial objects with respect to the axes of the spacecraft; observations on at least two such objects are needed to determine the orientation of the spacecraft on the lunar surface. The AOT provided six fields-of-view, each 60° wide, set on six discrete axes called detents, separated by 60°, and arranged at 45° to the vertical. In the case of Apollo 11, because of radiation from the sun and earth only two detents were usable and of these only one was deemed sufficiently near the center of the field of view to be accurate. Therefore it was decided to use the option of only one celestial sighting and replace the second one by the direction of the lunar gravity field, as determined from the accelerometers in the inertial measurement unit (IMU). After completion of the platform alignment the appropriate navigation program was entered into the primary navigation computer for the ascent to the command module. After checks of the reaction control system, the abort guidance system calibration, and another check of the platform alignment, the ascent fuel tanks were pressurized with helium, and the rendezvous radar placed in the antenna slew position. The lunar module was ready to leave the moon's surface leaving behind, an optical retroreflector for lunar ranging from the earth, a seismometer to monitor vibrations and a commemorative plaque to mark the visit of humans to another planet.

The ascent engine ignition was smooth and uneventful, as were the maneuvers of the module, while the actual path of the module conformed exactly to what was predicted. The module was inserted into an elliptical orbit about 87 × 18 km, as determined by the primary guidance system, and confirmed by the ground MSFN network. These orbital values agreed approximately with those obtained by the abort guidance computer, which, it should be pointed out, was quite different in design and manufacture from the primary guidance computer, with a different instruction set and hardware, and was even developed by a different laboratory. The next challenge is to rendezvous with the command module.

Meeting up with the command module means executing a programmed sequence of impulsive burns leading to the placement of the lunar module in sufficiently close proximity to the command module to effect the docking maneuver. This is unquestionably the most exacting and critical maneuver involving a complex problem in planetary motion, a maneuver that is going to test the ultimate capabilities and coordination between the crew and the ground support system. The astronauts were familiar with the procedure and the practical aspects of carrying out the maneuver; in fact it was Neil Armstrong with David Scott who manned the Gemini 8 mission in 1966 to dock with an unmanned Agena target vehicle and carry out extra-vehicular activities. That was the first *manned* docking operation, as opposed to just rendezvousing (if that is a word); the first Russian manned docking operation was between Soyuz 4 and Soyuz 5 in 1969, a mission in which crews were exchanged. The experience derived from the Gemini project provided invaluable information for the optimal lunar module design for the achievement of its rendezvous with the command module. From that experience two fundamental decisions were made (1) only one of the modules, the lunar module, would be maneuvered while the other remained passive and (2) to reduce, as far as possible, the operations of rendezvous and dock to a fixed routine with preplanned choice of variables to minimize the need for independent judgments by the astronauts.

The sequence of maneuvers adopted for Apollo11, chosen presumably for its energy efficiency, is given the name *coelliptic*. To make sense of the coelliptic approach, and how it achieves a rendezvous and dock with the control module, it is useful to recapitulate Kepler's planetary theory which we met in Chap. 3. We recall that Kepler's third law of planetary motion states that the *square* of the period of the orbit around a gravitating body is in direct proportion to the *cube* of its major axis, or in the case of a circular orbit, the cube of the radius. If T is the time to complete one revolution and R the radius of a circular orbit around a mass M, then the third law states:

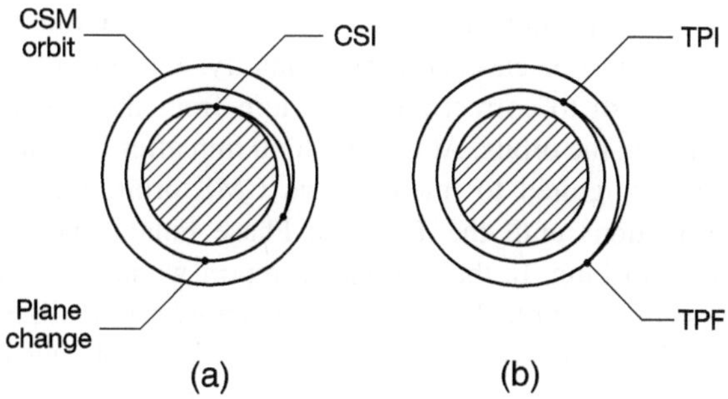

Fig. 17.4 Orbital coelliptical maneuvers of LM to dock with CSM

$$T^2 = \frac{4\pi^2}{GM}R^3, \tag{17.1}$$

where M is the mass of the gravitating body, in this case the moon, and G is the gravitational constant. It follows that the linear velocity V along the orbit is given by

$$V = \sqrt{\frac{GM}{R}}, \tag{17.2}$$

that is, the velocity is smaller in larger radii orbits than in smaller ones. It follows that in attempting to rendezvous two orbiting spacecraft, they obviously must approach the same orbit and therefore the same velocity. This raises the question as to how a spacecraft can be made to change the radius of its orbit, assuming for simplicity that it is circular. We already know how this can be done: we have already encountered the Hohmann maneuver. We recall that it consists of applying a calibrated forward impulse to the space craft in the direction of its orbital velocity by a short thruster burn; this results in an elliptical orbit with the major axis dependent on the magnitude of the impulse, with the farthest point (the apogee) at a diametrically opposite point. If another short forward impulse is applied at the apogee, a circular orbit can result with a larger radius than the original. Similarly if retrograde impulses are applied the spacecraft is transferred to a smaller orbit.

The coelliptic method achieves its goal of docking through a succession of steps allowing opportunity to carefully confirm and correct the results of each step and providing the possibility of alternative steps or even an abort, if that proved necessary. While the lunar module is executing these maneuvers, the command module maintains a 110 km circular orbit and constantly updates its own computer with navigational data and the position and velocity of the lunar module. The sequence of operations for the coelliptical method of lunar orbit rendezvous and dock is summarized in Fig. 17.4.

The defining phase in executing a coelliptical sequence is the placement of the active vehicle (LM) in an intermediate *circular* orbit 28 km below the circular orbit of the passive vehicle (CSM). This is achieved by the Coelliptic Sequence Initiation (CSI) burn that follows the initial launch from the moon's surface, placing the LM in the first elliptical orbit. After the CSI operation the planes of the two circular orbits must be checked for coplanarity, and corrected if necessary; this is followed by a burn to correct any departure from the planned 28 km altitude difference between the orbits. At about 2 h 40 min after launch, the Terminal Phase Initiation burn is ignited to intercept the CMS orbit, followed by two small midcourse corrections until the Terminal Phase Final, when braking is applied leading to final docking. In the astronauts' report we are told that in fact the abort guidance system, mentioned earlier, was in fact used to realign the platform during docking.

The success of this whole exercise hinged of course on the radar providing range and range rates as well as bearings, and no small part was also played by onboard computers, primitive by today's standards though they were. The subsequent transearth injection maneuver followed the jettison of the lunar module after all samples had been retrieved and stored in the command module. The injection into the homeward bound coast phase required the last firing of the service propulsion engine. The coast itself was relatively uncomplicated with two midcourse corrections, and the astronauts could relax since the remaining challenges were familiar, and thoughts of home buoyed their spirits. History had been made.

17.4 Navigation to Distant Planets: The Mariner Missions

The history of interplanetary exploration by the USA begins with the formation of an aeronautical engineering group under the Hungarian born American, Theodore von Kármán, at the California Institute of Technology (Caltech). He was a professor of aeronautical engineering who gave encouragement to a group of students dubbed the *rocket boys* who were the founding members of what he named the Jet Propulsion Laboratory (JPL). The first proof of the group's unique abilities was the building of the first U.S. spacecraft to be successfully launched into space; we know it as Explorer I. It was sent aloft successfully by the Army's Ballistic Missiles Agency in 1958. A dominant figure in the future development of JPL was the New Zealander William Pickering whose interests went far beyond rocketry per se and sought a broader mission to carry out scientific exploration in space. The success of Explorer I no doubt whetted the appetite for more missions of that type, extending the range to reach all the planets of the solar system. As director, he therefore pushed to have JPL be part of the newly formed space agency, NASA, but remaining part of Caltech, in the way that many applied physics laboratories in the USA are often attached to universities. Another example in the satellite business is the Applied Physics Laboratory of the Johns Hopkins University. The conduct of the Explorer program itself was eventually given over to the NASA Goddard Space Flight Center, in Maryland.

In the years between 1962 and 1973 JPL designed and built a series of spacecraft for a program of inner solar system exploration named Mariner. These were relatively small unmanned remotely controlled craft carrying instruments to observe and transmit back to ground the results of basic physical measurements such as radiation and magnetic field intensities, high energy particle flux densities, as well a photographic record of the planets' surfaces. They were totally controlled by telemetry over enormous distances (Mars is never less than about 79 million kilometers from Earth!) made possible through the use of giant radio-telescopes, which will be further discussed later.

The first Mariner space probe programmed to visit Venus was built on a hexagonal frame about 5 m wide and 3.6 m high and weighed about 200 kg. It was launched in 1962 using an Atlas-Agena rocket engine, but unfortunately veered off course and had to be destroyed. However a duplicate craft was successfully launched about 1 month later designated Mariner 2, which performed satisfactorily on its path to Venus for about 14 weeks. On the way it relayed measurements of radiation and particulate matter, principally the high energy cosmic rays originating from outside the solar system, but more interestingly the so-called solar wind, which is a continuous stream of charged particles emanating from the Sun into space, which are deflected by the Earth's magnetic field leading to their concentration in the polar regions and the familiar aurora displays. Experiments designed to analyze the composition of solar wind on later missions, for example Apollo 11, yielded the generally expected result that it is mainly protons (hydrogen nuclei) and electrons with traces of the ions of heavier elements. It is the collision processes between the high speed solar wind particles and air molecules in the upper atmosphere that produce the spectacular auroral displays. The light emitting collision processes are the same as produce a glow discharge in air (or any gas) at low pressure when a high voltage is applied.

The problem of computing and then executing an interplanetary trajectory that is optimal in terms of energy expenditure is a daunting one; it must begin from a platform, the Earth, moving in its orbit at about 106,000 km/h relative to the fixed stars, and travel through a gravitational field that varies in space and time because of the motions of the planets, and finally achieve its objective of landing on, or orbiting a moving target! This obviously puts a heavy burden on the computational power available, the ability to navigate on an interplanetary scale, and long-range communication and control. The computation of the admissible trajectories leads to certain severely limited windows for launching the space probe; the optimum trajectory would be a Hohmann orbit with the Earth at perigee and Venus at apogee which occurs periodically every 583 days, termed the synodic period. The Mariner spacecraft were not designed to actually land on the target planet and therefore their trajectory must be precisely placed close enough to the target to obtain useful observations but not actually impact its surface for fear of biological contamination. Thus there is a narrow target zone perhaps on the order of 1,000 km from the surface of the planet within which the spacecraft path must be restricted.

The scale of distances in these missions is vastly greater than even the lunar landing; for example, the trajectory to our nearest neighbor, Venus, extends to a distance of about 330 million kilometers, orders of magnitude greater the distance to the moon. It follows that the journey takes proportionately longer and therefore the onboard instruments must keep functioning satisfactorily for a much longer time. A mission to Mars would have a far longer trajectory of about 602 million kilometers, requiring on the order of 8 months to complete. Over these distances the

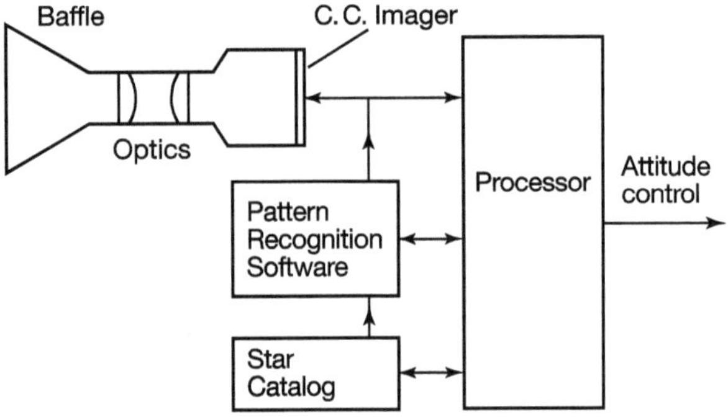

Fig. 17.5 The functional elements of a star tracker

environmental conditions are subject to great variation: for example, the solar constant, that is, the rate of incidence of radiant energy from the Sun per square meter may change by a factor of 2, requiring tight temperature control of sensitive electronic equipment. In addition, the spacecraft is constantly bombarded by energetic charged particles, cosmic dust, and meteorites. A variety of scientific instruments were carried on interplanetary missions including magnetometers, ionizing particle detectors, dust particle detectors, and radiometers for determining planetary surface temperatures.

The ability of a spacecraft to successfully pursue its course depends not only on being able to keep track of its progress but also to maintain the proper orientation in space. Thus the vital communication link between the ground tracking stations and the spacecraft critically depends on the onboard guidance and control system's ability to point the high gain antennas accurately back to the ground stations and to orient toward the Sun the solar panels on which the spacecraft depends for electrical power. But of particular interest to us is the navigational method used to keep track of the craft's position in order to determine when to apply midcourse corrections whether planned or as they become necessary. Through telemetry the navigational function was performed by a combination of a star tracker on board (see also Chap. 14) and the ground-based Deep Space Network (DSN) operated by JPL for NASA. Figure 17.5 shows the functional elements of a generic star tracker.

The star tracker used by the Mariner missions uniquely tracks the star Canopus in the southern constellation Carina with celestial coordinates of right ascension 6 h 24 min and declination $-52°41'$. It is the second brightest star in the sky after Sirius and has a relatively large observed angular diameter of 0.0066 in. measured by a special (intensity) stellar interferometer, a derivative of the Michelson stellar interferometer. The star is especially suitable as a reference direction because for trajectories in the ecliptic plane (orbital plane of the Earth) the sun is always nearly at right angles to Canopus, alleviating the problem of light shielding. Its location in the southern hemisphere makes it uniquely suitable as a reference for the roll attitude, that is, the angle about the craft's symmetry axis, with respect to the line of the sun. In a roll search of the tracker, the stars in an entire great circle belt may pass through the acquisition field of view; but since the sequence of stars in this belt is loaded in the DSN computers, it is possible to acquire Canopus by correlating it with the sequence telemetered down from the spacecraft.

The Deep Space Network is a facility of NASA operated by JPL. It presently consists of three deep space tracking and telecommunication stations distributed over the globe at roughly 120° apart in longitude for maximum length of baseline and coverage as the Earth rotates. One station is at Goldstone in the Mojave desert of California, one is near Madrid, Spain, and the third near Canberra, Australia. It provides the essential two-way communication for the telemetric control of unmanned spacecraft as they make their way to distant planets in the solar system. It tracks the position and velocity vector of the space craft and monitors the function of the systems on board, receiving data and transmitting commands both for corrective action or to pursue the planned flight path. Of course the purpose of the missions is to gather scientific data and therefore an important function of the network is to receive transmissions of data generated by the various instruments taken on a particular mission.

At each of the three locations there are at least five (Canberra and Madrid) or six (Goldstone) deep space observation complexes equipped with large parabolic antennas: one high efficiency antenna with a diameter of 34 m, another two or three 34 m antenna with a beam waveguide feed, and two more: 70 m, and 26 m diameter antennas. The *beam waveguide feed* of the second antenna mentioned is simply one in which the electromagnetic wave is not confined as in a standard waveguide where the walls impose boundary conditions on the wave, but rather a series of reflectors are used to direct the energy from one location to another. The advantage of this design is that it permits the placement of the electronics in a more convenient location. All the stations within a complex are remotely operated from one central Signal Processing Center; it is here that the telemetry data ARE processed and the navigational computations carried out and acted upon.

References

1. N. A. Armstrong, M. Collins, E. E. Aldrin, http://history.nasa.gov/alsj/a11/A11_MissionReport.pdf
2. R. Wheeler, Apollo Flight J., http://history.nasa.gov/afj/launchwindow/lw1.html

The Future of Navigation 18

18.1 Introduction

When we ponder what aspects of navigation may be radically transformed in the future, we are led to think about the most fundamental quantities navigation entails: to reach a given destination, we clearly require a means of measuring distances and directions in space as well as a way to integrate the path and project the expected arrival at the destination. Distances are now defined in terms of the propagation time of light, that is, the velocity of light is taken to be a constant and therefore time measurement has supplanted that of distance. It follows that future advances must be in the areas of the measurement of the propagation *time* of electromagnetic waves from one point to another, the measurement of *direction* with respect to an inertial coordinate system, and a rapid means of computing the integrated route. It is precisely in these three areas that fundamentally new developments are beginning to emerge that will transform navigation in the future. The first seemingly wild idea is to build a gyroscope using *matter* waves (de Broglie waves), rather than light waves as in the laser gyro, and the second is a new generation of atomic clocks using laser cooled atoms/ions, and lastly, in the field of computers we can look forward to *quantum* computers. Fundamental demonstrations of the feasibility of matter wave gyroscopes and quantum computing have already been published [1, 2] and laser cooled atoms in what are called *atomic fountains* are currently extant in several laboratory environments [3].

18.2 Matter Waves: The de Broglie Theory

Firstly let us recapitulate the de Broglie theory of matter waves. We recall in Chap. 8 we briefly traced Louis de Broglie's effort to reconcile the corpuscular behavior of atoms, acting like microscopic billiard balls in some circumstances, but with wavelike behavior in others. Light also clearly shows this duality of behavior: under some circumstances it exhibits the obvious properties of a wave as in

F.G. Major, *Quo Vadis: Evolution of Modern Navigation: The Rise of Quantum Techniques*, DOI 10.1007/978-1-4614-8672-5_18,

Young's double slit interference experiment, while showing a discrete particle nature in the photoelectric effect, where increasing the intensity of light falling on a photo-emitting surface does *not* lead to the emission of more energetic electrons, but rather a greater *number* of electrons. Incidentally Einstein was awarded the Nobel Prize in physics for his work on the photoelectric effect rather than relativity. It may be argued that perhaps light is a special case; after all, the success of the theory of relativity proves that it always travels with the same constant velocity, a unique property indeed. But the success of Bohr's theory of the hydrogen spectrum, in particular his hypothesis about stationary states, strongly suggests a resonance condition characteristic of wave behavior. Physics was passing through challenging times at the turn of the twentieth century. Using arguments based on Einstein's Special Theory of Relativity and Plank's concept of a quantum of energy, de Broglie was able to show that a description in which particle motion is represented by an accompanying wave can be constructed [4] using the same Lorentz transformation as Einstein. The material particle velocity is identified with the group velocity of the accompanying wave. He derived an expression that relates the momentum of a particle to the wavelength of its wave: the well-known formula $\lambda = h/mV$, where h is Plank's constant (6.626×10^{-34} J s), m is the particle mass, and V is its linear velocity, so that mV is the linear momentum. This formula may be written as $p = hk$, where $p = mV$ is the momentum vector and k is the wave vector. The de Broglie wavelength is extremely small even for a body as light as an elementary particle such as an electron; a quick calculation shows that for an electron of 1.0 eV energy the de Broglie wavelength is only about 1.2 nm, which, if it were radiation, would be in the X-ray region of the spectrum. Therefore the wave aspect of matter does not manifest itself in bodies on a human scale, where Newtonian theory still applies, but only on an atomic scale. The hypothesis of the particle-wave duality of matter was tested experimentally on electrons in a classic series of experiments by Davisson and Germer in 1927 in which they used an apparatus analogous to an X-ray diffraction experiment, as described in Chap. 7. A collimated beam of energetic electrons was directed at right angles on a plane metallic foil of nickel; they recorded the electrons that are scattered as they pass through the metal foil onto a photographic plate and found the resulting concentric ring pattern reminiscent of Debye X-ray diffraction patterns. In subsequent years similar diffraction patterns were obtained with other beams of atomic particles such as neutrons and helium. In the case of helium, Stern used a LiF crystal as the diffracting element because its relatively small lattice spacing gives rise to greater diffracting angles, which alleviated the collimation requirement on the helium beam. In all cases the wave behavior of material particles was confirmed and the de Broglie formula numerically validated.

Quantum theory has come a long way since the corpuscular-wave theory dispute, and the concept of a wave packet traveling at the group velocity determining particle motion conforms to modern quantum mechanics in a broad sense, and the de Broglie wavelength of a particle remains a valid quantity when discussing such wavelike behavior as the interference and diffraction of matter waves. We recall that the term interference in this context refers to the phenomenon observed when two waves overlap, resulting in a pattern of peaks and valleys that depends on the relative *phase* of the two waves at different points. In order to be able to observe interference in the present sense between two beams of atoms, each atomic wave must be *coherent,* that is, the waves associated with the atoms making up the beam must exhibit a deterministic, nonrandom, *phase* relationship in their oscillations. To generate such a coherent atomic beam presents a formidable task when you remember that a hydrogen atom for example at room temperature has a de Broglie wavelength on the order of 10^{-10} m; this is about twice the radius of a hydrogen atom itself! Clearly the de Broglie wavelength must be drastically extended by cooling the atoms to unprecedented levels of low temperature—to the microkelvin range. Indeed the reason we are even talking about atom interference is that we know that we have the techniques for achieving just such extreme temperatures.

We dealt at some length in Chap. 11 with the subject of using laser light scattering for the cooling of atoms and ions. We recall that Doppler cooling is effective down to a temperature T_{D} given by the following:

$$T_{\mathrm{D}} = \frac{h\Delta\nu_{\mathrm{n}}}{4\pi k_{\mathrm{B}}}, \tag{18.1}$$

where $\Delta\nu_{\mathrm{n}}$ is the natural frequency width of the atomic line, and k_{B} is the Boltzmann constant. Temperatures below the Doppler limit are attainable using optical polarization methods [5] with polarized light (see Chap. 11); quantitatively the degree of cooling of this mechanism is set by the shift in the atomic energy level by the action of the laser light, a shift proportional to the laser light intensity and inversely as the frequency detuning from the resonance line center. However any process involving the absorption or emission of single photons is ultimately limited by the recoil energy of the atom upon emitting or absorbing *one* photon, which amounts to the following:

$$\Delta E_{\mathrm{r}} = \left(\frac{h}{2\pi}\right)^2 \frac{k^2}{2M}. \tag{18.2}$$

For an alkali atom such as Rb this limit corresponds to a temperature on the order of a few hundred nano-Kelvin. This would seem to be an unbreachable fundamental limit; however using processes involving *two* photons each, typified by a technique using the Raman effect, even lower temperatures have been achieved by methods classified as sub-recoil cooling.

Before describing the sub-recoil cooling scheme called *velocity selective coherent population trapping* (VSCPT) let us recall the Raman light scattering phenomenon. First noted by the Indian spectroscopist Raman in 1928 in the scattering of light by molecules, this is a *nonlinear* light scattering process that differs from Rayleigh scattering by air molecules, in which the incident and scattered photons have the same wavelength, and differs from the resonance scattering of photons familiar in optical pumping, in which the incident wave must match an atomic transition frequency. In Raman scattering the frequency of the incident beam is not

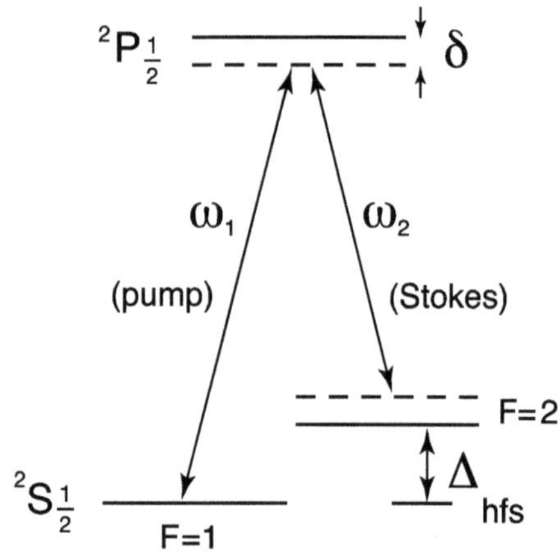

Fig. 18.1 A two-photon Raman interaction with three levels of a Rb atom

resonant with any absorbing transition, and the scattered photons are not all the same frequency as the incident photons. But what made Raman scattering so important to spectroscopists is the fact that the scattered photons contained not only the incident wavelength but also photons with wavelengths displaced from it by frequency intervals characteristic of the *scattering molecule*. In the language of electrical engineers it is as if the molecule modulated the incident light wave by its own molecular motion; in other words the base frequency of the molecular spectrum is shifted from its actual infrared frequency to the incident frequency, the hallmark of nonlinear behavior. This was of great practical importance since molecular spectra contain infrared wavelengths that were hard to measure directly.

Let us consider a two-photon stimulated Raman process applied to three energy levels of the Rb atom: the two hyperfine sublevels of the ground state and one excited state forming a Λ (lambda) configuration. Assume two coherent laser beams whose frequencies ω_1 and ω_2 correspond to transitions between the common upper energy level near the quantum P-state and the two lower levels, one coinciding with a hyperfine sublevel and the other detuned slightly, by an amount $\Delta\nu$, from the other hyperfine sublevel, as shown in Fig. 18.1. Note that the laser beams are *not* tuned to make real transitions between the ground states and the excited P-state.

It is observed that under the right conditions the rate of absorption/emission, that is, photon scattering, falls to *zero* at resonance when $\Delta\nu = 0$; the atom has been pumped into a coherent *superposition* of the two ground hyperfine states, and there is destructive interference between the transition amplitudes between the two states and the excited state. This quenching of the emission is an example of what is called *coherent population trapping*. To use this phenomenon to cool atoms, the frequency detuning $\Delta\nu$ in the above argument is made dependent on the velocity of the atom by placing it in two oppositely directed laser beams and selecting the laser frequencies so that $\Delta\nu = 0$ for an atom at rest. If the actual velocity of the atom is V then the opposing Doppler shifts of the two beams lead to a detuning of $\Delta\nu = (k_1 + k_2)V$. The way this approach circumvents the recoil energy limit on the achievable low temperature is by the suppression of further photon scattering at

$V = 0$. Quantum analysis of this cooling process [6] leads to the prediction that the velocity distribution that results has two sharp peaks at $+hk_1/2\pi$ and $-hk_2/2\pi$; the *widths* of these peaks correspond to temperatures more than 100 times lower than the recoil limit, down to the *nano-Kelvin* range.

Another atomic gas cooling technique that has been generally used to continue cooling beyond the limits of the laser techniques is called *evaporative cooling*, which is based on the thermodynamics of a gas. The principle of evaporative cooling is relatively simple: it consists of isolating a volume of atoms that have been cooled to the limit of laser cooling, and a forced evaporative cooling sequence is performed. This consists of allowing the isolated atoms to come to thermal equilibrium and then forcing the most energetic to *leave* the trap. This creates a nonequilibrium distribution of energy among the remaining atoms which subsequent collisions among them will restore to thermal equilibrium at a *lower temperature*. The crucial point is that the new thermal equilibrium requires a redistribution of energy among the atoms in which more atoms reach lower temperatures. Of course this procedure can only go on so far, eventually there will be very few atoms left, but they will be cold! Realistically one may start with say 10^{10} atoms at a temperature of 10 μK and end up with 10^7 atoms at a temperature of 100 nK, a reduction of temperature by a factor of 10^2.

Using this technique in 1995 the temperature of an atomic gas was cooled to a historic low point where the theoretically predicted Bose–Einstein condensation occurs. This thermodynamic quantum phenomenon was an early prediction by Einstein on the basis of the quantum statistical theory developed by the Indian theoretician Bose. According to quantum statistical mechanics a distinction must be made between a gas composed of particles whose wavefunction representing them must be symmetric (no change in sign) and antisymmetric (change in sign) with respect to an exchange of any two particles. As we saw in Chap. 8, particles with spin ½ such as electrons require an antisymmetric wavefunction, which leads to the Pauli exclusion principle that is central to the "building up" principle in atomic structure theory. But what are of interest here are particles that require the total wavefunction to be *symmetric* with respect to an exchange of particles; this applies to particles, such as photons, that have an integral total spin, called bosons. In their case not only is there no exclusion principle, but a given quantum state can have *any number* of particles in it. This manifests itself in the equilibrium thermal energy distribution law for the particles, first derived by Bose. It was Einstein who pointed out that as the temperature of a cloud of these bosons is lowered and the energy of each particle continuously reduced there will come a point where all the particles are in one state, the lowest state, now called a Bose–Einstein condensate (BEC).

The continued application of cooling techniques on a defined group of atoms to reach such extremely low temperatures clearly requires a method of isolating that group from its ambient warm environment. This can be accomplished with an arrangement of lasers and magnetic fields called a magneto-optical trap (MOT), illustrated in Fig. 18.2. We recall that when two counter-propagating laser beams are scattered from the same atom, the effect on the motion of the atom is like a force of resistance proportional to the velocity, opposing the motion of the atom in both

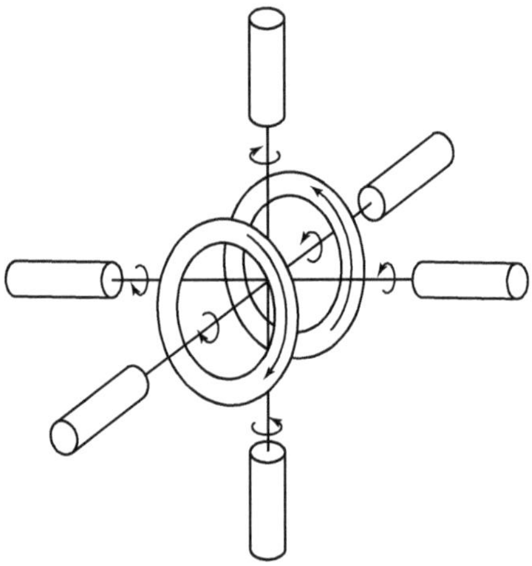

Fig. 18.2 Schematic drawing of a magneto-optical trap (MOT)

directions. The magneto-optical trap incorporates the same basic mechanism in a more elaborate system that uses *circularly polarized* laser beams and a magnetic field to both confine and cool the atoms. It consists of three mutually perpendicular counter-propagating laser beams which intersect at the center of the trap, and a quadrupole magnetic field whose axis is parallel with one of the laser beams. The field is commonly produced by passing a current in opposite directions through a pair of circular coils mounted coaxially with one of the laser beams, at equal distance from the center. The field has a constant gradient in intensity, passes through zero at the center of the trap, and then reverses direction. The function of the magnetic field is simply to shift the resonance frequency to compensate for the changing Doppler shift as the atoms slow down. To understand how this configuration cools and traps an atomic gas, we will analyze the function of the magnetic field with reference to Rb^{87} atoms. To do this let us recall the hyperfine structure of the ground state energy levels of that isotope, particularly the dependence of the magnetic sublevel energy on the magnetic field in the "weak field" region, as shown in Fig. 18.3.

We note again that the energies of the $|F,m_F>$ sublevels $|2, \pm2>$ and $|3, \pm3>$ vary approximately as linear functions of the applied magnetic field provided it is weak enough that it causes a magnetic splitting small compared with the nuclear hyperfine frequency. Therefore since the magnetic field in the magneto-optical trap is designed to vary linearly with distance along the axis of the trap, it follows that the transition frequency for example between $|2,2>$ and $|3,3>$ also varies linearly with *distance* from the center where the magnetic field is zero. Let us assume the axis of the magnetic field coils is the z-axis of a Cartesian coordinate system and assume the laser is tuned to resonance with the $|2,2> \rightarrow 3,3>$ transition at the origin, where the magnetic field is zero, then an atom with a positive z-coordinate moving in the positive z-direction will have a resonance frequency higher than the laser, and therefore in order to have the positive Doppler shift in the laser frequency necessary for resonance, the atom must interact with a laser beam traveling in the

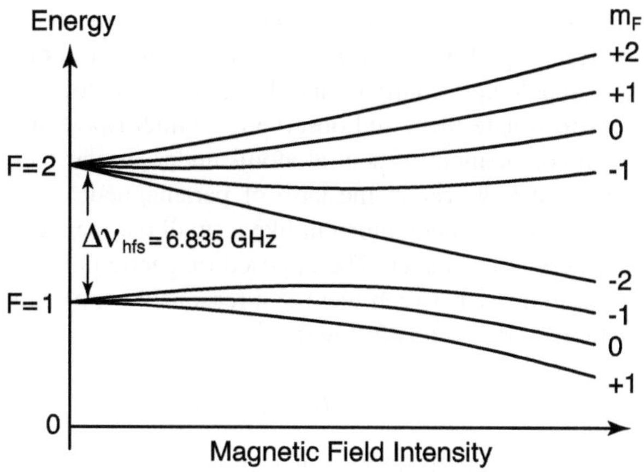

Fig. 18.3 The energy as a function of a weak magnetic field of some hyperfine levels of Rb^{87}

negative z-direction and be impelled toward the origin. The probability of a photon being actually absorbed and reemitted is determined by quantum selection rules for "allowed" transitions that involve the changes in the magnetic quantum number m_F and the sense (clockwise or counterclockwise) of the circular polarization σ_+ and $\sigma_=$ of the photon. The basis of these rules we recall is the law of the conservation of angular momentum applied to the atom–photon system. The photon has the unique property of having *one* unit of angular momentum $(h/2\pi)$ and yet has only *two* possible orientations with respect to a given axis, corresponding to clockwise (σ_+) and counterclockwise (σ_-) rotation (Only spin ½ particles such as electrons have just two orientations). Let us assume that the laser transitions occur between the ground electronic state represented by the spectroscopic notation $5^2S_{1/2}$ and the excited state $5^2P_{3/2}$ corresponding to a laser wavelength of $\lambda = 780$ nm. The conservation of angular momentum requires for example that an increase in m_F by one unit accompanies the absorption of light having σ_+ circular polarization; thus,

$$|F = 2, m_F = 2 > \rightarrow |F = 3, m_F = 3 >; \Delta m_F = +1(\sigma_+ \text{polarization}). \quad (18.3)$$

It follows that in the example described above an atom in the $/F = 2, m_F = 2>$ would interact with a σ_+ laser beam directed in the negative z-direction and make a transition to the $|F = 3, m_F = 3>$ state.

It should be noted that the cooling/confinement mechanism described above has a different dependence on the motion of the atom from Doppler cooling that we discussed in Chap. 11. In the present case the spatial variation of the retarding force is determined by the gradient in the magnetic field, which is designed to be constant, making the force depend on position rather than velocity as in the Doppler cooling. In the magneto-optical trap we have a restoring force that gives rise to containment, whereas in Doppler cooling the cooling force depends on velocity, more in the nature of a viscous force.

The discussion so far has concentrated on the cooling function of the magneto-optical trap. Equally important is the atom trapping properties of a magnetic quadrupole field configuration. In Chap. 10 the focus was on the trapping of atomic *ions* to isolate them and observe their microwave spectrum free of perturbations. Such confinement of neutral atoms under similar thermal conditions is out of the question; however in the case of paramagnetic atoms, such as Rb, that have a permanent magnetic moment like a small bar magnet, at microkelvin temperatures that is another matter. The opposed magnetic fields of the two coils with the null point midway between them is a reasonable approximation to a quadrupole magnetic field defined as follows:

$$B_r = Ar, B_\theta = 0, B_z = -2Az. \tag{18.4}$$

In describing the motion of a Rb atom in such a field, the physical conditions are such that we may make the assumption that the atom interacts with the magnetic field only through the magnetic moment of its outer electron. Moreover effective magnetic trapping of the atom requires that, as it moves through the spatially variable magnetic field of the trap, it maintains the same orientation, that is, the same magnetic quantum state, with respect to the magnetic field. Such a time-dependent interaction between atom and field that is so slow that it does not lead to transitions to other states is referred to as being *adiabatic*. This requires, using the vector model, that the atomic electron spin vector must precess around the magnetic field direction at a much higher frequency than the rate of change in the frequency itself. If that condition is violated then the electron spin direction can flip, that is, it may make a quantum transition to point in the opposite direction with a reversal of the force. Such transitions are named Majorana (pronounced "mayorana") transitions, named for a brilliant but tragic Italian physicist who lived during the period of early development of quantum theory. But the quadrupole field has a null point at the center where the precession frequency passes through zero; at that point the adiabatic condition would be violated even by a slow atom, and transitions would occur to the opposite spin state. To avoid that occurrence, a solution that has become standard is to introduce another time-varying *uniform* magnetic field that *rotates* about the quadrupole axis, referred to as a *time-orbiting potential* (TOP). Such a field of magnitude B_T rotating with angular velocity Ω has components given by the following:

$$B_x = B_T \cos(\Omega t); B_y = B_T \sin(\Omega t). \tag{18.5}$$

To be effective this field must have a frequency Ω that exceeds the frequency of rotation of the atoms in the x–y plane. Under the assumed adiabatic condition in which the magnetic moment remains at all times pointing in the opposite direction to the magnetic field, the motion of an atom is readily derived using the potential energy function given by

$$U = \mu B(x, y, z), \tag{18.6}$$

where μ is the magnetic moment of the atom, and B the magnitude of the magnetic field, given by

$$B^2 = A^2\left(x^2 + y^2 + 4z^2\right). \tag{18.7}$$

We can take advantage of the cylindrical symmetry of the field and write $(x^2 + y^2) = \rho^2$, and then the energy expression becomes

$$U = \pm\mu A\left(\rho^2 + 4z^2\right)^{1/2}, \tag{18.8}$$

and the equations of motion are simply given by the following (for spin antiparallel):

$$m\frac{d^2\rho}{dt^2} = -\frac{\partial U}{\partial \rho}, \quad m\frac{d^2 z}{dt^2} = -\frac{\partial U}{\partial z}. \tag{18.9}$$

The combined action of the static and rotating fields on an atom may be radically simplified by taking a time average approximation which reduces the problem to a time-independent Hamiltonian, thereby simplifying the solution in quantum mechanics. Indeed since these traps are intended to confine atoms under extremely low-temperature conditions, their motion must be treated using quantum theory. However the time dependence of the field cannot be totally ignored; in fact the atoms display a residual *micromotion* reminiscent of the Paul trap for ions. The simplification of assuming adiabatic conditions and time-independent average motion obscures the more detailed behavior, for example, an unexplained vertical shift of the center of the atomic cloud upon reversal of the sense of the rotation of the oscillating field.

18.3 Atom Interferometry: Diffraction Grating

The implicit requirement of interferometry is that there be two beams that are sufficiently coherent to interfere and produce a pattern from which useful knowledge can be gained concerning their relative propagation history. This is generally achieved by dividing, or *splitting,* one coherent beam into two coherent beams that can then be recombined and compared in phase after having traveled different paths. In the case of light, the act of splitting a beam while preserving coherence is almost trivial—all one needs is a polished flat plate of glass. However in the case of a coherent beam of atoms the available methods are considerably more complicated.

The classical diffraction grating used in spectroscopy, which consists of a large number of equally spaced lines or slits in a screen, does cause the separation of an input wave into many *coherent waves* diffracted in different directions; however, in the case of atoms the de Broglie wavelength, even at very low temperatures, is so small that it requires nanotechnology to machine a suitable grating. It is therefore a tribute to the experimental skills of the team at the Massachusetts Institute of Technology that a transmission grating was indeed successfully built and tested on a beam of sodium atoms; the de Broglie wavelength was 17 pm (17×10^{-12} m) and the spatial period of the grating was only 200 nm etched in a gold membrane.

18.4 The Kapitza–Dirac Effect

Because of the difficulty of dealing with such micro-gratings, attention turned to the use of lattices of scattering centers just as von Laue had argued for the use of crystal lattices for X-ray spectroscopy. One matter-wave technique applies a physical consequence of quantum theory first pointed out in 1933 by the anti-Soviet Russian physicist Kapitza, noted for his work in magnetism and low temperature physics, and Dirac, one of the titans of quantum theory [7]. It is known as the Kapitza–Dirac effect. Simply stated, the effect is the diffraction of an electron beam by the periodic optical field pattern of a standing light wave acting as a diffraction grating, as illustrated in Fig. 18.4. It is analogous to the diffraction of an X-ray beam by the periodic structure of a crystal which results in the von Laue patterns: a case of the roles of radiation and matter being interchanged. However a distinction is drawn between what has been described as a "thin" diffraction grating and a "thick" one; the former is applied to the Kapitza–Dirac diffraction which is characterized by many orders of diffraction from one scattering plane, while the latter is called Bragg diffraction involving more than one scattering plane. The difference arises from the manner in which they are observed: Kapitza–Dirac diffraction occurs with tightly focused laser beams and the simultaneous conservation of energy and momentum allows many diffraction angles, whereas in the case of Bragg diffraction only certain angles of incidence of a parallel laser beam lead to a reflection.

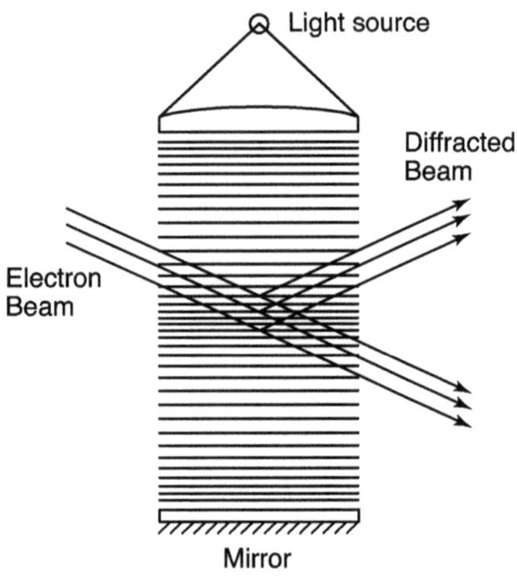

Fig. 18.4 The diffraction of an electron beam by a standing optical wave: the Kapitza–Dirac effect

A numerical estimate of the intensity of the diffracted free electron waves using a light field constructed from a conventional light source, such as a mercury arc, quickly shows that the effect is far too weak to be observable as originally envisioned. However the situation was radically changed with the advent of powerful coherent laser sources of light, and several laboratories were encouraged to try to demonstrate the effect. Success with an electron beam was achieved by Batelaan and his coworkers [8] at the University of Nebraska in the USA in 2001. But however great may be the interest in elementary particle–radiation interaction, or technically outstanding this achievement was, the problem at hand remains the coherent diffraction of *atomic* waves.

The experimental challenge in observing such scattering with atoms is the smallness of the angle of scattering; for example, a rough calculation for Rb at a temperature of 1°K using an optical laser wavelength yields a scattering angle on the order of only 10^{-6} rad. Nevertheless scattering of a supersonic beam of sodium atoms was observed by P. J. Martin et al. at MIT in 1988 [9]. An extended standing light wave pattern was used acting as a light diffraction grating. The maxima of the intensity of diffracted atoms were observed to occur in directions that conform to Bragg's law, namely,

$$n\lambda_{\mathrm{dB}} = 2d\sin\theta_{\mathrm{n}}, \tag{18.10}$$

where λ_{dB} is the de Broglie wavelength of the atom, d is the spacing of the light pattern, and θ_{n} the Bragg angle of reflection of the nth order. Substituting $\lambda_{\mathrm{dB}} = h/p$, $\sin\theta \approx \theta$, and $d = \lambda_{\mathrm{L}}/2 = \pi/k_{\mathrm{L}}$, the spacing of the standing wave maxima produced by the laser field, we find $\theta_n = n(hk_L/2\pi p)$. The integer n is called the order of the Bragg scattering and both first- and second-order reflections have been observed.

The experience of these experiments prepared the way for future atom interferometry to be conducted on atoms at the lowest possible temperature: the state in which *all the atoms* are in their lowest possible quantum state, that is, a Bose–Einstein condensate (BEC). In that state the velocity of the atoms is so low and their de Broglie wavelengths so long that the diffraction angles of *atoms* from an optical standing wave would be well within experimental reach. This circumstance has prompted the attempt to demonstrate atomic diffraction not by an atomic beam interacting with a stationary light field, but initially by the converse use of a "traveling" optical field scattered by a stationary BEC [10]. The traveling optical field can be created simply by having two laser beams propagating in opposite directions and slightly detuned from each other. Thus assume for example the following field composed of two running waves along the x-axis,

$$E(x,t) = E_0\cos(k_1x - \omega_1t) - E_0\cos(k_2x + \omega_2t), \tag{18.11}$$

where $(\omega_1 - \omega_2) = \Delta\omega$ and $k_1 - k_2 = \Delta k$ are small. We can rewrite $E(x,t)$ as follows:

Fig. 18.5 Laser scattering by a Bose Einstein condensate

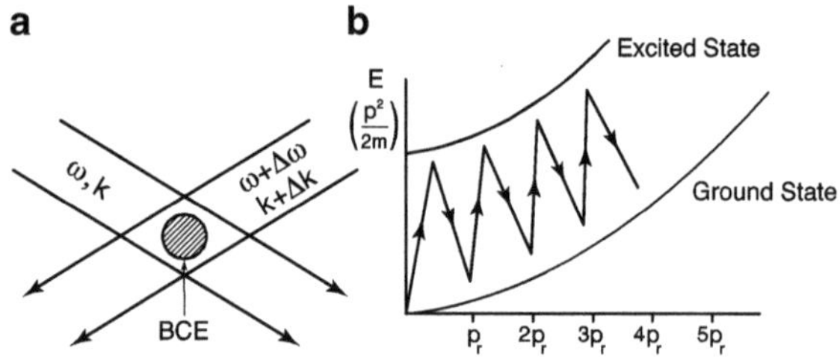

$$E(x,t) = 2\sin\left\{\frac{1}{2}(k_1 + k_2)x - \frac{1}{2}(\omega_1 - \omega_2)t\right\}\sin\left\{\frac{1}{2}(k_1 - k_2)x - \frac{1}{2}(\omega_1 + \omega_2)t\right\}, \tag{18.12}$$

which represents an amplitude modulated wave with the *envelope* traveling at a velocity of $\Delta\omega/k_{\text{ave}}$. Thus by tuning $\Delta\omega$ the velocity of the atoms relative to the optical lattice formed by the amplitude modulation is varied. Since the atoms scatter from a lattice moving with respect to the laboratory frame, the change in linear momentum is doubled.

A more general arrangement for the study of atomic scattering is depicted in Fig. 18.5. Two counter-propagating laser beams irradiate the BEC atoms; the atoms undergo stimulated two-photon Raman transitions in which they each absorb a photon from one beam and stimulated to emit it into the second beam with a net transfer of momentum and a change in the direction of motion. If the angle between the two laser directions is θ, then the net transfer of momentum to the atom is given by the following:

$$\Delta p = \frac{hk}{\pi}\sin\frac{\theta}{2}, \tag{18.13}$$

where the magnitude of the wave vector, k, is the same for both laser beams. The conservation of energy further requires that the following hold for a first order scattering involving the absorption/reemission of one photon:

$$\frac{(\Delta p)^2}{2M} = h\frac{\Delta\omega}{2\pi}. \tag{18.14}$$

It is this stimulated two-photon Raman scattering that is the basis of *beam splitting* in an atom wave interferometer. Thus by applying the laser excitation in the form of a pulse of appropriate duration the atom is left in a coherent mixture of two momentum states of the two ground hyperfine states. The use of pulsed laser fields raises the question of whether there is an optimum duration of the pulses and if so, its dependence on the laser intensity. Quantum transitions between two states induced by an oscillatory perturbation generally cause the system to oscillate between the upper and lower energy state; therefore the common way of specifying the strength of a pulsed excitation of defined duration is in terms of an angle; thus a π-pulse is one in which the system flips all the way from one state to the other. The imagery comes from the old magnetic resonance technique in which the direction of spins was involved. If the intensity and duration of the laser pulse is such as to leave the atom with *equal* probability of being in one sublevel as the other (½π–pulse) the atom beam is split into two directions because of the accompanying change in the momentum. If on the other hand the pulse length is doubled, leading to a π-pulse, the atom reverses momentum without changing state. The idea that a $\pi/2$ pulse can cause an atom to be in two different states at once is of course the same quantum mystery as Young's double slit experiment: there, in some sense, a photon can be at both slits at once. We will return to the topic of atom beam handling when we deal with interferometers in greater detail.

In experiments on Bragg diffraction by the group at NIST in Gaithersburg [10] a BEC is firstly formed in the way we described earlier, namely, by laser cooling in a magneto-optical trap an atomic gas, in this case sodium. The atoms are then confined in a combination of a static quadrupole and rotating *bias* magnetic fields where those in the magnetic quantum sublevel $F = 1$, $m_F = -1$ are further compressed and cooled by forced evaporation, until Bose–Einstein condensation occurs. Forced evaporation of the atoms is achieved by the application of radio frequency electromagnetic fields that, in conjunction with inter-atomic collisions, cause heating and loss of energetic atoms from the trap, leading to a subsequent thermal equilibrium at a lower temperature, but with fewer atoms. Starting with approximately 10^{10} sodium atoms, the final state was reported to consist of 10^6 atoms all in the condensed Bose–Einstein state, with characteristic frequencies along the coordinate x, y, z axes of 360, 250, and 180 Hz.

Having achieved Bose–Einstein condensation the next objective was to subject the atoms to scattering by the diffracting optical laser field. This was done by releasing the condensate from the trap, and letting it emerge in a beam. As already described, the diffracting field was produced by two laser pulses with frequencies ω and $\omega + \Delta\omega$ propagating in nearly opposite directions with a transitory overlap. To ensure that the two laser beams were coherent, they were derived from the same laser using acousto-optic modulators. As described earlier they were detuned from the resonant transition $(3S_{1/2}\, F = 1) \rightarrow 3(P_{3/2}\, F' = 2)$ to avoid resonance fluorescence and detuned from each other by $\Delta\omega/2\pi = -1.85$ GHz, which determines the fineness of the optical field "diffraction grating."

The method used by the NIST group to observe the atoms was to firstly optically pump them (see Chap. 8) into the $F = 0$ sublevel of the ground state and then form an image through absorption using a laser tuned to the optical transition from $F = 2$ in the ground state to $F' = 3$ in the first excited state. In this manner the liberated condensate can be followed as it moves as a unit. The essential point about this rather complicated procedure to produce a bunch of cold atoms is, as already emphasized, that the different orders of diffraction which emerge in different directions are *coherent* and therefore satisfy the essential requirement for interferometry.

18.5 An Atom Interferometer

By the time a patent was actually awarded for an atom interferometer back in 1977, there had already been expanding interest in exploiting its potential uses, including a new level of precision in gyroscopy, among other things. Many research groups around the world have been engaged in exploring new ways of improving the technique and expanding its application in new directions. The basic elements of a conventional interferometer whether optical or atomic are (a) a particle source, (b) a coherent beam splitter, (c) propagation test path, (d) phase correlator between the two waves, and (e) detector of the interference pattern of the recombined waves. In a sense the separated fields in a Ramsey cavity (see Chap. 8) could be thought of as an interferometer involving the hyperfine transitions of the atoms.

Perhaps the simplest interferometer configuration is an atomic version of the classical optical Mach–Zehnder type of interferometer. Mach is the famous Ernst Mach of the Mach's principle fame, a principle which stated that the dynamics of a body is affected by the action of all the matter in the universe. This point of view is thought to have influenced Einstein's thinking on the subject.

The classical Mach–Zehnder interferometer is illustrated in Fig. 18.6. Its basic function is similar to the more famous Michelson interferometer in which the amplitude of the light wave from the source is "split" into two waves ("arms") that travel similar paths before being brought together again to produce an interference pattern at the detector. Any small difference in the propagation delay between the two paths results in an observable shift in the interference pattern.

Of the many atom interferometer designs that have been studied at different laboratories we will begin with one published by Pritchard's group at MIT [11] that illustrates the essential features that typify the starting point for many elaborations. It uses three transmission gratings fabricated using nanotechnology to achieve the narrow slits required for the diffraction of de Broglie waves. It is notable in having achieved sufficient separation between the two beam paths to enable the introduction of phase-changing differences in the propagation along the two sides, for example, due to a rotation. The use of diffraction gratings properly designed can have the unintuitive property of producing a "white" fringe pattern that is independent of wavelength and therefore not widened by the atomic velocity distribution. A schematic drawing of this Mach–Zehnder type of atom interferometer is depicted in Fig. 18.7.

To achieve an intense narrow beam of atoms of sharp longitudinal velocity distribution a *supersonic source* was used. This is constructed by heating the subject element (in this case sodium) in a stainless steel chamber to a very high temperature that reaches 800°C, at which point the metal vapor pressure reaches 5 mmHg. High-pressure (2 atm) inert gas is introduced and the mixture is then allowed to undergo a supersonic expansion through a 70 μm diameter nozzle into vacuum, with the high-speed gas atoms acting as a carrier. A 500 μm diameter "skimmer" aperture removes divergent atoms, leaving atoms along the axis to propagate through the apparatus. A useful property of this type of source is that the velocity of the atoms depends in a calculable way on the *mass* of the carrier gas. Thus by using different

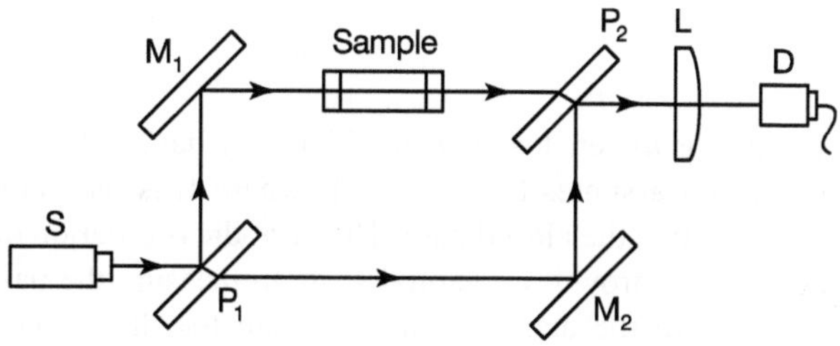

Fig. 18.6 Basic elements of an optical Mach–Zehnder interferometer

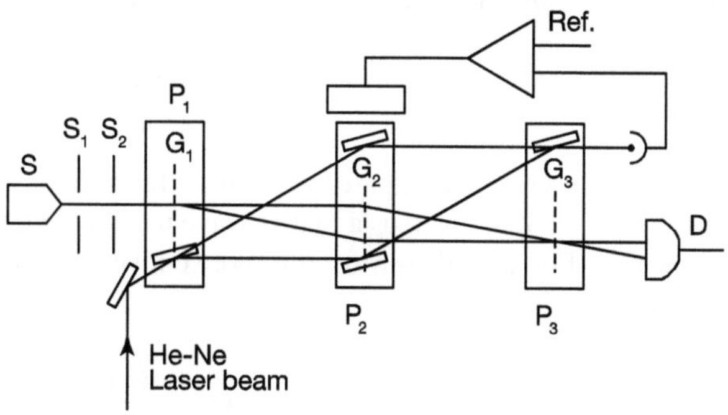

Fig. 18.7 Mach–Zehnder type of atom interferometer (taken from [10])

gases it is possible to realize the desired atomic velocity and therefore de Broglie wavelength.

Atoms from the source pass through collimating slits and strike the first diffraction grating which separates them into the zeroth and first-order diffracted beams that fall on the second grating; there they are diffracted reversing the deflections in momentum and forming an interference pattern on the third grating that acts like a screen. The gratings have slit spacing (period) of only 200 nm, or about one-third the wavelength of visible light. An auxiliary He–Ne laser is used as an optical interferometer designed to precisely fix the relative positions of the gratings.

18.6 An Atom Gyroscope

We recall that in Chap. 12 the subject of the Sagnac effect was introduced in the context of *laser* gyroscopes, which are in themselves a revolutionary change from the mechanical rotor type. There it was pointed out that since rotational motion defines a non-inertial system, a proper description requires the General Theory of Relativity; however, a classical treatment yields a correct first-order approximation. The result given there for the difference in time displayed by clocks traveling in the two possible senses (clockwise or counterclockwise) around a closed path in a frame of reference rotating with an angular velocity Ω was as follows:

$$\Delta t = \pm \frac{2\Omega}{c^2} A, \tag{18.15}$$

where A is the enclosed area, which may have any shape. As a first approximation this result assumes that $V/c \ll 1$, where V is the velocity with which the clocks move around the closed path. The periodic oscillation of a wave, whether optical or de Broglie, provides a natural scale to measure the passage of time, and therefore we can use the above result to obtain the difference in the phase $\Delta\varphi = 2\pi\nu\Delta t$ between matter waves going around the closed path in opposite directions. Thus using $h\nu = mc^2$ to obtain ν, we have the following:

$$\Delta\varphi = \frac{4\pi m\Omega A}{h}. \tag{18.16}$$

It is instructive to compare the phase difference for a light wave using the He–Ne laser at $\lambda = 632$ nm versus a matter wave using cold Cs atoms traveling around the same area. For an optical wave it is more convenient to express the phase difference in the following form:

$$\Delta\varphi = \frac{4\pi\Omega A}{\lambda c}. \tag{18.17}$$

Thus writing the phase shift in the form $\Delta\varphi = 4\pi\, C\, \Omega A$, we find for the He–Ne laser $C_L = 5.26 \times 10^{-3}$ and for cesium atoms $C_A = 3.3 \times 10^8$ an increase of a factor of 6.27×10^{10}! That is assuming equal area, which of course is totally unrealistic; on the other hand that large an improvement is not necessary to make it interesting—even a factor of 100 would be revolutionary.

This enormous potential advantage that atom interference has over laser gyros has spurred attempts by several research groups around the world to design systems that overcome some of the specific difficulties in realizing that advantage. Many have successfully experimented with different platforms and techniques. These may be categorized according to the operating temperature of the atoms, the way coherent atomic beams are formed, and the manner in which they are redirected to form a closed path in order to measure the Sagnac effect.

There are two groups at Stanford University actively engaged in atomic interferometry, one led by M. A. Kasevich devoted to rotation measurement, and the other concerned among other things with fundamental studies involving atomic fountains, led by the Nobel laureate S. Chu. In the year 2000 members of the Kasevich group were able to achieve sensitivity in rotation measurements reaching 3×10^{-8} degrees/s/Hz$^{1/2}$ using a long interference path. To be useful for vehicular navigation, ways must be found to increase the interference path without simply making the instrument longer. This requires large angle coherent beam splitters such has been investigated by Gattobigio et al. [12].

Finally we will consider one instrument designed as a compact gyroscope for mobile applications that was built by K.Takase at Stanford University [13]. It uses two vertically projected clouds of cold Cs atoms from magneto-optical traps in a

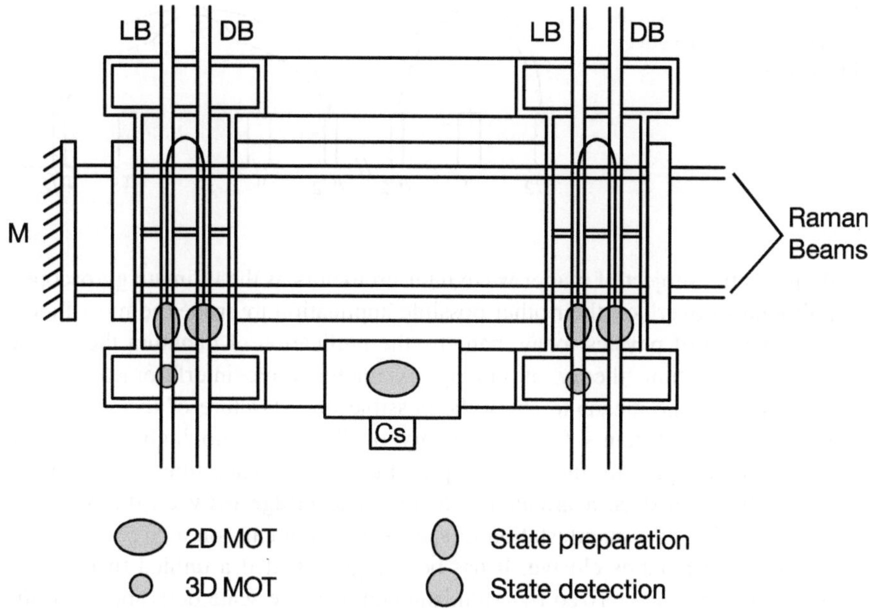

Fig. 18.8 Takase's mobile atomic interferometer for Rotation Measurement

manner reminiscent of the Cs fountain frequency standard. The Cs^{133} nucleus has a nuclear spin $I = 7/2$, and therefore the ground state nuclear hyperfine structure has two sublevels with total spin $F = 4$ and $F = 3$. The resonance frequency of a transition between these sublevels is *defined* as 9, 192,631,770 Hz: it *defines* the unit of time, the second. The atoms are subjected to two horizontal counter-propagating laser beams differing in frequency by 9.2 GHz, the hyperfine splitting between the two nuclear states to stimulate two-photon Raman transitions in a $\pi/2$–π–π–$\pi/2$ sequence of pulses, as shown in Fig. 18.8. The application of the Raman laser beams spatially manipulates the linear momentum vectors of atoms in these two states.

Under the action of gravity the atoms describe portions of parabolic arcs between changes induced by the Raman laser pulses and finally return to be detected by optical pumping at the bottom of their trajectory, as shown in Fig. 18.9.

From Fig. 18.9 we note how the application of the π and $\pi/2$ pulses affect atoms in these states. In the case of a $\pi/2$–π–$\pi/2$ sequence the atoms in the F–4 and $F = 3$ states describe a Fig. 18.8 with the Sagnac effect canceling out between them, whereas in the $\pi/2$–π–π–$\pi/2$ sequence the enclosed areas are different. Takase points out that imperfect π-pulses due to errors in their power or duration can lead to spurious trajectories with differing dependence on angular velocity leading to errors. Moreover such an error in the π-pulse makes the system sensitive to linear acceleration. The stability of the interferometer is given as within 6.6×10^{-3} degrees per hour.

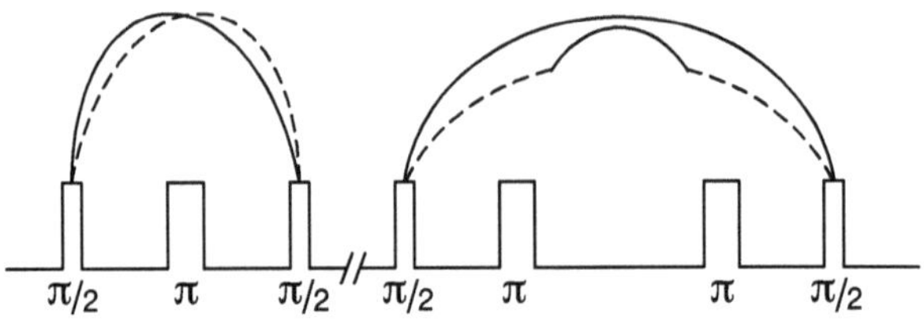

Fig. 18.9 Vertical arcs of the Cs atoms showing the effect of Raman pulses

While on the subject of matter wave interferometers as the ultimate gyroscopes, it is difficult to overlook their other possible application to probe the most fundamental problem of physics today, namely, the unification of quantum theory and Einstein's space-time theory of gravity. A matter wave interferometer would provide the ultimate means to study the possible fine structure of the space-time "continuum" of Einstein's General Theory of Relativity; no higher resolution method exists for probing the microscopic limits where space-time may manifest quantum effects, that is, a granular structure. The bridge between the relativity theory of gravity, as a manifestation of space-time geometry on a large scale, and quantum theory remains elusive. It has been argued that if a unified theory does exist, it must be characterized by a fundamental distance scale derivable from the known fundamental constants which characterize quantum theory, relativity, and gravity, namely, Planck's constant, the velocity of light, and Newton's gravitational constant, respectively. Following Planck, if we ask what is the magnitude of the fundamental scale of length that a unified theory might have, we find the value now called Planck's length, l_P, given by the formula [14]:

$$l_P = \sqrt{\frac{hG}{2\pi c^3}}.$$

This result may be obtained simply by using "dimensional analysis," that is, by expressing the units of h, G, and c in terms of the basic units of mass, length, and time and requiring that the resulting combination reduces to that of length. If we substitute numerical values we find l_P is about 10^{-26} nm! To all but the most indomitable physicists this result would halt any thoughts of trying to find experimental evidence of quantum effects in gravity. However proving the existence of quantization in a gravitational field does not require the measurement of Plank's length itself, but other large-scale consequences of it. Consider the analogous situation with the quantization of the electromagnetic field: even though the energy of a quantum of say sodium light is only about 3×10^{-19} J, the presence of quantization manifests itself in an easily detectable way through the black body spectrum, as discovered by Planck. In the case of a quantized gravity field, pictured as being granular on a Planck length scale, it would produce a random fluctuation of the phase of an atomic beam leading to a possibly observable effect. A comparison has been drawn between that fluctuation and the Brownian motion of a pollen particle suspended in water.

Several research groups have taken up the challenge to construct interferometers that can be carried aboard a freely orbiting earth satellite. This limitation to a satellite is essential for two reasons: first, a satellite is a "freely falling" system and therefore the gravitational field, aside from the "granularity" being sought, is zero, and second, a satellite environment can be made totally free of vibration.

18.7 Cesium Fountain Frequency Standard

We discussed in Chap. 8 the classical cesium beam frequency standard. We recall that the parameters that determine its accuracy and stability are (1) the total number of atoms observed and (2) the length of time during which the atoms are observed free from random perturbations. In that standard the atoms travel freely in the horizontal direction in high vacuum, and aside from rare collisions with background particles and a weak uniform magnetic field applied by design, the atoms are free of perturbations. To improve performance by increasing the number of atoms contributing to the signal would require that the vapor pressure in the source be raised, but this increases the velocity of the atoms, causing a reduction in the observation time for a given length of beam. In order to maintain the observation time as long as possible the length of the beam must be increased, but unfortunately there is not only an obvious practical limit to the length, there is also the action of gravity which would deflect and broaden the beam because of the velocity spread among the atoms from the thermal source. The conclusion is clear: the classical standard has its limits.

In the fountain standard the effect of gravity on a Cs beam is not only accommodated but in fact exploited by having a *vertical* beam. Long before the age of lasers, in fact as far back as 1954, Zacharias [15] had proposed just such a method of circumventing the effect of gravity on the standard horizontal beam. By directing the atoms in a vertical direction they would follow a narrow parabolic path reaching a height determined by their initial velocity and returning to the ground plane. Only one Ramsey cavity would be required to induce transitions since the atoms traverse it on the way up and again on the way down. Furthermore since the instrument has a finite height, only the atoms with initial velocities smaller than the limit set by its height contribute to the signal. At the time Zacharias made the proposal, the standard cesium sources typically operated in the 50°C range at which the *mean* height reached by an atom is computed to be about 1 km (!), but it was argued that the atoms at the low end of the Maxwell distribution of velocities, though small in number, would be able to complete the parabola within a practical height, and return to the detector. However in practice the technique proved unworkable because of the overwhelming presence of energetic atoms. In any event it was clear that the classical Cs oven source would have such a small number of atoms return to be detected that any gain in performance due to an increased observation time is more than offset by the loss of particles contributing to the signal. For an apparatus of height limit h the maximum source velocity V of atoms that will return is given by $(1/2)mV^2 = mgh$. In order to have a *mean* height on the order of, for example, 2 m, the temperature would have to be around 0.6 K.

Like so many other examples the situation was completely changed by the advent of the laser. As we have seen, a cloud of atoms can be cooled in a magneto-optical trap consisting of a quadrupole magnetic field, and three mutually perpendicular pairs of counter-propagating laser beams suitably tuned with respect to the atomic resonance. The effective retarding forces acting on the atoms are proportional to their distances from the center, provided these are small, causing confinement of their motion and concentrating them at the center. The cooling process needed to reach below the Doppler limit depends on the light polarization gradients which are effective even in the absence of the trapping magnetic field and result in the accumulation of atoms called an *optical molasses* [16] at the intersection of the beams with a temperature on the order of 30 μK.

The main technical choice remaining is how to project or "launch" this laser-cooled molasses in a narrowly defined vertical direction in such a way that the atoms that traverse the Ramsey cavity on the way up will return through it on the way down. In the classic thermal source the atoms simply effuse from a small opening in the oven; the equivalent action of simply releasing the molasses along one direction clearly would be inefficient at low temperatures since the expansion rate from the molasses, though higher than might be expected at such temperatures, is still very low. What is needed is the application somehow of a thrust to impart an initial impulse to the atoms. This is done by the same lasers that are used to cool the atoms. A *moving* light field is created, as discussed earlier, simply by introducing a negative frequency offset in the laser beams propagating in the downward direction and a positive offset in the upward beams. The resulting forces on the atoms result eventually in an equilibrium velocity of the atoms at which the Doppler frequency shifts bring the atoms into resonance with both laser beams. This is equivalent to referring the system to a frame of reference moving at a constant velocity determined according to the Doppler effect. After the atoms have been allowed to reach equilibrium the laser beams are blocked, allowing the atoms to continue their upward motion. This is more complicated than simply subjecting the molasses to a single upward laser pulse, but is significantly more effective in maintaining ordered motion.

The first published observation of an atomic fountain and the demonstration of its potential in high-resolution spectroscopy were by the Stanford group in 1989 [17]. The group demonstrated the formation of a Na atomic fountain by projecting the atoms vertically upwards using radiation pressure and saw the atoms fall back under gravity! The long observation time made possible by the low velocity of the atoms enabled the ground state hyperfine transition to be observed with a 2 Hz line width after a 1,000 s integration time, and the center of the microwave resonance frequency resolved to within ± 10 mHz. An excellent treatment of this subject can be found in the review article by S. R. Jefferts et al. [18].

Cesium fountain frequency standards have superseded the horizontal thermal beam machines and all major national standards laboratories now rely on this method of observing the cesium hyperfine frequency to establish the International

Atomic Time (French initials TAI) scale. These standards are large fixed installations whose primary design criteria are accuracy and reproducibility. Cesium fountain standards will no doubt in the future also be installed at all GNSS monitoring stations. These are not constrained of course in design with regard to size and weight, and have therefore achieved extraordinary accuracy and stability in frequency. For example at the French standards laboratory, Bureau National de Métrologie—Systèmes de Référence Temps Espace (BNM-SYRTE) results of stability tests in 2005 [19] on their FO1 and FO2 standards are given individually as $2.9 \times 10^{-14}\,\tau^{-1/2}$ for FO1 and $1.6 \times 10^{-14}\,\tau^{\,1/2}$ for FO2, but what is even more impressive is that the fractional difference in frequency between the two standards is only 4×10^{-16}. It is expressed in time, that is, about 1 μs in 1,000 years.

The actual design of a cesium fountain standard naturally divides into several distinct functions: (1) laser cooling of Cs vapor to form an optically induced molasses, concentrated into a bunch at the center, (2) the launch of the Cs using two oppositely directed vertical laser beams, (3) possible additional cooling stage to further reduce the spread in the velocity distribution of the atoms, (4) the selection of atoms in one of the hyperfine sublevel usually $|F = 3, m_F = 0$ □ by transitions from the $|F = 4$, $m_F = 0$ □ state into the $|F = 3, m_F = 0$ □, while atoms in the other |F = 4 □ sublevel are optically eliminated, leaving atoms in only the $|F = 3$ $m_F = 0$ □ state, (5) induce a coherent mixture of the two $m_F = 0$ sublevels by a resonant field in the Ramsey microwave cavity, (6) provide a perturbation-free vertical drift space (or "toss tube") typically 1 m long in a laboratory installation, in weak uniform magnetic field , And (7) detect by a laser probe the occurrence of transitions in the Ramsey cavity as the atoms emerge after their second passage through it. The sequence of operations is repeated at regular intervals, each time incrementing the frequency of the Ramsey cavity field. A schematic drawing of the FO1 Cs fountain standard at BNM-SYRTE is shown in Fig. 18.10.

For the purposes of navigation, the portability and stability of the standard are obviously important attributes, perhaps more than absolute accuracy; this requires that the Cs fountain be sufficiently compact and rugged to be usable in a mobile setting. In fact the justifications for developing a mobile standard actually go beyond navigation: such a standard would be useful in transferring precise time on a global scale, as well as mapping gravitational fields through the relativistic dependence of clock rates on the gravitational field.

The essential elements of such a compact Cs standard under development by the NIST group in Boulder [20] are shown schematically in Fig. 18.11.

The cold atom source in the compact standard has three mutually perpendicular counter-propagating pairs of laser beams to capture and cool Cs vapor, first in a magneto-optical trap which is then converted to a pure optical molasses. A moving light field generated by detuning the vertical beams serves to launch the atom bunch upward to a height of about 40 cm from the source. A rectangular microwave cavity mounted above the source provides the pulse at the hyperfine frequency to induce atoms in the $|F = 4, m_F = 0>$ state to transition to the desired $|F = 3, m_F> = 0$ state.

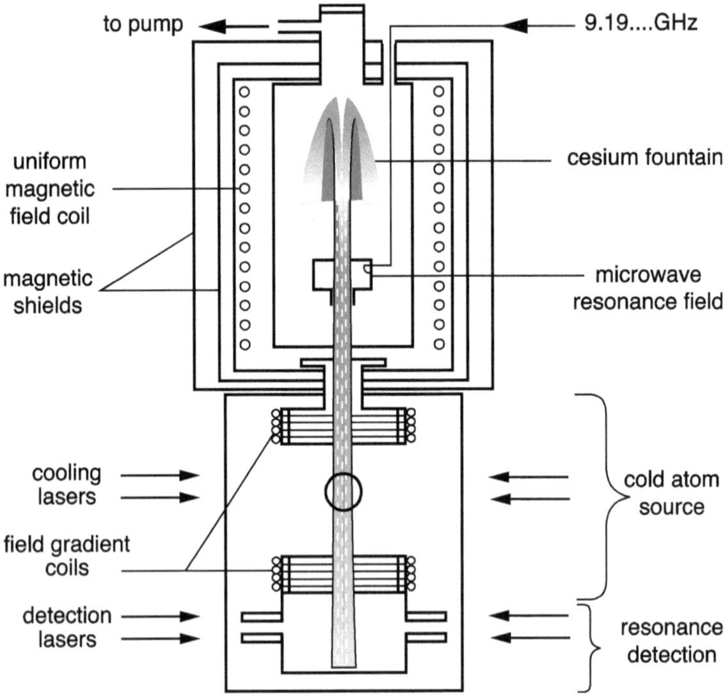

Fig. 18.10 The cesium fountain frequency standard FO1 at BNM-SYRTE

Fig. 18.11 The elements of a developmental NIST compact Cs fountain frequency standard

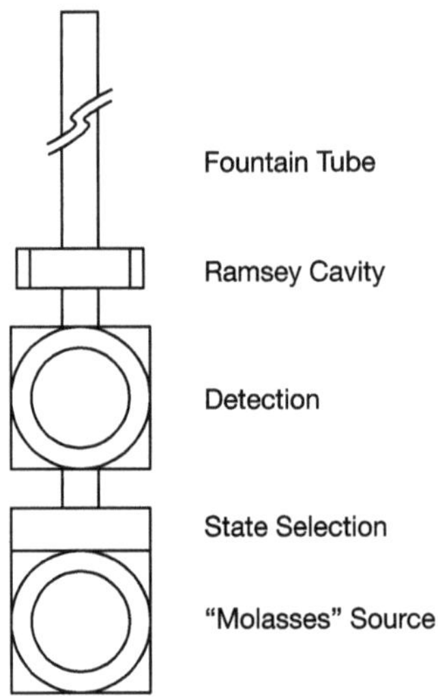

The remaining atoms in $|F = 4, m_F>$ substates are removed by optically inducing the transitions from all the $|F = 4, m_F>$ substates to all the $/F = 5, m_F>$ substates in the upper state.

The section for detecting the occurrence of transitions between the two hyperfine states with $m_F = 0$ is mounted above the state selection cavity and is capable of measuring the populations of both of these "field independent" states. The atoms descending from the Ramsey cavity, where the Cs standard transition frequency is established, first encounter a standing light wave which induces $|F = 4,\ m_F = 0> \rightarrow |F = 5, m_F>$ transitions resulting in the emission of photons proportional to the number of atoms in the $|F = 4,\ m_F = 0>$ state. Atoms in this state are then removed using a traveling wave at the $|F = 4,\ m_F> \rightarrow |F = 5,\ m>$ transition frequency. To obtain the emission of photons proportional to the number of atoms in the $|F = 3,\ m_F = 0>$ state, they are pumped into the $|F = 4,\ m_F = 0>$ using the standard hyperfine frequency microwave field, followed by the same sequence as before.

The Ramsey cavity is a copper cylinder with inner diameter 7 cm and a little under 2 cm high and operates in the TE_{011} mode at a frequency of 9.19... GHz, with an unloaded Q of 18,000. A uniform axial magnetic field 50 nT are used for the Ramsey cavity and fountain, but a stronger field 0.5 μT in the state selection and detection regions. The fountain reached 13.2 cm above the Ramsey microwave cavity, corresponding to a Ramsey time of 0.33 s and a correspondingly narrow linewidth.

Finally, a highly significant scientific program involving atomic timekeeping is in the early stages of development; it seeks to apply the extraordinary stability achievable in an atomic fountain clock on board a spacecraft to test the predictions of relativity theory. It is a joint program called PARCS (Primary Atomic Reference Clock in Space) whose objective is to develop a space hardened Cs fountain frequency standard in which, because of the reduced gravity aboard a distant satellite, the flight time of the Cs atoms in the fountain would be significantly longer than on the earth's surface. Moreover in a satellite freely orbiting in space with no onboard sources of vibration, there would be no resonance line broadening on that account. The program is a collaborative effort involving the National Institute of Standards and Technology, the Jet Propulsion Laboratory, and the University of Colorado. It is expected that this space-qualified clock will be placed on the International Space Station (ISS) at some date to be determined (TBD).

18.8 Quantum Computers

Until the last quarter century or so, the hardware for information processing and computation was inherently built up around two-state classical devices, whether mechanical or electronic. Numbers could be expressed using the binary system (base 2) as a series of *bits* each with a value of either one or zero, physically embodied as the on–off states of a switch, or two levels of an electrical potential in what are called registers. Various electronic circuits called gates were devised to apply logic in the performance of desired computations by the manipulation of binary numbers. The essence of this discrete system is that the value of each bit is unequivocally one or zero.

In the microscopic world where physical behavior requires a quantum description, a whole new physical form of representation and manipulation of mathematical and logical operations has emerged. It begins with the quantum treatment of the state of a physical system: a basic tenet is the principle of *superposition*, as in wave optics. When applied to material particles in the classical Young two-slit experiment it leads to counterintuitive results—in that experiment we recall that a *single* photon falling on the two slits emerges as a photon in a coherent superposition of *two* states as long as the passage through either slit is not detected; otherwise the photon "collapses" into one state. It thus appears that the passage of a photon through a given slit depends on whether or not the *other* slit is blocked! Another example of superposition is in the representation of a two-electron system. We recall that the wavefunction representing a collection of electrons (fermions) must be antisymmetric (change sign) with respect to an exchange of electrons. Thus for example a two-electron wavefunction is made antisymmetric by writing the total wavefunction as a product of a symmetrized (entangled) space function with a spin function of the opposite symmetry. For example, if the space function is symmetric, the spin function must be antisymmetric, that is, to use quantum symbolism, represented by $(|\uparrow>_1 |\downarrow>_2 — |\downarrow>_1 |\uparrow>_2)$ where the arrows indicate the directions of the two $s = \pm ½$ spin states with respect to a specified axis. A wavefunction involving the coordinates of two particles, which *cannot* be expressed as the product of functions each involving just one particle, is an example of what Schrödinger called *entanglement*. A wavefunction, which *can* be factored into functions each involving just one particle, can be shown to lead to two separate Schrödinger equations with independent solutions. On a more basic level, two quantum states of the *same* particle can be entangled; for example, consider a free spin ½ particle, which may be an electron or a nucleus initially in the state $m_S = +1/2$, placed in a magnetic field. Classically we have a picture of a magnetic field exerting a torque on the magnetic moment of the particle which, because of its spin, will cause it to precess like a gyroscope. If a weak alternating magnetic field, usually at radiofrequency, is applied to the particle transverse to the spin axis and resonant with its frequency of precession, it will cause the axis to tilt and eventually reverse direction. This rotation of the axis of spin between the spin-up and the spin-down state will continue as long as the transverse radiofrequency field persists. If it is on with the right amplitude and duration, the particle is put in a coherent equal mixture of spin-up, $m_S = +1/2$, and spin-down, $m_S = -1/2$, states. It is neither in one state nor the other; we think of it classically in this state as rotating in a plane perpendicular to the magnetic field, and hence the radiofrequency field is referred to as a 90°–*pulse*.

Just as the unit of classical information is called a *bit* and has just two values: 0 or 1, the unit of quantum information is called a *qubit* which can be represented as a coherent superposition of two spin-1/2 states; thus,

$$U = a_0|\uparrow> + a_1|\downarrow>; \ |a_0|^2 + |a_1|^2 = 1. \qquad (18.19)$$

The radically greater power of quantum computing derives from the fact that a register of N qubits occupies a space of 2^N dimensions, that is, it represents at once a coherent superposition of 2^N binary states. This means that computations are carried out effectively *in parallel*, to which is given the name *quantum parallelism.*

The first significant advance in the structural design of a computer based on the manipulation of quantum states was made by Deutsch in 1985. To appreciate his contribution we must recall the design of a classical computer. For example, the Turing machine is a conceptual computer which can be shown to be capable of carrying out all the classical computer operations. It is conceived as having an infinite tape divided into unit cells, each of which can contain a 0 or 1. It has a read-write head that can scan in either direction along the tape, and takes action through gates as determined by its current internal state and the reading of the tape. The action it takes consists in overwriting a new value on the tape and changing the internal state according to what has been read on the tape. Modern computers are not simple Turing machines, but their computational power can be expressed in terms of a Turing machine.

Again like classical computers, the operation of a quantum computer is built around various logic gates. Not surprisingly a great deal of effort has been made by several laboratories to experimentally demonstrate quantum gates. For example, as far back as 1995 the pioneer group in this field at NIST reported [21] on the successful demonstration of a controlled-NOT (CNOT) gate involving a *single* ion in an electromagnetic trap using the coupling of a magnetic hyperfine transition with the two lowest motional states of the ion. As a two-bit gate the controlled-NOT gate operates by flipping one bit if, and only if, the other bit is in a given state. The Boulder group had been engaged for many years in the applications of individual ions trapped and cooled to the lowest vibrational quantum states in the trapping field (see Chap. 10). The controlled-NOT logic gate was demonstrated on trapped individual $^9Be^+$ ions, which were cooled to the lowest two quantum levels of oscillation separated in frequency by 11 MHz in a harmonic trap. The ions have a $^2S_{1/2}$ ground state split into two magnetic hyperfine levels 1.25 GHz apart. The gate was constructed using the two center of mass oscillation states as one qubit and the two magnetic hyperfine states as another. They showed that by applying a single microwave π-pulse having a frequency equal to the sum of the hyperfine and vibrational frequency with the proper power and duration, they could flip the internal hyperfine states if and only if the ion was in the $n = 1$ vibrational level. In recognition of the significance of this experimental work by the Boulder group the 2012 Nobel Prize in physics was shared by its leader, David Wineland, with Serge Haroche, currently at the Collège de France in Paris. These studies in the manipulation of coherent quantum states represent significant advances at a fundamental level toward the realization of quantum computing. However, the ultimate success in building a quantum computer based on isolated ions will depend on its scalability, that is, the ability to expand the number and complexity of its constituents.

It is in principle possible to construct a large lattice of microscopic potential wells capable of storing individual ions that can be addressed by laser beams (see Chap. 10), but that is presumably somewhat in the future.

A highly promising form of such potential wells is already well known as semiconductor nanostructures known as *quantum dots* [22]. The sizes of these nanoparticles measure typically on the order between 100 nm to as small as 10 nm a side, about the width of 20 atoms. At this nanometer scale it is expected that the behavior of any confined particle must be described in quantum terms: thus the properties of quantum dots are different in character from those of the same material in bulk. The most striking of the differences is that when excited to emit photons the spectrum is characteristic not of the material itself but of the geometric *size* of the quantum dot. This is explained by the atom-like electron energy level structure in the potential well of the quantum dot. In fact quantum dots have been called "artificial atoms." The peak of the photon emission spectrum shifts from red toward the blue as the size of the quantum dot is made smaller; thus quantum dots can be manufactured to emit precisely the desired spectrum by controlling their core size. In fact the "tuning" range extends from the infrared to the ultraviolet ends of the spectrum. The physical form of an aggregate of quantum dots may be as a powder, a suspension in a liquid, or as a space lattice designed to embody qubits in a quantum computer using the spin of one electron in each dot. Like ions in a trap, the quantum states of an electron in a quantum dot can be manipulated by appropriately tuned laser beams.

In contrast to using the quantum states of confined single electrons or ions to represent qubits, another promising approach to building a quantum computer is based on using matter in bulk. It has recently been argued that nuclear magnetic resonance (NMR), the same NMR used as a medical diagnostic tool, has certain advantages with respect to scalability. The origin of the technique dates back to the 1960s with names like E.M Purcell and R.V. Pound associated with it. Most atoms, and therefore molecules formed from them, have nuclei with permanent magnetic moments associated with spin angular momentum. We will assume we are dealing with nuclei whose outer electronic structure around the nuclei has zero magnetic moment (no paramagnetism). When subjected to an external magnetic field the nuclear magnetic moments will precess about the field as axis at a frequency which depends on the strength of the field and the g-factor (see Chap. 8) of their nuclear moments. The phases of the nuclear precessions will be randomly distributed among the nuclei, and there is no observable transverse magnetic signal. Furthermore at ordinary temperatures the axial components of the nuclear moments are only weakly polarized with a preponderance of nuclei pointing in the direction of the field. However by applying a sufficiently strong axial magnetic field, in the 1–10 T range, there is an observable polarization of the nuclear spins. The technology of NMR has during the second half of the last century matured to the point of having a wide range of NMR spectrometers developed for specific applications in many areas of science and medicine. If a nucleus is chosen with a nuclear spin $I = 1/2$ then in a magnetic field it can exist in two possible states $m_I = \pm½$, and therefore can represent a qubit. These states can be manipulated by applying a transverse radiofrequency field which resonates with the nuclear precession frequency to cause transitions between the two magnetic sublevels. In the process there is energy absorption which can be detected using the classic radiofrequency absorption method of using a high-Q LC circuit to provide the resonant excitation field. Alternatively under the right conditions a radiofrequency inductive method can be used in which the detection field is in the form of a 90° radiofrequency pulse to create a magnetic moment precessing in a plane at right angles to the axis, causing an observable radiofrequency signal due to the precession of the nuclear moments.

To apply NMR to the construction of a quantum gate, assume we have a quantity of liquid consisting of *molecules* made up of atoms with nuclear spins of ½, so that in an external magnetic field the nuclei line up either parallel or antiparallel with the field, corresponding to the two values of a qubit. In a liquid the random thermal agitation of the molecules causes the magnetic field at one nucleus due to neighboring nuclei to randomly fluctuate and tend to average out; however within a molecule, a change in the state of one nucleus can affect another if it disturbs the electrons orbiting both. This circumstance was exploited by Chuang et al. [23] to construct a logic gate using the nuclear states of H^1 and C^{13} in chloroform ($CHCl_3$), each of which has a nuclear spin of ½. The chloroform liquid is placed in a strong axial magnetic field to orient the nuclear spins along the field. A 90° radiofrequency pulse is applied at the C^{13} resonance frequency causing the C^{13} nuclear spin axis to precess, in classical terms, in a plane at right angles to the axis. It happens that the frequency of that precession depends on the state, up or down, of the hydrogen nucleus (proton); therefore there will come a time when the difference in precession angles of the C^{13} nucleus for the two possible states of the H nucleus is 180°, that is, are in opposite directions. If now a 90° pulse is applied to the C^{13} nucleus, it will end up pointing down if the H nucleus was pointing up and vice versa. This constitutes in effect a controlled-NOT gate.

More recently a quantum controlled-NOT gate has been reported [24] which operates with radiation in the 1.55 μm telecommunications band. It is based on a fiber optic system and is able to circumvent the difficulty of working with photons for quantum computation due to the weakness of the interaction between two photonic qubits.

References

1. T.L. Gustavson et al., Classical Quant. Grav. **17**, 2385–2398 (2000)
2. M.A. Porter et al., Phys. Lett. A. **352**, 210–215 (2006)
3. P.D. Kunz et al., (NIST), 41st Annual PTTI Meeting, 2009
4. L. de Broglie, *Matter and Light* (Dover, New York, 1939)
5. C. Cohen-Tannoudji, Rev. Mod. Phys. **70**, 707 (1998)
6. B. Lounis, C. Cohen-Tannoudji, J. Phys. (France) **2**, 579 (1992)
7. P.L. Gould et al., Phys. Rev. Lett. **56**, 827 (1986)
8. D.L. Freimund, H. Batemaan, Phys. Rev. Lett. **89**, 283682 (2002)
9. P.J. Martin et al., Phys. Rev. Lett. **60**, 515 (1988)

10. M. Kozuma et al., Phys. Rev. Lett. **82**, 871 (1999)
11. D.E. Pritchard et al., Ann. Phys. (Leipzig) **10**, 35–54 (2001)
12. G.L. Gattobigio et al., Phys. Rev. Lett. **107**, 254104 (2011)
13. K. Takase, Ph.D. thesis, Stanford University, 2008
14. C. H-T. Wang, R. Bingham, J.T. Mendonça, arXiv:gr-qc/0603112v3, 26 Jul 2006
15. J.R. Zacharias, Phys. Rev. **94**, 751 (1954)
16. S. Chu et al., Phys. Rev. Lett. **55**, 48 (1985)
17. M.A. Kasevich, Phys. Rev. Lett. **63**, 612 (1989)
18. S.R. Jefferts et al., Proc. SPIE **6673**, 667309 (2007)
19. C. Vian et al., IEEE Trans. Instrum. Meas. **54**, 833 (2005)
20. T.P. Heavner et al., in *IEEE/EIA International Frequency Control Symposium*, 2002, p. 473
21. C. Munroe et al., Phys. Rev. Lett. **75**, 4714 (1995). and Phys. Rev.A, 55 R2489 (1997)
22. C. Kloeffel, D. Loss, Annu. Rev. Con. Mat. Phys. **4**, 51 (2013)
23. I.L. Chuang et al., Proc. R. Soc. London **454**, 447 (1998)
24. J. Chen et al., Phys. Rev. Lett. **100**, 133603 (2008)

Epilogue

If glimpses of the emerging quantum means of answering the question quo vadis (addressed not to God, but you as a navigator) have been made a little more intelligible in their historical context, albeit in a necessarily cursory fashion, it is hoped that sufficient interest has been aroused to inquire further into the subject and become a participant in its development or at least become an intelligent user of it. We owe it to Cher Ami, that determined avian navigator, to understand its seeming devotion to duty!

F.G. Major, *Quo Vadis: Evolution of Modern Navigation: The Rise of Quantum Techniques*, DOI 10.1007/978-1-4614-8672-5,

Index

F.G. Major, *Quo Vadis: Evolution of Modern Navigation: The Rise of Quantum Techniques*, DOI 10.1007/978-1-4614-8672-5,

MIX
Papier aus verantwortungsvollen Quellen
Paper from responsible sources
FSC® C105338

If you have any concerns about our products,
you can contact us on
ProductSafety@springernature.com

In case Publisher is established outside the EU,
the EU authorized representative is:
Springer Nature Customer Service Center GmbH
Europaplatz 3, 69115 Heidelberg, Germany

Printed by Libri Plureos GmbH
in Hamburg, Germany